DOPAMINE TRANSPORTERS

Wiley Series in Drug Discovery and Development

Binghe Wang, Series Editor

Drug Delivery: Principles and Applications
Edited by Binghe Wang, Teruna Siahaan, and Richard A. Soltero

Computer Applications in Pharmaceutical Research and Development
Edited by Sean Ekins

Glycogen Synthase Kinase-3 (GSK-3) and Its Inhibitors: Drug Discovery and Development
Edited by Ana Martinez, Ana Castro, and Miguel Medina

Drug Transporters: Molecular Characterization and Role in Drug Disposition
Edited by Guofeng You and Marilyn E. Morris

Aminoglycoside Antibiotics: From Chemical Biology to Drug Discovery
Edited by Dev P. Arya

Drug-Drug Interactions in Pharmaceutical Development
Edited by Albert P. Li

Dopamine Transporters: Chemistry, Biology, and Pharmacology
Edited by Mark L. Trudell and Sari Izenwasser

DOPAMINE TRANSPORTERS

Chemistry, Biology, and Pharmacology

Edited by

MARK L. TRUDELL
University of New Orleans

SARI IZENWASSER
University of Miami Miller School of Medicine

WILEY

A JOHN WILEY & SONS, INC., PUBLICATION

CONTENTS

CONTRIBUTORS   vii

PREFACE   xi

PART I BIOCHEMISTRY AND MOLECULAR BIOLOGY   1

1   **The Dopamine Transporter: An Anatomical Perspective**   3
Hilary R. Smith, Thomas J. Beveridge, Colleen A. Hanlon, and Linda J. Porrino

2   **Dopamine Transporter, Disease States, and Pathology**   29
Deborah C. Mash

3   **Cloning and Genetic Analysis of Dopamine Transporters**   47
David J. Vandenbergh

4   **Molecular Structure and Composition of Dopamine Transporters**   73
M. Laura Parnas and Roxanne A. Vaughan

5   **Electrochemical Characterization of Dopamine Transporters**   97
Evgeny A. Budygin and Sara R. Jones

PART II MEDICINAL CHEMISTRY   123

6   **Tropane-Based Dopamine Transporter–Uptake Inhibitors**   125
Scott P. Runyon and F. Ivy Carroll

7   **The Benztropines: Atypical Dopamine-Uptake Inhibitors That Provide Clues About Cocaine's Mechanism at the Dopamine Transporter**   **171**

    Amy Hauck Newman and Jonathan L. Katz

8   **Structure–Activity Relationship of GBR 12909 Ligands**   **211**

    Thomas E. Prisinzano and Kenner C. Rice

9   **Structure–Activity Relationship Study of Piperidine Derivatives for Dopamine Transporters**   **233**

    Prashant S. Kharkar, Aloke K. Dutta, and Maarten E. A. Reith

10 **Non-Nitrogen-Containing Dopamine Transporter–Uptake Inhibitors**   **265**

    Peter C. Meltzer

11 **Dopamine-Releasing Agents**   **305**

    Bruce E. Blough

PART III PHARMACOLOGY   **321**

12 **PET/SPECT Imaging Studies of the Plasma Membrane Dopamine Transporter**   **323**

    Paul Cumming, Weiguo Ye, and Dean F. Wong

13 **In Vitro Studies of Dopamine Transporter Function and Regulation**   **347**

    Brian R. Hoover, Bruce H. Mandt, and Nancy R. Zahniser

14 **In Vivo Studies of Dopamine Transporter Function**   **391**

    Jane B. Acri

INDEX   **439**

CONTRIBUTORS

Jane B. Acri, Addiction Treatment Discovery Program, Division of Pharmacotherapies and Medical Consequences of Drug Abuse, National Institute on Drug Abuse, Bethesda, Maryland

Thomas J. Beveridge, Center for the Neurobiological Investigation of Drugs of Abuse, Wake Forest University Health Sciences, Winston-Salem, North Carolina

Bruce E. Blough, Center for Organic and Medicinal Chemistry, Research Triangle Institute, Research Triangle Park, North Carolina

Evgeny A. Budygin, Department of Physiology and Pharmacology, Wake Forest University Health Sciences, Winston-Salem, North Carolina

F. Ivy Carroll, Center for Organic and Medicinal Chemistry, Research Triangle Institute, Research Triangle Park, North Carolina

Paul Cumming, Department of Radiology, Ludwig Maximillian University, Munich, Germany

Aloke K. Dutta, Department of Pharmaceutical Sciences, Wayne State University, Detroit, Michigan

Colleen A. Hanlon, Center for the Neurobiological Investigation of Drugs of Abuse, Wake Forest University Health Sciences, Winston-Salem, North Carolina

Brian R. Hoover, Department of Neurology, University of Minnesota, Minneapolis, Minnesota

Sari Izenwasser, Department of Psychiatry and Behavioral Sciences, University of Miami Miller School of Medicine, Miami, Florida

Sara R. Jones, Department of Physiology and Pharmacology, Wake Forest University Health Sciences, Winston-Salem, North Carolina

Jonathan L. Katz, Medications Discovery Research Branch, National Institute on Drug Abuse Intramural Research Program, National Institutes of Health, Bethesda, Maryland

Prashant S. Kharkar, Department of Pharmaceutical Sciences, Wayne State University, Detroit, Michigan

Bruce H. Mandt, Department of Pharmacology, University of Colorado Denver, Aurora, Colorado

Deborah C. Mash, Departments of Neurology and Molecular and Cellular Pharmacology, University of Miami Miller School of Medicine, Miami, Florida

Peter C. Meltzer, Organix Inc., Woburn, Massachusetts

Amy Hauck Newman, Medications Discovery Research Branch, National Institute on Drug Abuse Intramural Research Program, National Institutes of Health, Baltimore, Maryland

M. Laura Parnas, Department of Biochemistry and Molecular Biology, University of North Dakota School of Medicine and Health Sciences, Grand Forks, North Dakota

Linda J. Porrino, Center for the Neurobiological Investigation of Drugs of Abuse, Wake Forest University Health Sciences, Winston-Salem, North Carolina

Thomas E. Prisinzano, Department of Medicinal Chemistry, The University of Kansas, Lawrence, Kansas

Maarten E. A. Reith, Department of Psychiatry, New York University, New York, New York

Kenner C. Rice, Chemical Biology Research Branch, National Institute on Drug Abuse, National Institutes of Health, Bethesda, Maryland

Scott P. Runyon, Center for Organic and Medicinal Chemistry, Research Triangle Institute, Research Triangle Park, North Carolina

Hilary R. Smith, Center for the Neurobiological Investigation of Drugs of Abuse, Wake Forest University Health Sciences, Winston-Salem, North Carolina

Mark L. Trudell, Department of Chemistry, University of New Orleans, New Orleans, Louisiana

David J. Vandenbergh, Department of Biobehavioral Health, Center for Developmental and Health Genetics, and Neuroscience Institute, The Pennsylvania State University, University Park, Pennsylvania

Roxanne A. Vaughan, Department of Biochemistry and Molecular Biology, University of North Dakota School of Medicine and Health Sciences, Grand Forks, North Dakota

Dean F. Wong, The Russell H. Morgan Department of Radiology and Radiological Science, Department of Psychiatry, and Department of Environmental Health Sciences—Bloomberg School of Public Health, Johns Hopkins University, Baltimore, Maryland

Weiguo Ye, The Russell H. Morgan Department of Radiology and Radiological Science, Johns Hopkins University, Baltimore, Maryland

Nancy R. Zahniser, Department of Pharmacology and Neuroscience Program, University of Colorado Denver, Aurora, Colorado

PREFACE

Dopamine neurotransmission has been the subject of research for nearly half a century. However, it was not until 1987, when it was implicated with the reinforcing effects of cocaine, that the dopamine transporter (DAT) was thrust into the scientific spotlight and became a pharmacological target for the development of a potential medication for one of the most addictive drugs known. At this time a surge in biological, chemical, and pharmacological investigations directed toward understanding the DAT occurred, spearheaded primarily by the National Institute on Drug Abuse. To date, the development of a clinically useful therapeutic agent for cocaine addiction has been elusive, but tremendous scientific advances have been made toward understanding the DAT and its role in a variety of neurological disease states and disorders. This includes Parkinson's disease, schizophrenia, attention-deficit/hyperactivity disorder, Tourette's syndrome, and drug abuse. Although, according to SciFinder Scholar the annual number of research publications focused on dopamine-uptake inhibitors peaked in 2001, the annual output of research publications concerning all aspects of the DAT is increasing continually. This clearly illustrates the importance of the DAT in neuroscience and as a pharmacological target for neurotherapeutic agents.

Like many similar substances, our interest in the DAT began in the early 1990s. Our focus was directed toward the synthesis and development of selective dopamine-uptake inhibitors as potential medications for cocaine addiction. During the course of our studies we have learned a great deal about the structure–activity relationships of the DAT. In addition, we have serendipitously gained a greater understanding of both the serotonin and norepinephrine transporters. Moreover, recent discoveries which suggest that the DAT may be affected indirectly by other drugs of abuse (e.g., nicotine and canabanoids) have led to additional investigations into the neurochemistry of these systems.

Overall, our understanding of the DAT and dopaminergic systems has increased dramatically over the past decade, and that level of understanding is being elevated by new discoveries every day. When training new researchers in such a rapidly developing field, it has been difficult to find background materials that would provide a foundation that was sufficiently broad but detailed enough to furnish a basic understanding of the biological, chemical, and pharmacological aspects of the DAT. Although there were excellent reviews of specific subject material in many areas, a single comprehensive source of DAT biology, chemistry, and pharmacology was not available. This spawned the idea that a book that reviewed significant advances in the various fields of the DAT would be highly valuable to new as well as experienced researchers in the field.

The book is organized to summarize the significant biological, chemical, and pharmacological developments that have led to own current understanding of the DAT. Most of the research and discoveries presented here are derived from efforts directed toward understanding the pharmacological mechanisms of cocaine as well as medication development for cocaine dependence. However, these studies have contributed significantly toward the elucidation of general DAT function. In Part I the focus is on important biological developments. Chapters 1 to 5 are authoritative reviews written by eminent scientists in the areas of DAT function, pathological conditions, DAT cloning, DAT molecular structure, and DAT electrochemical characterization. Part II deals with the medicinal chemistry of DAT inhibitory ligands and substrates. Chapters 6 to 11 are written by pioneering medicinal chemists in this area. Chemical syntheses and various structure–activity relationships of important classes of dopamine-uptake inhibitors and dopamine releasers are discussed. Finally, in Part III, the reader is directed to pharmacological discoveries that have provided extraordinary insight into the role of the DAT in normal neurological function as well as in disease states. Chapters 12 to 14 are written by leading researchers in the fields of DAT imaging, in vitro studies, and in vivo behavioral studies.

ACKNOWLEDGMENTS

The editors dedicate the book to all the postdoctoral researchers, graduate students, technicians, and undergraduate students who have, and will, contribute to the research efforts of our laboratories. Their hard work and dedicated efforts have been, and will continue to be, vital to the success of our research program. The editors would also like to thank Dr. Stacey Lomenzo for technical assistance in assembling the manuscript.

On August 29, 2005, Hurricane Katrina dealt our program a devastating blow. Our University of New Orleans laboratories were shut down for nearly eight months as electricity was restored, mold was eradicated, and repairs were made. Many of our personnel lost their homes and were scattered around the country. At a time when careers were in jeopardy with no clear view of recovery, it was a great comfort to have the sympathy and support of both the scientific and administrative communities of the National Institutes of Health and the National Institute on Drug Abuse. It is with sincere appreciation that we acknowledge the National Institute on Drug Abuse for financial support of our research (DA11528).

MARK L. TRUDELL
SARI IZENWASSER

PART I

BIOCHEMISTRY AND
MOLECULAR BIOLOGY

1

THE DOPAMINE TRANSPORTER: AN ANATOMICAL PERSPECTIVE

HILARY R. SMITH, THOMAS J. BEVERIDGE, COLLEEN A. HANLON, AND LINDA J. PORRINO

Center for the Neurobiological Investigation of Drugs of Abuse, Wake Forest University Health Sciences, Winston-Salem, North Carolina

1.1 Introduction   3
1.2 Anatomy of the DAT   6
    1.2.1 DAT mRNA   6
    1.2.2 DAT Protein Distribution   8
    1.2.3 Changes in DAT Distribution with Age   14
1.3 Visualizing the DAT in Humans   15
1.4 Conclusions   17
Acknowledgments   18
References   18

1.1 INTRODUCTION

Until the mid-1950s, dopamine was considered exclusively to be an intermediate in the biosynthesis of norepinephrine and adrenaline. In 1958, Arvid Carlsson and his colleagues discovered that dopamine was present in high concentrations in the rat corpus striatum and that it could be depleted by administration of the drug reserpine [1]. Following the reversal of reserpine-induced hypokinesia by L-dihydroxyphenyl-alanine (L-dopa), Carlsson subsequently proposed a biological role for dopamine in

Dopamine Transporters: Chemistry, Biology, and Pharmacology. Edited by Mark L. Trudell and Sari Izenwasser

neurotransmission, independent of its function as a precursor for norepinephrine biosynthesis [2]. This achievement was in part the basis of the Nobel Prize in Medicine awarded to Arvid Carlsson in 2000.

A major breakthrough in our understanding of the role of dopamine in the central nervous system, however, came with the application of histochemical techniques to visualize the cellular location of dopamine with a fluorescence microscope. These techniques were developed by Falck and co-workers [3]. The technique consisted of treating tissues with formaldehyde vapor to produce catecholamine condensation products called isoquinolines. The specific compound formed from dopamine and formaldehyde vapor condensation, 3,4-dihydroisoquinoline, emits a bright green fluorescent glow when illuminated by ultraviolet light. Once dopamine could be identified consistently, many experiments were devoted to the precise cellular localization and mapping of dopaminergic neuronal pathways [4]. Some of the earliest studies using this approach showed that a lesion of the substantia nigra (SN) resulted in the disappearance of green fluorescent nerve terminals from the rat neostriatum, confirming the neuronal location of dopamine [5]. The first systematic mapping studies of dopaminergic pathways were performed by Dahlström and Fuxe in 1964 [6]. They showed the concentration of dopaminergic cell bodies in the ventral midbrain as well as the widespread distribution of dopamine within the striatum. Demonstration that the highest levels of dopamine were present within the basal ganglia led to the hypothesis that it might be involved in motor control and that decreased striatal dopamine could be the cause of the extrapyramidal symptoms characteristic of Parkinson's disease. Subsequent mapping studies by Bjorklund, among others, in the rodent [7,8], and by Sladek and others in the primate [9–11], showed that dopamine had a much wider distribution than simply the basal ganglia, which suggested a more complex role in behavior. Today, dopamine is thought to make critical contributions to reward processing, attention, memory, and cognitive function. In addition, Parkinson's disease, schizophrenia, addiction, and attention-deficit disorder are known or suspected to be the result of dysregulation of dopamine processing in the central nervous system (CNS).

A key to understanding the function of dopamine and dopamine transmission is knowledge of its anatomy and of its distribution within both the cell and the brain pathways. An important way in which to gain insight into the anatomy of the dopamine system is through the dopamine transporter (DAT). The DAT is a plasma membrane transport protein that is responsible for regulation of the concentration of extracellular dopamine. Removal of neurotransmitters from the synapse via degradation by enzymes or uptake into presynaptic terminals represents the final step in the process of neurotransmission. The catecholamine nerve terminals possess high-affinity uptake sites, which serve to terminate transmitter action and maintain transmitter homeostasis. Within the dopamine system, uptake is accomplished by the DAT, which transports dopamine into, and sometimes out of, the presynaptic terminal.

It was hypothesized originally that the structure of the DAT would be similar to that of the norepinephrine transporter, due to the structural relationship of their substrates. In fact, the norepinephrine transporter, together with the γ-aminobutyric acid

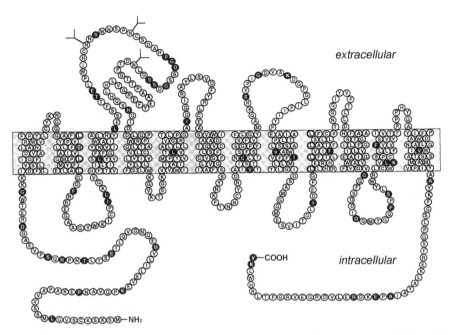

Figure 1-1   Primary amino acid sequence of the human dopamine transporter. Amino acid residues that differ between the rat and human sequence are shown in black. Y-shaped symbols represent potential N-glycosylation sites. (Adapted with permission from Giros and Caron [161].)

(GABA) transporter, was first found to belong to a novel and unique family of sodium- and chloride-dependent transporters [12,13]. Structural analysis predicted that these novel transporters would be made up of 12 transmembrane domains connected by alternating extracellular and intracellular loops (Fig. 1-1). Based on these ideas, Giros et al. (1991) isolated and cloned the gene encoding the DAT from rat SN and inserted it into mammalian cells [14]. They then applied dopamine to these cells and measured the uptake, confirming the identity of the transporter. During the same year, three other groups also isolated and characterized the DAT, due in part to the development of a series of potent selective dopamine-uptake inhibitors [15–17].

Great advances have been made in recent years in further determining the structure of the DAT, with its cloning in several species, including *Caenorhabditis elegans*, mouse, bovine, rat, monkey, and human [14–21]. Between these species, the sequence of the DAT is highly conserved; for example, the human DAT is 98% identical to the monkey DAT, 92% identical to the rat DAT, and 88% identical to the bovine DAT [20,22]. Additionally, the DNA encoding the DAT exhibits high sequence similarity with the previously cloned norepinephrine ($\sim$67%), serotonin ($\sim$49%), and GABA ($\sim$45%) transporters in the rat [16]. The goal of this chapter is to consider localization of the DAT from both a cellular and a systems perspective.

We focus on the more expansive distribution and potentially expanding importance of dopamine in nonhuman and human primates.

1.2 ANATOMY OF THE DAT

Coincident with our expanding knowledge of DAT structure and function was the introduction of selective probes for investigating its anatomical profile of expression in the brain. As the DAT is expressed selectively in somata, dendrites, and axons of dopaminergic neurons, it can be used as a precise and specific marker for cell bodies, axon pathways, and terminal projection fields. The availability of various types of DAT markers made feasible mapping the anatomy of the dopamine system at multiple levels, from localization of gene expression to ultrastructural and regional expression of protein.

1.2.1 DAT mRNA

The distribution of DAT mRNA-positive cells in the rodent brain has been described extensively using in situ hybridization histochemistry, RNAse protection, and Northern blot analyses [23–33]. DAT mRNA in the rat is confined primarily to cell bodies of the ventral midbrain which are immunopositive for tyrosine hydroxylase (TH), the rate-limiting enzyme of the dopamine synthetic pathway, and which correspond to the A8, A9, and A10 monoaminergic cell groups described by Dahlström and Fuxe in 1964 [6]. Within these structures of the ventral midbrain, the source of the majority of the brain's dopamine, DAT mRNA is expressed most abundantly in calbindin-negative cells [32] of the substantia nigra pars compacta and pars lateralis (SNc, SNl; A9), ventral tegmental area (VTA; A10), and retrorubral area (A8) (Fig. 1-2a). The absolute levels of DAT mRNA within these midbrain structures are variable, such that the densest expression is within cells of the medial SNc, moderate labeling is observed in the middle and lateral extent of the SNc as well as the parabrachial pigmented and intrafascicular nuclei of the VTA, and more diffuse expression can be observed in the retrorubral field. In contrast to the extensive labeling of cells in the SNc and VTA, DAT mRNA in the substantia nigra pars reticulata

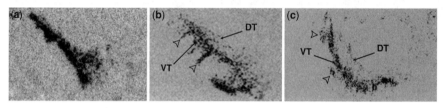

Figure 1-2 Autoradiographic images of the distribution of dopamine transporter mRNA in the midbrain dopamine cell regions of the rat (a), rhesus monkey (b), and human (c). Arrowheads indicate the cell columns of the ventral tier. DT, dorsal tier; VT, ventral tier. (Image of human mRNA contributed by Yasmin Hurd.)

(SNr) is confined to a few scattered islands of dopamine cells, which are also positive for TH immunoreactivity. In more dorsal areas of the midbrain, cells of the A11 group, which arises rostrally in the posterior hypothalamus and extends caudally into the ventral portions of the periaqueductal gray, as well as the caudal zona incerta (A13), have also been reported to express mRNA for the DAT. Although the overwhelming majority of neurons in these midbrain cell groups express both TH and DAT mRNA, there appear to be populations of cells within these areas in which the overlap is only partial, and examples of TH-positive cells that do not express DAT mRNA, as well as DAT mRNA-positive/TH-negative cells, have been reported [23,24,28,29]. This is particularly true in subdivisions of the A10 cell group, where subpopulations of TH-positive cells express little or no measurable DAT mRNA [23,24,28–30,33].

In addition to the dense labeling seen in the midbrain dopamine neurons, DAT mRNA has also been identified in some, although not all, of the dopaminergic cells of the hypothalamus [26,28–31]. For example, DAT mRNA-containing cell bodies have been described in the zona incenta (A13), the periventricular nucleus (A14), and the dorsomedial portions of the arcuate nucleus (A12), which contribute dopamine to the tuberoinfundibular pathway, amygdala, and locally to other hypothalamic nuclei [34,35]. Using the sensitive RNAse-protection method, Maggos and colleagues were also able to detect DAT mRNA in quantities equal to or greater than those measured in the hypothalamus, in the amygdala, caudate, and nucleus accumbens of the rodent brain, areas previously thought to be devoid of DAT gene expression [30]. Finally, DAT mRNA is also expressed in low quantities in the glomerular layer of the olfactory bulb (A16) and the retina [26,29].

Multiple studies in the monkey and human have confirmed that DAT gene expression in the primate midbrain is similar in topography to that of the rodent, with labeling occurring almost exclusively in the dopaminergic nuclei of the SN and VTA [36–42] (Fig. 1-2b and c). In the primate brain these cells have been regrouped into the dorsal and ventral tiers, based on their shared neurochemical characteristics, connectivity patterns, and differential regulation [43–48]. The dorsal tier, which is calbindin-positive and contributes projections preferentially to ventral striatum, amygdala, and cortex via the mesolimbic and mesocortical pathways, consists of dorsal SNc, VTA (parabrachial pigmented and paranigral nuclei), and the retrorubral cell groups. The ventral tier, which is calbindin-negative and projects via the nigrostriatal pathway to the more dorsal sensorimotor striatum, consists of the ventral portions of the SNc, including columns of cells that penetrate with fingerlike extensions into the SNr. The heaviest labeling of DAT mRNA in the primate brain occurs within the ventral tier, where dense gene expression is localized to the ventral SNc and cell columns, while the dorsal tier, including the VTA, is much more diffusely labeled [27,37,38,40] (Fig. 1-2b and c). This differential pattern of gene expression is consistent not only with the greater cell density of the ventral tier, but also with the relative abundance of DAT protein in the nigrostriatal terminal fields compared to mesocortical and mesolimbic areas [49–59].

In the case of the primate, DAT gene expression outside the confines of the ventral midbrain remains, for the most part, a mystery. The presence or absence of DAT mRNA in the primate hypothalamus, striatum, and amygdala, for example, has yet to be characterized thoroughly, although DAT gene expression within the human cerebellar vermis has been reported [60]. Full characterization of the DAT gene outside the primate midbrain will help to crystallize the role of dopamine and its regulation, particularly in brain disorders.

1.2.2 DAT Protein Distribution

In contrast to the restricted nature of the topography of expression of DAT mRNA, the distribution of DAT protein in the brain is distinctly heterogeneous. It is localized predominantly to such areas as the caudate, putamen, and nucleus accumbens, which are known to be innervated by dopaminergic terminals (Fig. 1-3), or to sites such as the SN and VTA which are populated by dopaminergic cell bodies. The distribution of the DAT protein has been described extensively at subcellular and regional levels in rodent and primate species utilizing immunohistochemical [27,28,50,52,57,61–74] and autoradiographic [49,51,53–56,58,59,75–87] techniques.

Ultrastructural Localization of DAT Protein   On dopaminergic axons and terminals in the rat striatum the DAT is located primarily on extrasynaptic plasma membranes [57,67,88] (Fig. 1-4a), although a limited amount of DAT immunogold labeling occurs intracellularly, where it is associated with vesicular membranes [57,67]. Localization of the DAT to membranes at the periphery of synaptic areas and not directly over synaptic densities (Fig. 1-4b) suggests that the DAT's primary role in dopamine clearance lies in limiting the spatial diffusion of dopamine into the neuropil following tonic and phasic release. In the SN and VTA the DAT is localized most prominently on plasma membranes and smooth endoplasmic reticulum in dendrites [57,67,71] (Fig. 1-5a), while somatic labeling is cytoplasmic,

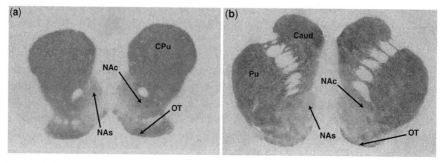

Figure 1-3   Binding of a tropane radioligand, [³H]2β-propanoyl-3β-(4-tolyl)tropane, to dopamine transporters in the rat (a) and monkey (b) striatum. Caud, caudate; CPu, caudate–putamen; NAc, nucleus accumbens core; NAs, nucleus accumbens shell; OT, olfactory tubercle; Pu, putamen.

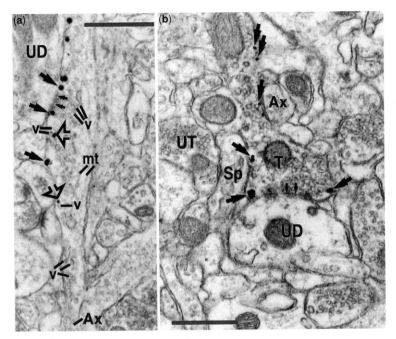

Figure 1-4 Immunogold labeling of dopamine transporters in TH-immunoreactive axons of the dorsal striatum of the rat. (a) Immunogold labeling of dopamine transporters (large closed arrows) can be seen on the plasma membrane of a varicose axon that forms a synaptic junction (small closed arrows) with an adjacent unlabeled dendrite (UD). Some gold particles are also associated with unlabeled pleomorphic vesicles (v) (open arrows). (b) Immunogold labeling for DAT (large arrows) can be seen along the plasma membrane of a TH-labeled axon terminal (T) which contacts a large unlabeled dendrite (UD). Immunogold labeling for the DAT can be seen lateral to the site of contact (small arrows). Ax, axon. (Adapted with permission from Nirenberg et al. [57]. Copyright © 1996 Society for Neuroscience.)

occurs almost exclusively in association with tubulovesicles, rough endoplasmic reticulum, and Golgi apparatus [57,67,71], and no doubt reflects sites of DAT synthesis and transport (Fig. 1-5b). Plasmalemmal labeling in the dendrites is both synaptic and nonsynaptic; in fact, the densest DAT labeling in the SN occurs predominantly at sites distant from synaptic junctions (Fig. 1-5a), as well as in dendritic pines, in which synapses are rarely seen [57,71]. Interestingly, the same characteristic of DAT immunolabeling distant from synaptic release sites has been observed in the prelimbic cortex of the rat, where extracellular dopamine diffusion has been reported to be greater than in dorsal striatum [89–92]. The fact that dopamine is released locally from nigral cell dendrites has been well established [93–100], and synaptic vesicles and intrinsic dopaminergic axon terminals are relatively uncommon within the SN. Furthermore, despite its more frequent localization to extrasynaptic membranes, DAT immunoreactivity has also been visualized at the site of dendrodendritic synapses in the SN [67]. Localization of the DAT to both synaptic and nonsynaptic

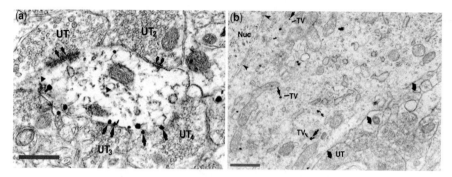

Figure 1-5 Immunogold labeling of dopamine transporters in the substantia nigra of the rat. (a), Immunogold-labeled dopamine transporters (large arrows) in a dendrite diffusely immunoperoxidase labeled for tyrosine hydroxylase (arrowheads). Two unlabeled terminals (UT_1, UT_2) form synaptic junctions (small arrows) onto portions of the dendrite devoid of immunogold labeling for the DAT, while other terminals (UT_3, UT_4) contact portions of the dendrite on which several gold particles for DAT can be seen. (b) Immunogold labeling of the DAT is localized to internal membranes of a cell body in the SN. Immunogold particles for DAT (straight arrows) are associated with tubulovesicular membranes (TV), but not with the nuclear (arrowheads) or plasma (curved arrows) membranes. Nuc, nucleus; UT, unlabeled terminal. (Adapted from Nirenberg et al. [57]. Copyright © 1996 Society for Neuroscience.)

loci therefore suggests a multifunctional role for the DAT in the SN involving reuptake of diffused transmitter at extrasynaptic sites, as may be the case in the cortex, as well as local clearance of dopamine from the dendrodendritic synaptic space. In addition, consistent with the lack of conventional vesicular release of dopamine and the placement of DATs directly at the synapse, there is a body of evidence which suggests that dopamine release from SN dendrites may occur via reverse transport by the DAT under certain abnormal physiological conditions [101–104] and may even be a routine means of dopamine transmission and dendrodendritic inhibition in the midbrain [105].

Regional Distribution of DAT Protein in the Rodent Brain Early autoradiographic mapping studies employed nonselective radioligands that targeted not only the DAT but also serotonin and norepinephrine transporters [77,79,86,106–109] or unrelated sites [58,110]. The development of highly selective ligands such as [^{3}H]WIN 35,428 and [^{125}I]RTI-121, however, provided superior tools for visualizing the regional distribution of DAT-binding sites [49,59,75,85]. In the rodent brain, DAT binding accumulates most heavily in the caudate and putamen, with more moderate levels of binding sites in the nucleus accumbens core and olfactory tubercle, and light accumulation in the nucleus accumbens shell, SN, VTA, and subthalamic nucleus [49,55,78,87] (Fig. 1-3a). Increasing medial–lateral gradients of DAT binding density within the caudate and putamen have been reported [78,79,87]; however, the literature is somewhat inconsistent on the existence and direction of rostral–caudal gradients.

While immunochemical analyses of DAT protein distribution are largely consistent with the findings from binding studies, the superior sensitivity and spatial resolution of these methods have resulted in identification of DAT immunoreactivity in brain regions that do not express the protein in amounts sufficient to be detected by standard autoradiographic techniques. Immunochemical analyses confirm the robust expression of DAT protein in the midbrain dopaminergic perikarya as well as in the neuropil of terminal fields such as striatum [50,72,74,111], where the lateral–medial gradient seen in receptor autoradiography is mentioned [27], as is a pattern of patchy expression [50,52] consistent with the striosome–matrix concept of cellular compartmentation in the striatum [112,113]. The nucleus accumbens is populated heterogeneously with DAT immunoreactivity; the labeling of the core compartment is more or less continuous with that of the adjacent ventral caudate–putamen, while immunolabeling in the shell is much lighter and distinctly patchy in appearance [50,52]. DAT immunoreactivity has also been detected in axonal fibers and arborizations in the medial forebrain bundle, lateral habenula, entopeduncular nucleus, anterior cingulate and prelimbic cortices, claustrum–dorsal endopiriform nucleus, bed nucleus of the stria terminalis, intermediate lateral septum, zona incerta, median eminence, glomerular region of the olfactory bulb, and amygdala [50,72,74]. DAT protein in the amygdala is expressed heterogeneously, with labeling occurring most predominantly in the central nucleus and the intercalated nuclei. The basal and medial nuclei, however, seem to be unfailingly DAT-negative [50,72]. Conspicuously absent from any of the descriptions of DAT immunolocalization are the dopaminergic cells of the hypothalamus. With the exception of one report of extremely diffuse labeling of scattered cells in the arcuate nucleus [72], none of the hypothalamic dopamine cell groups that are DAT mRNA-positive appear to express somatic DAT protein. This prominent lack of protein expression within mRNA-positive hypothalamic cell bodies, along with the robust labeling of DATs at their terminals site in the median eminence, suggests that the processes of DAT production, storage, and transport in the tuberoifundibular system may operate via different mechanisms than those in the midbrain.

Regional Distribution of DAT Protein in the Primate Brain The distribution of DAT protein in the primate brain largely parallels its appearance in the rodent, with the expected densities of labeling in many of the dopaminergic terminal field structures. Receptor autoradiography studies in both monkeys and humans therefore describe high accumulations of radioligand in caudate and putamen along with moderate binding in nucleus accumbens, olfactory tubercle, and SN [51,53,54,59,75,76,80,82,86,112] (Fig. 1-3b). Subregional analyses of the monkey caudate and putamen have revealed decreasing dorsal–ventral and anterior–posterior gradients [51,54,82], although a separate analysis of the ventral caudate would no doubt reveal an opposite gradient due to the more uniform pattern of binding at post-commissural levels. The heterogeneous binding characteristics of striosome–matrix compartments are confined largely to ventral portions of the caudate, while the putamen and dorsal caudate are labeled somewhat more uniformly [82]. Low, but detectable levels of binding to the DAT have also been described in the median eminence, hypothalamus, thalamus, bed nucleus of the stria terminalis, globus pallidus, zona incerta, central, basolateral, and basomedial nuclei of the amygdala, hippocampus, and VTA of monkeys [51,54].

Similar to the subregional striatal pattern observed in the monkey, human studies have noted lower binding in the ventral caudate and nucleus accumbens core than in dorsal areas; however, analyses of anterior–posterior binding gradients in human striatum have described both increasing and decreasing levels in the caudate and putamen [59,84,107]. These discordant findings are probably due to the use of a variety of both ligands and anterior–posterior brain levels. Studies with the selective ligand [^{125}I]RTI-121, however, have indicated that unlike the pattern seen in the monkey and rodent, the anterior-to-posterior gradient *increases* in density in the human striatum [59]. As is the case in the monkey striatum, binding in the human has been observed to be greater in matrix than in striosomal compartments [114]. There are few descriptions of extrastriatal binding of selective ligands to the human DAT, however, Staley and colleagues (1995) have reported very low binding levels in the cortex, amygdala, and globus pallidus (1995), while Little et al. (1995) focused on the temporal lobe and reported low but measurable quantities of DAT in the hippocampus and amygdala [56]. Both of these studies were undertaken using [^{125}I]RTI-121; however, Staley et al. detected DATs in the central and lateral nuclei of the amygdala, while Little et al. reported that DAT protein was restricted to the basolateral nucleus. These discrepancies are due, perhaps, to the limitations inherent in resolution of the autoradiographic method as well as interpretation of data obtained from a region of the brain where binding levels are only marginally above background.

Immunohistochemical examination of DAT distribution in the primate brain has not been extensive. Aside from the findings of Ciliax et al. (1999), who undertook a thorough description of DAT immunoreactivity in the human brain [62], the layout of DAT labeling in the primate must be pieced together from numerous studies that have looked at the DAT in restricted areas of the brain [61,63–66,68,69,73,115]. DAT immunoreactivity is found in established dopamine mesostriatal, mesolimbic, and mesocortical areas, and conforms largely to the topography expected based on rodent and autoradiographic findings.

As in the rodent, robustly labeled DAT-positive cells are found in the A9 and A10 cells of the human ventral mesencephalon; however, Ciliax and colleagues do not mention the A8 cell group [62], in which DAT immunoreactivity has been identified in rat studies [50,52]. This study employed double labeling for the DAT and TH, and interestingly, although mesencephalic DAT and TH immunoreactivities were almost identical in appearance, small subsets of labeled neurons were observed to be either TH-positive and DAT-negative, or TH-negative and DAT-positive, reminiscent of the similar findings with DAT mRNA and TH double-labeling studies in rodents [23,24,28,29]. Immunoreactivity in the human striatum is intense, with denser labeling of fibers in the matrix than in patch compartments [62], which again is consistent with the autoradiographic and rodent immunohistochemical findings. Aside from the expected labeling of axon fibers, there is some evidence that the primate striatum, particularly in ventral zones, may contain a scattered population of intrinsic

DAT-positive dopamine neurons [61,63,64] that respond to dopamine denervation by proliferating—perhaps as a compensatory mechanism for dopamine depletion. The amygdala appears to be innervated by dopamine more extensively in the human than in the rodent, as DAT immunoreactive axons can be visualized not only in the central nucleus as in the rat, but also in the lateral, basolateral, and basomedial nuclei. The densest overall labeling in the human amygdala occurs in the basolateral nucleus, while in the central nucleus DAT immunoreactivity is heterogeneous and occurs only in dense patches. The dopaminergic cell groups of the human hypothalamus, like those of the rodent, are devoid of DAT immunoreactive cell bodies; however, DAT-positive axons are plentiful, particularly on the medial border of the hypothalamus, adjacent to the third ventricle [62]. The sparse labeling that occurs in both internal and external segments of the globus pallidus can be attributed to labeling of the medial forebrain bundle and other ascending dopaminergic fibers.

Two important areas of considerable divergence of DAT densities between species are the thalamus and the cortex. In 1991, Berger et al. pointed out the apparent increase in the density of dopaminergic fibers in the cortex as one ascends the phylogenetic tree from rodent to primate [116]. In the rat, for example, DAT-positive fibers are restricted to anterior cingulate and prelimbic cortices [50,74], while in the monkey, the range of innervation by DAT-positive fibers is far more extensive spatially and is heterogeneous in both density and laminar distribution. The densest labeling occurs in primary motor and premotor cortical regions, intermediate levels in association regions, and lowest levels in primary sensory areas [115]. In addition to extensive DAT presence in the neocortex, the primate entorhinal cortex contains relatively dense immunoreactivity [65], and DAT-positive axons form compact bands in the molecular and polymorphic layers of the monkey dentate gyrus [115]. In rodents the primary monoaminergic input to the hippocampus is noradrenergic; however, [^{3}H]nisoxetine binding to norepinephrine transporters in the monkey hippocampus is extremely sparse [117]. It is quite likely, therefore, that dopamine has a greater role in modulating hippocampal processing in the primate than it does in lower species. Like that of the monkey, the human cortex is rich in DAT-positive fibers; in fact, Ciliax and colleagues found immunoreactive axons in all regions of the cortex in which they looked, although the relative abundancies were heterogeneous both topographically and in density [62]. The greatest accumulation of labeled fibers can be found in the premotor cortical areas, while the least dense labeling occurs in the visual cortex. As is the case in the cell bodies of ventral midbrain, double labeling for TH and DAT reveals some fibers in both rodent and primate cortices that stain positive for TH and either negative or variable in intensity for the DAT. It is probable that a subset of these TH-positive and DAT-negative fibers are noradrenergic; however, in the human, their morphological characteristics indicate that they are predominantly of dopaminergic rather than noradrenergic origin [62,118]. These subpopulations of dopaminergic fibers in the cortex may correspond to the populations of mesocortical neurons in the VTA that show similar differential TH and DAT profiles.

Dopaminergic innervation of the rodent thalamus is sparse; however, there is substantial evidence of the presence of dopamine in the primate thalamus (see, e.g., [119–121]). In fact, the primate thalamus is the site of substantial innervation by DAT immunoreactive fibers [66,69,73]. The profile of DAT immunolabeling in the monkey thalamus is extensive, and although the densest labeling occurs in the mediodorsal, lateral posterior, posterior portions of ventrolateral, and the rostral central lateral nuclei, moderate to weak labeling can be observed in almost all the remaining nuclei. The density of DAT immunopositive fibers in the human thalamus is as great as, or even greater than, in the monkey. Although the topography of areas of densest labeling in the human does not overlap completely with that of the monkey, DAT immunolabeling in both species is extensive, and as is the case in the nonhuman primate, virtually all human thalamic nuclei are positive for DAT labeling [66,73]. Retrograde tract tracing reveals that dopaminergic innervation of the primate thalamus arises from multiple sources rather than any single conventional dopamine pathway. Furthermore, the distribution of DAT immunoreactivity in the primate brain does not correspond completely with dopamine immunoreactivity. Dopamine-positive fibers tend to demonstrate a more restricted distribution, concentrated more densely in the midline nuclei than elsewhere in the thalamus, while as mentioned above, DAT immunolabeling is far more extensive. According to Sanchez-Gonzalez et al. (2005), two broadly defined dopaminergic domains can be identified in the primate thalamus: the midline region, which is rich in dopamine and poor in DAT and whose source of dopamine is primarily hypothalamic, and a nonmidline region, low in dopamine and dense in DAT, which receives its dopaminergic input preferentially from the midbrain A8 to A11 cell groups [73].

Finally, one other primate brain region that has been demonstrated to contain dopaminergic elements is the cerebellum. Although the cerebellum is not typically thought of as the site of dopaminergic transmission, Purkinje cells have, in fact, been shown to express dopamine receptors [122]. Consistent with the concept of dopamine transmission and uptake in the cerebellum, relatively high densities of DAT and TH-immunoreactive fibers have been localized to lobules II, III, and IV of the vermis [68].

1.2.3 Changes in DAT Distribution with Age

In concluding a description of DAT distribution in the brain, we would be remiss if we failed to mention that any descriptions contained in this chapter can only be generalized to the young adult brain, since as the brain ages there is a precipitous decline in DAT mRNA and protein expression [36,37,81,123–132]. Declines in the level of human DAT mRNA can eventually eliminate the regional heterogeneity in levels of labeling that are evident in the young, but mature midbrain. Not only do profound losses of gene expression occur (up to 95% in some areas), but they occur in such a way that the regions with the greatest initial levels of DAT mRNA experience the greatest decrements, while dorsal tier areas such as the retrorubral fields, which have the lowest levels of expression initially, lose the least, resulting in equivalent levels of mRNA across both dorsal and ventral tier structures [37]. The DAT protein in the midbrain suffers similar losses. Declines of up to 73% in the nigrostriatal domains of the SN and 66% in the mesolimbic VTA have been measured autoradiographically in the rat [123], and in the human, reductions in the number of DAT-immunoreactive neurons in the SN were of similar magnitude [126].

Examination of the dopamine terminal fields in the aging rodent brain reveals a parallel loss of DAT protein in nigrostriatal and mesolimbic target areas [123]; however, in the monkey, there appears to be a much greater decline in DAT protein in nigrostriatal territories, such as caudate and putamen, than in mesolimbic terminal areas such as nucleus accumbens and olfactory tubercle [81]. This also appears to be true of the human striatum, where loss of radioligand binding to the DAT has been demonstrated to be greater in dorsal striatal domains than in nucleus accumbens [129]. Although these findings should probably be treated with caution, as they were obtained using small numbers of subjects, they are consistent with the replicated mRNA data, which show preferential losses in nigrostriatal neurons [37].

1.3 VISUALIZING THE DAT IN HUMANS

Although the distribution of the DAT can be revealed in postmortem studies, and nonhuman primate studies of animal models can complement this work, in vivo imaging tools are required to fully realize the nature of the changes in the anatomy and function of the dopamine system in human psychiatric and neurological disorders. Because DAT distribution in the CNS coincides with dopaminergic innervation, DAT ligands have been developed for use in neuroimaging as in vivo markers of dopaminergic systems. Positron emission tomography (PET) and single-photon emission computed tomography (SPECT) are molecular imaging techniques used to image ongoing biological processes via radiolabeled ligands.

The PET technique utilizes radiotracers labeled with relatively short-lived positron-emitting radionuclides such as [^{18}F] or [^{76}Br], or very short-lived radionuclides such as [^{11}C] and [^{15}O]. Due to their high specific radioactivity, very small amounts of radiolabeled tracers may be used with this technique. Minimizing both the amount and the duration of exposure to radioactivity is crucial for studying ongoing clinical disease, which makes the specific, short-lived radiotracers available for PET advantageous when investigating clinical populations (e.g., [^{11}C] has a half-life of 20 minutes). Given the short half-lives of the PET ligands, however, they must be synthesized in a cyclotron on-site and administered to patients immediately. This limits the widespread use and availability of PET.

In the absence of PET imaging, SPECT is often used. SPECT is closely related to PET, but the isotopes utilized have longer half-lives (e.g., [^{123}I] has a half-life of 13 hours) and may be stored on site. SPECT was developed 10 to 20 years earlier than PET, and its resolution is much lower, yielding less detailed images which require more time to acquire. The experimental and clinical applications of SPECT are similar to those of PET, with the exception of functional imaging research, where rapid observation of changes is required. SPECT has been used, in particular, to study the neurotransmitter dopamine. Thus, whereas PET has the advantage of more quantitative instrumentation and more versatile choice of radionuclides, SPECT requires less apparatus and its use is therefore more widespread.

Consequently, for use in large clinical trials, SPECT maintains an advantage over PET, due to wider accessibility and ready availability of its radioligands.

Historically, it was not until the 1980s that we were able to visualize the scope of the dopamine system in the live human brain. Using a technology traditionally reserved for oncological research, Garnett et al. (1983) synthesized and administered dopamine radiolabeled with ^{18}F to five participants while PET recorded the uptake volume and distribution of the radiolabeled tracer [133]. This highly publicized study was followed by several other landmark investigations into the regional distribution of dopamine receptors in the human brain [134,135] as well as attempts to characterize the role of altered dopamine binding in ongoing disease [136]. Using 6-fluoro-dopa, Calne et al. (1985) provided the first direct evidence that nigrostriatal dopamine impairment may exist in the absence of overt clinical symptoms [137]. These early studies of in vivo dopamine imaging in patients opened a fresh avenue for an examination of the role of dopamine in healthy humans as well as in pathological conditions.

Over the last decade, a large number of radiotracers have been developed to image and quantify transporter availability with PET or SPECT [138,139]. As outlined by Piccini in 2003, several characteristics guide development of radiotracers for DAT imaging; for example, the ligand must cross the blood–brain barrier and have high affinity and selectivity for DAT sites and low nonspecific binding, and the metabolites of the ligand must not interfere with quantification of DAT binding [140]. Finally, for quantification purposes a ligand must possess the characteristics that would allow it to be labeled with positron- or single-photon-emitting radionuclides.

The first compound available to image the DAT in vivo was [^{11}C]cocaine [141,142]. The uptake and clearance of [^{11}C]cocaine in the human brain, however, is very fast, with peak striatal activity reached within 10 minutes. Although [^{11}C]cocaine is useful for investigating the pharmacokinetics of cocaine, it is not an ideal probe to quantify DAT binding, due to its rapid kinetics and similar affinity for the other catecholamine transporters [143]. To address the issues of kinetics and selectivity, significant advances were made in the 1990s in attempting to modify the structure of cocaine [139]. A family of tropane analogs of cocaine was shown to be highly selective for the DAT. The tropanes are currently the most common radioligands used for both PET and SPECT imaging in humans. β-CIT [2β-carbomethoxy-3β-(4-iodophenyl)tropane] was the first cocaine derivative developed for DAT imaging in the tropane family. The kinetics of [^{123}I]β-CIT, however, were not practical for imaging patient populations, as it required a 24-hour period between injection of the tracer and acquisition of the imaging data.

Subsequently, several other tropanes were developed with faster kinetics. The β-CIT derivatives [^{123}I]CIT-FE and [^{123}I]CIT-FP, as well as the cocaine derivative CFT [2β-carbomethoxy-3β-(4-fluorophenyl)tropane], can be labeled with either ^{11}C or ^{18}F to measure DAT binding using PET. β-CIT was developed for use with SPECT cameras [144,145] and to date, ^{123}I-labeled β-CIT, along with [^{99m}Tc]TRODAT-1 [146–149], have been widely used for SPECT imaging of DAT density in the human brain. In addition to the tropane analogs,

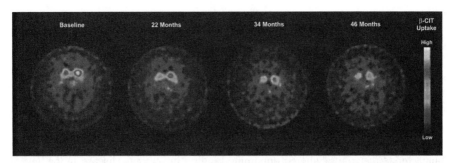

Figure 1-6 In Parkinson's disease DAT binding in the caudate and putamen declines with disease progression, as revealed by [^{123}I]β-CIT SPECT imaging in an early-stage PD patient. (Replicated with permission from the Parkinson Study Group [160].) (*See insert for color representation of figure.*)

[^{11}C]*d*-threo-methylphenidate has also been used successfully to imaging the DAT in humans [150–153].

Tracers such as these have advanced our ability to diagnose and follow disease progression (Fig. 1-6). Consequently, DAT imaging in patient populations has been used as a tool for detecting early-stage Parkinson's disease [154] and has been used to assess the efficacy of therapeutic interventions on neural functioning in Parkinson's patients [155–157]. In addition, clinical signs similar to those of PD patients are observed in a number of other disorders, such as Wilson's disease [158]. Use of PET imaging with DAT tracers has greatly aided the differential diagnosis of Parkinson's disease from similar movement disorders [159], thereby improving the treatment of these disorders.

One of the key advances in the development of these tracers, however, has been the ability to visualize the DAT in regions beyond the striatum. Many human psychiatric and neurological disorders are characterized by dopamine dysfunction in areas such as the cortex. Therefore, those tracers that allow visualization of the distribution and concentration of the DAT throughout the brain will provide the greatest insights into human disease and help to clarify the role of dopamine in these disorders.

1.4 CONCLUSIONS

In this chapter we have reviewed and attempted to synthesize reports investigating the distribution and localization of the DAT in rodents and primates from both a cellular and a systems perspective. In the rodent brain the DAT is located on somatic, dendritic, and axonal membranes in the dopaminergic cell groups of the ventral mesencephalon, as well as on terminal fibers in the basal ganglia, hypothalamus, central amygdala, and a highly restricted region of the cortex. Different subpopulations of midbrain dopamine neurons have been identified which express varying levels of the DAT. These groups of neurons may correspond to, and be the sources of, similar subpopulations of downstream dopaminergic fibers, particularly in extrastriatal areas such as cortex, which also express levels of the DAT differentially. At dopaminergic terminals the DAT tends to be located adjacent to, rather than directly over, synaptic densities, indicating that clearance of dopamine typically occurs only after diffusion from the synaptic cleft. In some locations, particularly in cortex, the DAT is found some distance from the synapse, pointing to the potential for greater spatial diffusion and temporal differences in dopamine transmission.

In the primate, the distribution of DAT protein has a profile similar to that in the rodent, particularly in the dopaminergic cell body nuclei; however, it has a greater anatomical range of expression in the dopamine terminal fields. In the primate brain, for example, the DAT is found in a greater number of nuclei within the amygdala and is present in areas in which it is relatively scant in the rodent (if present at all), such as the thalamus, hippocampus, and in the case of the human, virtually all of the cortical mantle. Compared to the restricted nature of DAT expression in these areas in the rodent, the widespread nature of DAT distribution in the thalamus and cortex of the human points to a far greater role for dopamine in the modulation of cortical activity and suggests an exquisitely fine level of dopaminergic control over higher-order processing. By developing ever more sophisticated probes for in vivo imaging of the DAT, we are gaining not only an appreciation of the contribution of dopamine to this type of processing, but also an expanding base of knowledge regarding its role in neurological and psychiatric disorders.

ACKNOWLEDGMENTS

This work was supported by NIH grants DA06634 and DA009085.

REFERENCES

1. Carlsson, A., Lindqvist, M., Magnusson, T., and Waldeck, B. (1958). On the presence of 3-hydroxytyramine in brain. *Science, 127*, 471.

2. Carlsson, A. (1959). The occurrence, distribution and physiological role of catecholamines in the nervous system. *Pharmacol Rev, 11*, 490–493.

3. Falck, B., Hillarp, N. A., Thieme, G., and Torp, A. (1962). Fluorescence of catecholamines and related compounds with formaldehyde. *J Histochem Cytochem, 10*, 348–354.

4. Carlsson, A., Falck, B., and Hillarp, N. A. (1962). Cellular localization of brain monoamines. *Acta Physiol Scand Suppl, 56*, 1–28.

5. Anden, N. E., Carlsson, A., Dahlstroem, A., Fuxe, K., Hillarp, N. A., and Larsson, K. (1964). Demonstration and mapping out of nigro-neostriatal dopamine neurons. *Life Sci, 3*, 523–530.

6. Dahlström, A., and Fuxe, K. (1964). Evidence for the existence of monoamine-containing neurons in the central nervous system: I. Demonstration of monoamines in the cell bodies of brain stem neurons. *Acta Physiol Scand Suppl, 232*, 231–255.

7. Bjorklund, A., Lindvall, O., and Nobin, A. (1975). Evidence of an incerto-hypothalamic dopamine neurone system in the rat. *Brain Res, 89*, 29–42.

8. Bjorklund, A., and Nobin, A. (1973). Fluorescence histochemical and microspectro-fluorometric mapping of dopamine and noradrenaline cell groups in the rat diencephalon. *Brain Res, 51,* 193–205.

9. Felten, D. L., and Sladek, J. R., Jr. (1983). Monoamine distribution in primate brain. V: Monoaminergic nuclei—anatomy, pathways and local organization. *Brain Res Bull, 10,* 171–284.

10. Hoffman, G. E., Felten, D. L., and Sladek, J. R., Jr. (1976). Monoamine distribution in primate brain. III: Catecholamine-containing varicosities in the hypothalamus of *Macaca mulatta. Am J Anat, 147,* 501–513.

11. Sladek, J. R., Jr., Garver, D. L., and Cummings, J. P. (1982). Monoamine distribution in primate brain. IV: Indoleamine-containing perikarya in the brain stem of *Macaca arctoides. Neuroscience, 7,* 477–493.

12. Guastella, J., Nelson, N., Nelson, H., Czyzyk, L., Keynan, S., Miedel, M. C., Davidson, N., Lester, H. A., and Kanner, B. I. (1990). Cloning and expression of a rat brain GABA transporter. *Science, 249,* 1303–1306.

13. Pacholczyk, T., Blakely, R. D., and Amara, S. G. (1991). Expression cloning of a cocaine- and antidepressant-sensitive human noradrenaline transporter. *Nature, 350,* 350–354.

14. Giros, B., el Mestikawy, S., Bertrand, L., and Caron, M. G. (1991). Cloning and functional characterization of a cocaine-sensitive dopamine transporter. *FEBS Lett, 295,* 149–154.

15. Kilty, J. E., Lorang, D., and Amara, S. G. (1991). Cloning and expression of a cocaine-sensitive rat dopamine transporter. *Science, 254,* 578–579.

16. Shimada, S., Kitayama, S., Lin, C. L., Patel, A., Nanthakumar, E., Gregor, P., Kuhar, M., and Uhl, G. (1991). Cloning and expression of a cocaine-sensitive dopamine transporter complementary DNA. *Science, 254,* 576–578.

17. Usdin, T. B., Mezey, E., Chen, C., Brownstein, M. J., and Hoffman, B. J. (1991). Cloning of the cocaine-sensitive bovine dopamine transporter. *Proc Natl Acad Sci U S A, 88,* 11168–11171.

18. Giros, B., el Mestikawy, S., Godinot, N., Zheng, K., Han, H., Yang-Feng, T., and Caron, M. G. (1992). Cloning, pharmacological characterization, and chromosome assignment of the human dopamine transporter. *Mol Pharmacol, 42,* 383–390.

19. Jayanthi, L. D., Apparsundaram, S., Malone, M. D., Ward, E., Miller, D. M., Eppler, M., and Blakely, R. D. (1998). The *Caenorhabditis elegans* gene T23G5.5 encodes an antidepressant- and cocaine-sensitive dopamine transporter. *Mol Pharmacol, 54,* 601–609.

20. Miller, G. M., Yatin, S. M., De La Garza, R., 2nd, Goulet, M., and Madras, B. K. (2001). Cloning of dopamine, norepinephrine and serotonin transporters from monkey brain: relevance to cocaine sensitivity. *Brain Res Mol Brain Res, 87,* 124–143.

21. Wu, X., and Gu, H. H. (1999). Molecular cloning of the mouse dopamine transporter and pharmacological comparison with the human homologue. *Gene, 233,* 163–170.

22. Chen, N., and Reith, M. E. (2000). Structure and function of the dopamine transporter. *Eur J Pharmacol, 405,* 329–339.

23. Augood, S. J., Westmore, K., McKenna, P. J., and Emson, P. C. (1993). Co-expression of dopamine transporter mRNA and tyrosine hydroxylase mRNA in ventral mesencephalic neurones. *Brain Res Mol Brain Res, 20,* 328–334.

24. Blanchard, V., Raisman-Vozari, R., Vyas, S., Michel, P. P., Javoy-Agid, F., Uhl, G., and Agid, Y. (1994). Differential expression of tyrosine hydroxylase and membrane dopamine transporter genes in subpopulations of dopaminergic neurons of the rat mesencephalon. *Brain Res Mol Brain Res, 22,* 29–38.

25. Burchett, S. A., and Bannon, M. J. (1997). Serotonin, dopamine and norepinephrine transporter mRNAs: heterogeneity of distribution and response to "binge" cocaine administration. *Brain Res Mol Brain Res, 49,* 95–102.

26. Cerruti, C., Walther, D. M., Kuhar, M. J., and Uhl, G. R. (1993). Dopamine transporter mRNA expression is intense in rat midbrain neurons and modest outside midbrain. *Brain Res Mol Brain Res, 18,* 181–186.

27. Gonzalez-Hernandez, T., Barroso-Chinea, P., De La Cruz Muros, I., Del Mar Perez-Delgado, M., and Rodriguez, M. (2004). Expression of dopamine and vesicular monoamine transporters and differential vulnerability of mesostriatal dopaminergic neurons. *J Comp Neurol, 479,* 198–215.

28. Hoffman, B. J., Hansson, S. R., Mezey, E., and Palkovits, M. (1998). Localization and dynamic regulation of biogenic amine transporters in the mammalian central nervous system. *Front Neuroendocrinol, 19,* 187–231.

29. Lorang, D., Amara, S. G., and Simerly, R. B. (1994). Cell-type-specific expression of catecholamine transporters in the rat brain. *J Neurosci, 14,* 4903–4914.

30. Maggos, C. E., Spangler, R., Zhou, Y., Schlussman, S. D., Ho, A., and Kreek, M. J. (1997). Quantitation of dopamine transporter mRNA in the rat brain: mapping, effects of "binge" cocaine administration and withdrawal. *Synapse, 26,* 55–61.

31. Meister, B., and Elde, R. (1993). Dopamine transporter mRNA in neurons of the rat hypothalamus. *Neuroendocrinology, 58,* 388–395.

32. Sanghera, M. K., Manaye, K. F., Liang, C. L., Lacopino, A. M., Bannon, M. J., and German, D. C. (1994). Low dopamine transporter mRNA levels in midbrain regions containing calbindin. *Neuroreport, 5,* 1641–1644.

33. Shimada, S., Kitayama, S., Walther, D., and Uhl, G. (1992). Dopamine transporter mRNA: dense expression in ventral midbrain neurons. *Brain Res Mol Brain Res, 13,* 359–362.

34. Bodoky, M., and Rethelyi, M. (1977). Dendritis arborization and axon trajectory of neurons in the hypothalamic arcuate nucleus of the rat. *Exp Brain Res, 28,* 543–555.

35. Zaborszky, L., and Makara, G. B. (1979). Intrahypothalamic connections: an electron microscopic study in the rat. *Exp Brain Res, 34,* 201–215.

36. Bannon, M. J., Poosch, M. S., Xia, Y., Goebel, D. J., Cassin, B., and Kapatos, G. (1992). Dopamine transporter mRNA content in human substantia nigra decreases precipitously with age. *Proc Natl Acad Sci U S A, 89,* 7095–7099.

37. Bannon, M. J., and Whitty, C. J. (1997). Age-related and regional differences in dopamine transporter mRNA expression in human midbrain. *Neurology, 48,* 969–977.

38. Haber, S. N., Ryoo, H., Cox, C., and Lu, W. (1995). Subsets of midbrain dopaminergic neurons in monkeys are distinguished by different levels of mRNA for the dopamine transporter: comparison with the mRNA for the D2 receptor, tyrosine hydroxylase and calbindin immunoreactivity. *J Comp Neurol, 362,* 400–410.

39. Harrington, K. A., Augood, S. J., Kingsbury, A. E., Foster, O. J., and Emson, P. C. (1996). Dopamine transporter (DAT) and synaptic vesicle amine transporter (VMAT2)

gene expression in the substantia nigra of control and Parkinson's disease. *Brain Res Mol Brain Res, 36,* 157–162.

40. Hurd, Y. L., Pristupa, Z. B., Herman, M. M., Niznik, H. B., and Kleinman, J. E. (1994). The dopamine transporter and dopamine D2 receptor messenger RNAs are differentially expressed in limbic- and motor-related subpopulations of human mesencephalic neurons. *Neuroscience, 63,* 357–362.

41. Uhl, G. R., Walther, D., Mash, D., Faucheux, B., and Javoy-Agid, F. (1994). Dopamine transporter messenger RNA in Parkinson's disease control substantia nigra neurons. *Ann Neurol, 35,* 494–498.

42. Counihan, T. J., and Penney, J. B., Jr. (1998). Regional dopamine transporter gene expression in the substantia nigra from control and Parkinson's disease brains. *J Neurol Neurosurg Psychiatry, 65,* 164–169.

43. Fallon, J. H., Koziell, D. A., and Moore, R. Y. (1978). Catecholamine innervation of the basal forebrain. II: Amygdala, suprarhinal cortex and entorhinal cortex. *J Comp Neurol, 180,* 509–532.

44. Fallon, J. H., and Moore, R. Y. (1978). Catecholamine innervation of the basal forebrain. IV: Topography of the dopamine projection to the basal forebrain and neostriatum. *J Comp Neurol, 180,* 545–580.

45. Gerfen, C. R., Baimbridge, K. G., and Thibault, J. (1987). The neostriatal mosaic. III: Biochemical and developmental dissociation of patch-matrix mesostriatal systems. *J Neurosci, 7,* 3935–3944.

46. Gerfen, C. R., Herkenham, M., and Thibault, J. (1987). The neostriatal mosaic. II: Patch- and matrix-directed mesostriatal dopaminergic and non-dopaminergic systems. *J Neurosci, 7,* 3915–3934.

47. Lynd-Balta, E., and Haber, S. N. (1994). The organization of midbrain projections to the striatum in the primate: sensorimotor-related striatum versus ventral striatum. *Neuroscience, 59,* 625–640.

48. Lynd-Balta, E., and Haber, S. N. (1994). The organization of midbrain projections to the ventral striatum in the primate. *Neuroscience, 59,* 609–623.

49. Boja, J. W., Cadet, J. L., Kopajtic, T. A., Lever, J., Seltzman, H. H., Wyrick, C. D., Lewin, A. H., Abraham, P., and Carroll, F. I. (1995). Selective labeling of the dopamine transporter by the high affinity ligand 3β-(4-[^{125}I]iodophenyl)tropane-2β-carboxylic acid isopropyl ester. *Mol Pharmacol, 47,* 779–786.

50. Ciliax, B. J., Heilman, C., Demchyshyn, L. L., Pristupa, Z. B., Ince, E., Hersch, S. M., Niznik, H. B., and Levey, A. I. (1995). The dopamine transporter: immunochemical characterization and localization in brain. *J Neurosci, 15,* 1714–1723.

51. De La Garza, R., 2nd, Meltzer, P. C., and Madras, B. K. (1999). Non-amine dopamine transporter probe [(3)H]tropoxene distributes to dopamine-rich regions of monkey brain. *Synapse, 34,* 20–27.

52. Freed, C., Revay, R., Vaughan, R. A., Kriek, E., Grant, S., Uhl, G. R., and Kuhar, M. J. (1995). Dopamine transporter immunoreactivity in rat brain. *J Comp Neurol, 359,* 340–349.

53. Hall, H., Halldin, C., Guilloteau, D., Chalon, S., Emond, P., Besnard, J., Farde, L., and Sedvall, G. (1999). Visualization of the dopamine transporter in the human brain post-mortem with the new selective ligand [^{125}I]PE2I. *Neuroimage, 9,* 108–116.

54. Kaufman, M. J., Spealman, R. D., and Madras, B. K. (1991). Distribution of cocaine recognition sites in monkey brain. I: In vitro autoradiography with [^{3}H]CFT. *Synapse, 9*, 177–187.

55. Letchworth, S. R., Smith, H. R., Porrino, L. J., Bennett, B. A., Davies, H. M., Sexton, T., and Childers, S. R. (2000). Characterization of a tropane radioligand, [(3)H]2β-propanoyl-3β-(4-tolyl)tropane ([(3)H]PTT), for dopamine transport sites in rat brain. *J Pharmacol Exp Ther, 293*, 686–696.

56. Little, K. Y., Carroll, F. I., and Cassin, B. J. (1995). Characterization and localization of [^{125}I]RTI-121 binding sites in human striatum and medial temporal lobe. *J Pharmacol Exp Ther, 274*, 1473–1483.

57. Nirenberg, M. J., Vaughan, R. A., Uhl, G. R., Kuhar, M. J., and Pickel, V. M. (1996). The dopamine transporter is localized to dendritic and axonal plasma membranes of nigrostriatal dopaminergic neurons. *J Neurosci, 16*, 436–447.

58. Richfield, E. K. (1991). Quantitative autoradiography of the dopamine uptake complex in rat brain using [^{3}H]GBR 12935: binding characteristics. *Brain Res, 540*, 1–13.

59. Staley, J. K., Boja, J. W., Carroll, F. I., Seltzman, H. H., Wyrick, C. D., Lewin, A. H., Abraham, P., and Mash, D. C. (1995). Mapping dopamine transporters in the human brain with novel selective cocaine analog [^{125}I]RTI-121. *Synapse, 21*, 364–372.

60. Hurley, M. J., Mash, D. C., and Jenner, P. (2003). Markers for dopaminergic neurotransmission in the cerebellum in normal individuals and patients with Parkinson's disease examined by RT-PCR. *Eur J Neurosci, 18*, 2668–2672.

61. Betarbet, R., Turner, R., Chockkan, V., DeLong, M. R., Allers, K. A., Walters, J., Levey, A. I., and Greenamyre, J. T. (1997). Dopaminergic neurons intrinsic to the primate striatum. *J Neurosci, 17*, 6761–6768.

62. Ciliax, B. J., Drash, G. W., Staley, J. K., Haber, S., Mobley, C. J., Miller, G. W., Mufson, E. J., Mash, D. C., and Levey, A. I. (1999). Immunocytochemical localization of the dopamine transporter in human brain. *J Comp Neurol, 409*, 38–56.

63. Cossette, M., Lecomte, F., and Parent, A. (2005). Morphology and distribution of dopaminergic neurons intrinsic to the human striatum. *J Chem Neuroanat, 29*, 1–11.

64. Cossette, M., Levesque, D., and Parent, A. (2005). Neurochemical characterization of dopaminergic neurons in human striatum. *Parkinsonism Relat Disord, 11*, 277–286.

65. Erickson, S. L., Akil, M., Levey, A. I., and Lewis, D. A. (1998). Postnatal development of tyrosine hydroxylase- and dopamine transporter–immunoreactive axons in monkey rostral entorhinal cortex. *Cereb Cortex, 8*, 415–427.

66. Garcia-Cabezas, M. A., Rico, B., Sanchez-Gonzalez, M. A., and Cavada, C. (2007). Distribution of the dopamine innervation in the macaque and human thalamus. *Neuroimage, 34*, 965–984.

67. Hersch, S. M., Yi, H., Heilman, C. J., Edwards, R. H., and Levey, A. I. (1997). Subcellular localization and molecular topology of the dopamine transporter in the striatum and substantia nigra. *J Comp Neurol, 388*, 211–227.

68. Melchitzky, D. S., and Lewis, D. A. (2000). Tyrosine hydroxylase- and dopamine transporter-immunoreactive axons in the primate cerebellum: evidence for a lobular- and laminar-specific dopamine innervation. *Neuropsychopharmacology, 22*, 466–472.

69. Melchitzky, D. S., and Lewis, D. A. (2001). Dopamine transporter–immunoreactive axons in the mediodorsal thalamic nucleus of the macaque monkey. *Neuroscience, 103*, 1033–1042.

70. Mengual, E., and Pickel, V. M. (2004). Regional and subcellular compartmentation of the dopamine transporter and tyrosine hydroxylase in the rat ventral pallidum. *J Comp Neurol, 468*, 395–409.

71. Nirenberg, M. J., Chan, J., Vaughan, R. A., Uhl, G. R., Kuhar, M. J., and Pickel, V. M. (1997). Immunogold localization of the dopamine transporter: an ultrastructural study of the rat ventral tegmental area. *J Neurosci, 17*, 5255–5262.

72. Revay, R., Vaughan, R., Grant, S., and Kuhar, M. J. (1996). Dopamine transporter immunohistochemistry in median eminence, amygdala, and other areas of the rat brain. *Synapse, 22*, 93–99.

73. Sanchez-Gonzalez, M. A., Garcia-Cabezas, M. A., Rico, B., and Cavada, C. (2005). The primate thalamus is a key target for brain dopamine. *J Neurosci, 25*, 6076–6083.

74. Sesack, S. R., Hawrylak, V. A., Matus, C., Guido, M. A., and Levey, A. I. (1998). Dopamine axon varicosities in the prelimbic division of the rat prefrontal cortex exhibit sparse immunoreactivity for the dopamine transporter. *J Neurosci, 18*, 2697–2708.

75. Canfield, D. R., Spealman, R. D., Kaufman, M. J., and Madras, B. K. (1990). Autoradiographic localization of cocaine binding sites by [^{3}H]CFT ([^{3}H]WIN 35,428) in the monkey brain. *Synapse, 6*, 189–195.

76. Chinaglia, G., Alvarez, F. J., Probst, A., and Palacios, J. M. (1992). Mesostriatal and mesolimbic dopamine uptake binding sites are reduced in Parkinson's disease and progressive supranuclear palsy: a quantitative autoradiographic study using [^{3}H]mazindol. *Neuroscience, 49*, 317–327.

77. Cline, E. J., Scheffel, U., Boja, J. W., Mitchell, W. M., Carroll, F. I., Abraham, P., Lewin, A. H., and Kuhar, M. J. (1992). In vivo binding of [^{125}I]RTI-55 to dopamine transporters: pharmacology and regional distribution with autoradiography. *Synapse, 12*, 37–46.

78. Coulter, C. L., Happe, H. K., Bergman, D. A., and Murrin, L. C. (1995). Localization and quantification of the dopamine transporter: comparison of [^{3}H]WIN 35,428 and [^{125}I]RTI-55. *Brain Res, 690*, 217–224.

79. Fujita, M., Shimada, S., Fukuchi, K., Tohyama, M., and Nishimura, T. (1994). Distribution of cocaine recognition sites in rat brain: in vitro and ex vivo autoradiography with [^{125}I]RTI-55. *J Chem Neuroanat, 7*, 13–23.

80. Goulet, M., Morissette, M., Grondin, R., Falardeau, P., Bedard, P. J., Rostene, W., and Di Paolo, T. (1999). Neurotensin receptors and dopamine transporters: effects of MPTP lesioning and chronic dopaminergic treatments in monkeys. *Synapse, 32*, 153–164.

81. Kaufman, M. J., and Madras, B. K. (1993). [^{3}H]CFT ([^{3}H]WIN 35,428) accumulation in dopamine regions of monkey brain: comparison of a mature and an aged monkey. *Brain Res, 611*, 322–325.

82. Letchworth, S. R., Nader, M. A., Smith, H. R., Friedman, D. P., and Porrino, L. J. (2001). Progression of changes in dopamine transporter binding site density as a result of cocaine self-administration in rhesus monkeys. *J Neurosci, 21*, 2799–2807.

83. Mennicken, F., Savasta, M., Peretti-Renucci, R., and Feuerstein, C. (1992). Autoradiographic localization of dopamine uptake sites in the rat brain with ^{3}H-GBR 12935. *J Neural Transm Gen Sect, 87*, 1–14.

84. Piggott, M. A., Marshall, E. F., Thomas, N., Lloyd, S., Court, J. A., Jaros, E., Costa, D., Perry, R. H., and Perry, E. K. (1999). Dopaminergic activities in the human

striatum: rostrocaudal gradients of uptake sites and of D1 and D2 but not of D3 receptor binding or dopamine. *Neuroscience, 90*, 433–445.

85. Scheffel, U., Pogun, S., Stathis, M., Boja, J. W., and Kuhar, M. J. (1991). In vivo labeling of cocaine binding sites on dopamine transporters with [³H]WIN 35,428. *J Pharmacol Exp Ther, 257*, 954–958.

86. Staley, J. K., Basile, M., Flynn, D. D., and Mash, D. C. (1994). Visualizing dopamine and serotonin transporters in the human brain with the potent cocaine analogue [¹²⁵I]RTI-55: in vitro binding and autoradiographic characterization. *J Neurochem, 62*, 549–556.

87. Wilson, J. M., Nobrega, J. N., Carroll, M. E., Niznik, H. B., Shannak, K., Lac, S. T., Pristupa, Z. B., Dixon, L. M., and Kish, S. J. (1994). Heterogeneous subregional binding patterns of ³H-WIN 35,428 and ³H-GBR 12,935 are differentially regulated by chronic cocaine self-administration. *J Neurosci, 14*, 2966–2979.

88. Svingos, A. L., Clarke, C. L., and Pickel, V. M. (1999). Localization of the delta-opioid receptor and dopamine transporter in the nucleus accumbens shell: implications for opiate and psychostimulant cross-sensitization. *Synapse, 34*, 1–10.

89. Cass, W. A., and Gerhardt, G. A. (1995). In vivo assessment of dopamine uptake in rat medial prefrontal cortex: comparison with dorsal striatum and nucleus accumbens. *J Neurochem, 65*, 201–207.

90. Garris, P. A., Collins, L. B., Jones, S. R., and Wightman, R. M. (1993). Evoked extra-cellular dopamine in vivo in the medial prefrontal cortex. *J Neurochem, 61*, 637–647.

91. Garris, P. A., and Wightman, R. M. (1994). Different kinetics govern dopaminergic trans-mission in the amygdala, prefrontal cortex, and striatum: an in vivo voltammetric study. *J Neurosci, 14*, 442–450.

92. Lee, T. H., Gee, K. R., Ellinwood, E. H., and Seidler, F. J. (1996). Combining "caged-dopamine" photolysis with fast-scan cyclic voltammetry to assess dopamine clearance and release autoinhibition in vitro. *J Neurosci Methods, 67*, 221–231.

93. Bernardini, G. L., Gu, X., Viscardi, E., and German, D. C. (1991). Amphetamine-induced and spontaneous release of dopamine from A9 and A10 cell dendrites: an in vitro electrophysiological study in the mouse. *J Neural Transm Gen Sect, 84*, 183–193.

94. Bjorklund, A., and Lindvall, O. (1975). Dopamine in dendrites of substantia nigra neurons: suggestions for a role in dendritic terminals. *Brain Res, 83*, 531–537.

95. Cheramy, A., Leviel, V., and Glowinski, J. (1981). Dendritic release of dopamine in the substantia nigra. *Nature, 289*, 537–542.

96. Paden, C., Wilson, C. J., and Groves, P. M. (1976). Amphetamine-induced release of dopamine from the substantia nigra in vitro. *Life Sci, 19*, 1499–1506.

97. Robertson, G. S., Damsma, G., and Fibiger, H. C. (1991). Characterization of dopamine release in the substantia nigra by in vivo microdialysis in freely moving rats. *J Neurosci, 11*, 2209–2216.

98. Rosales, M. G., Flores, G., Hernandez, S., Martinez-Fong, D., and Aceves, J. (1994). Activation of subthalamic neurons produces NMDA receptor–mediated dendritic dopa-mine release in substantia nigra pars reticulata: a microdialysis study in the rat. *Brain Res, 645*, 335–337.

99. Geffen, L. B., Jessell, T. M., Cuello, A. C., and Iversen, L. L. (1976). Release of dopa-mine from dendrites in rat substantia nigra. *Nature, 260*, 258–260.

100. Nieoullon, A., Cheramy, A., and Glowinski, J. (1977). Nigral and striatal dopamine release under sensory stimuli. *Nature, 269*, 340–342.

101. Eshleman, A. J., Henningsen, R. A., Neve, K. A., and Janowsky, A. (1994). Release of dopamine via the human transporter. *Mol Pharmacol, 45*, 312–316.

102. Pifl, C., Drobny, H., Reither, H., Hornykiewicz, O., and Singer, E. A. (1995). Mechanism of the dopamine-releasing actions of amphetamine and cocaine: plasmalemmal dopamine transporter versus vesicular monoamine transporter. *Mol Pharmacol, 47*, 368–373.

103. Sitges, M., Reyes, A., and Chiu, L. M. (1994). Dopamine transporter mediated release of dopamine: role of chloride. *J Neurosci Res, 39*, 11–22.

104. Sulzer, D., Maidment, N. T., and Rayport, S. (1993). Amphetamine and other weak bases act to promote reverse transport of dopamine in ventral midbrain neurons. *J Neurochem, 60*, 527–535.

105. Falkenburger, B. H., Barstow, K. L., and Mintz, I. M. (2001). Dendrodendritic inhibition through reversal of dopamine transport. *Science, 293*, 2465–2470.

106. Donnan, G. A., Kaczmarczyk, S. J., McKenzie, J. S., Kalnins, R. M., Chilco, P. J., and Mendelsohn, F. A. (1989). Catecholamine uptake sites in mouse brain: distribution determined by quantitative [3H]mazindol autoradiography. *Brain Res, 504*, 64–71.

107. Donnan, G. A., Kaczmarczyk, S. J., Paxinos, G., Chilco, P. J., Kalnins, R. M., Woodhouse, D. G., and Mendelsohn, F. A. (1991). Distribution of catecholamine uptake sites in human brain as determined by quantitative [3H]mazindol autoradiography. *J Comp Neurol, 304*, 419–434.

108. Javitch, J. A., Strittmatter, S. M., and Snyder, S. H. (1985). Differential visualization of dopamine and norepinephrine uptake sites in rat brain using [3H]mazindol autoradiography. *J Neurosci, 5*, 1513–1521.

109. Scatton, B., Dubois, A., Dubocovich, M. L., Zahniser, N. R., and Fage, D. (1985). Quantitative autoradiography of 3H-nomifensine binding sites in rat brain. *Life Sci, 36*, 815–822.

110. Dawson, T. M., Gehlert, D. R., and Wamsley, J. K. (1986). Quantitative autoradiographic localization of the dopamine transport complex in the rat brain: use of a highly selective radioligand: [3H]GBR 12935. *Eur J Pharmacol, 126*, 171–173.

111. Jansson, A., Goldstein, M., Tinner, B., Zoli, M., Meador-Woodruff, J. H., Lew, J. Y., Levey, A. I., Watson, S., Agnati, L. F., and Fuxe, K. (1999). On the distribution patterns of D1, D2, tyrosine hydroxylase and dopamine transporter immunoreactivities in the ventral striatum of the rat. *Neuroscience, 89*, 473–489.

112. Graybiel, A. M., and Moratalla, R. (1989). Dopamine uptake sites in the striatum are distributed differentially in striosome and matrix compartments. *Proc Natl Acad Sci U S A, 86*, 9020–9024.

113. Graybiel, A. M., and Ragsdale, C. W., Jr. (1978). Histochemically distinct compartments in the striatum of human, monkeys, and cat demonstrated by acetylthiocholinesterase staining. *Proc Natl Acad Sci U S A, 75*, 5723–5726.

114. Little, K. Y., Kirkman, J. A., Carroll, F. I., Breese, G. R., and Duncan, G. E. (1993). [125I]RTI-55 binding to cocaine-sensitive dopaminergic and serotonergic uptake sites in the human brain. *J Neurochem, 61*, 1996–2006.

115. Lewis, D. A., Melchitzky, D. S., Sesack, S. R., Whitehead, R. E., Auh, S., and Sampson, A. (2001). Dopamine transporter immunoreactivity in monkey cerebral cortex: regional, laminar, and ultrastructural localization. *J Comp Neurol, 432*, 119–136.

116. Berger, B., Gaspar, P., and Verney, C. (1991). Dopaminergic innervation of the cerebral cortex: unexpected differences between rodents and primates. *Trends Neurosci, 14*, 21–27.

117. Smith, H. R., Beveridge, T. J., and Porrino, L. J. (2006). Distribution of norepinephrine transporters in the non-human primate brain. *Neuroscience, 138*, 703–714.

118. Gaspar, P., Berger, B., Febvret, A., Vigny, A., and Henry, J. P. (1989). Catecholamine innervation of the human cerebral cortex as revealed by comparative immunohisto-chemistry of tyrosine hydroxylase and dopamine-β-hydroxylase. *J Comp Neurol, 279*, 249–271.

119. Goldman-Rakic, P. S., and Brown, R. M. (1981). Regional changes of monoamines in cerebral cortex and subcortical structures of aging rhesus monkeys. *Neuroscience, 6*, 177–187.

120. Rieck, R. W., Ansari, M. S., Whetsell, W. O., Jr., Deutch, A. Y., and Kessler, R. M. (2004). Distribution of dopamine D2-like receptors in the human thalamus: autoradiographic and PET studies. *Neuropsychopharmacology, 29*, 362–372.

121. Wang, G. J., Volkow, N. D., Fowler, J. S., Ding, Y. S., Logan, J., Gatley, S. J., MacGregor, R. R., and Wolf, A. P. (1995). Comparison of two pet radioligands for imaging extrastriatal dopamine transporters in human brain. *Life Sci, 57*, PL187–PL191.

122. Khan, Z. U., Gutierrez, A., Martin, R., Penafiel, A., Rivera, A., and De La Calle, A. (1998). Differential regional and cellular distribution of dopamine D2-like receptors: an immunocytochemical study of subtype-specific antibodies in rat and human brain. *J Comp Neurol, 402*, 353–371.

123. Hebert, M. A., Larson, G. A., Zahniser, N. R., and Gerhardt, G. A. (1999). Age-related reductions in [^{3}H]WIN 35,428 binding to the dopamine transporter in nigrostriatal and mesolimbic brain regions of the Fischer 344 rat. *J Pharmacol Exp Ther, 288*, 1334–1339.

124. Himi, T., Cao, M., and Mori, N. (1995). Reduced expression of the molecular markers of dopaminergic neuronal atrophy in the aging rat brain. *J Gerontol A Biol Sci Med Sci, 50*, B193–B200.

125. Leroux-Nicollet, I., and Costentin, J. (1994). Comparison of the subregional distributions of the monoamine vesicular transporter and dopamine uptake complex in the rat striatum and changes during aging. *J Neural Transm Gen Sect, 97*, 93–106.

126. Ma, S. Y., Ciliax, B. J., Stebbins, G., Jaffar, S., Joyce, J. N., Cochran, E. J., Kordower, J. H., Mash, D. C., Levey, A. I., and Mufson, E. J. (1999). Dopamine transporter-immunoreactive neurons decrease with age in the human substantia nigra. *J Comp Neurol, 409*, 25–37.

127. Meng, S. Z., Ozawa, Y., Itoh, M., and Takashima, S. (1999). Developmental and age-related changes of dopamine transporter, and dopamine D1 and D2 receptors in human basal ganglia. *Brain Res, 843*, 136–144.

128. Mozley, P. D., Acton, P. D., Barraclough, E. D., Plossl, K., Gur, R. C., Alavi, A., Mathur, A., Saffer, J., and Kung, H. F. (1999). Effects of age on dopamine transporters in healthy humans. *J Nucl Med, 40*, 1812–1817.

129. Tupala, E., Hall, H., Bergstrom, K., Mantere, T., Rasanen, P., Sarkioja, T., Hiltunen, J., and Tiihonen, J. (2003). Different effect of age on dopamine transporters in the dorsal and ventral striatum of controls and alcoholics. *Synapse, 48*, 205–211.

130. Volkow, N. D., Ding, Y. S., Fowler, J. S., Wang, G. J., Logan, J., Gatley, S. J., Hitzemann, R., Smith, G., Fields, S. D., and Gur, R. (1996). Dopamine transporters decrease with age. *J Nucl Med, 37*, 554–559.

131. Volkow, N. D., Fowler, J. S., Wang, G. J., Logan, J., Schlyer, D., MacGregor, R., Hitzemann, R., and Wolf, A. P. (1994). Decreased dopamine transporters with age in health human subjects. *Ann Neurol, 36*, 237–239.

132. Zelnik, N., Angel, I., Paul, S. M., and Kleinman, J. E. (1986). Decreased density of human striatal dopamine uptake sites with age. *Eur J Pharmacol, 126*, 175–176.

133. Garnett, E. S., Firnau, G., and Nahmias, C. (1983). Dopamine visualized in the basal ganglia of living man. *Nature, 305*, 137–138.

134. Baron, J. C., Comar, D., Zarifian, E., Agid, Y., Crouzel, C., Loo, H., Deniker, P., and Kellershohn, C. (1985). Dopaminergic receptor sites in human brain: positron emission tomography. *Neurology, 35*, 16–24.

135. Inoue, Y., Wagner, H. N., Jr., Wong, D. F., Links, J. M., Frost, J. J., Dannals, R. F., Rosenbaum, A. E., Takeda, K., Di Chiro, G., and Kuhar, M. J. (1985). Atlas of dopamine receptor images (PET) of the human brain. *J Comput Assist Tomogr, 9*, 129–140.

136. Ziporyn, T. (1985). PET scans "relate clinical picture to more specific nerve function." *JAMA, 253*, 943–945, 949.

137. Calne, D. B., Langston, J. W., Martin, W. R., Stoessl, A. J., Ruth, T. J., Adam, M. J., Pate, B. D., and Schulzer, M. (1985). Positron emission tomography after MPTP: observations relating to the cause of Parkinson's disease. *Nature, 317*, 246–248.

138. Elsinga, P. H., Hatano, K., and Ishiwata, K. (2006). PET tracers for imaging of the dopaminergic system. *Curr Med Chem, 13*, 2139–2153.

139. Laruelle, M., Slifstein, M., and Huang, Y. (2002). Positron emission tomography: imaging and quantification of neurotransporter availability. *Methods, 27*, 287–299.

140. Piccini, P. P. (2003). Dopamine transporter: basic aspects and neuroimaging. *Mov Disord,, 18*(Suppl 7), S3–S8.

141. Fowler, J. S., Volkow, N. D., Wolf, A. P., Dewey, S. L., Schlyer, D. J., Macgregor, R. R., Hitzemann, R., Logan, J., Bendriem, B., and Gatley, S. J., et al. (1989). Mapping cocaine binding sites in human and baboon brain in vivo. *Synapse, 4*, 371–377.

142. Logan, J., Fowler, J. S., Volkow, N. D., Wolf, A. P., Dewey, S. L., Schlyer, D. J., MacGregor, R. R., Hitzemann, R., Bendriem, B., and Gatley, S. J., et al. (1990). Graphical analysis of reversible radioligand binding from time-activity measurements applied to $[N\text{-}^{11}C\text{-methyl}]\text{-}(-)$-cocaine PET studies in human subjects. *J Cereb Blood Flow Metab, 10*, 740–747.

143. Ritz, M. C., Cone, E. J., and Kuhar, M. J. (1990). Cocaine inhibition of ligand binding at dopamine, norepinephrine and serotonin transporters: a structure–activity study. *Life Sci, 46*, 635–645.

144. Haaparanta, M., Bergman, J., Laakso, A., Hietala, J., and Solin, O. (1996). $[^{18}F]$CFT ($[^{18}F]$WIN 35,428), a radioligand to study the dopamine transporter with PET: biodistribution in rats. *Synapse, 23*, 321–327.

145. Madras, B. K., Gracz, L. M., Fahey, M. A., Elmaleh, D., Meltzer, P. C., Liang, A. Y., Stopa, E. G., Babich, J., and Fischman, A. J. (1998). Altropane, a SPECT or PET imaging probe for dopamine neurons. III: Human dopamine transporter in postmortem normal and Parkinson's diseased brain. *Synapse, 29*, 116–127.

146. Chou, K. L., Hurtig, H. I., Stern, M. B., Colcher, A., Ravina, B., Newberg, A., Mozley, P. D., and Siderowf, A. (2004). Diagnostic accuracy of [^{99m}Tc]TRODAT-1 SPECT imaging in early Parkinson's disease. *Parkinsonism Relat Disord, 10*, 375–379.

147. Johannsen, B., and Pietzsch, H. J. (2002). Development of technetium-99 m-based CNS receptor ligands: Have there been any advances? *Eur J Nucl Med Mol Imaging, 29*, 263–275.

148. Krause, K. H., Dresel, S. H., Krause, J., la Fougere, C., and Ackenheil, M. (2003). The dopamine transporter and neuroimaging in attention deficit hyperactivity disorder. *Neurosci Biobehav Rev, 27*, 605–613.

149. Wang, J., Jiang, Y. P., Liu, X. D., Chen, Z. P., Yang, L. Q., Liu, C. J., Xiang, J. D., and Su, H. L. (2005). ^{99m}Tc-TRODAT-1 SPECT study in early Parkinson's disease and essential tremor. *Acta Neurol Scand, 112*, 380–385.

150. Gatley, S. J., Pan, D., Chen, R., Chaturvedi, G., and Ding, Y. S. (1996). Affinities of methylphenidate derivatives for dopamine, norepinephrine and serotonin transporters. *Life Sci, 58*, 231–239.

151. Lee, C. S., Samii, A., Sossi, V., Ruth, T. J., Schulzer, M., Holden, J. E., Wudel, J., Pal, P. K., de la Fuente-Fernandez, R., Calne, D. B., and Stoessl, A. J. (2000). In vivo positron emission tomographic evidence for compensatory changes in presynaptic dopaminergic nerve terminals in Parkinson's disease. *Ann Neurol, 47*, 493–503.

152. Sossi, V., Holden, J. E., Chan, G., Krzywinski, M., Stoessl, A. J., and Ruth, T. J. (2000). Analysis of four dopaminergic tracers kinetics using two different tissue input function methods. *J Cereb Blood Flow Metab, 20*, 653–660.

153. Volkow, N. D., Wang, G., Fowler, J. S., Logan, J., Gerasimov, M., Maynard, L., Ding, Y., Gatley, S. J., Gifford, A., and Franceschi, D. (2001). Therapeutic doses of oral methylphenidate significantly increase extracellular dopamine in the human brain. *J Neurosci, 21*, RC121.

154. Marshall, V., and Grosset, D. (2003). Role of dopamine transporter imaging in routine clinical practice. *Mov Disord, 18*, 1415–1423.

155. Bannon, M. J. (2005). The dopamine transporter: role in neurotoxicity and human disease. *Toxicol Appl Pharmacol, 204*, 355–360.

156. Brooks, D. J. (1997). Advances in imaging Parkinson's disease. *Curr Opin Neurol, 10*, 327–331.

157. Brooks, D. J., Frey, K. A., Marek, K. L., Oakes, D., Paty, D., Prentice, R., Shults, C. W., and Stoessl, A. J. (2003). Assessment of neuroimaging techniques as biomarkers of the progression of Parkinson's disease. *Exp Neurol, 184*(Suppl 1), S68–S79.

158. Barthel, H., Sorger, D., Kuhn, H. J., Wagner, A., Kluge, R., and Hermann, W. (2001). Differential alteration of the nigrostriatal dopaminergic system in Wilson's disease investigated with [^{123}I]β-CIT and high-resolution SPECT. *Eur J Nucl Med, 28*, 1656–1663.

159. Ilgin, N., Zubieta, J., Reich, S. G., Dannals, R. F., Ravert, H. T., and Frost, J. J. (1999). PET imaging of the dopamine transporter in progressive supranuclear palsy and Parkinson's disease. *Neurology, 52*, 1221–1226.

160. Parkinson Study Group (2002). Dopamine transporter brain imaging to assess the effects of pramipexole vs. levodopa on Parkinson disease progression. *JAMA, 28*, 1653–1661.

161. Giros, B., and Caron, M.G. (1993). Molecular characterization of the dopamine transporter. *Trends Pharmacol Sci, 14*, 43–49.

2

DOPAMINE TRANSPORTER, DISEASE STATES, AND PATHOLOGY

DEBORAH C. MASH

Departments of Neurology and Molecular and Cellular Pharmacology, University of Miami Miller School of Medicine, Miami, Florida

2.1	Introduction	29
2.2	DAT Expression and Radioligand Interactions	30
2.3	In Vitro and In Vivo Imaging	32
2.4	Neurotoxins and the Effects of Aging	34
2.5	DAT and Parkinson's Disease	35
2.6	DAT and Cocaine Abuse	36
2.7	DAT and Attention-Deficit/Hyperactivity Disorder	39
2.8	Conclusions	39
	References	40

2.1 INTRODUCTION

The dopamine transporter (DAT) is a plasma membrane transport protein that functions to regulate dopamine (DA) neurotransmission. Essential for normal central nervous system function, DA signaling mediates physiological functions as diverse as movement, motivation, and reward. The DAT is involved in terminating DA signaling by removing the DA from nerve synapses via reuptake into the releasing neurons [1–3]. DAT binds amphetamine, cocaine, and other psychostimulants,

Dopamine Transporters: Chemistry, Biology, and Pharmacology. Edited by Mark L. Trudell and Sari Izenwasser
Copyright © 2008 John Wiley & Sons, Inc.

which inhibit DA reuptake. Amphetamine also stimulates the release of DA through interaction with DAT. Abnormal concentrations of DA in synapses leads to a hyper-dopaminergic state, which initiates a series of events associated with the behavioral effects of these drugs.

The DAT is a major target for various pharmacologically active drugs and environmental toxins. The development of selective radioligands has provided information on the role of DAT structure and function in mediating neurotoxicity and its role in drug abuse, attention-deficit disorder, and Parkinson's disease. Binding domains for DA and various blocking drugs, including cocaine, occur by interaction with multiple amino acid residues, some of which are separate in the primary structure but lie close together in the still unknown tertiary structure [3,4]. Chimera and site-directed mutagenesis studies suggest the involvement of both overlapping and separate domains in the interaction with substrates and blockers, whereas recent findings with sulfhydryl reagents selectively targeting cysteine residues support a role for conformational changes in the binding of blockers such as cocaine. The DAT can transport DA in an efflux mode, and recent mutagenesis experiments show different structural requirements for inward and outward transport.

DA uptake is inhibited with a distinct pharmacological profile by a variety of structurally diverse compounds. Identification and cloning of the gene for the DAT has provided insight into the molecular mechanism of DA reuptake inhibition by drugs that bind to the transport carrier [5–8]. DA transport inhibitors can be divided into two groups of reuptake inhibitors: inhibitors such as cocaine, which produce euphoria and addiction in humans, and inhibitors such as mazindol and buproprion, which do not produce euphoria and have low abuse liability. The underlying assumption is that the second class of inhibitors interacts differently than cocaine at sites on the DAT. Molecular biological and radioligand binding studies have delineated discrete domains within the structure of the DAT protein for substrate, cocaine, and antidepressant interactions [9–11], raising the possibility that it may be feasible to design cocaine antagonists that are devoid of uptake blockade for the clinical management of cocaine addiction. Although strong evidence for DAT domains selectively influencing binding of DA or cocaine analogs has not yet emerged, the development of a cocaine antagonist at the level of the transporter remains a possibility [4].

2.2 DAT EXPRESSION AND RADIOLIGAND INTERACTIONS

The neuronal DAT has been identified and studied in human brain postmortem by radioligand-binding techniques using novel cocaine congeners and transport inhibitors. In vitro binding and autoradiography of radioligands afford systematic visualization of the distribution and regulation of binding sites on the DAT. These studies in postmortem human brain are a counterpart to the noninvasive techniques of in vivo brain imaging and are important for establishing adaptations in DAT function that result from exposure to abused substances and neurotoxins.

Radioligand binding to the DAT has been characterized best with the cocaine congeners [^{3}H]WIN 35,428 and [^{125}I]RTI-121 [12]. In contrast to the classic DA transport inhibitors ([³H]mazindol, [³H]GBR 12935, and [³H]nomifensine), the cocaine congeners ([³H]WIN 35,428, [¹²⁵I]RTI-55, and [¹²⁵I]RTI-121) label multiple sites with a pharmacological profile characteristic of the DAT in primate and human brain [7,13–15]. Radioligand binding to COS cells transfected with the cloned cocaine-sensitive DAT demonstrates that [³H]WIN 35,428 identified two sites for binding to the protein expressed from a single cDNA [16]. Whether the multiple cocaine-binding sites represent altered functional forms or states of the DAT or discrete recognition sites on the DAT protein remains unclear. Pharmacological studies have demonstrated a lack of correspondence between DA transport function and ligand binding [7]. These findings provide additional support for the incomplete correspondence of pharmacologically overlapping sites for [³H]WIN 35,428, [³H]GBR 12935, and [³H]mazindol labeling of the DAT in native membrane preparations from rat brain [17,18]. Pharmacological heterogeneity of the cloned and native human DAT was suggested further by the dissociation of [³H]WIN 35,428 and [³H]GBR 12935 binding sites [7]. Interestingly, the proportion of high- and low-affinity [³H]WIN 35,428–binding sites observed differs across studies using cloned [5,7,16] or native membranes [13,17,18]. The binding of [³H]WIN 35,428 to cloned human DAT demonstrates the recognition of multiple sites with only one that is functionally correlated with that of the cloned DA-uptake process [7,16].

Saturation binding of the cocaine congeners [³H]WIN 35,428 [19,20], [¹²⁵I]RTI-55 [14,21], and [¹²⁵I]RTI-121 [15,22] is biphasic, indicating the presence of multiple cocaine-recognition sites on the DAT. The binding site densities in human striatum estimated from saturation isotherms for the cocaine congeners ([³H]WIN 35,428 [20], [¹²⁵I]RTI-55 [14], and [¹²⁵I]RTI-121 [15] (approximately 150 to 200 pmol/g original tissue) are in agreement with that observed for the noncocainelike transport inhibitors [³H]mazindol and [³H]GBR 12935 [23]. The total binding densities corresponding to both high- and low-affinity cocaine-recognition sites were comparable to those for noncocainelike transport inhibitors, which recognize a single class of sites associated with the DAT. Multiple cocaine-recognition sites associated with the DAT have not been reported consistently in human striatum [7,24]. The binding density corresponding to a single high-affinity site is significantly lower than that comprised by both high- and low-affinity cocaine-recognition sites in human brain [7,14,15,20,21,24]. In some instances, the existence of only a single component for WIN 35,428 [7,24] or RTI-121 [15] may be due to different conditions of the binding assay. RTI-121 saturation curves performed in sucrose–phosphate buffer reveal high- and low-affinity cocaine-recognition sites with a total density approximately equal to 200 pmol/g tissue. The heterogeneity of radioligand binding suggests that substrate and inhibitor binding is not mediated by a single class of recognition sites. The functional significance of multiple cocaine-recognition sites associated with the DAT remains unclear. The two cocaine-recognition sites may reflect binding interactions with distinct domains of a single DAT polypeptide or with different conformations of the DAT, which are recognized with equal affinities by non-cocainelike DA transport inhibitors. Reorientation of the DAT from inside to outside or structural folding of the protein that favors outward- vs. inward-directed residues may account for the pharmacological heterogeneity seen with different classes of radioligands that bind to the DAT.

2.3   IN VITRO AND IN VIVO IMAGING

The DAT in the human brain has been mapped in vitro using ligand binding and autoradiography on postmortem brain sections and visualized in vivo using positron emission tomography (PET) and single-photon emission computed tomography (SPECT). The first detailed regional maps of the distribution of DAT in postmortem human brain were generated using [³H]mazindol [25]. When binding to the norepinephrine transporter was occluded with desipramine, [³H]mazindol labeling was most evident in the striatum, with local gradients corresponding to the pattern of DAergic projections. Moderate to low [³H]mazindol labeling was visualized over DA cell body fields, including the substantia nigra compacta, ventral tegmental area, and the cell group in the retrorubral field [25].

Autoradiographic mapping of the DAT in human brain using [³H]cocaine [26] and the cocaine congeners [³H]WIN 35,428 [19,20], [¹²⁵I]RTI-55 [14], and [¹²⁵I]RTI-121 [15] have demonstrated high densities of labeling over the striatum, with moderate labeling over DAergic cell body fields. Autoradiographic visualization of the labeling of the radiolabeled cocaine congeners demonstrates distinct regional distribution patterns (Fig. 2-1). Although these radioligands have been useful for determining the regional distribution of ligand binding sites associated with the DAT, the overlapping affinities for the serotonin and/or the norepinephrine transporter makes it

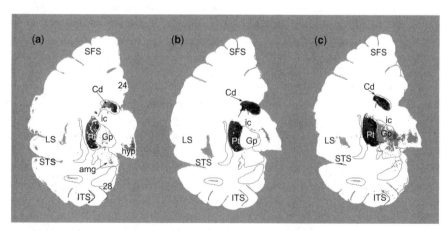

Figure 2-1 In vitro autoradiography of radiolabeled cocaine congeners in human brain. Gray-scale images of (a) [¹²⁵I]RTI-55 (50 pM) in the presence of 500 nM citalopram, (b) [³H]WIN 35,428 (2 nM), and (c) [¹²⁵I]RTI-121 (20 pM). Amg, amygdala; Cd, caudate; Gp, globus pallidus; hyp, hypothalamus; ic, internal capsule; ITS, inferior temporal sulcus; LS, lateral sulcus; Pt, putamen; SFS, superior frontal sulcus; STS, superior temporal sulcus; th, thalamus; 28, Brodman area 28, entorhinal cortex.

difficult to study the mapping of regulatory changes in the DAT by cocaine [12,14,21,26,27]. The distribution of [^{125}I]RTI-55 binding is very widespread, with binding sites prevalent throughout the cerebral cortex, striatum, thalamus, hypothalamus, and amygdala (Fig. 2-1a). This pattern of labeling correlates with the known distribution of monoaminergic nerve terminals and is consistent with binding of [^{125}I]RTI-55 to both serotonin (5-HT) and DA transporters [14,27]. Selective visualization of the DAT may be achieved by occluding the binding of [^{125}I]RTI-55 to the 5-HT transporter with the serotonin-reuptake inhibitors paroxetine or citalopram [14,27] (Fig. 2-1a). The DAT has been labeled selectively in vitro using the novel cocaine congener [^{125}I]RTI-121 [15]. The anatomical distribution of [^{125}I]RTI-121 binding (Fig. 2-1c) is much more restricted and correlates more closely with the known distribution and density of DA projection systems than does either [^{125}I]RTI-55 or [^{3}H]WIN 35,428 (Fig. 2-1b). Interestingly, the anatomical distribution of [^{3}H]WIN 35,428 labeling is more highly correlated than [^{125}I]RTI-55 with the pattern of [^{3}H]cocaine labeling [28].

The DAT has been labeled in vivo in human brain with [^{11}C]nomifensine [29,30], [^{11}C]dl-threo-methylphenidate [31], [^{11}C]cocaine [32,33], [^{11}C]WIN 35,428 [34], [^{11}C]β-CIT (also called RTI-55 [27]), [^{123}I]-β-CIT [35–37], [^{123}I]β-CIT-FE [38], and [^{123}I]β-CIT-FP [39]. In the living human brain, the regional distribution of [^{11}C]cocaine and [^{11}C]dl-threo-methylphenidate is heterogeneous with the high radiotracer labeling apparent in the striatum and the low labeling seen throughout the cerebral cortex. The psychostimulants [^{11}C]cocaine and [^{11}C]dl-threo-methylphenidate bind to the DAT with low in vivo occupancy of norepinephrine (NE) and 5-HT transporters [31,32]. Administration of DA transport inhibitors significantly decreased striatal [^{11}C]cocaine and [^{11}C]dl-threo-methylphenidate uptake, further demonstrating that these radioligands label the DAT selectively in vivo.

The cocaine congeners [^{11}C]WIN 35,428 [90], [^{11}C]β-CIT [27], and [^{123}I]β-CIT [35–37] also demonstrated marked radiotracer accumulation in the striatum, with low levels of binding to the thalamus, hypothalamus, midbrain, and pons. Binding of the radiotracer to the extrastriatal regions were occluded by pretreatment with mazindol and citalopram, indicating that the labeling in these regions was primarily to the 5-HT and NE transporters, not to the DAT [27,36,37]. The binding of [^{123}I]β-CIT to the thalamus, hypothalamus, midbrain, and pons was reduced significantly in depressed patients treated with the selective 5-HT-reuptake inhibitor citalopram [36]. The cocaine congeners ([^{123}I]β-CIT-FP, [^{123}I]β-CIT-FE) demonstrate greater selectivity for binding to the DAT [38,39]. In vivo imaging of [^{123}I]β-CIT-FP or [^{123}I]β-CIT-FE was not detected in the thalamus, hypothalamus, or cerebral cortical areas, and the specific binding measured in the striatum was approximately 10% less than that observed for [^{123}I]β-CIT. Considering that the ratio of 5-HT to DA terminals in the human striatum is approximately 1 : 9, the 10% decrease probably corresponds to the decreased labeling of striatal serotonin transporters. In vivo imaging of the DAT in neocortical regions known to receive DA projections, including the amygdala, hippocampus, frontal, cingulate and entorhinal cortices is often difficult. The lack of measurable DAT binding in corticolimbic regions may be due to the much lower densities of DATs in these brain areas. However, understanding the effects of antidepressant drugs on corticolimbic DA reuptake may prove important for clarifying the role of the DAT in a variety of neuropsychiatric disease states.

2.4 NEUROTOXINS AND THE EFFECTS OF AGING

Reductions in DA and DA-related synaptic markers in the striatum are known to contribute to the cognitive and motor deficits associated with normal aging. In the postmortem human striatum, a progressive decrease in DAT density with age has been demonstrated using [^{3}H]GBR 12935 [40–43]. Decreases in DAT density of 75 and 65% were reported for subjects ranging from 19 to 100 years [40] and 18 to 88 years [41], respectively. In vivo imaging of the DAT with cocaine and cocaine congeners ([^{11}C]cocaine [44] and [^{123}I]β-CIT [45]) and "classical" DA transport inhibitors ([^{11}C]nomifensine [46]) also demonstrated a decline in DA transporters with increasing age. Using [^{11}C]cocaine, a gradual decline in the density of cocaine recognition sites was detected over the age range 21 to 63 years [44]. Using [^{123}I]β-CIT, a 51% decline in DAT density was observed over the age range 18 to 83 years [45]. A decrease in the [^{11}C]nomifensine striatum/cerebellum ratios was observed over the age range 24 to 81 years [46]. Taken together, in vivo imaging with a variety of radiotracers demonstrate age-related declines in DAT density that occur at a rate of approximately 10% per decade.

In keeping with the marked decline in DATs with normal aging, studies of the mRNA encoding the DAT demonstrated a profound loss of DAT gene expression in DA-containing substantia nigra neurons with increasing age [47]. Although a precipitous age-related decline (>95% in subjects >57 years age) was reported for DAT mRNA, the mRNA for tyrosine hydroxylase (another phenotypic marker of DA neurons) decreased linearly with age [48]. The abrupt decline in the DAT at the end of the fifth decade was surprising considering that the decrease in DAT density was linear over the lifespan. This difference may reflect differential regulation of DAT mRNA and protein with normal aging [49].

The decline in DAT density may reflect a loss of DA nerve terminals or a decrease in the number of DATs expressed by aging DAergic neurons. Comparable changes in DAergic pre- and postsynaptic markers and DAT densities suggest that the decrease in DAT labeling may be due to reduced integrity of DAergic projections [41]. Age-related decreases have been shown for tyrosine hydroxylase [48], striatal DA content [50,51], and in dopamine D_1 and D_2 receptors [41]. The loss of DA synaptic markers suggests that degenerative changes occur in DA neurons. Alternatively, if the DAT is regulated by synaptic DA content, the decrease in DAT density may be a compensatory response to the age-related decline in neuronal DA content with correspondingly lower rates of DA turnover.

MPTP (1-methyl-4-phenyl-1,2,3,6-tetrahydropyridine) is well known to damage the nigrostriatal DAergic pathway as seen in Parkinson's disease. The function of the DAT is important to the appearance of MPTP neurotoxicity because to be neurotoxic, an MPTP metabolite must first gain access to the DAergic neurons by reuptake mechanisms. In fact, DAT is a mandatory factor for expression of MPTP neurotoxicity and may explain the selective neuronal damage in the substantia nigra in MPTP toxicity [52,53]. Overexpression of DAT in transgenic mice enhances the neurotoxicity of MPTP exposures [54,55]. Therefore, DAT is thought to play an important role in the MPTP neurotoxic process, and specific blockade of DAT with high-affinity inhibitors in neurodegenerative diseases such as Parkinson's disease, where the effective levels of DA are markedly reduced, may have beneficial consequences. Exposures to MPTP or other DAT-binding toxins may play a contributory role in some cases of nigrostriatal damage, including Parkinson's disease.

2.5   DAT AND PARKINSON'S DISEASE

Parkinson's disease is a common neurodegenerative disorder characterized by progressive loss of DAergic neurons and the presence of Lewy bodies in the substantia nigra. The estimated prevalence of Parkinson's disease is 0.2% but increases with age to 1 to 3% in persons over 80 years [40]. The diagnosis of possible and probable Parkinson's disease is based on clinical criteria, with few studies addressing actual

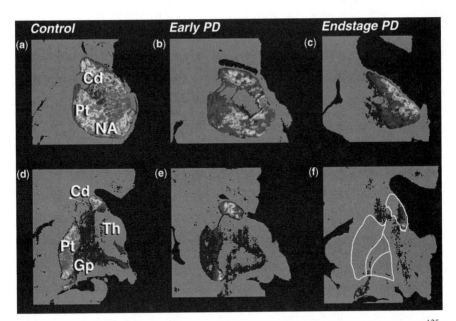

Figure 2-2   Mapping DAT in Parkinson's disease. In vitro autoradiographic maps of [^{125}I] RTI-121 labeling of the DAT in coronal sections of the striatum from a representative age-matched control subject (W, M, 73 years; panels a and d) and an early-stage Parkinson's disease patient (W, M, 69 years; panels b and e) and an advanced end-stage Parkinson's disease patient (W, M, 76 years; panels c and f). The DAT distribution and density are shown in a pseudocolor scale. Red-orange, high densities; yellow-green, intermediate; blue-purple, low). Cd, caudate; GP, globus pallidus; NA, nucleus accumbens; Th, thalamus; Pt, putamen. (*See insert for color representation of figure.*)

reliability or diagnostic certainty by disease stage. In clinical pathological studies, the clinical diagnosis is often not confirmed at postmortem examination [56,57]. Since Parkinsonian patients manifest symptoms when more than 50% of neurons have degenerated, clinical diagnosis fails to identify persons who would benefit from early diagnosis and treatment to slow disease progression. Molecular neuroimaging using PET and SPECT affords the sensitivity and specificity necessary to evaluate patients with nigrostriatal degeneration.

Postmortem studies have confirmed an association between the distribution and density of striatal DAT binding and Parkinson's disease. In end-stage Parkinson's disease there is a severe loss of the DAT over the anterior to posterior extent of the striatum (Fig. 2-2). The posterior putamen is severely depleted of DAT-binding sites, with selective involvement also seen with the more anterior sectors. Molecular imaging approaches with [^{18}F]dopa and PET have been considered the gold standard for evaluating nigral degeneration in vivo [58,59]. Tropane derivative studies using [^{11}C]CFT and [^{123}I]β-CIT have shown a direct correlation between decreases in DAT and Parkinson's disease symptoms [10,60]. However, most studies to date have examined patients with more advanced disease progression [61]. Although DAT tracers afford certain advantages over [^{18}F]dopa, tropane derivatives have not been fully validated in vivo. DAT tracers present certain problems for imaging disease progression, because DAT binding site density (up- or down-regulation) is regulated by drug treatments that affect DAergic transmission. Despite this limitation, early DA neuron loss can be reliably quantified with DAT tracers using PET or SPECT imaging. This radiotracer approach is warranted because neuroprotective therapies will probably benefit Parkinsonian patients in the future.

2.6 DAT AND COCAINE ABUSE

Since the DAT is a key regulator of DA neurotransmission, cocaine-mediated alterations in DAT numbers or function would lead to profound effects on DAergic signaling. Understanding the sites of cocaine's action in the brain and the acute and long-term neurobiological effects may help shed light on the clinical relevance of regulatory changes in DAT function to the behavioral effects of chronic cocaine use. The cocaine congeners [^{3}H]WIN 35,428 and [^{125}I]RTI-55 have been used to visualize the regional density of DAT postmortem in the human brain from cocaine abusers. Radioligand binding and autoradiographic studies demonstrated significant increases in DAT densities using the cocaine congeners [^{125}I]RTI-55 and [^{3}H]WIN 35,428 throughout the caudate, putamen, and nucleus accumbens from cocaine abusers [20,62]. Saturation binding analysis confirmed the increase in [^{3}H]WIN 35,428 binding to putamen membranes of cocaine users compared to drug-free and age-matched control subjects [20,62]. The increase in [^{3}H]WIN 35,428 binding observed in cocaine abusers was due to an elevation in the apparent density of the high-affinity cocaine recognition site on the DAT [20,62,63].

The elevations in DAT densities demonstrated in the postmortem brain from human cocaine abusers have been confirmed by in vivo SPECT imaging in human cocaine-dependent subjects [64]. The striatal uptake of $[^{123}I]\beta$-CIT (also called RTI-55) was elevated significantly (25%) in acutely abstinent ($\geq$96 hour) cocaine-dependent subjects. These studies suggest that the high-affinity cocaine-binding site may up-regulate in the human striatum with chronic cocaine abuse as a compensatory response to elevated synaptic levels of DA. We have shown corresponding increases postmortem in high-affinity $[^{3}H]$WIN 35,428–binding sites on the human DAT and functional DA uptake in cocaine abusers [63]. The increased ability of the protein to transport DA may help to explain the addictive liability of cocaine. As the transporter up-regulates the density in the nerve terminal to transport DA more efficiently, more cocaine will be needed to experience cocaine's reinforcing effects and euphoria. Since the transporter up-regulates with chronic use, abrupt withdrawal from cocaine will result in a net DA deficit in the synapse of cocaine abusers. This regulatory change in the DAT may explain the reports of anhedonia associated with the postcocaine crash.

In keeping with these findings, in vivo SPECT measurements of DAT densities in cocaine abusers vary depending on the time since the last cocaine administration. When the DAT was imaged in vivo after a period of drug abstinence, $[^{123}I]\beta$-CIT measures were still elevated, but decreased in their level of statistical significance [64]. Initial imaging studies with $[^{123}I]\beta$-CIT in cocaine-dependent subjects abstinent for 3 to 18 months demonstrated a significant trend toward a return to baseline densities measured in drug-free control subjects. In another study, chronic cocaine abusers had significantly lower $[^{11}C]$cocaine uptake in the basal ganglia and thalamus when screened 10 to 90 days after the last use of cocaine [65]. The uptake of $[^{11}C]$cocaine was negatively correlated with "cocaine craving" and with depressive symptoms, suggesting an association between withdrawal symptoms and DAT densities. The results of human and rodent studies suggest that the DAT up-regulates in response to acute "binges" of cocaine administration, but may gradually normalize or decrease with long periods of drug abstinence. The decrease in DAT densities in the striatal reward centers suggests that decreased DAT densities may reflect lower DAergic tone. A hypodopaminergic state may be one of the triggers of cocaine craving during abstinence that causes the addict to relapse to previous patterns of drug use.

Unlike methamphetamine, cocaine does not lead to neurotoxic damage of DAergic neurons [3]. However, one study reported that in contrast to the up-regulation in DAT-binding sites, immunoblots of DAT protein were reduced in cocaine abusers compared to control subjects [66]. The observed reduction in DAT protein would suggest a neurotoxic effect of cocaine in some addicted populations. Although some studies have reported a decrease in the vesicular monoamine transporter in DA terminals [66,67], other studies have reported no change in this presynaptic marker in cocaine addicts who had correspondingly elevated DAT densities [68]. It is interesting to point out that α-synuclein has been implicated in DAergic degeneration in Parkinson's disease [69,70]. In cocaine abusers, α-synuclein protein levels are increased in cocaine abusers in the nigrostriatal pathway [71,72]. The overexpression of α-synuclein may occur as a protective response to changes in DA turnover, vesicular storage, and increased oxidative stress resulting from cocaine abuse. However, the accumulation of α-synuclein protein with long-term cocaine exposure may put addicts at increased risk for motor and cognitive deficits with advanced age. Further studies are needed to confirm an interaction between overexpression of α-synuclein and increased DAT function to determine their contribution to possible neurotoxic effects in aging cocaine addicts.

A subgroup of cocaine abusers who died following a syndrome of excited delirium (ED) was first described in 1985 [73]. DAT regulation in cocaine abusers that came to autopsy has been compared to victims of cocaine ED using ligand binding and in vitro autoradiography. Excited delirium victims have a similar pattern of cocaine abuse and have positive drug screens for cocaine and benzoylecgonine at death [74–76]. In contrast to other cocaine abusers, autoradiographic mapping with a single concentration of [^{3}H]WIN 35,428 and [^{125}I]RTI-55 failed to demonstrate an elevation in the apparent density of the DAT in the striatum of ED victims [62,77]. The lack of a compensatory increase in DA transport function in the ED subgroup may indicate a diminished capacity for DA reuptake during a cocaine challenge or binge use [63]. Since the concentration of synaptic DA is controlled by the reuptake mechanism(s), the lack of compensatory increase in cocaine-recognition sites could be the molecular defect that explains the paranoia and agitation associated with this syndrome. Paranoia in the context of cocaine abuse is common, and several lines of evidence suggest that this phenomenon may be related to the function of the DAT protein [78]. Genetic differences in the makeup of persons who abuse cocaine may underlie some of these differences in susceptibility to adverse neuro-psychiatric effects of chronic cocaine abuse.

PET and SPECT imaging allows for drug interactions within the brain to be assessed while simultaneously monitoring the behavioral effects and plasma bioavail-ability of the drug [79]. In vivo imaging with [^{11}C]cocaine has demonstrated that the time course of cocaine binding to the striatum correlates with the onset of the eupho-ric experience [32,80]. These studies support the hypothesis that the "high" or euphoria induced by psychostimulants is related to the rapid occupancy of the DAT and the corresponding rise in the synaptic concentration of DA [79]. Similar studies with [^{11}C]*dl*-threo-methylphenidate also demonstrated rapid uptake, which correlates with the onset of the behavioral effects. [^{11}C]*dl*-Threo-methylphenidate exhibited slower clearance from the striatum than did [^{11}C]cocaine. Since methyl-phenidate has lower abuse liability than cocaine, it is possible that kinetic differences in the rates of occupancy of the DAT by [^{11}C]*dl*-threo-methylphenidate may explain the lack of the euphoriant "rush" and lower arousal [31]. Rothman and [81] has suggested that DAT inhibitors with slow rates of entry into the brain and slow onset of action may be candidate drugs for treating cocaine dependence.

Although there is still controversy regarding the long-term sequelae of cocaine abuse, pharmacological treatment strategies have been based on the assumption that chronic cocaine abuse leads to neuroadaptive changes in the sensitivity of DA neurons, with increased DA turnover and resulting DA depletion [82,83]. The com-pensatory increase in high-affinity cocaine recognition sites on the human DAT may indicate an enhanced ability of cocaine to inhibit DA transport in chronic cocaine users. An elevation in [^{3}H]methylphenidate binding to the DAT has been shown to occur as early as 1 hour after the acute administration of cocaine to drug-naive rats [84], indicating that DAT may undergo rapid regulatory changes in the membrane to regulate synaptic levels of neurotransmitter. This time course suggests that there may be alterations in subcellular distributions and trafficking of the DAT that occur with acute changes in synaptic DA.

2.7 DAT AND ATTENTION-DEFICIT/HYPERACTIVITY DISORDER

DAT has emerged as an important neuropsychiatric target for attention-deficit/hyperactivity disorder (ADHD) in children and adults. There is converging evidence that DAT binding is increased in adults and children with ADHD [85]. Genetic studies have found an association between ADHD and the 40-base-pair allele of the DAT VNTR (variable number of tandem repeats) polymorphism located in the 3' untranslated region [86]. A haplotype study of three polymorphisms at the DAT locus confirm linkage to ADHD [87]. Using different radiotracers, six studies have demonstrated an increase of 20 to 70% in DAT binding in unmedicated patients [88]. However, two studies reported no change in the overall binding in vivo in either adolescent or young adults [89,90], and some controversy exists regarding the regulatory effects of methylphenidate exposures [91,92]. These conflicting observations in ADHD are reminiscent of DAT regulation in cocaine abusers, which underscore that careful clinical descriptions of the cohort, time of drug effects, and choice of tracer for imaging and measurement are needed to disentangle DAT occupancy from up- or down-regulation.

ADHD stimulant medications affect DAT function, with good therapeutic outcomes [85,93]. Volkow et al. [94,95] have stressed the importance of the relatively low abuse potential of methylphenidate compared to cocaine, due to its slow rate of uptake after oral administration, combined with its slow rate of clearance. Thus, drugs that target the DAT with slow kinetics of association and dissociation offer therapeutic advantages for developing drugs to normalize DAergic synaptic tone through an interaction with the DAT. In mice, DAT knockout leads to behavioral hyperactivity that is responsive to stimulant therapy [96,97]. The mechanism by which stimulants decrease hyperactivity in the DAT knockout mouse is unclear. This apparent hyperactivity paradox in mouse models may suggest an involvement of the noradrenergic system in ADHD.

2.8 CONCLUSIONS

DA uptake through the DAT is essential for maintaining normal DA homeostasis in the brain by affecting the intensity and duration of DA signaling at receptor sites. The DAT is an important molecular target for treatment of ADHD and for development of medications to block the euphoriant effects of psychostimulants. Neurotoxins selectively target DAergic neurons through interaction with the DAT, and DAT radiotracers are useful for assessing degeneration of DAergic neurons in Parkinson's disease. Genetic association studies have reported a variable number of tandem repeat polymorphisms in the 3'-noncoding region of the DAT gene, implicating this protein in the development of various neuropsychiatric disorders.

REFERENCES

1. Uhl, G. R. (2003). Dopamine transporter: basic science and human variation of a key molecule for dopaminergic function, locomotion, and parkinsonism. *Mov Disord*, *18*(Suppl 7), S71–S80.

2. Mortensen, O. V., and Amara, S. G. (2003). Dynamic regulation of the dopamine transporter. *Eur J Pharmacol*, *479*, 159–170.

3. Bannon, M. J. (2005). The dopamine transporter: role in neurotoxicity and human disease. *Toxicol Appl Pharmacol*, *204*(3), 355–360.

4. Zhen, J., Chen, N., and Reith, M. E. (2005). Differences in interactions with the dopamine transporter as revealed by diminishment of Na(+) gradient and membrane potential: dopamine versus other substrates. *Neuropharmacology*, *49*(6), 769–779.

5. Eshleman, A. J., Neve, R. L., Janowsky, A., and Neve, K. A. (1995). Characterization of a recombinant human dopamine transporter in multiple cell lines. *J Pharmacol Exp Ther*, *274*(1), 276–283.

6. Giros, B., el Mestikawy, S., Godinot, N., Zheng, K., Han, H., Yang-Feng, T., and Caron, M. G. (1992). Cloning, pharmacological characterization, and chromosome assignment of the human dopamine transporter. *Mol Pharmacol*, *42*(3), 383–390.

7. Pristupa, Z. B., Wilson, J. M., Hoffman, B. J., Kish, S. J., and Niznik, H. B. (1994). Pharmacological heterogeneity of the cloned and native human dopamine transporter: dissociation of [^{3}H]WIN 35,428 and [^{3}H]GBR 12,935 binding. *Mol Pharmacol*, *45*(1), 125–135.

8. Vandenbergh, D. J., Persico, A. M., and Uhl, G. R. (1992). A human dopamine transporter cDNA predicts reduced glycosylation, displays a novel repetitive element and provides racially-dimorphic TaqI RFLPs. *Brain Res Mol Brain Res*, *15*(1–2), 161–166.

9. Giros, B., Wang, Y. M., Suter, S., McLeskey, S. B., Pifl, C., and Caron, M. G. (1994). Delineation of discrete domains for substrate, cocaine, and tricyclic antidepressant interactions using chimeric dopamine–norepinephrine transporters. *J Biol Chem*, *269*(23), 15985–15988.

10. Kitayama, S., Shimada, S., Xu, H., Markham, L., Donovan, D. M., and Uhl, G. R. (1992). Dopamine transporter site-directed mutations differentially alter substrate transport and cocaine binding. *Proc Natl Acad Sci U S A*, *89*(16), 7782–7785.

11. Kitayama, S., Wang, J. B., and Uhl, G. R. (1993). Dopamine transporter mutants selectively enhance MPP+ transport. *Synapse*, *15*(1), 58–62.

12. Boja, J. W., Vaughen, R., Patel, A., Shaya, E. K., and Kuhar, M. J. (1994). The dopamine transporter. In *Dopamine Transporters and Receptors*, Niznik, H. B., Ed. Marcel Dekker, New York, pp. 611–644.

13. Madras, B. K., Spealman, R. D., Fahey, M. A., Neumeyer, J. L., Saha, J. K., and Milius, R. A. (1989). Cocaine receptors labeled by [^{3}H]2β-carbomethoxy-3β-(4-fluorophenyl)tropane. *Mol Pharmacol*, *36*(4), 518–524.

14. Staley, J. K., Basile, M., Flynn, D. D., and Mash, D. C. (1994). Visualizing dopamine and serotonin transporters in the human brain with the potent cocaine analogue [^{125}I]RTI-55: in vitro binding and autoradiographic characterization. *J Neurochem, 62*(2), 549–556.

15. Staley, J. K., Boja, J. W., Carroll, F. I., Seltzman, H. H., Wyrick, C. D., Lewin, A. H., Abraham, P., and Mash, D. C. (1995). Mapping dopamine transporters in the human brain with novel selective cocaine analog [^{125}I]RTI-121. *Synapse, 21*(4), 364–372.

16. Boja, J. W., Markham, L., Patel, A., Uhl, G., and Kuhar, M. J. (1992). Expression of a single dopamine transporter cDNA can confer two cocaine binding sites. *Neuroreport, 3*(3), 247–248.

17. Reith, M. E., de Costa, B., Rice, K. C., and Jacobson, A. E. (1992). Evidence for mutually exclusive binding of cocaine, BTCP, GBR 12935, and dopamine to the dopamine transporter. *Eur J Pharmacol, 227*(4), 417–425.

18. Reith, M. E., and Selmeci, G. (1992). Radiolabeling of dopamine uptake sites in mouse striatum: comparison of binding sites for cocaine, mazindol, and GBR 12935. *Naunyn Schmiedebergs Arch Pharmacol, 345*(3), 309–318.

19. Kaufman, M. J., and Madras, B. K. (1991). Severe depletion of cocaine-recognition sites associated with the dopamine transporter in Parkinson's-diseased striatum. *Synapse, 9*(1), 43–49.

20. Staley, J. K., Hearn, W. L., Ruttenber, A. J., Wetli, C. V., and Mash, D. C. (1994). High affinity cocaine-recognition sites on the dopamine transporter are elevated in fatal cocaine overdose victims. *J Pharmacol Exp Ther, 271*(3), 1678–1685.

21. Little, K. Y., Kirkman, J. A., Carroll, F. I., Breese, G. R., and Duncan, G. E. (1993). [^{125}I]RTI-55 binding to cocaine-sensitive dopaminergic and serotonergic uptake sites in the human brain. *J Neurochem, 61*(6), 1996–2006.

22. Boja, J. W., Carroll, F. I., Vaughan, R. A., Kopajtic, T., and Kuhar, M. J. (1998). Multiple binding sites for [^{125}I]RTI-121 and other cocaine analogs in rat frontal cerebral cortex. *Synapse, 30*(1), 9–17.

23. Marcusson, J., and Eriksson, K. (1988). [^{3}H]GBR-12935 binding to dopamine uptake sites in the human brain. *Brain Res, 457*(1), 122–129.

24. Little, K. Y., Kirkman, J. A., Carroll, F. I., Clark, T. B., and Duncan, G. E. (1993). Cocaine use increases [^{3}H]WIN 35428 binding sites in human striatum. *Brain Res, 628*(1–2), 17–25.

25. Donnan, G. A., Kaczmarczyk, S. J., Paxinos, G., Chilco, P. J., Kalnins, R. M., Woodhouse, D. G., and Mendelsohn, F. A. (1991). Distribution of catecholamine uptake sites in human brain as determined by quantitative [^{3}H]mazindol autoradiography. *J Comp Neurol, 304*(3), 419–434.

26. Biegon, A., Dillon, K., Volkow, N. D., Hitzemann, R. J., Fowler, J. S., and Wolf, A. P. (1992). Quantitative autoradiography of cocaine binding sites in human brain postmortem. *Synapse, 10*(2), 126–30.

27. Farde, L., Halldin, C., Muller, L., Suhara, T., Karlsson, P., and Hall, H. (1994). PET study of [^{11}C]β-CIT binding to monoamine transporters in the monkey and human brain. *Synapse, 16*(2), 93–103.

28. Madras, B. K., and Kaufman, M. J. (1994). Cocaine accumulates in dopamine-rich regions of primate brain after i.v. administration: comparison with mazindol distribution. *Synapse, 18*(3), 261–275.

29. Aquilonius, S. M., Bergstrom, K., Eckernas, S. A., Hartvig, P., Leenders, K. L., Lundquist, H., Antoni, G., Gee, A., Rimland, A., and Uhlin, J. (1987). In vivo evaluation of striatal dopamine reuptake sites using ^{11}C-nomifensine and positron emission tomography. *Acta Neurol Scand*, *76*(4), 283–287.

30. Salmon, E., Brooks, D. J., Leenders, K. L., Turton, D. R., Hume, S. P., Cremer, J. E., Jones, T., and Frackowiak, R. S. (1990). A two-compartment description and kinetic procedure for measuring regional cerebral ^{11}C]nomifensine uptake using positron emission tomography. *J Cereb Blood Flow Metab*, *10*(3), 307–316.

31. Ding, Y. S., Fowler, J. S., Volkow, N. D., Gatley, S. J., Logan, J., Dewey, S. L., Alexoff, D., Fazzini, E., and Wolf, A. P. (1994). Pharmacokinetics and in vivo specificity of ^{11}C]*dl*-threo-methylphenidate for the presynaptic dopaminergic neuron. *Synapse*, *18*(2), 152–160.

32. Fowler, J. S., Volkow, N. D., Wolf, A. P., Dewey, S. L., Schlyer, D. J., Macgregor, R. R., Hitzemann, R., Logan, J., Bendriem, B., and Gatley, S. J. (1989). Mapping cocaine binding sites in human and baboon brain in vivo. *Synapse*, *4*(4), 371–377.

33. Gatley, S. J., Volkow, N. D., Fowler, J. S., Dewey, S. L., and Logan, J. (1995). Sensitivity of striatal ^{11}C]cocaine binding to decreases in synaptic dopamine. *Synapse*, *20*(2), 137–144.

34. Wong, D. F., Yung, B., Dannals, R. F., Shaya, E. K., Ravert, H. T., Chen, C. A., Chan, B., Folio, T., Scheffel, U., Ricaurte, G. A., et al. (1993). In vivo imaging of baboon and human dopamine transporters by positron emission tomography using ^{11}C]WIN 35,428. *Synapse*, *15*(2), 130–142.

35. Neumeyer, J. L., Wang, S., Gao, Y., Milius, R. A., Kula, N. S., Campbell, A., Baldessarini, R. J., Zea-Ponce, Y., Baldwin, R. M., and Innis, R. B. (1994). *N*-ω-fluoroalkyl analogs of (1*R*)-2β-carbomethoxy-3β-(4-iodophenyl)-tropane (β-CIT): radiotracers for positron emission tomography and single photon emission computed tomography imaging of dopamine transporters. *J Med Chem*, *37*(11), 1558–1561.

36. Pirker, W., Asenbaum, S., Kasper, S., Walter, H., Angelberger, P., Koch, G., Pozzera, A., Deecke, L., Podreka, I., and Brucke, T. (1995). β-CIT SPECT demonstrates blockade of 5HT-uptake sites by citalopram in the human brain in vivo. *J Neural Transm Gen Sect*, *100*(3), 247–256.

37. Seibyl, J. P., Wallace, E., Smith, E. O., Stabin, M., Baldwin, R. M., Zoghbi, S., Zea-Ponce, Y., Gao, Y., Zhang, W. Y., Neumeyer, J. L., et al. (1994). Whole-body biodistribution, radiation absorbed dose and brain SPECT imaging with iodine-123-β-CIT in healthy human subjects. *J Nucl Med*, *35*(5), 764–770.

38. Kuikka, J. T., Akerman, K., Bergstrom, K. A., Karhu, J., Hiltunen, J., Haukka, J., Heikkinen, J., Tiihonen, J., Wang, S., and Neumeyer, J. L. (1995). Iodine-123 labelled *N*-(2-fluoroethyl)-2β-carbomethoxy-3 β-(4-iodophenyl)nortropane for dopamine transporter imaging in the living human brain. *Eur J Nucl Med*, *22*(7), 682–686.

39. Kuikka, J. T., Bergstrom, K. A., Ahonen, A., Hiltunen, J., Haukka, J., Lansimies, E., Wang, S., and Neumeyer, J. L. (1995). Comparison of iodine-123 labelled 2β-carbomethoxy-3β-(4-iodophenyl)tropane and 2β-carbomethoxy-3β-(4-iodophenyl)-*N*-(3-fluoropropyl)-nortropane for imaging of the dopamine transporter in the living human brain. *Eur J Nucl Med*, *22*(4), 356–360.

40. Allard, P., and Marcusson, J. O. (1989). Age-correlated loss of dopamine uptake sites labeled with ^{3}H]GBR-12935 in human putamen. *Neurobiol Aging*, *10*(6), 661–664.

41. De Keyser, J., Ebinger, G., and Vauquelin, G. (1990). Age-related changes in the human nigrostriatal dopaminergic system. *Ann Neurol*, *27*(2), 157–161.

42. Hitri, A., Casanova, M. F., Kleinman, J. E., Weinberger, D. R., and Wyatt, R. J. (1995). Age-related changes in [^{3}H]GBR 12935 binding site density in the prefrontal cortex of controls and schizophrenics. *Biol Psychiatry*, *37*(3), 175–182.

43. Zelnik, N., Angel, I., Paul, S. M., and Kleinman, J. E. (1986). Decreased density of human striatal dopamine uptake sites with age. *Eur J Pharmacol*, *126*(1–2), 175–176.

44. Volkow, N. D., Fowler, J. S., Wang, G. J., Logan, J., Schlyer, D., MacGregor, R., Hitzemann, R., and Wolf, A. P. (1994). Decreased dopamine transporters with age in health human subjects. *Ann Neurol*, *36*(2), 237–9.

45. van Dyck, C. H., Seibyl, J. P., Malison, R. T., Laruelle, M., Wallace, E., Zoghbi, S. S., Zea-Ponce, Y., Baldwin, R. M., Charney, D. S., and Hoffer, P. B. (1995). Age-related decline in striatal dopamine transporter binding with iodine-123-β-CITSPECT. *J Nucl Med*, *36*(7), 1175–1181.

46. Tedroff, J., Aquilonius, S. M., Hartvig, P., Lundqvist, H., Gee, A. G., Uhlin, J., and Langstrom, B. (1988). Monoamine re-uptake sites in the human brain evaluated in vivo by means of ^{11}C-nomifensine and positron emission tomography: the effects of age and Parkinson's disease. *Acta Neurol Scand*, *77*(3), 192–201.

47. Bannon, M. J., Poosch, M. S., Xia, Y., Goebel, D. J., Cassin, B., and Kapatos, G. (1992). Dopamine transporter mRNA content in human substantia nigra decreases precipitously with age. *Proc Natl Acad Sci U S A*, *89*(15), 7095–7099.

48. McGeer, P. L., McGeer, E. G., and Suzuki, J. S. (1977). Aging and extrapyramidal function. *Arch Neurol*, *34*(1), 33–35.

49. Ma, S. Y., Ciliax, B. J., Stebbins, G., Jaffar, S., Joyce, J. N., Cochran, E. J., Kordower, J. H., Mash, D. C., Levey, A. I., and Mufson, E. J. (1999). Dopamine transporter–immunoreactive neurons decrease with age in the human substantia nigra. *J Comp Neurol*, *409*(1), 25–37.

50. Adolfsson, R., Gottfries, C. G., Roos, B. E., and Winblad, B. (1979). Post-mortem distribution of dopamine and homovanillic acid in human brain, variations related to age, and a review of the literature. *J Neural Transm*, *45*(2), 81–105.

51. Hornykiewicz, O. (1983). Dopamine Changes in the aging human brain. In *Aging Brain and Ergot Alkaloids*, Agnoli, A., Grepaldi, G., Spno, P. F., and Trabucchi, M., Eds. Raven Press, New York, pp. 9–14.

52. Uhl, G. R., Javitch, J. A., and Snyder, S. H. (1985). Normal MPTP binding in parkinsonian substantia nigra: evidence for extraneuronal toxin conversion in human brain. *Lancet*, *1*(8435), 956–957.

53. Richfield, E. K., Thiruchelvam, M. J., Cory-Slechta, D. A., Wuertzer, C., Gainetdinov, R. R., Caron, M. G., Di Monte, D. A., and Federoff, H. J. (2002). Behavioral and neurochemical effects of wild-type and mutated human α-synuclein in transgenic mice. *Exp Neurol*, *175*(1), 35–48.

54. Gainetdinov, R. R., Fumagalli, F., Jones, S. R., and Caron, M. G. (1997). Dopamine transporter is required for in vivo MPTP neurotoxicity: evidence from mice lacking the transporter. *J Neurochem*, *69*(3), 1322–1325.

55. Donovan, D. M., Miner, L. L., Perry, M. P., Revay, R. S., Sharpe, L. G., Przedborski, S., Kostic, V., Philpot, R. M., Kirstein, C. L., Rothman, R. B., Schindler, C. W., and Uhl, G. R. (1999). Cocaine reward and MPTP toxicity: alteration by regional variant dopamine transporter overexpression. *Brain Res Mol Brain Res*, *73*(1–2), 37–49.

56. Gelb, D. J., Oliver, E., and Gilman, S. (1999). Diagnostic criteria for Parkinson disease. *Arch Neurol*, *56*(1), 33–39.

57. Agid, Y. (1991). Parkinson's disease: pathophysiology. *Lancet*, *337*(8753), 1321–1324.

58. Brooks, D. J., Ibanez, V., Sawle, G. V., Quinn, N., Lees, A. J., Mathias, C. J., Bannister, R., Marsden, C. D., and Frackowiak, R. S. (1990). Differing patterns of striatal ^{18}F-dopa uptake in Parkinson's disease, multiple system atrophy, and progressive supranuclear palsy. *Ann Neurol*, *28*(4), 47–55.

59. Leenders, K. L., Palmer, A. J., Quinn, N., Clark, J. C., Firnau, G., Garnett, E. S., Nahmias, C., Jones, T., and Marsden, C. D. (1986). Brain dopamine metabolism in patients with Parkinson's disease measured with positron emission tomography. *J Neurol Neurosurg Psychiatry*, *49*(8), 853–860.

60. Frost, J. J., Rosier, A. J., Reich, S. G., Smith, J. S., Ehlers, M. D., Snyder, S. H., Ravert, H. T., and Dannals, R. F. (1993). Positron emission tomographic imaging of the dopamine transporter with ^{11}C-WIN 35,428 reveals marked declines in mild Parkinson's disease. *Ann Neurol*, *34*(3), 423–431.

61. Shih, M. C., Hoexter, M. Q., Andrade, L. A., and Bressan, R. A. (2006). Parkinson's disease and dopamine transporter neuroimaging: a critical review. *Sao Paulo Med J*, *124*(3), 168–175.

62. Staley, J. K., Basile, M., Wetli, C. V., Hearn, W. L., Flynn, D. D., Ruttenber, A. J., and Mash, D. C. (1994). *Differential regulation of the DAT in Cocaine Overdose Deaths*. Monographs, N. R., Ed. NIH, Bethesda, MD, p. 32.

63. Mash, D. C., Pablo, J., Ouyang, Q., Hearn, W. L., and Izenwasser, S. (2002). Dopamine transport function is elevated in cocaine users. *J Neurochem*, *81*(2), 292–300.

64. Malison, R. T. (1995). SPECT Imaging of DA Transporters in Cocaine Dependence with ^{123}I]β-CIT, Monographs, N. R., Ed. NIH, Bethesda, MD, p. 60.

65. Volkow, N. D., Wang, G. J., Fowler, J. S., Logan, J., Hitzemannn, R., Gatley, S. J., MacGregor, R. R., and Wolf, A. P. (1996). Cocaine uptake is decreased in the brain of detoxified cocaine abusers. *Neuropsychopharmacology*, *14*(3), 159–168.

66. Wilson, J. M., Levey, A. I., Bergeron, C., Kalasinsky, K., Ang, L., Peretti, F., Adams, V. I., Smialek, J., Anderson, W. R., Shannak, K., Deck, J., Niznik, H. B., and Kish, S. J. (1996). Striatal dopamine, dopamine transporter, and vesicular monoamine transporter in chronic cocaine users. *Ann Neurol*, *40*(3), 428–439.

67. Little, K. Y., McLaughlin, D. P., Zhang, L., McFinton, P. R., Dalack, G. W., Cook, E. H., Jr., Cassin, B. J., and Watson, S. J. (1998). Brain dopamine transporter messenger RNA and binding sites in cocaine users: a postmortem study. *Arch Gen Psychiatry*, *55*(9), 793–799.

68. Staley, J. K., Talbot, J. Z., Ciliax, B. J., Miller, G. W., Levey, A. I., Kung, M. P., Kung, H. F., and Mash, D. C. (1997). Radioligand binding and immunoautoradiographic evidence for a lack of toxicity to dopaminergic nerve terminals in human cocaine overdose victims. *Brain Res*, *747*(2), 219–229.

69. Eriksen, J. L., Dawson, T. M., Dickson, D. W., and Petrucelli, L. (2003). Caught in the act: α-synuclein is the culprit in Parkinson's disease. *Neuron*, *40*(3), 453–456.

70. Maries, E., Dass, B., Collier, T. J., Kordower, J. H., and Steece-Collier, K. (2003). The role of α-synuclein in Parkinson's disease: insights from animal models. *Nat Rev Neurosci*, *4*(9), 727–738.

71. Mash, D. C., Ouyang, Q., Pablo, J., Basile, M., Izenwasser, S., Lieberman, A., and Perrin, R. J. (2003). Cocaine abusers have an overexpression of α-synuclein in dopamine neurons. *J Neurosci*, *23*(7), 2564–2571.

72. Qin, Y., Ouyang, Q., Pablo, J., and Mash, D. C. (2005). Cocaine abuse elevates α-synuclein and dopamine transporter levels in the human striatum. *Neuroreport*, *16*(13), 1489–1493.

73. Wetli, C. V., and Fishbain, D. A. (1985). Cocaine-induced psychosis and sudden death in recreational cocaine users. *J Forensic Sci*, *30*(3), 873–880.

74. Ruttenber, A. J., Lawler-Heavner, J., Yin, M., Wetli, C. V., Hearn, W. L., and Mash, D. C. (1997). Fatal excited delirium following cocaine use: epidemiologic findings provide new evidence for mechanisms of cocaine toxicity. *J Forensic Sci*, *42*(1), 25–31.

75. Karch, S. B., and Stephens, B. G. (1999). Drug abusers who die during arrest or in custody. *J R Soc Med*, *92*(3), 110–113.

76. Karch, S. B., and Wetli, C. V. (1995). Agitated delirium versus positional asphyxia. *Ann Emerg Med*, *26*(6), 760–761.

77. Staley, J. K., Wetli, C. V., Ruttenber, A. J., Hearn, W. L., and Mash, D. C. (1995). *Altered Dopaminergic Synaptic Markers in Cocaine Psychosis and Sudden Death*, Monographs, N. R., Ed. NIH, Bethesda, MD, p. 491.

78. Gelernter, J., Kranzler, H. R., Satel, S. L., and Rao, P. A. (1994). Genetic association between dopamine transporter protein alleles and cocaine-induced paranoia. *Neuropsychopharmacology*, *11*(3), 195–200.

79. Volkow, N. D., Ding, Y. S., Fowler, J. S., Wang, G. J., Logan, J., Gatley, J. S., Dewey, S., Ashby, C., Liebermann, J., Hitzemann, R., et al. (1995). Is methylphenidate like cocaine? Studies on their pharmacokinetics and distribution in the human brain. *Arch Gen Psychiatry*, *52*(6), 456–463.

80. Cook, C. E., Jeffcoat, A. R., and Perez-Reyes, M. (1985). Pharmacokinetic studies of cocaine and phencyclidine in man. In *Pharmacokinetics and Pharmacodynamics of Psychoactive Drugs*, Barnett, G., and Chiang, C. N., Eds. Biomedical Publications, Foster City, CA, pp. 48–74.

81. Rothman, R. B. (1990). High affinity dopamine reuptake inhibitors as potential cocaine antagonists: a strategy for drug development. *Life Sci*, *46*(20), PL17–PL21.

82. Dackis, C. A., and Gold, M. S. (1985). Bromocriptine as treatment of cocaine abuse. *Lancet*, *1*(8438), 1151–1152.

83. Gawin, F. H., and Ellinwood, E. H., Jr. (1988). Cocaine and other stimulants: actions, abuse, and treatment. *N Engl J Med*, *318*(18), 1173–1182.

84. Schweri, M. M. (1993). Rapid increase of stimulant binding to the DAT after acute cocaine administration: physiological basis of drug craving, Soc Neurosci Abstr 936, presented at the Society for Neuroscience 19th Annual Meeting.

85. Spencer, T. J., Biederman, J., Madras, B. K., Faraone, S. V., Dougherty, D. D., Bonab, A. A., and Fischman, A. J. (2005). In vivo neuroreceptor imaging in attention-deficit/hyper-activity disorder: a focus on the dopamine transporter. *Biol Psychiatry*, *57*(11), 1293–1300.

86. Faraone, S. V., Perlis, R. H., Doyle, A. E., Smoller, J. W., Goralnick, J. J., Holmgren, M. A., and Sklar, P. (2005). Molecular genetics of attention-deficit/hyperactivity disorder. *Biol Psychiatry*, *57*(11), 1313–1323.

87. Barr, C. L., Feng, Y., Wigg, K. G., Schachar, R., Tannock, R., Roberts, W., Malone, M., and Kennedy, J. L. (2001). 5′-Untranslated region of the dopamine D4 receptor gene and attention-deficit hyperactivity disorder. *Am J Med Genet, 105*(1), 84–90.

88. Dougherty, D. D., Bonab, A. A., Spencer, T. J., Rauch, S. L., Madras, B. K., and Fischman, A. J. (1999). Dopamine transporter density in patients with attention deficit hyperactivity disorder. *Lancet, 354*(9196), 2132–2133.

89. van Dyck, C. H., Quinlan, D. M., Cretella, L. M., Staley, J. K., Malison, R. T., Baldwin, R. M., Seibyl, J. P., and Innis, R. B. (2002). Unaltered dopamine transporter availability in adult attention deficit hyperactivity disorder. *Am J Psychiatry, 159*(2), 309–312.

90. Jucaite, A., Fernell, E., Halldin, C., Forssberg, H., and Farde, L. (2005). Reduced midbrain dopamine transporter binding in male adolescents with attention-deficit/hyperactivity disorder: association between striatal dopamine markers and motor hyperactivity. *Biol Psychiatry, 57*(3), 229–238.

91. Dresel, S., Krause, J., Krause, K. H., LaFougere, C., Brinkbaumer, K., Kung, H. F., Hahn, K., and Tatsch, K. (2000). Attention deficit hyperactivity disorder: binding of [^{99}mTc]TRODAT-1 to the dopamine transporter before and after methylphenidate treatment. *Eur J Nucl Med, 27*(10), 1518–1524.

92. Vles, J. S., Feron, F. J., Hendriksen, J. G., Jolles, J., van Kroonenburgh, M. J., and Weber, W. E. (2003). Methylphenidate down-regulates the dopamine receptor and transporter system in children with attention deficit hyperkinetic disorder (ADHD). *Neuropediatrics, 34*(2), 77–80.

93. Madras, B. K., Miller, G. M., and Fischman, A. J. (2005). The dopamine transporter and attention-deficit/hyperactivity disorder. *Biol Psychiatry, 57*(11), 1397–1409.

94. Volkow, N. D., Fowler, J. S., Wang, G. J., Ding, Y. S., and Gatley, S. J. (2002). Role of dopamine in the therapeutic and reinforcing effects of methylphenidate in humans: results from imaging studies. *Eur Neuropsychopharmacol, 12*(6), 557–566.

95. Volkow, N. D., Wang, G. J., Fowler, J. S., Logan, J., Franceschi, D., Maynard, L., Ding, Y. S., Gatley, S. J., Gifford, A., Zhu, W., and Swanson, J. M. (2002). Relationship between blockade of dopamine transporters by oral methylphenidate and the increases in extracellular dopamine: therapeutic implications. *Synapse, 43*(3), 181–187.

96. Gainetdinov, R. R., Jones, S. R., and Caron, M. G. (1999). Functional hyperdopaminergia in dopamine transporter knock-out mice. *Biol Psychiatry, 46*(3), 303–311.

97. Giros, B., Jaber, M., Jones, S. R., Wightman, R. M., and Caron, M. G. (1996). Hyperlocomotion and indifference to cocaine and amphetamine in mice lacking the dopamine transporter. *Nature, 379*(6566), 606–612.

3

CLONING AND GENETIC ANALYSIS OF DOPAMINE TRANSPORTERS

DAVID J. VANDENBERGH

Department of Biobehavioral Health, Center for Developmental and Health Genetics, and Neuroscience Institute, The Pennsylvania State University, University Park, Pennsylvania

3.1   Introduction   47
3.2   cDNA Cloning   48
3.3   Genomic DNA Cloning   51
3.4   Genetics of *DAT1* and Dopamine-Related Phenotypes   55
3.5   Genetic Analysis of Transporter Activity in Human and Nonhuman Primates   60
3.6   Conclusions   62
References   62

3.1   INTRODUCTION

There is a single dopamine transporter gene (*DAT1*, or *DAT*) in all species checked to date that use dopamine as a neurotransmitter. The gene is officially termed the Solute Ligand Carrier Family 6A, Member 3 (*SLC6A3*), with the number 3 indicating that it was the third gene in the family of neurotransmitter transporter genes to be cloned, after the γ-aminobutyric acid (GABA) and norepinephrine transporters. The details of gene nomenclature may be found in the reports of the Human Genome Organization (HUGO, www.gene.ucl.ac.uk/nomenclature/index.html) and the International Committee on Standardized Genetic Nomenclature for Mice (www.informatics.jax.org/mgihome/nomen/). In this chapter the standardized

Dopamine Transporters: Chemistry, Biology, and Pharmacology. Edited by Mark L. Trudell and Sari Izenwasser

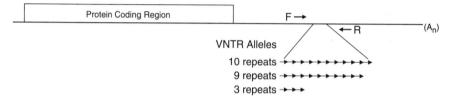

Figure 3-1 Schematic of the sequence of the *DAT1* cDNA. A thin line represents the 5′- and 3′-untranslated regions (UTR), while the box represents the coding region of the gene, and the polyA tail (A$_n$) is shown. The VNTR in the 3′-UTR is indicated below the cDNA with the two most common alleles (9- and 10-copy) and the smallest human allele (3-copy) for comparison. These alleles can be amplified by PCR of genomic DNA using primers (arrows labeled F and R).

nomenclature is utilized, with the gene symbol in italic type when referring to the gene (*DAT1*) and in regular text type when referring to the protein (DAT1). Symbols in all-capital letters indicate the human gene or protein, and those with a single capital letter indicate that of other species.

The chapter is organized into sections that largely follow a chronological sequence from the cloning of the cDNA, to isolating the genomic clones, then to identification of polymorphisms within and around the gene. A wealth of information was gleaned from the cDNA and genomic DNA clones, based on their nucleotide sequences that fueled research in pharmacology, physiology, and behavior (Fig. 3-1). The polymorphisms allowed genetic tests of the role of *DAT1* in behaviors, disorders, and conditions that are heritable traits and are related to dopamine function.

3.2 cDNA CLONING

The rat cDNA (complementary DNA) was the first dopamine transporter cDNA to be cloned [1–3], followed shortly by the bovine cDNA [4]. These reports utilized the information provided by cloning of the transporter cDNAs for GABA [5] and norepinephrine [6]. Alignment of the cDNA sequences of these two transporters revealed regions of similarity that were likely to be present in all family members. These conserved regions were used as probes to clone the *DAT1* cDNA from the human [7,8] as well as other cDNAs for members of the neurotransmitter transporter gene family, including transporters for serotonin [9,10], glycine [11,12], taurine [13,14], and proline [15], as well as transporters with unidentified ligands.

The first experiments that tested the function of the protein encoded by the cloned cDNA were carried out to ensure that the cDNA did, in fact, encode the dopamine transporter. Thus, it was shown that cells that could not accumulate dopamine naturally would do so when transfected with the cDNA. This transport occurred in a sodium- and chloride-dependent manner, as predicted from the literature (e.g., [1]). Furthermore, the rank order of potency of drugs known to bind to DAT1 matched that expected from studies of the native protein (e.g., [8]). These data also indicated that the dopamine transporter was inserted properly into the plasma membrane and did not require other proteins unique to dopaminergic cells for dopamine uptake. It was clear, however, that drug affinity profiles measured from cells transfected with *DAT1* were not as high as those of the native protein [1,8], suggesting that DAT1 might need posttranslational modifications, or accessory proteins, unique to dopaminergic neurons, for full activity.

DAT1 sequence data cDNAs revealed a great amount of information about the potential for posttranslational modification of the protein product without ever having to study the protein directly. This indirect route to learning about the protein was possible given the exact relationship between codons in the cDNA (more exactly, the mRNA from which the cDNA was reverse-transcribed) and the amino acids they encode. The amino acid sequence revealed that the most highly conserved cDNA sequences were regions of high hydrophobicity and suggested that they span the cell membrane when the protein is exported to the cell surface. There are 12 of these transmembrane regions in every transporter of this family. The absence of an export signal [16] in the protein sequence indicated that the protein's amino-terminal sequences are located inside the cell, and given that there are an even number of transmembrane domains, the carboxy terminal is located internally as well. In addition to this general topology scheme, the amino acid sequence revealed consensus sequences representing sites for phosphorylation and glycosylation of the proteins. Previous work had shown that the transporter was glycosylated [17], and now the exact positions of attachment of glycosyl groups to the protein were apparent. The potential for phosphorylation to alter transporter activity [18] was confirmed shortly thereafter, when it was shown that exposure to phorbol esters, which act to stimulate protein kinase C activity, reduced the uptake velocity of Dat1 in COS cells [19]. These cells do not express their native copy of the *Dat1* gene, but had been transfected with a plasmid that allowed expression of the cDNA cloned into the plasmid. Subsequent work demonstrated increased phosphorylation of Dat1 in parallel with a decrease in the V_{max} value of the protein in cultured cells, confirming that phosphorylation might regulate Dat1 function [20,21]. In a set of experiments that utilized synaptosomes from rat brain, it was shown that Dat1's reuptake activity was not altered by forskolin and 8-bromoadenosine 3,5-cyclic monophosphate, which act to stimulate phosphorylation by protein kinase A [22]. These researchers went on to show that the GABA transporter was inhibited with these drugs, indicating that the two related transporters might be regulated by different kinases [22]. It is worth noting that all the putative phosphorylation sites on the protein were found in domains predicted to be internal to the cell membrane, and all the putative glycosylation sites were located in cell-external domains, confirming the topology scheme that was predicted from the cDNA sequence.

Subsequent experiments addressed an interesting and long-standing question about the protein's cocaine-binding activity. As early as 1987 it had been suggested that under particular conditions, there were two binding sites for cocaine in synaptosomal preparations from brain regions rich in dopamine transporters, such as the striatum [23]. The two sites also differed in the density of sites per milligram of total protein, an additional hint that there might be two proteins [24,25]. Synaptosomal membrane preparations contain many proteins, so were the two sites located on one or two proteins, and by extension, was there one gene or two? This concern with the possibility of multiple genes led to the original naming of the gene as *DAT1* rather than *DAT*. Comparison of the clones from many laboratories isolated from one species of animal revealed only one cDNA with cocaine- and dopamine-binding activity, and similarly, comparison of clones from many species identified one cDNA, albeit taking into account the expectation of minor nucleotide differences across species. These data suggested that there was only one dopamine transporter gene. The proof that one protein is encoded by this gene was presented in work published by Boja and colleagues. They found pharmacological characteristics expected of both binding sites when the dopamine transporter was expressed from the *Dat1* cDNA [26]. Thus, the low-affinity site represents a different conformation of, or different site on, one protein, not a second, distinct protein. Expression of Dat1 in cells that could be cultured continuously opened the door to many further pharmacological studies to determine the structure–activity relationship of this protein, which is the focus of several chapters in this book.

An important experiment directed toward development of a model for Parkinson's disease utilized the cloned cDNA. Expression of Dat1 from cDNA transfected into SK-N-MC cells, which normally do not express this protein, conferred cytotoxicity to MPP^+ (1-methyl-4-phenylpiridinium) [27,28]. MPP^+ is the active metabolite of MPTP (1-methyl-4-phenyl-1,2,3,6-tetrahydropyridine), a contaminant found in some illicit preparations of an analog of the opiate, meperidine. The trace amounts of MPP^+ that were generated from the MPTP caused the selective death of dopaminergic neurons in drug users, who then developed cases of a Parkinson's-like syndrome [29]. Thus, the cytotoxicity induced in cells normally resistant to the chemical when the cells expressed Dat1 indicated that the transporter was probably serving to concentrate the chemical in dopaminergic neurons in vivo. Using site-directed mutagenesis, amino acid residues critical for Dat1's ability to transport dopamine and MPP^+ were identified in the first and seventh transmembrane domains [30]. In subsequent work, both the affinity and the rate of uptake of MPP^+ were increased by selective mutations in the seventh and eleventh transmembrane domains [31].

The technique of site-directed mutagenesis depends completely on the availability of a protein's cDNA, due to the ease with which a mutation can be introduced into the cDNA, resulting in a desired replacement of one amino acid for another in the resulting protein [32]. Further indications of the selectivity of MPP^+ for the dopamine transporter were shown in experiments that utilized the *NET* (norepinephrine transporter) cDNA in combination with the *Dat1* cDNA. Despite the high degree of similarity between the two genes (and their proteins), NET does not transport MPP^+ efficiently, and cells expressing *NET* are not killed by MPP^+. Buck and Amara took advantage of this difference to construct a series of chimeric cDNAs that made proteins with various combinations of portions of *DAT1* and *NET* [33]. In doing so, they demonstrated that the portion of DAT1 containing the fourth through eighth transmembrane amino acids was important for uptake of MPP^+ [33]. These studies that combined aspects of pharmacology and toxicology were able to take advantage of opportunities to study DAT1 provided by cloning the cDNA that were not possible when studying the protein in its natural state. There was another advantage in addition to the single amino acid replacement by Kitayama et al. [30,31] or the mixing of entire portions of two different transporters by Buck and Amara [33]. Cells in culture that do not express DAT1 provide a very low, consistent, negative control condition for comparison to the signal of the same cells transfected with *DAT1* cDNA.

In addition to enabling study of the DAT1 function, cloning the cDNA also provided a means to identify proteins that interact with DAT1. With the *DAT1* sequence in hand it was possible to construct a yeast two-hybrid selection scheme to find the proteins that bind to DAT1. The adaptor protein, Hic-5, was found to bind to Dat1 through a LIM domain in Hic-5, and the binding inhibits transporter activity [34]. Additional work showed that the interaction identified in cultured cells in vitro also occurred in the brain, and that Hic-5 is expressed in dopamine projections, supporting the idea that this activity of Hic-5 may be important for accurate dopamine uptake [34]. A complete understanding of the identity and manner of interaction of these Dat1-interacting proteins may provide new avenues to an understanding of the molecular basis of dopamine transporter–related diseases and disorders. For example, genetic variants of these associated proteins might alter the nature of their interactions with Dat1, and thus dopamine uptake. This type of alteration, known as epistasis, may be an important source of variation in dopamine-related phenotypes.

3.3 GENOMIC DNA CLONING

In addition to initial data concerning the protein's structure and pharmacological characteristics, the cDNA provided a tool to reveal the gene's position in the genome. The gene was shown to be located on the tip of the short arm of chromosome 5 at position 5p15.3 in the human [8,35] (Fig. 3-2) and chromosome 13 in the mouse [36]. The tip of chromosome 5p is the critical region deleted in cri-du-chat, but it was soon clear that although *DAT1* was deleted in some cases, cri-du-chat deletions were often proximal to *DAT1* on this arm of the chromosome [37], leaving *DAT1* intact.

It also became apparent as additional transporter genes were mapped to chromosome positions that this gene family was widely scattered across many chromosomes rather than being located in a single locus (Fig. 3-2). For exact nucleotide positions of specific transporters, the human genome browser at www.genome.ucsc.edu can be searched with a transporter gene's name or symbol. Given the completion of the human genome, it is unlikely that more than the current 19 gene family members, numbered 1 to 20, will be found. Gene 10, which is a pseudogene on chromosome 16 with sequence similarity to the creatine transporter on X, is not shown. The small molecules carried by several of these transporters have yet to be identified, but the gene family is not limited to transporting neurotransmitters, as many carry amino acids in nonneuronal tissues [38].

A natural step subsequent to the cloning of the cDNA and positioning of the gene on chromosome 5 was cloning of the gene in its entirety. The human and mouse genes were described in a single report [39]. The sequence data revealed by *DAT1*

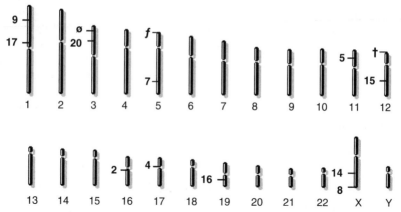

Figure 3-2 Ideograms indicating chromosomal position of the human SLC6A family of transporter genes. Each gene's location is indicated by a tick mark to the left of the chromosomes. Where multiple genes are located close to each other, the gene number was replaced by a superscript symbol ($^\emptyset$, ^{f}, †). The gene numbering is: $^\emptyset$1, *GABA* (GAT1), 2, *norepinephrine*, ^{f}3, *dopamine*, 4, *serotonin*, 5, *glycine* (GLYT2), $^\emptyset$6, *taurine*, 7, *proline* (PROT), 8, *creatine* (CT1), 9, *glycine*, $^\emptyset$11, *GABA* (GAT3), †12, *betaine/GABA* (BGT1), †13, *GABA* (GAT2), 14, amino acid, 15, unknown, 16, unknown, 17, unknown, ^{f}18, unknown, ^{f}19, neutral amino acid, 20, *proline* (IMINO type).

genomic clones opened the door to further study of both expression of the gene and its role in dopamine-related behaviors. The genomic clones revealed that the gene was divided into 15 exons, with parts of the coding region present on exons 2 to 15, although the last exon, containing nearly half of the nucleotides of the mature mRNA, is comprised largely of the 3' untranslated region (UTR) (Fig. 3-3). The size of the human gene was determined to be approximately 52,000 base

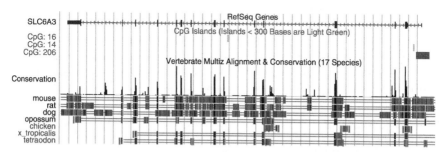

Figure 3-3 Structure of *DAT1*. The gene is displayed with boxes for each of the 15 exons, connected by a thin line representing the introns. The arrowheads within the introns represent the direction of transcription with the first exon on the right. This orientation is based on the convention of displaying the chromosome with the telomere of the p arm to the left. Only a few of the tracks of information from the Genome Browser are shown in this image, including the CpG islands, and conserved regions of sequence with the species noted to the left. (From the UCSC Genome Browser, www.genome.ucsc.edu [125].) (*See insert for color representation of figure.*)

pairs (52 kb), which is now known to be near the average of all genes in the genome [40,41], and differed only minimally from the genes from the other family members for GABA (46 kb) [42], norepinephrine (47 kb) [43], and serotonin (38 kb) [44].

Control of gene expression is an important part of gene function, and the developmental timing, tissue specificity, and quantity of transcripts are critical to proper expression of the protein. *DAT1* is expressed starting with the stage of brain development when the neurogenesis is thought to be initiated (embryonic day 14 or 15 in the mouse) [45] and remains in a transcriptionally active state throughout life. This expression is restricted to a very small number of neurons, specifically those that use dopamine as a neurotransmitter in the substantia nigra, the ventral tegmental area, and a few other areas of the brain [46]. An example of this regional specificity of expression is shown in Figure 3-4.

The cloned genomic DNA from *DAT1* contained flanking sequences, which allowed for identification of sequences involved in controlling transcription of the gene. The initiation of transcription was mapped precisely to one primary and several secondary sites within a few nucleotides of each other [39]. Alignment of the sequences immediately upstream of the start sites of *DAT1* with the flanking sequence of other transporter genes previously cloned identified the promoter as G+C rich and lacking canonical TATAA and CAAT boxes [39]. This combination is found frequently in genes that are expressed in a tissue-specific manner. The G+C-rich sequence at the 5′-end of *DAT1* contains an abundance of CG pairs of nucleotides, known as a CpG island, which is indicated by the thick bar below exon 1 and extending to the right (upstream of the gene) in Figure 3-3. The C of

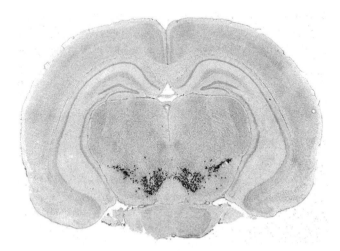

Figure 3-4 Expression of *DAT1* (black dots) is limited to cells of the substantia nigra (SN) and the ventral tegmental area (VTA) in the ventral midbrain as detected by in situ hybridization of an antisense cDNA probe. (Image used with permission from the Allen Brain Atlas, www.brain-map.org, [47], accessed March 2007. Copyright © 2004 Allen Institute for Brain Science.)

these pairs is often methylated as a means of restricting transcription in tissues where the protein is not required. A potential silencer element was identified in intron 1 that plays a role in expression of *DAT1* [48]. More recent genetic analysis indicated that sequences in introns 9, 12, and 14 might contribute to as much as twofold differences in the expression of mRNA from *DAT1* [49]. The fact that these polymorphisms are not codon altering, and there are no codon-altering polymorphisms in exons in linkage disequilibrium with the intron polymorphisms, suggested that the introns harbored enhancerlike sequences. Thus, Greenwood and colleagues tested plasmids constructed with various portions of the 5' *DAT1* promoter sequences in combination with intron sequences to drive expression of a luciferase reporter gene. By using a dopaminergic cell line (SN4741), they ensured, to the greatest extent possible, that the native transcription factors that control *DAT1* expression were present. In addition, appropriate controls for variability in transfection of these cells were included to provide reliable results. These results need to be followed up by genetic studies of *DAT1* expression in human beings that focus on these polymorphisms.

The promoter sequences in the 5'-flank of the gene were studied by cloning parts of the sequences in front of a reporter gene [50]. These experiments demonstrated that the core elements (the CpG island), proximal to the start of transcription served as a powerful, nonspecific promoter and that more distal elements served to regulate the core, mostly by repression [50]. Cotransfection of a plasmid containing *Nurr1* along with the *DAT1*-reporter gene plasmid elevated expression from the reporter gene, suggesting that *Nurr1* was able to act via the *DAT1* sequences [50]. That Nurr1 binds to the *DAT1* promoter and enhances its expression was shown in a subsequent report [51], but it did not bind to sequences that had been shown to be its binding site in previous studies of other promoters. This report was the first to suggest that Nurr1 activates transcription by binding to noncannonical sequences [51,52]. Note that the experiments were carried out using a mouse transcription factor and a human gene's promoter. The conservation of sequence in *Nurr1* and *DAT1* is sufficiently high to allow this type of cross-species experiment to work routinely. Consistent with a possible role for this protein in regulating *DAT1*, analysis of NURR1 in cocaine abusers showed that its expression was down-regulated [53]. Additional work will be necessary to determine the manner in which cocaine administration alters NURR1 expression. Subsequently, another transcription factor, ZFP161, was shown to bind to the *DAT1* promoter [54], and SP1 and SP3, two widely expressed transcription factors, were shown to activate expression of *DAT1* [55]. The latter two proteins aare known to act via G+C-rich sequences, similar to those found in the 5' proximal promoter, as mentioned above.

Control of *DAT1* expression appears to be linked intimately to proper development of dopaminergic neurons. It has been found that expression of *Nurr1* in mouse embryonic stem cells promotes differentiation to a dopamine cell type [56], which may be a direct result of interaction with *DAT1* or with other genes important for correct differentiation. These results may be important in the development of cell-based therapies to treat Parkinson's disease [57]. Further knowledge about DNA elements that control *Dat1* function may also allow one to target expression of other genes in dopamine neurons, which may provide additional protection for these neurons in Parkinson's disease [58]. These studies have not revealed everything that needs to be known about *DAT1* expression in order to explain the high degree of tissue specificity and the developmental timing of expression, but work in this area is ongoing [59].

Given the difficulty in identifying promoter and enhancer sequences that regulate *DAT1* expression, other, more comprehensive methods may illuminate important elements in ensuring that the correct amount of DAT1 is produced in the correct tissues and at the correct stage of development. One such method is quantitative trait locus (QTL) analysis, which employs two parental strains of mice, frequently C57BL/6J and DBA/2J, and recombinant inbred (RI) mice generated by crossing the two strains. In a QTL analysis of this type, the amount of Dat1 protein on the surface of neurons was shown to be controlled by a gene located on mouse chromosome 19 [60]. This region of chromosome 19 is not the site of any of the transcription factors mentioned in the previous paragraphs, indicating that there are more genes to be identified that activate or repress the expression of *Dat1*.

In addition to gaining an understanding of the regulation of genes in dopamine neurons, knowledge of the genomic DNA has allowed for deletion of the gene in order to study the effect of loss of the gene in behavior and pharmacology. This subject is covered in a subsequent chapter, but brief mention here is worthwhile because of the relevance of the information generated to human genetics described below. The first knockout mice created without functional dopamine transporters survived into adulthood and displayed levels of activity which suggested that they would be a model of attention-deficit/hyperactivity disorder (ADHD) [61]. Although this initial report also demonstrated that the knockout mice were indifferent to cocaine and amphetamine, use of other drug-reward measures indicated that they were not completely unresponsive [62] and that serotonin function was an important factor in psychostimulant-related behaviors. The idea that a second neurotransmitter system is involved in behaviors that were attributed to a single (dopamine) neurotransmitter system is well documented in animal studies but has not been incorporated in human genetic work to the extent that it could be.

3.4 GENETICS OF *DAT1* AND DOPAMINE-RELATED PHENOTYPES

The inferred protein sequence was not the only source of new information about the dopamine transporter; the DNA sequence revealed important data in the form of genetic polymorphisms within the exons of *DAT1*. The use of the cDNA as a probe revealed additional polymorphisms that were not in the exons but were physically linked to them in the introns and flanking regions of the chromosome. The identification of sites in the DNA that varied from person to person led to an explosion of reports examining the relationship between *DAT1* and a number of disorders and diseases of varying degrees of relatedness to dopamine function.

The first polymorphism reported for the human *DAT1* was a site identified due to differential cutting with the restriction endonuclease *Taq*I and named *Taq*IA [7]. The two alleles at this site were detected by the presence of DNA fragments of different sizes on Southern blots and thus termed a restriction fragment length polymorphism (RFLP). This original polymorphism was subsequently shown to be located in the fourth intron (Vandenbergh, unpublished data), which greatly diminished the likelihood that there would be a functional change in transporter activity of the protein, and it was not used widely. Newer techniques that are faster and require less starting genomic DNA [i.e., Polymerase Chain Reaction (PCR)] are used to detect this type of polymorphism, and the general name single nucleotide polymorphism (SNP) is now used. In fact, using high-throughput methods, the site was "rediscovered" and is identified by the name rs461753 on chromosome 5 at nucleotide position 1,484,214 (from the March 2006 assembly, on the Genome Browser, www.genome.ucsc.edu).

The second polymorphic site to be identified was in the 3′-untranslated region (3′-UTR) of *DAT1* cDNA, not in an intron [35], and proved to be of much greater interest. The polymorphism was shown to be a tandem repeat of 40 nucleotides, present in head-to-tail fashion with a total of 10 copies in the cDNA. The number of copies of the repeat varied from person to person [35], so the polymorphism was called a variable number of tandem repeats (VNTR). As opposed to an SNP, which usually exists only as one of two alleles, VNTRs can be much more variable. For the *DAT1* VNTR, there are only two common alleles, a 10-copy form at a frequency of roughly 70%, and a 9-copy form at a frequency of roughly 25% in most populations. The remaining 5% of the alleles are rare and sizes range from 3 to 13 copies [63,64]. The 9-copy form has been reported to be present at a reduced frequency (17%) in African-Americans, with a concomitant increase in frequency of rare alleles [63].

A number of studies published shortly after discovery of the VNTR examined the genetic relationship between *DAT1* and neuropsychiatric disorders thought to be related to dopamine. The first phenotypes to be tested were related to illicit drug use, abuse, and dependence. The VNTR polymorphism and the *Taq*I polymorphism described earlier were compared to drug use as measured by the *Diagnostic and Statistical Manual III-R* (DSM-III-R) or a drug use survey that assessed the quantity and frequency of use at peak lifetime use, but neither polymorphism was associated with drug use [65]. An association was found between the 9-copy allele of the VNTR and cocaine-induced paranoia, although there was no association with the broader measure of cocaine use [66]. The underlying mechanism that might explain this relationship between alleles of *DAT1* and a drug-induced state of paranoia is not clear; however, recent work on DAT1 protein levels and function in the brains of cocaine users may suggest potential avenues for future work. The use of WIN 35,428, a DAT1 ligand, revealed that the levels of DAT1 do not change in people who experience excited delirium when using cocaine, whereas DAT1 levels were increased in other cocaine users [67]. The contradictory results concerning the relationship between genotype and DAT1 protein levels using positron emission tomography (PET) imaging (described in Section 3.5) might be due to small effects in normal individuals, but those effects could be exacerbated by the presence of cocaine. In an attempt to find additional polymorphic sites in the *DAT1* gene that might be used in these types of studies it was found that the gene is highly conserved and that no polymorphism that alters an amino acid was present at a frequency of

>5% (the standard cutoff for rare vs. common alleles) [68]. These authors screened exons of the gene by sequencing DNA from persons free of psychiatric disorders and those with Tourette's syndrome, ethanol dependence, or attention-deficit/ hyperactivity disorder to bias toward finding polymorphic sites that might only be present in dopamine-related disorders; however, no relationship between genotype and trait status was found [68]. In a separate study, an allele of a VNTR found in intron 8 of *DAT1* [39] was shown to be associated with cocaine abuse in a Brazilian sample [69]. The authors went on to test for a possible functional difference between the two common alleles at this site by use of a *Renilla* luciferase reporter gene system. In this system, an intron found within the reporter gene is replaced with an intron of the gene of interest (*DAT1*) to test the effects of intronic sequences on a heterologous gene. The two most common intron 8 VNTR alleles (five and six copies of the repeat, labeled as alleles 2 and 3, respectively) were inserted and luciferase expression was tested in a dopaminergic cell line. The vector containing the five-repeat allele generated more expression than the vector with the six-repeat allele, although no numerical comparison was given and both alleles generated less than 50% of the expression of the unmodified vector [69]. In general, analyses have shown that there are very few polymorphisms that alter amino acids of DAT1 [68], and analyses of *DAT1* polymorphisms must look to the less obvious, and harder to assay, functional outcomes, such as mRNA stability or alteration of transcription rate, as opposed to steady-state mRNA levels.

Alcoholism was another disorder to be tested for a genetic relationship to *DAT1* shortly after discovery of the gene. One of the earliest reports utilized a more sophisticated approach than was common at the time, with the aim of understanding alcoholism as a complex disorder. In this report, Japanese alcoholics and controls were selected according to their genotype at the Aldehyde Dehydrogenase 2 (*ALDH2*) gene before analyzing the relationship between the *DAT1* VNTR and alcoholism. The goal was to generate a more homogeneous population of alcoholics because the mutant allele, *ALDH2*2*, is a well-recognized protective factor against alcoholism. It was found that alcoholics with the *2 allele were significantly more likely to have the seven-repeat allele of *DAT1* than were nonalcoholic individuals with the *ALDH2*2* allele [70]. Unfortunately, this increase in frequency (5.6% in alcoholics compared to 2.1% in controls) focuses on a relatively rare allele that is unlikely to contribute to alcoholism to any great extent in the general population. Additional studies presented conflicting reports on a relationship between *DAT1* and alcoholism or specific traits related to alcoholism, such as withdrawal seizures or delirium, with some studies finding an association [71] and some not [72].

Genetic assessment of a role for *DAT1* and smoking came much later than for other types of drug use. In 1999, two back-to-back papers examined the role of *DAT1* VNTR alleles and smoking status (non-, current, and formersmoker). Lerman and colleagues reported that the 9-copy allele was found more frequently in nonsmokers [73]. This group also found a gene–gene interaction (epistasis) with alleles at the dopamine receptor D2 gene (*DRD2*), such that the effect of the 9-copy allele was stronger in persons carrying the *Taq*I-A2 allele in the receptor gene. In the accompanying report, Sabol and colleagues confirmed the association with nonsmoking status and showed that the effect was related to both late initiation of smoking and length of quitting attempts [74]. The 9-copy allele was associated with low levels of novelty seeking, suggesting that these individuals had a lower need for novelty and reward (both traits related to dopamine release) and thus were less likely to seek cigarettes [74]. Subsequent analyses did not confirm these results [68,75], but one of these studies [68] did find a genetic effect related to never-smokers. These persons have never tried smoking, and in standard classifications, based on U.S. government definitions of smokers, they are grouped with those who have smoked fewer than 100 cigarettes in the nonsmoker category. Without having smoked a cigarette, these people have had no exposure to the pharmacological effects of nicotine, and it is not clear how they would respond to smoking; thus, these people may form a genetically distinct group based on their avoidance of a common behavior. Interestingly, these never-smokers were found to have a higher frequency of the 10-copy allele [68], which is opposite of what would be expected based on the Lerman and Sabol studies. Lerman and colleagues extended their studies by examining the role of alleles of *DAT1* with more narrowly defined smoking phenotypes, with the intent of understanding aspects of smoking-related behaviors that might be "closer" to the function of the transporter, thereby gaining an understanding of the associations. In one study there was evidence for a significant interaction of *DAT1* and *DRD2* on prolonged smoking abstinence and time to relapse at the end of treatment (10 weeks), but these effects were no longer significant at six months [76]. These results were important because they provided the first evidence of a genetic component to pharmacological treatment for smoking. These findings were supported in an analysis of short-term (1 week) success at smoking cessation with either nicotine replacement therapy or bupropion treatment, which showed modest association to alleles of *DAT1*; however, analogous to the publication of Lerman, the effect dissipated by the fourth week [77].

In addition to drug abuse phenotypes, early studies examined several traits related to dopamine function. Despite initial hopes of uncovering new genetic information that would clarify the relationship between the transporter and a disease, disorder, or personality, little in the way of consistent results appeared. Thus, some positive associations were found between *DAT1* and Parkinson's disease [78,79], but subsequent work was unable to extend these findings.

In the case of schizophrenia, another disorder treated by dopamine-related drugs, two early reports excluded linkage to *DAT1* in multiple pedigrees. In the first report, a highly polymorphic site(s) detected by *Taq*I restriction enzyme digestion and Southern blotting [80] detected no linkage under an assumption of autosomal dominant or recessive models when the disorder was ascertained by the research diagnostic criteria [81]. In an analysis of schizophrenia and schizophrenia spectrum disorders based on DSM-III-R criteria, Persico and colleagues detected no linkage [82]. Negative findings were presented in analyses of genetic association, rather than linkage, of *DAT1* and schizophrenia followed in studies of patients and controls from China [83], the United Kingdom [84], and France [85]. A recent study of schizophrenic psychosis has suggested that there may be an association with polymorphisms in the 5′ flank of *DAT1* [86], which may rekindle interest in *DAT1* as a genetic contributor to the disorder, but most attention has focused on other genes, such as disrupted-in-schizophrenia (*DISC1*), dysbindin (*DTNBP1*), and neuregulin (*NRG1*) [87].

Genetic analyses of manic-depression or bipolar disorder have provided a complex set of outcomes. Early studies excluded linkage [88] and found no association [89] for *DAT1*. Subsequent work by Greenwood and colleagues suggested that there might be linkage disequilibrium between haplotypes of SNPs in *DAT1* and manic depression [90,91]. A separate study found a rare missense SNP that alters the glutamic acid at amino acid position 602 to a glycine (E602G) that was inherited by a patient from an affected father [92]. Further tests by this group of the two forms of the protein in an in vitro cell system showed that the mutant allele blocks insertion of the protein into the cell membrane. The latter study raises the interesting possibility that a rare mutation in *DAT1* might cause manic depression in some families, and may expand our understanding of the mechanism by which altered dopamine reuptake contributes to manic depression, but it probably does not explain the majority of cases of this disorder.

The strongest genetic relationship between *DAT1* and a neuropsychiatric disorder is with attention-deficit/hyperactivity disorder (ADHD), with over 160 related publications as of 2007, and several recent reviews cover some of the details of ADHD and *DAT1* [93,94]. The original work in 1995 demonstrated association and linkage between ADHD and the 10-copy allele of the *DAT1* VNTR [95]. This study used a form of transmission disequilibrium analysis that tests for transmission of an allele at a frequency greater than expected based on the parents' genotypes to their offspring with ADHD. This type of test has an advantage in that it is using information within families, which diminishes the chance of spurious association due to differences in allele frequency in ethnic groups, and thus the results are thought to be more convincing. The 10-copy allele was found in these ADHD children, despite it being the most common allele at the VNTR. A subsequent replication [96] in an independent sample encouraged others to examine the relationship. For example, an analysis of subtypes of ADHD demonstrated that the 10-copy allele was associated with levels of hyperactive-impulsive symptoms, but not inattentive symptoms [97]. This and earlier work demonstrating high heritability of ADHD in twin [98] and adoption [99] studies, and the role of drugs that act via DAT1, such as methylphenidate and amphetamine, to treat ADHD, led to the hypothesis that the 10-copy allele would be associated with responsiveness to methylphenidate treatment. This idea was confirmed in a study of 119 Irish children [100], but not in the work of others [101,102]. In general, some studies have replicated a genetic association [103] and others have not [104]. With an eye toward analyzing larger sample sizes of ADHD children, metaanalyses have assessed the relationship by merging the data from multiple studies. All three of the metaanalyses that have been conducted to date have found no significant association [105–107]. This result appears incongruous with the high heritability demonstrated for ADHD, but high heritability does not guarantee that it will be easy to find the underlying genes. In fact, one of the metaanalyses [106] detected significant phenotypic heterogeneity between the samples in the 13 studies that were merged. It was realized as early as 1998 that the various subtypes of ADHD may reflect the actions of different genes [97], and it is possible that these subclassifications themselves are insufficiently precise phenotypes to reveal the gene effects in a reliable way. It is also quite likely that subtle differences in assessment of ADHD through the various questionnaires, self-reports, teacher reports, and so on, used, may be tapping into inherent differences in the disorder.

Other studies have approached the problem differently than the metaanalysis route by examining other neurophysiological aspects that might be associated with ADHD. Thus, one study assessed persons with different VNTR alleles on an attention task administered to the research volunteers, and measured methylphenidate-related electroencephalograph (EEG) changes. Children homozygous for the 10-repeat performed more poorly on the attention task, and they displayed different EEG responses when adminstered methylphenidate than did those who had one or two copies of the 9-repeat allele [108]. Another study analyzed the relationship between prenatal exposure to smoke and subsequent ADHD symptoms based on VNTR genotype. This work indicated that children who were homozygous for the 10-repeat allele had significantly higher hyperactivity–impulsivity symptoms, but only if they had been exposed to maternal prenatal smoking [109]. Given the difficulty in making clear associations with metaanalyses of DSM-like diagnoses as a phenotype, analyses that examine other aspects of the disorder, or interacting environments, may provide important advances in understanding the role of *DAT1* in ADHD.

3.5 GENETIC ANALYSIS OF TRANSPORTER ACTIVITY IN HUMAN AND NONHUMAN PRIMATES

The *DAT1* VNTR is present in nonhuman primates but not in other mammals or nonmammalian animals, at least for those checked to date. Cloning of the gene from rhesus (*Macaca mulatta* and *M. fasicularis*) and South American squirrel (*Saimiri sciureus*) monkeys demonstrated that the repeat sequence was present in species other than human beings [110]. Unfortunately, although the VNTR is present in apes and in Old and New World monkeys, other primate sequences have not been reported, so the point at which this sequence appeared evolutionarily is not clear. The site is polymorphic in the species that have been checked, but usually only as two common alleles, being found as single- and double-repeat alleles in chimpanzees, 5- and 11-repeat alleles in cynomolgus monkeys, or 5- and 12-repeat alleles in African green monkeys [111].

The work of several groups tested for a functional consequence of the VNTR on DAT1 protein or mRNA levels. Experiments that used a reporter gene fused to sequence from the 3′-UTR of the human gene demonstrated significantly higher expression if the 3′-UTR contained a 9-repeat rather than a 10-repeat sequence [110]. Similarly, Bannon and colleagues presented evidence that the 9-repeat allele enhanced *DAT1* transcription when introduced into an immortalized dopaminergic cell line or in neonatal dopamine neurons [52]. At approximately the same time, a separate group reported conflicting evidence that expression was highest with the 10-repeat fused to a reporter gene [112]. These reports suggested that concentrations of DAT1 in dopaminergic neurons might vary with allele status in the $3'$-UTR, but there was disagreement as to which allele was responsible for elevated expression. A potential modulator of expression was identified in the protein Hesr1/Hey1, which acts through binding to the VNTR and down-regulating gene expression [113], but this work utilized a reporter gene in a transfection assay of cells in culture, and binding of Hesr1 to the VNTR has not been demonstrated to occur in vivo. Interestingly, knockout mice deficient in Hesr1 have an increase in *Dat1* expression and a decrease in locomotor activity, suggestive of a *Hesr1–Dat1* interaction [114].

In an attempt to circumvent the problems associated with in vitro assays, Miller and colleagues tested levels of *DAT1* mRNA by quantitative reverse transcriptase–polymerase chain reaction (qRT-PCR) from the cerebellum and temporal lobe of autopsy samples of known VNTR genotype [115]. Although there was only one sample with a 9/9 genotype, expression of *DAT1* was highest in samples with the 10/10 genotype compared to the 9/10 genotype [115]. Unfortunately, this study did not test expression in the brain regions that are thought to possess functional levels of the transporter (e.g., substantia nigra and ventral tegmental area), and expression specific to each allele was not tested. For an example of such an allele-specific expression analysis, see work on the serotonin transporter in which a poly-morphism within the transcript was used to test the role of a promoter polymorphism by exploiting haplotypes of the two polymorphisms [116]. It is possible that accurate assessment of differences in expression induced by these alleles will require a system that more closely approximates the in vivo conditions in human neurons, and the possibility that different life histories might cause variation in *DAT1* gene expression cannot be ruled out.

The suggestion that expression of *DAT1* might depend on the allele present in the VNTR was also examined in human beings with PET imaging studies, and although imaging studies are addressed in a separate chapter, it is worth noting here those that included a genetic analysis. One study of recently detoxified cocaine users and healthy controls found that 9-repeat carriers (9/9 and 9/10 genotypes) had approximately 13% higher binding to $[^{123}I]\beta$-CIT than did 10/10 homozygotes [117]. In contrast, a study using the same ligand but a different population (abstinent alcoholics and healthy controls) found a 22% decrease in persons with a 9/10 genotype compared to those with a 10/10 genotype [118]. The absence of persons with a 9/9 genotype is due to the fact that this rare genotype was not represented in the sample studied; however, the absence of this genotype in the analysis is unlikely to account for a finding in the opposite direction from that of Jacobsen and colleagues. Martinez et al. found no association of genotype with transporter binding to $[^{123}I]\beta$-CIT by single-photon emission computerized tomography (SPECT) [119]; however, a relatively large study of 96 persons conducted by the same group demonstrated an increased dopamine transporter availability with the 9-repeat allele [120]. This study included age as a covariate to account for known decreases in transporter density with age. Subsequently, in a study of smokers, the presence of at least one 9-repeat allele, along with specific genotypes at several other dopamine-related genes, was correlated with greater decreases in DAT1 binding potential when smoking than those with the alternate genotypes [121]. This result was not supported by the work of Krause et al., who found no increase in dopamine transporter in persons homozygous for the 10-repeat allele [122]. These conflicting reports suggest that the effect of VNTR status on transporter density is modest in humans, or at least not so strong as to be detected above the variability in density that may arise from environmental effects such as disease, stress, and drug use, both licit and illicit. These types of studies may reveal more consistent results as new PET and SPECT ligands are synthesized and instrumentation is improved.

3.6   CONCLUSIONS

Isolation of cDNAs opened new doors for studies in many fields, including neurobiology, physical chemistry, and pharmacology, because the cDNAs provided critical reagents in large quantities. Many studies described in this and other chapters would not have been possible without the cDNA, genomic DNA, and protein that became available after the original publications in 1991 and 1992. In fact, publication of cDNA sequences promoted the spread of these reagents by making it possible for anyone with molecular biology skills to produce their own cDNA or to amplify smaller regions by PCR. Many reviews on *DAT1* are available in the literature, and some have been cited. For further information on topics not covered here, two recent reviews that cover neurotoxicology related to the dopamine transporter [123] and the short historical view from a transporter researcher from the earliest days [124], are well worth reading.

REFERENCES

1. Shimada, S., Kitayama, S., Lin, C. L., Patel, A., Nanthakumar, E., Gregor, P., Kuhar, M., and Uhl, G. (1991). Cloning and expression of a cocaine-sensitive dopamine transporter complementary DNA. *Science*, *254*(5031), 576–578.

2. Kilty, J. E., Lorang, D., and Amara, S. G. (1991). Cloning and expression of a cocaine-sensitive rat dopamine transporter. *Science*, *254*(5031), 578–579.

3. Giros, B., el Mestikawy, S., Bertrand, L., and Caron, M. G. (1991). Cloning and functional characterization of a cocaine-sensitive dopamine transporter. *FEBS Lett*, *295*(1–3), 149–154.

4. Usdin, T. B., Mezey, E., Chen, C., Brownstein, M. J., and Hoffman, B. J. (1991). Cloning of the cocaine-sensitive bovine dopamine transporter. *Proc Natl Acad Sci U S A*, *88*(24), 11168–11171.

5. Guastella, J., Nelson, N., Nelson, H., Czyzyk, L., Keynan, S., Miedel, M. C., Davidson, N., Lester, H. A., and Kanner, B. I. (1990). Cloning and expression of a rat brain GABA transporter. *Science*, *249*(4974), 1303–1306.

6. Pacholczyk, T., Blakely, R. D., and Amara, S. G. (1991). Expression cloning of a cocaine- and antidepressant-sensitive human noradrenaline transporter. *Nature*, *350*(6316), 350–354.

7. Vandenbergh, D. J., Persico, A. M., and Uhl, G. R. (1992). A human dopamine transporter cDNA predicts reduced glycosylation, displays a novel repetitive element and provides racially-dimorphic TaqI RFLPs. *Brain Res Mol Brain Res*, *15*(1–2), 161–166.

8. Giros, B., el Mestikawy, S., Godinot, N., Zheng, K., Han, H., Yang-Feng, T., and Caron, M. G. (1992). Cloning, pharmacological characterization, and chromosome assignment of the human dopamine transporter. *Mol Pharmacol*, *42*(3), 383–390.

9. Hoffman, B. J., Mezey, E., and Brownstein, M. J. (1991). Cloning of a serotonin transporter affected by antidepressants. *Science*, *254*(5031), 579–580.

10. Blakely, R. D., Berson, H. E., Fremeau, R. T., Jr., Caron, M. G., Peek, M. M., Prince, H. K., and Bradley, C. C. (1991). Cloning and expression of a functional serotonin transporter from rat brain. *Nature*, *354*(6348), 66–70.

11. Smith, K. E., Borden, L. A., Hartig, P. R., Branchek, T., and Weinshank, R. L. (1992). Cloning and expression of a glycine transporter reveal colocalization with NMDA receptors. *Neuron*, *8*(5), 927–935.

12. Liu, Q. R., Nelson, H., Mandiyan, S., Lopez-Corcuera, B., and Nelson, N. (1992). Cloning and expression of a glycine transporter from mouse brain. *FEBS Lett*, *305*(2), 110–114.

13. Smith, K. E., Borden, L. A., Wang, C. H., Hartig, P. R., Branchek, T. A., and Weinshank, R. L. (1992). Cloning and expression of a high affinity taurine transporter from rat brain. *Mol Pharmacol*, *42*(4), 563–569.

14. Liu, Q. R., Lopez-Corcuera, B., Nelson, H., Mandiyan, S., and Nelson, N. (1992). Cloning and expression of a cDNA encoding the transporter of taurine and β-alanine in mouse brain. *Proc Natl Acad Sci U S A*, *89*(24), 12145–12149.

15. Velaz-Faircloth, M., Guadano-Ferraz, A., Henzi, V. A., and Fremeau, R. T., Jr. (1995). Mammalian brain-specific L-proline transporter: neuronal localization of mRNA and enrichment of transporter protein in synaptic plasma membranes. *J Biol Chem*, *270*(26), 15755–15761.

16. Blobel, G. (1980). Intracellular protein topogenesis. *Proc Natl Acad Sci U S A*, *77*(3), 1496–1500.

17. Lew, R., Grigoriadis, D., Wilson, A., Boja, J. W., Simantov, R., and Kuhar, M. J. (1991). Dopamine transporter: deglycosylation with exo- and endoglycosidases. *Brain Res*, *539*(2), 239–246.

18. Surratt, C. K., Wang, J. B., Yuhasz, S., Amzel, M., Kwon, H. M., Handler, J. S., and Uhl, G. R. (1993). Sodium- and chloride-dependent transporters in brain, kidney, and gut: lessons from complementary DNA cloning and structure–function studies. *Curr Opin Nephrol Hypertens*, *2*(5), 744–760.

19. Kitayama, S., Dohi, T., and Uhl, G. R. (1994). Phorbol esters alter functions of the expressed dopamine transporter. *Eur J Pharmacol*, *268*(2), 115–119.

20. Huff, R. A., Vaughan, R. A., Kuhar, M. J., and Uhl, G. R. (1997). Phorbol esters increase dopamine transporter phosphorylation and decrease transport V_{max}. *J Neurochem*, *68*(1), 225–232.

21. Vaughan, R. A., Huff, R. A., Uhl, G. R., and Kuhar, M. J. (1997). Protein kinase C–mediated phosphorylation and functional regulation of dopamine transporters in striatal synaptosomes. *J Biol Chem*, *272*(24), 15541–15546.

22. Tian, Y., Kapatos, G., Granneman, J. G., and Bannon, M. J. (1994). Dopamine and gamma-aminobutyric acid transporters: differential regulation by agents that promote phosphorylation. *Neurosci Lett, 173*(1–2), 143–146.

23. Calligaro, D. O., and Eldefrawi, M. E. (1987). High affinity stereospecific binding of [^{3}H]cocaine in striatum and its relationship to the dopamine transporter. *Membr Biochem, 7*(2), 87–106.

24. Madras, B. K., Spealman, R. D., Fahey, M. A., Neumeyer, J. L., Saha, J. K., and Milius, R. A. (1989). Cocaine receptors labeled by [^{3}H]2β-carbomethoxy-3β-(4-fluorophenyl)-tropane. *Mol Pharmacol, 36*(4), 518–524.

25. Ritz, M. C., and Kuhar, M. J. (1989). Relationship between self-administration of amphetamine and monoamine receptors in brain: comparison with cocaine. *J Pharmacol Exp Ther, 248*(3), 1010–1017.

26. Boja, J. W., Markham, L., Patel, A., Uhl, G., and Kuhar, M. J. (1992). Expression of a single dopamine transporter cDNA can confer two cocaine binding sites. *Neuroreport, 3*(3), 247–248.

27. Kitayama, S., Shimada, S., and Uhl, G. R. (1992). Parkinsonism-inducing neurotoxin MPP$^+$: uptake and toxicity in nonneuronal COS cells expressing dopamine transporter cDNA. *Ann Neurol, 32*(1), 109–111.

28. Pifl, C., Giros, B., and Caron, M. G. (1993). Dopamine transporter expression confers cytotoxicity to low doses of the parkinsonism-inducing neurotoxin 1-methyl-4-phenyl-pyridinium. *J Neurosci, 13*(10), 4246–4253.

29. Langston, J. W., Ballard, P., Tetrud, J. W., and Irwin, I. (1983). Chronic Parkinsonism in humans due to a product of meperidine-analog synthesis. *Science, 219*(4587), 979–980.

30. Kitayama, S., Shimada, S., Xu, H., Markham, L., Donovan, D. M., and Uhl, G. R. (1992). Dopamine transporter site-directed mutations differentially alter substrate transport and cocaine binding. *Proc Natl Acad Sci U S A, 89*(16), 7782–7785.

31. Kitayama, S., Wang, J. B., and Uhl, G. R. (1993). Dopamine transporter mutants selectively enhance MPP$^+$ transport. *Synapse, 15*(1), 58–62.

32. Winter, G., Fersht, A. R., Wilkinson, A. J., Zoller, M., and Smith, M. (1982). Redesigning enzyme structure by site-directed mutagenesis: tyrosyl tRNA synthetase and ATP binding. *Nature, 299*(5885), 756–758.

33. Buck, K. J., and Amara, S. G. (1994). Chimeric dopamine–norepinephrine transporters delineate structural domains influencing selectivity for catecholamines and 1-methyl-4-phenylpyridinium. *Proc Natl Acad Sci U S A, 91*(26), 12584–12588.

34. Carneiro, A. M., Ingram, S. L., Beaulieu, J. M., Sweeney, A., Amara, S. G., Thomas, S. M., Caron, M. G., and Torres, G. E. (2002). The multiple LIM domain-containing adaptor protein Hic-5 synaptically colocalizes and interacts with the dopamine transporter. *J Neurosci, 22*(16), 7045–7054.

35. Vandenbergh, D. J., Persico, A. M., Hawkins, A. L., Griffin, C. A., Li, X., Jabs, E. W., and Uhl, G. R. (1992). Human dopamine transporter gene (*DAT1*) maps to chromosome 5p15.3 and displays a VNTR. *Genomics, 14*(4), 1104–1106.

36. Lossie, A. C., Vandenbergh, D. J., Uhl, G. R., and Camper, S. A. (1994). Localization of the dopamine transporter gene, *Dat1*, on mouse chromosome 13. *Mamm Genome, 5*(2), 117–118.

37. Overhauser, J., Huang, X., Gersh, M., Wilson, W., McMahon, J., Bengtsson, U., Rojas, K., Meyer, M., and Wasmuth, J. J. (1994). Molecular and phenotypic mapping

of the short arm of chromosome 5: sublocalization of the critical region for the cri-du-chat syndrome. *Hum Mol Genet*, *3*(2), 247–252.

38. Broer, S. (2006). The SLC6 orphans are forming a family of amino acid transporters. *Neurochem Int*, *48*(6–7), 559–567.

39. Donovan, D. M., Vandenbergh, D. J., Perry, M. P., Bird, G. S., Ingersoll, R., Nanthakumar, E., and Uhl, G. R. (1995). Human and mouse dopamine transporter genes: conservation of 5′-flanking sequence elements and gene structures. *Brain Res Mol Brain Res*, *30*(2), 327–335.

40. Lander, E. S., Linton, L. M., Birren, B., Nusbaum, C., Zody, M. C., Baldwin, J., Devon, K., Dewar, K., Doyle, M., FitzHugh, W., et al. (International Human Genome Sequencing Consortium) (2001). Initial sequencing and analysis of the human genome. *Nature*, *409*(6822), 860–921.

41. Venter, J. C., Adams, M. D., Myers, E. W., Li, P. W., Mural, R. J., Sutton, G. G., Smith, H. O., Yandell, M., Evans, C. A., Holt, R. A., et al. (2001). The sequence of the human genome. *Science*, *291*(5507), 1304–1351.

42. Lam, D. M., Fei, J., Zhang, X. Y., Tam, A. C., Zhu, L. H., Huang, F., King, S. C., and Guo, L. H. (1993). Molecular cloning and structure of the human (*GABATHG*) GABA transporter gene. *Brain Res Mol Brain Res*, *19*(3), 227–232.

43. Porzgen, P., Bonisch, H., and Bruss, M. (1995). Molecular cloning and organization of the coding region of the human norepinephrine transporter gene. *Biochem Biophys Res Commun*, *215*(3), 1145–1150.

44. Lesch, K. P., Balling, U., Gross, J., Strauss, K., Wolozin, B. L., Murphy, D. L., and Riederer, P. (1994). Organization of the human serotonin transporter gene. *J Neural Transm Gen Sect*, *95*(2), 157–162.

45. Nakai, J., and Fujita, S. (1994). Early events in the histo- and cytogenesis of the vertebrate CNS. *Int J Dev Biol*, *38*(2), 175–183.

46. Shimada, S., Kitayama, S., Walther, D., and Uhl, G. (1992). Dopamine transporter mRNA: dense expression in ventral midbrain neurons. *Brain Res Mol Brain Res*, *13*(4), 359–362.

47. Lein, E. S., Hawrylycz, M. J., Ao, N., Ayres, M., Bensinger, A., Bernard, A., Boe, A. F., Boguski, M. S., Brockway, K. S., Byrnes, E. J., et al. (2007). Genome-wide atlas of gene expression in the adult mouse brain. *Nature*, *445*(7124), 168–176.

48. Kouzmenko, A. P., Pereira, A. M., and Singh, B. S. (1997). Intronic sequences are involved in neural targeting of human dopamine transporter gene expression. *Biochem Biophys Res Commun*, *240*(3), 807–811.

49. Greenwood, T. A., and Kelsoe, J. R. (2003). Promoter and intronic variants affect the transcriptional regulation of the human dopamine transporter gene. *Genomics*, *82*(5), 511–520.

50. Sacchetti, P., Brownschidle, L. A., Granneman, J. G., and Bannon, M. J. (1999). Characterization of the 5′-flanking region of the human dopamine transporter gene. *Brain Res Mol Brain Res*, *74*(1–2), 167–174.

51. Sacchetti, P., Mitchell, T. R., Granneman, J. G., and Bannon, M. J. (2001). *Nurr1* enhances transcription of the human dopamine transporter gene through a novel mechanism. *J Neurochem*, *76*(5), 1565–1572.

52. Bannon, M. J., Michelhaugh, S. K., Wang, J., and Sacchetti, P. (2001). The human dopamine transporter gene: gene organization, transcriptional regulation, and potential involvement in neuropsychiatric disorders. *Eur Neuropsychopharmacol, 11*(6), 449–455.

53. Bannon, M. J., Pruetz, B., Manning-Bog, A. B., Whitty, C. J., Michelhaugh, S. K., Sacchetti, P., Granneman, J. G., Mash, D. C., and Schmidt, C. J. (2002). Decreased expression of the transcription factor *NURR1* in dopamine neurons of cocaine abusers. *Proc Natl Acad Sci U S A, 99*(9), 6382–6385.

54. Lee, K. H., Kwak, Y. D., Kim, D. H., Chang, M. Y., and Lee, Y. S. (2004). Human zinc finger protein 161, a novel transcriptional activator of the dopamine transporter. *Biochem Biophys Res Commun, 313*(4), 969–976.

55. Wang, J., and Bannon, M. J. (2005). Sp1 and Sp3 activate transcription of the human dopamine transporter gene. *J Neurochem, 93*(2), 474–482.

56. Chung, S., Sonntag, K. C., Andersson, T., Bjorklund, L. M., Park, J. J., Kim, D. W., Kang, U. J., Isacson, O., and Kim, K. S. (2002). Genetic engineering of mouse embryonic stem cells by *Nurr1* enhances differentiation and maturation into dopaminergic neurons. *Eur J Neurosci, 16*(10), 1829–1838.

57. Freed, C. R., Leehey, M. A., Zawada, M., Bjugstad, K., Thompson, L., and Breeze, R. E. (2003). Do patients with Parkinson's disease benefit from embryonic dopamine cell transplantation? *J Neurol, 250*(Suppl 3), III44–III46.

58. Zhuang, X., Masson, J., Gingrich, J. A., Rayport, S., and Hen, R. (2005). Targeted gene expression in dopamine and serotonin neurons of the mouse brain. *J Neurosci Methods, 143*(1), 27–32.

59. Kitayama, S., Sogawa, C., Sogawa, N., Mitsuhata, C., Kozai, K., Morita, K., and Dohi, T. (2006). Nicotinic modulation of the transcriptional expression of the human dopamine transporter gene, presented at the Society for Neuroscience, Annual Meeting, Atlanta, GA.

60. Janowsky, A., Mah, C., Johnson, R. A., Cunningham, C. L., Phillips, T. J., Crabbe, J. C., Eshleman, A. J., and Belknap, J. K. (2001). Mapping genes that regulate density of dopamine transporters and correlated behaviors in recombinant inbred mice. *J Pharmacol Exp Ther, 298*(2), 634–643.

61. Giros, B., Jaber, M., Jones, S. R., Wightman, R. M., and Caron, M. G. (1996). Hyperlocomotion and indifference to cocaine and amphetamine in mice lacking the dopamine transporter. *Nature, 379*(6566), 606–612.

62. Sora, I., Hall, F. S., Andrews, A. M., Itokawa, M., Li, X. F., Wei, H. B., Wichems, C., Lesch, K. P., Murphy, D. L., and Uhl, G. R. (2001). Molecular mechanisms of cocaine reward: combined dopamine and serotonin transporter knockouts eliminate cocaine place preference. *Proc Natl Acad Sci U S A, 98*(9), 5300–5305.

63. Doucette-Stamm, L. A., Blakely, D. J., Tian, J., Mockus, S., and Mao, J. I. (1995). Population genetic study of the human dopamine transporter gene (*DAT1*). *Genet Epidemiol, 12*(3), 303–380.

64. Gelernter, J., Kranzler, H., and Lacobelle, J. (1998). Population studies of polymorphisms at loci of neuropsychiatric interest (tryptophan hydroxylase (TPH), dopamine transporter protein (SLC6A3), D3 dopamine receptor (DRD3), apolipoprotein E (APOE), μ-opioid receptor (OPRM1), and ciliary neurotrophic factor (CNTF)). *Genomics, 52*(3), 289–297.

65. Persico, A. M., Vandenbergh, D. J., Smith, S. S., and Uhl, G. R. (1993). Dopamine transporter gene polymorphisms are not associated with polysubstance abuse. *Biol Psychiatry*, *34*(4), 265–267.

66. Gelernter, J., Kranzler, H. R., Satel, S. L., and Rao, P. A. (1994). Genetic association between dopamine transporter protein alleles and cocaine-induced paranoia. *Neuropsychopharmacology*, *11*(3), 195–200.

67. Mash, D. C., Pablo, J., Ouyang, Q., Hearn, W. L., and Izenwasser, S. (2002). Dopamine transport function is elevated in cocaine users. *J Neurochem*, *81*(2), 292–300.

68. Vandenbergh, D. J., Thompson, M. D., Cook, E. H., Bendahhou, E., Nguyen, T., Krasowski, M. D., Zarrabian, D., Comings, D., Sellers, E. M., Tyndale, R. F., et al. (2000). Human dopamine transporter gene: coding region conservation among normal, Tourette's disorder, alcohol dependence and attention-deficit hyperactivity disorder populations. *Mol Psychiatry*, *5*(3), 283–292.

69. Guindalini, C., Howard, M., Haddley, K., Laranjeira, R., Collier, D., Ammar, N., Craig, I., O'Gara, C., Bubb, V. J., Greenwood, T., et al. (2006). A dopamine transporter gene functional variant associated with cocaine abuse in a Brazilian sample. *Proc Natl Acad Sci U S A*, *103*(12), 4552–4557.

70. Muramatsu, T., and Higuchi, S. (1995). Dopamine transporter gene polymorphism and alcoholism. *Biochem Biophys Res Commun*, *211*(1), 28–32.

71. Sander, T., Harms, H., Podschus, J., Finckh, U., Nickel, B., Rolfs, A., Rommelspacher, H., and Schmidt, L. G. (1997). Allelic association of a dopamine transporter gene polymorphism in alcohol dependence with withdrawal seizures or delirium. *Biol Psychiatry*, *41*(3), 299–304.

72. Franke, P., Schwab, S. G., Knapp, M., Gansicke, M., Delmo, C., Zill, P., Trixler, M., Lichtermann, D., Hallmayer, J., Wildenauer, D. B., et al. (1999). *DAT1* gene polymorphism in alcoholism: a family-based association study. *Biol Psychiatry*, *45*(5), 652–654.

73. Lerman, C., Caporaso, N. E., Audrain, J., Main, D., Bowman, E. D., Lockshin, B., Boyd, N. R., and Shields, P. G. (1999). Evidence suggesting the role of specific genetic factors in cigarette smoking. *Health Psychol*, *18*(1), 14–20.

74. Sabol, S. Z., Nelson, M. L., Fisher, C., Gunzerath, L., Brody, C. L., Hu, S., Sirota, L. A., Marcus, S. E., Greenberg, B. D., Lucas, F. R.T., et al. (1999). A genetic association for cigarette smoking behavior. *Health Psychol*, *18*(1), 7–13.

75. Jorm, A. F., Henderson, A. S., Jacomb, P. A., Christensen, H., Korten, A. E., Rodgers, B., Tan, X., and Easteal, S. (2000). Association of smoking and personality with a polymorphism of the dopamine transporter gene: results from a community survey. *Am J Med Genet*, *96*(3), 331–334.

76. Lerman, C., Shields, P. G., Wileyto, E. P., Audrain, J., Hawk, L. H., Jr., Pinto, A., Kucharski, S., Krishnan, S., Niaura, R., Epstein, L. H. (2003). Effects of dopamine transporter and receptor polymorphisms on smoking cessation in a bupropion clinical trial. *Health Psychol*, *22*(5), 541–548.

77. O'Gara, C., Stapleton, J., Sutherland, G., Guindalini, C., Neale, B., Breen, G., and Ball, D. (2007). Dopamine transporter polymorphisms are associated with short-term response to smoking cessation treatment. *Pharmacogenet Genomics*, *17*(1), 61–67.

78. Le Couteur, D. G., Leighton, P. W., McCann, S. J., and Pond, S. (1997). Association of a polymorphism in the dopamine-transporter gene with Parkinson's disease. *Mov Disord*, *12*(5), 760–763.

79. Leighton, P. W., Le Couteur, D. G., Pang, C. C., McCann, S. J., Chan, D., Law, L. K., Kay, R., Pond, S. M., and Woo, J. (1997). The dopamine transporter gene and Parkinson's disease in a Chinese population. *Neurology, 49*(6), 1577–1579.

80. Byerley, W., Hoff, M., Holik, J., Caron, M. G., and Giros, B. (1993). VNTR polymorphism for the human dopamine transporter gene (*DAT1*). *Hum Mol Genet, 2*(3), 335.

81. Byerley, W., Coon, H., Hoff, M., Holik, J., Waldo, M., Freedman, R., Caron, M. G., and Giros, B. (1993). Human dopamine transporter gene not linked to schizophrenia in multigenerational pedigrees. *Hum Hered, 43*(5), 319–322.

82. Persico, A. M., Wang, Z. W., Black, D. W., Andreasen, N. C., Uhl, G. R., and Crowe, R. R. (1995). Exclusion of close linkage of the dopamine transporter gene with schizophrenia spectrum disorders. *Am J Psychiatry, 152*(1), 134–136.

83. Li, T., Yang, L., Wiese, C., Xu, C. T., Zeng, Z., Giros, B., Caron, M. G., Moises, H. W., and Liu, X. (1994). No association between alleles or genotypes at the dopamine transporter gene and schizophrenia. *Psychiatry Res, 52*(1), 17–23.

84. Daniels, J., Williams, J., Asherson, P., McGuffin, P., and Owen, M. (1995). No association between schizophrenia and polymorphisms within the genes for debrisoquine 4-hydroxylase (CYP2D6) and the dopamine transporter (DAT). *Am J Med Genet, 60*(1), 85–87.

85. Bodeau-Pean, S., Laurent, C., Campion, D., Jay, M., Thibaut, F., Dollfus, S., Petit, M., Samolyk, D., d'Amato, T., Martinez, M., et al. (1995). No evidence for linkage or association between the dopamine transporter gene and schizophrenia in a French population. *Psychiatry Res, 59*(1–2), 1–6.

86. Stober, G., Sprandel, J., Jabs, B., Pfuhlmann, B., Moller-Ehrlich, K., and Knapp, M. (2006). Family-based study of markers at the 5′-flanking region of the human dopamine transporter gene reveals potential association with schizophrenic psychoses. *Eur Arch Psychiatry Clin Neurosci, 256*(7), 422–427.

87. Ross, C. A., Margolis, R. L., Reading, S. A., Pletnikov, M., and Coyle, J. T. (2006). Neurobiology of schizophrenia. *Neuron, 52*(1), 139–153.

88. De bruyn, A., Souery, D., Mendelbaum, K., Mendlewicz, J., and Van Broeckhoven, C. (1996). A linkage study between bipolar disorder and genes involved in dopaminergic and GABAergic neurotransmission. *Psychiatr Genet, 6*(2), 67–73.

89. Gomez-Casero, E., Perez de Castro, I., Saiz-Ruiz, J., Llinares, C., and Fernandez-Piqueras, J. (1996). No association between particular DRD3 and DAT gene polymorphisms and manic-depressive illness in a Spanish sample. *Psychiatr Genet, 6*(4), 209–212.

90. Greenwood, T. A., Alexander, M., Keck, P. E., McElroy, S., Sadovnick, A. D., Remick, R. A., and Kelsoe, J. R. (2001). Evidence for linkage disequilibrium between the dopamine transporter and bipolar disorder. *Am J Med Genet, 105*(2), 145–151.

91. Greenwood, T. A., Schork, N. J., Eskin, E., and Kelsoe, J. R. (2006). Identification of additional variants within the human dopamine transporter gene provides further evidence for an association with bipolar disorder in two independent samples. *Mol Psychiatry, 11*(2), 125–133, 115.

92. Horschitz, S., Hummerich, R., Lau, T., Rietschel, M., and Schloss, P. (2005). A dopamine transporter mutation associated with bipolar affective disorder causes inhibition of transporter cell surface expression. *Mol Psychiatry, 10*(12), 1104–1109.

93. Mazei-Robinson, M. S., and Blakely, R. D. (2006). ADHD and the dopamine transporter: Are there reasons to pay attention? *Handb Exp Pharmacol*, (175), 373–415.

94. Waldman, I. D., and Gizer, I. R. (2006). The genetics of attention deficit hyperactivity disorder. *Clin Psychol Rev*, *26*(4), 396–432.

95. Cook, E. H., Jr., Stein, M. A., Krasowski, M. D., Cox, N. J., Olkon, D. M., Kieffer, J. E., and Leventhal, B. L. (1995). Association of attention-deficit disorder and the dopamine transporter gene. *Am J Hum Genet*, *56*(4), 993–998.

96. Gill, M., Daly, G., Heron, S., Hawi, Z., and Fitzgerald, M. (1997). Confirmation of association between attention deficit hyperactivity disorder and a dopamine transporter polymorphism. *Mol Psychiatry*, *2*(4), 311–313.

97. Waldman, I. D., Rowe, D. C., Abramowitz, A., Kozel, S. T., Mohr, J. H., Sherman, S. L., Cleveland, H. H., Sanders, M. L., Gard, J. M., and Stever, C. (1998). Association and linkage of the dopamine transporter gene and attention-deficit hyperactivity disorder in children: heterogeneity owing to diagnostic subtype and severity. *Am J Hum Genet*, *63*(6), 1767–1776.

98. Gillis, J. J., Gilger, J. W., Pennington, B. F., and DeFries, J. C. (1992). Attention deficit disorder in reading-disabled twins: evidence for a genetic etiology. *J Abnorm Child Psychol*, *20*(3), 303–315.

99. Alberts-Corush, J., Firestone, P., and Goodman, J. T. (1986). Attention and impulsivity characteristics of the biological and adoptive parents of hyperactive and normal control children. *Am J Orthopsychiatry*, *56*(3), 413–423.

100. Kirley, A., Lowe, N., Hawi, Z., Mullins, C., Daly, G., Waldman, I., McCarron, M., O'Donnell, D., Fitzgerald, M., and Gill, M. (2003). Association of the 480 bp *DAT1* allele with methylphenidate response in a sample of Irish children with ADHD. *Am J Med Genet B Neuropsychiatr Genet*, *121*(1), 50–54.

101. Winsberg, B. G., and Comings, D. E. (1999). Association of the dopamine transporter gene (*DAT1*) with poor methylphenidate response. *J Am Acad Child Adolesc Psychiatry*, *38*(12), 1474–1477.

102. Roman, T., Szobot, C., Martins, S., Biederman, J., Rohde, L. A., and Hutz, M. H. (2002). Dopamine transporter gene and response to methylphenidate in attention-deficit/ hyperactivity disorder. *Pharmacogenetics*, *12*(6), 497–499.

103. Barr, C. L., Xu, C., Kroft, J., Feng, Y., Wigg, K., Zai, G., Tannock, R., Schachar, R., Malone, M., Roberts, W., et al. (2001). Haplotype study of three polymorphisms at the dopamine transporter locus confirm linkage to attention-deficit/hyperactivity disorder. *Biol Psychiatry*, *49*(4), 333–339.

104. Palmer, C. G., Bailey, J. N., Ramsey, C., Cantwell, D., Sinsheimer, J. S., Del'Homme, M., McGough, J., Woodward, J. A., Asarnow, R., Asarnow, J., et al. (1999). No evidence of linkage or linkage disequilibrium between *DAT1* and attention deficit hyperactivity disorder in a large sample. *Psychiatr Genet*, *9*(3), 157–160.

105. Maher, B. S., Marazita, M. L., Ferrell, R. E., and Vanyukov, M. M. (2002). Dopamine system genes and attention deficit hyperactivity disorder: a meta-analysis. *Psychiatr Genet*, *12*(4), 207–215.

106. Purper-Ouakil, D., Wohl, M., Mouren, M. C., Verpillat, P., Ades, J., and Gorwood, P. (2005). Meta-analysis of family-based association studies between the dopamine transporter gene and attention deficit hyperactivity disorder. *Psychiatr Genet*, *15*(1), 53–59.

107. Li, D., Sham, P. C., Owen, M. J., and He, L. (2006). Meta-analysis shows significant association between dopamine system genes and attention deficit hyperactivity disorder (ADHD). *Hum Mol Genet*, *15*(14), 2276–2284.

108. Loo, S. K., Specter, E., Smolen, A., Hopfer, C., Teale, P. D., and Reite, M. L. (2003). Functional effects of the *DAT1* polymorphism on EEG measures in ADHD. *J Am Acad Child Adolesc Psychiatry*, *42*(8), 986–993.

109. Kahn, R. S., Khoury, J., Nichols, W. C., and Lanphear, B. P. (2003). Role of dopamine transporter genotype and maternal prenatal smoking in childhood hyperactive-impulsive, inattentive, and oppositional behaviors. *J Pediatr*, *143*(1), 104–110.

110. Miller, G. M., Yatin, S. M., De La Garza, R., 2nd, Goulet, M., and Madras, B. K. (2001). Cloning of dopamine, norepinephrine and serotonin transporters from monkey brain: relevance to cocaine sensitivity. *Brain Res Mol Brain Res*, *87*(1), 124–143.

111. Inoue-Murayama, M., Adachi, S., Mishima, N., Mitani, H., Takenaka, O., Terao, K., Hayasaka, I., Ito, S., and Murayama, Y. (2002). Variation of variable number of tandem repeat sequences in the 3′-untranslated region of primate dopamine transporter genes that affects reporter gene expression. *Neurosci Lett*, *334*(3), 206–210.

112. Fuke, S., Suo, S., Takahashi, N., Koike, H., Sasagawa, N., and Ishiura, S. (2001). The VNTR polymorphism of the human dopamine transporter (*DAT1*) gene affects gene expression. *Pharmacogenom J*, *1*(2), 152–156.

113. Fuke, S., Sasagawa, N., and Ishiura, S. (2005). Identification and characterization of the *Hesr1/Hey1* as a candidate trans-acting factor on gene expression through the 3′ non-coding polymorphic region of the human dopamine transporter (*DAT1*) gene. *J Biochem (Tokyo)*, *137*(2), 205–216.

114. Fuke, S., Minami, N., Kokubo, H., Yoshikawa, A., Yasumatsu, H., Sasagawa, N., Saga, Y., Tsukahara, T., and Ishiura, S. (2006). *Hesr1* knockout mice exhibit behavioral alterations through the dopaminergic nervous system. *J Neurosci Res*, *84*(7), 1555–1563.

115. Miller, G. M., and Madras, B. K. (2002). Polymorphisms in the 3′-untranslated region of human and monkey dopamine transporter genes affect reporter gene expression. *Mol Psychiatry*, *7*(1), 44–55.

116. Lim, J. E., Papp, A., Pinsonneault, J., Sadee, W., and Saffen, D. (2006). Allelic expression of serotonin transporter (SERT) mRNA in human pons: lack of correlation with the polymorphism SERTLPR. *Mol Psychiatry*, *11*(7), 649–662.

117. Jacobsen, L. K., Staley, J. K., Zoghbi, S. S., Seibyl, J. P., Kosten, T. R., Innis, R. B., and Gelernter, J. (2000). Prediction of dopamine transporter binding availability by genotype: a preliminary report. *Am J Psychiatry*, *157*(10), 1700–1703.

118. Heinz, A., Goldman, D., Jones, D. W., Palmour, R., Hommer, D., Gorey, J. G., Lee, K. S., Linnoila, M., and Weinberger, D. R. (2000). Genotype influences in vivo dopamine transporter availability in human striatum. *Neuropsychopharmacology*, *22*(2), 133–139.

119. Martinez, D., Gelernter, J., Abi-Dargham, A., van Dyck, C. H., Kegeles, L., Innis, R. B., and Laruelle, M. (2001). The variable number of tandem repeats polymorphism of the dopamine transporter gene is not associated with significant change in dopamine transporter phenotype in humans. *Neuropsychopharmacology*, *24*(5), 553–560.

120. van Dyck, C. H., Malison, R. T., Jacobsen, L. K., Seibyl, J. P., Staley, J. K., Laruelle, M., Baldwin, R. M., Innis, R. B., and Gelernter, J. (2005). Increased dopamine transporter availability associated with the 9-repeat allele of the SLC6A3 gene. *J Nucl Med*, *46*(5), 745–751.

121. Brody, A. L., Mandelkern, M. A., Olmstead, R. E., Scheibal, D., Hahn, E., Shiraga, S., Zamora-Paja, E., Farahi, J., Saxena, S., London, E. D., et al. (2006). Gene variants of

brain dopamine pathways and smoking-induced dopamine release in the ventral caudate/ nucleus accumbens. *Arch Gen Psychiatry*, *63*(7), 808–816.

122. Krause, J., Dresel, S. H., Krause, K. H., La Fougere, C., Zill, P., and Ackenheil, M. (2006). Striatal dopamine transporter availability and *DAT-1* gene in adults with ADHD: no higher DAT availability in patients with homozygosity for the 10-repeat allele. *World J Biol Psychiatry*, *7*(3), 152–157.

123. Bannon, M. J. (2005). The dopamine transporter: role in neurotoxicity and human disease. *Toxicol Appl Pharmacol*, *204*(3), 355–360.

124. Iversen, L. (2006). Neurotransmitter transporters and their impact on the development of psychopharmacology. *Br J Pharmacol*, *147*(Suppl 1), S82–S88.

125. Kent, W. J., Sugnet, C. W., Furey, T. S., Roskin, K. M., Pringle, T. H., Zahler, A. M., and Haussler, D. (2002). The human genome browser at UCSC. *Genome Res, 12*,(6), 996–1006.

4

MOLECULAR STRUCTURE AND COMPOSITION OF DOPAMINE TRANSPORTERS

M. Laura Parnas and Roxanne A. Vaughan

Department of Biochemistry and Molecular Biology, University of North Dakota School of Medicine and Health Sciences, Grand Forks, North Dakota

4.1	Introduction	74
4.2	DAT Structure and Molecular Characterization	74
	4.2.1 Primary Sequence and Topology	74
	4.2.2 Bacterial Leucine Transporter and Relationship to the DAT	75
4.3	Posttranslational Modifications	77
	4.3.1 Glycosylation	77
	4.3.2 Disulfide Bonds	78
	4.3.3 Phosphorylation	79
	4.3.4 Ubiquitylation	80
4.4	DAT Tertiary and Quaternary Structure	81
	4.4.1 DAT Oligomerization	81
	4.4.2 Zinc-Binding Site	81
	4.4.3 Protein–Protein Interactions and Functional Motifs	82
4.5	Substrate- and Ligand-Binding Sites	83
	4.5.1 Chimera Studies	83
	4.5.2 Affinity Labeling	84
	4.5.3 Site-Directed Mutagenesis	85
	4.5.4 SCAM Studies	86
4.6	Conclusions	87
References		87

Dopamine Transporters: Chemistry, Biology, and Pharmacology. Edited by Mark L. Trudell and Sari Izenwasser
Copyright © 2008 John Wiley & Sons, Inc.

4.1 INTRODUCTION

Neurotransmission mediated by dopamine (DA) is involved in the control of numerous processes, including locomotion, reward, emotion, and sympathetic regulation. Malfunctions in the dopaminergic system have been implicated in several psychiatric and mood disorders, such as Parkinson's disease (PD), schizophrenia, drug abuse, depression, and attention-deficit/hyperactivity disorder. Neurotransmission initiated by presynaptic DA release is terminated by transmitter reuptake through the dopamine transporter (DAT), a protein localized to the plasma membrane of synaptic terminals, axons, and dendrites of limbic and midbrain dopaminergic neurons. The key role of the DAT in dopaminergic transmission is highlighted by conditions such as drug abuse and genetic knockouts in which transport is inhibited, leading to elevated DA levels associated with psychomotor stimulation and addiction [1,2].

The DAT is a major target for several abused drugs, including cocaine, amphetamine (AMPH), and methamphetamine (METH), and for such therapeutic agents as methylphenidate (Ritalin) and bupropion (Wellbutrin). Cocaine and related psychostimulants bind to the DAT and inhibit its activity but are not transported, while AMPH and other compounds with structural similarity to DA are carried by the DAT. Within the brain these exogenous substrates compete with DA for transport, which reduces the rate of DA reuptake and induces nonexocytotic efflux of DA from the neuron into the synapse via reversal of transport (reviewed in [3]). The net effect of these drugs is to increase the concentration and duration of DA in the synapse, leading to supraphysiological activation of postsynaptic neurons that results in euphoria and psychomotor stimulation. In addition to these effects, many of the compounds transported are neurotoxic, due to their ability to enter the cytoplasm. At sufficiently high levels, DA and oxidized DA metabolites such as 6-hydroxydopamine, AMPH, METH, and other compounds are capable of inducing oxidative damage that can culminate in neuronal death (reviewed in [4,5]) providing the basis for one hypothesis for the progressive loss of dopaminergic neurons characteristic of PD. In this chapter we present an overview of the current state of knowledge of the molecular characteristics and composition of the DAT, including its structure, posttranslational modifications, and active sites, which have implications for its functions in these normal and disease conditions.

4.2 DAT STRUCTURE AND MOLECULAR CHARACTERIZATION

4.2.1 Primary Sequence and Topology

The DAT belongs to the SLC6 family of Na^+/Cl^--dependent neurotransmitter transporters, also known as neurotransmitter sodium symporters (NSSs), which includes the related cocaine-sensitive norepinephrine transporter (NET) and serotonin transporter (SERT) as well as carriers for γ-aminobutyric acid, glycine, proline, taurine, betaine, and creatine (reviewed in [6]). These proteins are secondary active transporters that mediate substrate translocation through coupling of Na^+ and Cl^-

movement down their electrochemical gradients to uphill movement of solute. Several members of the SLC6 family were cloned in the 1990s, including the DATs from human, monkey, rat, mouse, and cow [7–13]. In addition, nonmammalian DATs have been identified from *Caenorhabditis elegans* [14]. *Drosophila melanogaster* [15]. *Bombyx mori* [16], and *Eloria noyesi* [16].

DAT proteins contain between 610 and 630 amino acids, with the rat and human isoforms, containing 619 and 620 residues, being studied most extensively. Hydrophobicity analysis of the amino acid sequences revealed the presence of 12 transmembrane spanning domains (TMs), presumed to consist primarily of α-helical structure. The TMs are linked through intracellular and extracellular loops (ILs and ELs), with the N- and C-termini facing the cytoplasm (Figs. 4-1 and 4-2). The mammalian DATs are highly homologous, with the highest degree of similarity found within the TMs and the lowest found in the loops and cytoplasmic tails. EL2, which connects TMs 3 and 4, is characteristic of this transporter family, being the largest of the connecting loops and possessing extensive glycosylation and disulfide bonding. Some of the other loops (e.g., EL1, IL2, and EL5) contain only a few amino acids, spatially constraining the ends of the connected TMs.

Transport is not fully understood at the molecular level, but is believed to occur by an alternating access mechanism [17]. In this process the protein undergoes conformational movements that result in sequential exposure of the extracellular and intracellular portions of the substrate permeation pathway to alternate sides of the membrane. In the forward transport direction, DA and ions bind at the outside face and are transitioned through the substrate pathway, and once released into the cell interior, the empty transporter returns to its outward-facing form. Molecular gates that open and close at specific points in the transport cycle restrict compound access and ensure appropriate vectorial movement of substrates. The kinetics of transport (<1 DA/s) are slow relative to ion channels and ATP-driven pumps, and are thought to reflect the rate of return of the empty transporter to the outward-facing conformation. DAT also displays reverse transport, in which conditions such as extracellular AMPH induce the efflux of DA from the cytoplasm to the extracellular medium [3].

Whereas ions are moved stoichiometrically with a ratio of $1DA : 2Na^+ : 1Cl^-$ per transport cycle [18], DATs also move ions in a nonstoichiometric manner similar to movement in ion channels [19–21], indicating that the permeation pathway is not perfectly sealed and that slippage occurs in the gates or other structures. Channel-like ion movements occur by both substrate-dependent and spontaneous "leak" mechanisms, and currents produced may be of sufficient quantity to affect membrane potential and neuronal activity [20]. Efflux induced by AMPH also occurs through a channel-like mechanism that allows many DA molecules to move simultaneously [22].

4.2.2 Bacterial Leucine Transporter and Relationship to the DAT

One of the major challenges to understanding how the DAT and related neurotransmitter transporters function has been the limited knowledge of the protein structure. The sites of action for DA and ions, the nature of the conformational rearrangements

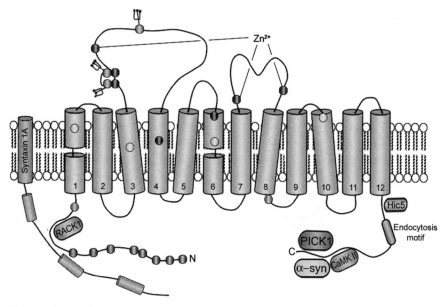

Figure 4-1 DAT posttranslational modifications and protein interaction sites. Schematic diagram of human DAT showing topological organization, transmembrane spanning domains 1 to 12, with TMs 1 and 6 shown as broken helices separated by short regions of random structure, and intracellularly oriented N- and C-termini. Specific sites and modifications represented are: consensus N-linked glycosylation sites at N181, N188, and N205 (orange circles) conjugated to complex carbohydrates (branched structures) terminated by sialic acids (dots); disulfide-bonded cysteines C180 and C189 in EL2 (connected purple circles); zinc-binding site residues H193, H375, and E396 (red circles); dimer and tetramer cross-linking sites at C243 and C306 (purple circles); putative extracellular gate residues R86, Y156, F320, and D471 (yellow circles); putative intracellular gate residues R60 and D436 (green circles); phosphorylation site cluster at serines 2, 4, 7, 12, and 13 (pink circles); and ubiquitylated lysines 19, 27, and 35 (blue circles). N-terminal tail interaction domains are indicated for syntaxin 1A and RACK1. C-terminal tail interaction domains are indicated for PICK1, α-synuclein, CaMKII, Hic-5, and the residue 587–596 endocytosis motif. (*See insert for color representation of figure.*)

that drive transport, and the mechanisms by which cocaine and other blockers bind to DAT and prevent transport are poorly understood. None of the NSS proteins has been crystallized, and there is currently only limited experimental evidence pertaining to their three-dimensional structures. A breakthrough in this area was made in 2005 by Gouaux and co-workers, who determined a high-resolution crystal structure of LeuT$_{Aa}$, a Na$^+$-dependent 12-TM bacterial leucine transporter from *Aquifex aeolicus* [23]. Although only distantly related in amino acid sequence, this transporter is functionally homologous to the NSSs, and it is currently expected to provide an important template for interpretation and direction of DAT structure–function studies.

In LeuT$_{Aa}$, TMs 1 to 5 and 6 to 10 form structural cassettes oriented to each other by a pseudo twofold axis, with TMs 1, 3, 6, and 8 forming a central core that creates a binding pocket for leucine and Na$^+$. TMs 1 and 6 are adjacent to each other and are not continuous helices but are separated into halves by a short stretch of random structure, allowing participation of main-chain atoms in coordination of Na^+ and substrate binding. TMs 3 and 8 are long helices of >30 residues tilted by $\sim 50°$ relative to the plane of the membrane. They are positioned next to each other and face TMs 1 and 6. The other TMs are arranged to the outside of this central core, apparently functioning to support the proper structure of the active site TMs. The leucine- and Na^+-binding sites are found about halfway across the membrane and are in close proximity, consistent with the substrates moving as a unit. In the crystallized structure these binding sites are occluded from the aqueous medium by the external and internal gates, suggesting that the transport mechanism involves a minimum of three conformations: open extracellularly, closed at both ends, and open intracellularly. Residues implicated in the external gate are the probable ion pair R30 and D404 in TMs 1 and 10, and Y108 and F253 in TMs 3 and 6, with the analogous residues in the DAT (R86/D471 and Y156/F320) strictly conserved (Fig. 4-1). The intracellular gate is composed of substantial protein structure from the inner halves of TMs 1, 6, and 8, as well as another ion pair, R5 and D369, which corresponds to DAT residues R60 and D436 at the intracellular ends of TMs 1 and 8 (Fig. 4-1). In the crystallized form, $LeuT_{Aa}$ is present as a symmetrical dimer formed by interactions between TMs 9 and 12, but the significance of this is unknown.

Despite many similarities and what is sure to be the importance of $LeuT_{Aa}$ in guiding future studies on NSSs, the mammalian transporters also possess significant structural and functional differences from $LeuT_{Aa}$, including the sizes of some of the loops and cytoplasmic tails, the presence of posttranslational modifications, and inhibition by cocaine and other exogenous compounds. Some current data on the DAT that do not conform easily to the $LeuT_{Aa}$ structure already suggest the presence of differences between the two proteins related to substrate selectivity and other properties [24]. In addition, the form of $LeuT_{Aa}$ that was crystallized is only one of what is likely to be numerous conformations that occur during the transport cycle; thus many unknowns remain about the molecular basis of the transport process.

4.3 POSTTRANSLATIONAL MODIFICATIONS

The DAT possesses numerous posttranslational modifications that regulate various aspects of its function (Fig. 4-1). These modifications, particularly glycosylation and disulfide bonding on EL2, and phosphorylation and ubiquitylation on the long N-terminal tail, may represent some of the most significant structural departures from $LeuT_{Aa}$. The domains containing these modifications in eukaryotic transporters may thus be evolutionary additions that confer regulatory properties not found in bacterial homologs.

4.3.1 Glycosylation

The DAT contains extensive glycosylation, originally demonstrated on the rat striatal protein by analysis with exo- and endoglycosidases. Native DAT displays a mass of $\sim 80\,kDa$ on SDS-PAGE (sodium dodecyl sulfate–polyacrylamide gel electrophoresis) gels that is reduced by ~25 or ~5 kDa by treatment with N-glycanase or neuraminidase, respectively [25–27], demonstrating the presence of N-linked carbohydrates and sialic acids (*N*-acetylneuraminic acid). Sialic acids are typically found on glycosyl chain termini (Fig. 4-1), although it is not known for the DAT if every N-linked chain is modified. The DAT mass is not affected by α-mannosidase [27], indicating that mannose is not a constituent of the glycosyl structure. Solubilized DATs also bind to wheat germ agglutinin resin, which complexes with sialic acid but not to concanavalin A resin, which binds high mannose [27], supporting the enzymatic deglycosylation findings. With these exceptions the composition of DAT carbohydrate structures is unknown. On SDS-PAGE gels the immature form of DAT electrophoreses as a narrow band at ~55 kDa, while the mature 80-kDa form displays a broad migration pattern that often resolves into multiple components, potentially indicating the presence of heterogeneous glycosylation isoforms. DATs from different mammalian species and brain regions show similar overall patterns of N-linked and sialic acid glycosylation [28], although slight molecular mass differences attributed to glycosylation are seen in the DAT from rat striatum and nucleus accumbens [29–31]. In the rat, glycosylation of the DAT is developmentally regulated, with little glycosylation apparent in early postnatal development and increasing amounts occurring through later stages up to adulthood [32], but the significance of this finding is unknown.

Multiple canonical N-linked glycosylation sequences (N-X-S/T) are present on the DAT EL2 (Fig. 4-1), with the human and rat DAT forms containing three and four of these sites, respectively. The functional role of glycosylation was explored by subjecting the consensus hDAT N-glycosylation sites at N181, N188, and N205 to sequential mutation [33]. This caused mass reductions of ~10 kDa per N-modified residue, consistent with utilization of all three sites. Analysis of the mutants with cell surface biotinylation and confocal microscopy revealed that reductions in N-glycosylation led to lessened DAT cell surface levels resulting from increased endocytosis, but that glycosylation was not absolutely required for biosynthetic maturation or plasma membrane expression. Nonglycosylated DATs also displayed reduced transport V_{max} and increased potency for cocaine-based drugs [33,34], suggesting a further role for glycosylation in proper protein structure. Treatment of native DAT with neuraminidase to remove only sialic acids did not affect cocaine analog–binding affinity [30], leaving the function of this modification unknown. The rat DAT isoform contains the analogous glycosylation sites at N181, N188, and N204 and a fourth site at N196, but the significance of the additional site and whether it is utilized are not known.

4.3.2 Disulfide Bonds

The DAT contains numerous cysteine residues throughout its primary sequence. Two of these in EL2, C180, and C189 (Fig. 4-1) were originally proposed to be disulfide bonded, based on loss of transport activity upon mutation [35], and this has been supported more directly in a recent study [36]. However, while reducing agents suppress transport activity of wild-type DAT, they have no effect on a mutant in which all cysteines except C180 and C189 were removed [36]. This suggests that the disulfide bond supports proper folding of DAT during maturation or trafficking, but that once the fully mature protein is expressed, maintenance of the bond is not necessary for transport activity. Interestingly, two of the N-glycosylated residues (N181 and N188) are directly adjacent to the disulfide-bonded cysteines (C180 and C189), and a nearby residue (H193) is part of a zinc-binding site. These residues are found just past the extracellular end of the TM 3 helix [24], which in LeuT$_{Aa}$ forms part of the active site and in the DAT is implicated by increasing evidence as participating in substrate transport and cocaine binding. These modifications may thus contribute to proper orientation of this helix, affecting both the active site structure and the transporter biosynthesis and trafficking processes.

4.3.3 Phosphorylation

Several aspects of DAT function are affected by a variety of phosphorylation pathways, including protein kinase C (PKC), calcium-calmodulin-dependent protein kinase (CaMK), mitogen-activated kinases (MAPKs), tyrosine kinases, and protein phosphatases 1 and 2A (PP1/PP2A) (reviewed in [37–39]). One of the best studied of these is membrane trafficking, which regulates the density of DAT surface expression and overall level of transport activity (reviewed in [40]). Substrate efflux and possibly protein–protein interactions are also regulated by phosphorylation [41,42], and many questions remain as to how the various kinase and phosphatase pathways are integrated with respect to these processes and whether the DAT or associated proteins are the enzyme targets.

Human and rat isoforms of the DAT are phosphorylated in vivo in both rat brain tissue and heterologous expression systems. Under resting conditions the DAT displays a constitutive level of phosphorylation that is rapidly increased by PKC activators and PP1/PP2A inhibitors [43–45]. Most of the phosphate incorporated is recovered as phosphoserine, with a minor amount found as phosphothreonine [44,46], but to date, phosphotyrosine has not been detected. DAT phosphorylation is also stimulated by AMPH and METH via a PKC-dependent mechanism [47], but is not affected by cocaine or DA [48]. The majority of constitutive and stimulated phosphorylation sites in human and rat DAT have been localized to the first 21 residues at the N-terminus of the protein [45,47,49]. In both isoforms this domain contains a cluster of five serines at positions 2, 4, 7, 12, and 13 (Fig. 4-1), with a sixth serine present at position 21 in that DAT. There are no threonines in this sequence, and the identity of the phosphorylated threonine is not known. Other unknowns that remain with respect to phosphorylation include the stoichiometry, identity of the sites utilized, and whether the same or different sites are utilized in basal and various stimulated conditions.

It is not yet clear if PKC phosphorylates DAT directly, but this idea is supported by recent findings that a recombinant DAT sequence containing N-terminal residues 1 to 64 can serve as an in vitro substrate for PKCα, PKCβI, and PKCβII [48]. In addition, the potential for inputs from other kinases is also suggested, as this sequence shows robust in vitro phosphorylation by PKA, PKG, CaMKII, ERK1/2, and casein kinase [48], and similarly, both CaMKII and PKCα phosphorylate a peptide consisting of residues 1 to 27 [50]. Phosphorylation of DAT mediated by kinases other than PKC has not been demonstrated in vivo, but there is some suggestive evidence for this in that basal DAT phosphorylation is not blocked by PKC inhibitors [44] but may be suppressed by MAPK inhibitors [51], and AMPH-induced DA efflux, which is increased by N-terminal phosphorylation, is suppressed by removal of a DAT CaMKII-binding site [50]. The only activity currently linked to DAT phosphorylation is AMPH-stimulated DA efflux, with phosphorylation suggested to promote an efflux favoring conformation of the protein [52]. Although PKC and AMPH both stimulate DAT endocytosis and down-regulation concomitantly with DAT phosphorylation in the wild-type protein, removal of phosphorylation sites has no apparent effect on these processes [45,47], and the phosphorylation targets for these events have not been identified. There is also some evidence to suggest a link between the SNARE protein syntaxin 1A and DAT phosphorylation [53], but the functional significance of this is not known.

4.3.4 Ubiquitylation

Ubiquitin is a 76-amino acid protein that undergoes ATP-dependent ligation to target proteins through the formation of an isopeptide bond between the ubiquitin C-terminal glycine carboxyl group and the ε-amino group of an internal lysine on the target protein (reviewed in [54]). This conjugation process is mediated by a three-step cascade of enzymatic reactions [55]. The best characterized of these processes is polyubiquitylation, which involves the attachment of a chain of four or more ubiquitin molecules and is associated primarily with the targeting of intracellular proteins for degradation by the 26S proteasome. Other processes involve the attachment of one or a shorter chain of ubiquitin molecules, which is associated with activities including endocytosis and lysosomal protein degradation. Thus, ubiquitin plays a key role in the modulation of plasma membrane protein trafficking by serving as an internalization signal for endocytosis and by regulating the activity of the endocytotic machinery [54,56].

Ubiquitylation and associated degradation of the DAT were first demonstrated to occur via the action of Parkin [57], a ubiquitin ligase for which ligase-inhibiting mutations are linked to the development of PD [58]. Inefficient removal of damaged DAT proteins in response mutant Parkin has thus been suggested to result in dominant-negative inhibition of transport activity leading to aberrant DA clearance potentially related to PD development [57]. Using mass spectrometry, Sorkin and co-workers demonstrated that the DAT undergoes constitutive and PKC-stimulated modification with single and short-chain ubiquitin complexes on N-terminal tail lysines 19, 27, and 35 [59,60] (Fig. 4-1). The ubiquitylated transporter was localized to the plasma membrane, and PKC-stimulated ubiquitylation caused transporter movement into early and then late endosomes. This suggests that PKC-induced ubiquitylation targets DATs for degradation through the lysosomal pathway [59] via involvement with the ubiquitin ligase Nedd4-2 [61].

4.4 DAT TERTIARY AND QUATERNARY STRUCTURE

4.4.1 DAT Oligomerization

Until relative recently it was presumed that dopamine transporters functioned as monomers, but evidence now strongly supports the presence of oligomeric DAT complexes (reviewed in [62]). This was first suggested from in situ radiation inactivation studies that indicated the presence of DAT complexes with dimeric or tetrameric masses [63,64]. Direct verification of this idea was obtained in a study showing formation of DAT homodimers by chemical cross-linking of C306 at the top of TM 6 (Fig. 4-1), with dimerization suggested to occur through a conserved oligomerization motif (GVXXGVXX) present in the cytoplasmic half of TM 6 [65]. An additional symmetrical interface at C243 in TM 4 (Fig. 4-1) was found to participate in the formation of DAT tetramers formed as dimers of dimers [66]. However, although DAT monomers can be cross-linked exogenously through these cysteines, there is no evidence for endogenous intersubunit disulfide bonding, as DAT shows the same mass on reducing and nonreducing SDS-PAGE gels.

The presence of noncovalent DAT complexes was demonstrated by Torres and co-workers using co-immunoprecipitation methods. In these studies coexpression of wild-type DAT with trafficking-defective mutants led to reduced transporter surface expression via a dominant negative effect, indicating that oligomeric interactions promote proper transporter trafficking and biosynthesis [34]. DAT oligomer formation in living cells has also been demonstrated through fluorescence resonance energy transfer (FRET) analysis [67]. Intracellular DAT complexes detected by FRET were seen throughout the cell interior, and FRET signals were lost by coexpression of wild-type DAT with endoplasmic reticulum export-defective forms [67]. Several independent lines of evidence thus indicate that the DAT constitutively forms oligomers and that this promotes proper intracellular sorting and cell surface expression. A question that remains is whether or not DAT oligomerization is necessary for substrate transport or cocaine binding, or if the monomers within these complexes function independently.

4.4.2 Zinc-Binding Site

The divalent cation zinc (Zn^{2+}) is known to modulate many central nervous system functions with extensive physiological roles in several brain areas [68]. Potential regulation of dopaminergic neurotransmission by Zn^{2+} was suggested by studies in which DA uptake in rat brain tissue was noncompetitively inhibited in vitro by physiological concentrations of the ion [69,70]. Zinc was also found to increase cocaine analog–binding affinity. How these opposite effects on transport and cocaine binding were achieved was elucidated by Gether and co-workers, who identified an endogenous Zn^{2+}-binding site on the extracellular side of the DAT, composed of H193 in EL2, H375 at the top of TM 7, and E396 at the top of TM 8 [71,72] (Fig. 4-1). Because Zn^{2+} coordination residues must be within $\sim$4 Å of each other [73], these findings provided direct evidence for the three-dimensional proximity of these amino acids. In these studies it was found that Zn^{2+} binding to this site suppresses inward transport but enhances amphetamine-induced DA efflux [74]. Zn^{2+} binding also increases DA- and AMPH-induced ion currents [75,76] and facilitates an uncoupled Cl^- current [76]. Based on these results, it was proposed that Zn^{2+} binding to DAT inhibits uptake and promotes efflux by altering the membrane electrochemical potential needed to drive substrate movement [77].

The geometry of Zn^{2+} binding and its inhibition of DA uptake have also provided valuable tools for the study of DAT structural and conformational properties. Exogenous Zn^{2+}-binding sites engineered into the DAT have served to delineate secondary and tertiary structure at the extracellular end of TMs 7 and 8 [71,72,78,79], with inhibition of DA-uptake functioning as a readout for investigation of residues involved in conformational switches during substrate translocation [80,81].

4.4.3   Protein–Protein Interactions and Functional Motifs

In recent years it has become clear that DAT does not function in isolation and is modulated via interaction with a variety of proteins (reviewed in [82]). These findings have been obtained by yeast two-hybrid co-immunoprecipitation mass spectrometry and FRET analyses, and where known, interactions have been localized to the cytoplasmically oriented N- and C-terminal tails (Fig. 4-1). The N-terminal tail, which contains the sites of PKC-mediated phosphorylation and ubiquitylation, interacts with several proteins related to kinase and phosphatase action. DAT co-immunoprecipitates with PKCβI and PKCβII [83], which regulate AMPH-induced DA efflux and can phosphorylate the recombinant DAT N-terminus in vitro [48], and with PP2A [84], which dephosphorylates DAT in vivo and in vitro [49]. Although the co-immunoprecipitation sites on DAT for these enzymes have not been mapped, the enzymatic interaction clearly occurs on the distal end of the N-terminal tail. The ligases that catalyze constitutive and PKC-dependent ubiquitylation of lysines 19, 27, and 35 must also interact with this domain. Yeast two-hybrid studies identified in vitro DAT N-terminal tail interactions with RACK1 (receptor associated with C-kinase), a soluble PKC adaptor protein, and syntaxin 1A, a plasma membrane SNARE protein [85] (Fig. 4-1). The RACK1 interaction has been suggested, although not yet demonstrated, to be involved with regulating PKC-dependent processes of the native protein. DAT-syntaxin 1A–binding interactions have also not yet been demonstrated to occur with native protein, but treatment of rat striatal tissue with botulinum neurotoxin C, which proteolyzes syntaxin 1A and separates its cytoplasmic and membrane domains, results in increased DA uptake and decreased DAT phosphorylation [53], suggesting a functional interaction with the DAT consistent with the N-terminal tail.

Proteins that interact with the DAT C-terminal tail include PICK1, Hic-5, α-synuclein, and CaMKII (Fig. 4-1). The final four residues (617 to 620) of the DAT C-terminal tail consist of a PDZ-binding domain that interacts with the PDZ protein PICK1 [86]. Residues 615 to 617 just upstream of this site mediate surface membrane targeting, but in a PICK1-independent manner [87], leaving the function of DAT-PICK1 interaction unknown. Hic-5, a focal adhesion adaptor protein, interacts with residues 571 to 580 at the membrane proximal region of the C-terminus, and its overexpression reduces DAT plasma membrane levels [88]. α-Synuclein, a protein associated with familial PD, interacts with DAT at the last 15 amino acids (606 to 620) of the C-terminal tail [89]. This interaction affects DAT membrane expression, but the mechanism is not clear, as opposing results have been obtained in different studies [89,90]. CaMKII binds to residues 612 to 617 on the C-terminal tail of DAT, resulting in regulation of AMPH-stimulated efflux, potentially through N-terminal tail phosphorylation [50], suggesting the interesting possibility that DAT N-terminal modifying enzymes may be positioned by C-terminal tail binding. Another functional region identified in the C-terminal tail (Fig. 4-1) is a novel endocytosis motif between residues 587 and 596 involved in constitutive and PKC-induced internalization [91]. This property suggests that this region may interact with clathrin endocytosis machinery, although the proteins involved with this motif have not yet been identified. Several other DAT-interacting proteins, including synapsin I, dynamin I, Brca2, neurocam, and the Kv2.1 potassium channel, have been identified by proteomic approaches [92], but the significance of these interactions has not been elucidated. It is thus clear that DAT functions within the context of an extensive network of proteins that have the potential for regulation of many transport properties, and determining how these interactions are integrated and regulated will undoubtedly be of major importance for our understanding of normal and aberrant DAT function.

4.5 SUBSTRATE- AND LIGAND-BINDING SITES

In the years since the cloning of DAT and related transporters, extensive efforts have been made to identify the interaction sites for substrates and uptake blocking compounds. These studies have significant implications for the mechanistic understanding of DAT activity and for potential development of pharmacological agents to treat drug abuse and dopaminergic diseases. An idea that has emerged from these studies, although far from fully demonstrated, is that active sites for substrates and blockers are located primarily in TM domains and that connecting loops function more in structural roles and/or as access gates.

4.5.1 Chimera Studies

Several investigators have analyzed DAT function using DAT-NET and DAT cross-species chimeras. As the parental proteins possess distinct affinities for various substrates and uptake blockers within a background of overall high-sequence homology, these studies have provided excellent indications of domains likely to contribute directly to DAT actions. Using this approach with DAT/NET chimeras Giros and co-workers [93] and Buck and Amara [94,95] identified substrate transport- and inhibitor-binding functions within the DAT TM 1–3 and 4–8 regions. Subsequently, Lee and co-workers found that by crossing bovine DAT, which has a low cocaine affinity, with human DAT, which has high-affinity

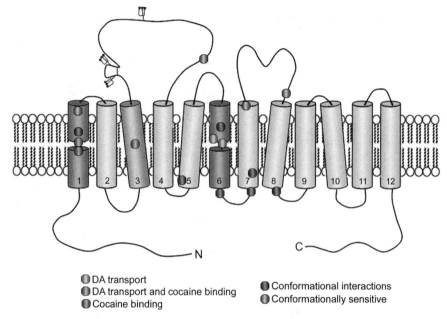

Figure 4-2 DAT residues and domains involved in transport, cocaine binding, and conformational changes. TMs 1 and 6 (pink cylinders) indicate domains irreversibly labeled by cocaine photoaffinity analogs, and TM 3 (orange cylinder) represents a domain essential for high-affinity cocaine binding identified through cross-species chimeras. Residues with support for presence within active sites are shown for dopamine transport (green circles), cocaine binding (red circle), and both transport and cocaine binding (yellow circles). Blue circles indicate residues that are essential for protein structural rearrangements during transport, and orange circles indicate residues that are not essential for activity but undergo conformational movements in the presence of substrates or transport inhibitors. Specific residues indicated are (TM1) F76, D79, W84, C90; (EL2) R219; (TM3) V152; (TM5) K264; (TM6) D313, F320, G323, Y335; (TM7) C342, D345, M371; (EL4) A399; (TM8) D436. (*See insert for color respresentation of figure.*)

cocaine binding, required the simultaneous presence of the human TM 3 and TM 6–8 regions [96,97], implicating the involvement of multiple domains in this function (Fig. 4-2). DAT/NET chimeras have also been used to map sites for Na^+ dependence to TMs 1 and 2, the junction between TMs 5 and 6, and TMs 9 to 12, and Cl^- dependence to TMs 9 to 12 [98]. Many of these results correspond well with what is known about the contributions of TMs 1, 3, 6, and 8 to the substrate pocket in LeuT$_{Aa}$.

4.5.2 Affinity Labeling

Active sites on DAT have also been investigated using irreversible affinity labeling, in which binding sites for cocaine, benztropine, and GBR classes of uptake blockers have been mapped using ligands that contain a reactive group that attaches covalently to the protein. All of the ligands analyzed to date have been shown to become incorporated in the N-terminal (TMs 1 and 2) and/or central (TMs 4 to 7) regions of the protein, suggesting a certain level of commonality in their binding domains [99–103]. Within the TM 4–7 domain, the attachment site of one of the cocaine analogs has now been further localized to TM6 [103], while a related cocaine analog with an identical pharmacophore but different reactive group orientation becomes attached within or close to TM 1 [104] (Fig. 4-2). This strongly supports the three-dimensional proximity of these TMs and provides additional evidence that cocaine interacts with multiple regions of the primary sequence. These results are also remarkably consistent with many of the chimera results, and together the two approaches strongly implicate the three-dimensional proximity and contribution of TMs 1, 3, and 6 for cocaine binding (Fig. 4-2).

4.5.3 Site-Directed Mutagenesis

To date, over 250 of the 619/620 residues in rat and human DAT have been examined by site-directed mutagenesis, and the results of these studies have recently been reviewed extensively [105,106]. Mutation of many of these residues leads to modest reductions in cocaine binding or transport, consistent with indirect effects on active site structure, although distinguishing between direct and indirect effects is difficult based on single findings. There is also good evidence that several mutations that affect cocaine affinity do so by inhibiting ion-binding or ion-transport conformational changes rather than by directly affecting ligand- or DA-binding sites [107–109]. Thus identification of residues within the substrate or ligand interaction sites is challenging and requires thorough analysis using multiple approaches, such as ion dependence, chimeric and cross-species switches, aqueous accessibility, and homology to active sites in LeuT$_{Aa}$ and related NSSs.

The first DAT residue shown to be essential for transport was D79, found about halfway across TM 1 [110] (Fig. 4-2). This residue is conserved in the biogenic amine NSSs, and its mutation also strongly inhibits cocaine binding. Its importance for transport and inhibitor binding has been supported by similar results from SERT [111,112] and by the finding that the analogous residue is present in the LeuT$_{Aa}$ Na$^+$ binding site [23]. It was proposed originally that the negatively charged carboxyl group on D79 coordinated with the positively charged DA amine and cocaine tropane ring nitrogen [110]. Subsequent work by Wang and colleagues [113] questioned this mechanism, and in LeuT$_{Aa}$ the analogous glycine residue coordinates a Na$^+$ that forms an ion pair with the leucine carboxyl group [23]. This suggests that D79 is present within the DAT active site and compensates for the lack of a negative charge on DA by coordinating one of the cotransported Na$^+$ ions [23].

A nearby TM1 residue with multiple lines of support for active site involvement is F76 (Fig. 4-2), which results in lowered cocaine affinity when mutated [114]. This residue corresponds to N21 in LeuT$_{Aa}$, which participates in substrate binding [23], and Y95 in SERT, which is implicated in substrate and ligand binding [115,116], supporting the importance of this position in the NSS binding pocket. V152 in TM 3 (Fig. 4-2) also has good support for direct involvement in transport and cocaine binding, based on chimeric switches, multiple substitution analysis, and homology to a SERT inhibitor contact point at I172 [115,116] and a LeuT$_{Aa}$ leucine contact point at V104 [117]. Within TM 6, two residues, rDAT F319 (which corresponds to hDAT F320) and G323 (Fig. 4-2), which are analogous to LeuT$_{Aa}$ TM 6 active-site phenylalanine and serine residues, cause the loss of essentially all transport activity when mutated [65,114], consistent with roles in transport. However, these residues are less well characterized and G323 has also been implicated as part of an oligomerization motif.

Simultaneous mutation of three TM 2 residues, L104, F105, and A109, leads to loss of cocaine binding with maintenance of DA transport [118,119]. However, LeuT$_{Aa}$ structure and SCAM (substituted cysteine accessibility method) analysis of DAT [120] support the interpretation that this TM is found outside the active-site core and acts as a structural support for TM 6, suggesting the likelihood that the TM 2 mutations affect cocaine binding indirectly by leading to misalignment of TM 6. These studies highlight the importance of the overall transporter structure in maintenance of proper function, the apparent ease with which indirect effects can be transmitted to active sites, and the challenges in determining DAT structure–function relationships from mutagenesis.

Several residues involved in conformational changes occurring during the transport cycle have been identified by the Reith and Gether labs. Mutation of W84 in TM 1, K264 in TM 5, D313, Y335, and D345 in TMs 6 and 7, and D436 in TM 8 (Fig. 4-2) leads to loss of transport activity and alterations in cocaine binding [107–109]. Analysis of the ionic and Zn^{2+} dependence of these mutants supports an interpretation that the mutations suppress transport by stabilizing particular structures of the protein, indicating that the wild-type residues participate in intermolecular interactions necessary for conformational rearrangements during the transport cycle. In these studies it was also found that DATs blocked in outward-facing conformations displayed increased cocaine affinity, whereas those blocked in inward-facing conformations displayed decreased cocaine affinity [107,108], consistent with the cocaine-binding pocket being formed by an extracellular-facing form of the protein.

4.5.4 SCAM Studies

The aqueous accessibility of DAT domains has been evaluated by Javitch and co-workers using SCAM, which determines the sensitivity of endogenous and exogenous cysteine residues to reactivity with sulfhydryl-modifying reagents [121]. In this way it is possible to identify aqueously accessible regions in interhelical loops and DA or ligand access pathways, and nonaccessible regions adjacent to membrane lipids or other protein structure. Although only a fraction of the DAT sequence has been analyzed in this way, the results obtained to date are consistent with the 12 TM topology shown in Figures 4-1 and 4-2. TM 2 has been shown to have low aqueous accessibility throughout its entire length [120], suggesting that it is unlikely to interact directly with DA or cocaine. SCAM has also been used to detect conformational changes in the DAT. Residues that are not required for activity but show altered sulfhydryl reagent accessibility during DA transport and cocaine binding include C90 in TM 1, R219 in EL2, C342 and M371 in TM 7, A399 in EL4, and D436 in TM 8 [78,122–124]

(Fig. 4-2). These studies have revealed that ions, substrates, and blockers induce distinct conformations in various parts of the transporter in complex manners that are not fully understood with respect to the transport cycle.

4.6 CONCLUSIONS

Although significant progress has been made in recent years with respect to the elucidation of DAT structure and molecular composition, we are far from fully understanding the mechanisms responsible for transport and transport inhibition, or the significance of the many DAT modifications in regulation of transporter activity, location, and subcellular distribution. These properties are crucial for normal dopamine clearance, and mutations or dysregulations in these processes are likely to play important roles in the development of disorders associated with inappropriate dopaminergic function.

REFERENCES

1. Kuhar, M. J., Ritz, M. C., and Boja, J. W. (1991). The dopamine hypothesis of the reinforcing properties of cocaine. *Trends Neurosci, 14*, 299–302.

2. Giros, B., Jaber, M., Jones, S. R., Wightman, R. M., and Caron, M. G. (1996). Hyperlocomotion and indifference to cocaine and amphetamine in mice lacking the dopamine transporter. *Nature, 379*, 606–612.

3. Sulzer, D., Sonders, M. S., Poulsen, N. W., and Galli, A. (2005). Mechanisms of neurotransmitter release by amphetamines: a review. *Prog Neurobiol, 75*, 406–433.

4. Storch, A., Ludolph, A. C., and Schwarz, J. (2004). Dopamine transporter: involvement in selective dopaminergic neurotoxicity and degeneration. *J Neural Transm, 111*, 1267–1286.

5. Fleckenstein, A. E., Volz, T. J., Riddle, E. L., Gibb, J. W., and Hanson, G. R. (2007). New insights into the mechanism of action of amphetamines. *Annu Rev Pharmacol Toxicol, 47*, 681–698.

6. Chen, N. H., Reith, M. E., and Quick, M. W. (2004). Synaptic uptake and beyond: the sodium- and chloride-dependent neurotransmitter transporter family SLC6. *Pflugers Arch, 447*, 519–531.

7. Shimada, S., Kitayama, S., Lin, C. L., Patel, A., Nanthakumar, E., Gregor, P., Kuhar, M., and Uhl, G. (1991). Cloning and expression of a cocaine-sensitive dopamine transporter complementary DNA. *Science, 254*, 576–578.

8. Giros, B., el Mestikawy, S., Godinot, N., Zheng, K., Han, H., Yang-Feng, T., and Caron, M. G. (1992). Cloning, pharmacological characterization, and chromosome assignment of the human dopamine transporter. *Mol Pharmacol, 42*, 383–390.

9. Kilty, J. E., Lorang, D., and Amara, S. G. (1991). Cloning and expression of a cocaine-sensitive rat dopamine transporter. *Science, 254*, 578–579.

10. Usdin, T. B., Mezey, E., Chen, C., Brownstein, M. J., and Hoffman, B. J. (1991). Cloning of the cocaine-sensitive bovine dopamine transporter. *Proc Natl Acad Sci U S A, 88*, 11168–11171.

11. Giros, B., el Mestikawy, S., Bertrand, L., and Caron, M. G. (1991). Cloning and functional characterization of a cocaine-sensitive dopamine transporter. *FEBS Lett*, *295*, 149–154.

12. Wu, X., and Gu, H. H. (1999). Molecular cloning of the mouse dopamine transporter and pharmacological comparison with the human homologue. *Gene*, *233*, 163–170.

13. Miller, G. M., Yatin, S. M., De La Garza, R., 2nd, Goulet, M., and Madras, B. K. (2001). Cloning of dopamine, norepinephrine and serotonin transporters from monkey brain: relevance to cocaine sensitivity. *Brain Res Mol Brain Res*, *87*, 124–143.

14. Jayanthi, L. D., Apparsundaram, S., Malone, M. D., Ward, E., Miller, D. M., Eppler, M., and Blakely, R. D. (1998). The *Caenorhabditis elegans* gene T23G5.5 encodes an anti-depressant- and cocaine-sensitive dopamine transporter. *Mol Pharmacol*, *54*, 601–609.

15. Porzgen, P., Park, S. K., Hirsh, J., Sonders, M. S., and Amara, S. G. (2001). The antidepressant-sensitive dopamine transporter in *Drosophila melanogaster*: a primordial carrier for catecholamines. *Mol Pharmacol*, *59*, 83–95.

16. Chen, R., Wu, X., Wei, H., Han, D. D., and Gu, H. H. (2006). Molecular cloning and functional characterization of the dopamine transporter from *Eloria noyesi*, a caterpillar pest of cocaine-rich coca plants. *Gene*, *366*, 152–160.

17. Rudnick, G. (2002). Mechanisms of biogenic amine neurotransmitter transporters. In Neurotransmitter Transporters: Structure, Function, and Regulation, Reith, M. E., Ed., Humana Press Totowa, NJ, pp. 25–52.

18. Gu, H., Wall, S. C., and Rudnick, G. (1994). Stable expression of biogenic amine transporters reveals differences in inhibitor sensitivity, kinetics, and ion dependence. *J Biol Chem*, *269*, 7124–7130.

19. Sonders, M. S., Zhu, S. J., Zahniser, N. R., Kavanaugh, M. P., and Amara, S. G. (1997). Multiple ionic conductances of the human dopamine transporter: the actions of dopamine and psychostimulants. *J Neurosci*, *17*, 960–974.

20. Ingram, S. L., Prasad, B. M., and Amara, S. G. (2002). Dopamine transporter–mediated conductances increase excitability of midbrain dopamine neurons. *Nat Neurosci*, *5*, 971–978.

21. Carvelli, L., McDonald, P. W., Blakely, R. D., and Defelice, L. J. (2004). Dopamine transporters depolarize neurons by a channel mechanism. *Proc Natl Acad Sci U S A*, *101*, 16046–16051.

22. Kahlig, K. M., Binda, F., Khoshbouei, H., Blakely, R. D., McMahon, D. G., Javitch, J. A., and Galli, A. (2005). Amphetamine induces dopamine efflux through a dopamine transporter channel. *Proc Natl Acad Sci U S A*, *102*, 3495–3500.

23. Yamashita, A., Singh, S. K., Kawate, T., Jin, Y., and Gouaux, E. (2005). Crystal structure of a bacterial homologue of Na^+/Cl^--dependent neurotransmitter transporters. *Nature*, *437*, 215–223.

24. Beuming, T., Shi, L., Javitch, J. A., and Weinstein, H. (2006). A comprehensive structure-based alignment of prokaryotic and eukaryotic neurotransmitter/Na^+ symporters (NSS) aids in the use of the LeuT structure to probe NSS structure and function. *Mol Pharmacol*, *70*, 1630–1642.

25. Grigoriadis, D. E., Wilson, A. A., Lew, R., Sharkey, J. S., and Kuhar, M. J. (1989). Dopamine transport sites selectively labeled by a novel photoaffinity probe: ^{125}I-DEEP. *J Neurosci*, *9*, 2664–2670.

26. Sallee, F. R., Fogel, E. L., Schwartz, E., Choi, S. M., Curran, D. P., and Niznik, H. B. (1989). Photoaffinity labeling of the mammalian dopamine transporter. *FEBS Lett*, *256*, 219–224.

27. Lew, R., Grigoriadis, D., Wilson, A., Boja, J. W., Simantov, R., and Kuhar, M. J. (1991). Dopamine transporter: deglycosylation with exo- and endoglycosidases. *Brain Res*, *539*, 239–246.

28. Vaughan, R. A., Brown, V. L., McCoy, M. T., and Kuhar, M. J. (1996). Species- and brain region–specific dopamine transporters: immunological and glycosylation characteristics. *J Neurochem*, *66*, 2146–2152.

29. Lew, R., Vaughan, R., Simantov, R., Wilson, A., and Kuhar, M. J. (1991). Dopamine transporters in the nucleus accumbens and the striatum have different apparent molecular weights. *Synapse*, *8*, 152–153.

30. Lew, R., Patel, A., Vaughan, R. A., Wilson, A., and Kuhar, M. J. (1992). Microheterogeneity of dopamine transporters in rat striatum and nucleus accumbens. *Brain Res*, *584*, 266–271.

31. Patel, A., Uhl, G., and Kuhar, M. J. (1993). Species differences in dopamine transporters: postmortem changes and glycosylation differences. *J Neurochem*, *61*, 496–500.

32. Patel, A. P., Cerruti, C., Vaughan, R. A., and Kuhar, M. J. (1994). Developmentally regulated glycosylation of dopamine transporter. *Brain Res Dev Brain Res*, *83*, 53–58.

33. Li, L. B., Chen, N., Ramamoorthy, S., Chi, L., Cui, X. N., Wang, L. C., and Reith, M. E. (2004). The role of N-glycosylation in function and surface trafficking of the human dopamine transporter. *J Biol Chem*, *279*, 21012–21020.

34. Torres, G. E., Carneiro, A., Seamans, K., Fiorentini, C., Sweeney, A., Yao, W. D., and Caron, M. G. (2003). Oligomerization and trafficking of the human dopamine transporter: mutational analysis identifies critical domains important for the functional expression of the transporter. *J Biol Chem*, *278*, 2731–2739.

35. Wang, J. B., Moriwaki, A., and Uhl, G. R. (1995). Dopamine transporter cysteine mutants: second extracellular loop cysteines are required for transporter expression. *J Neurochem*, *64*, 1416–1419.

36. Chen, R., Wei, H., Hill, E. R., Chen, L., Jiang, L., Han, D. D., and Gu, H. H. (2006). Direct evidence that two cysteines in the dopamine transporter form a disulfide bond. *Mol Cell Biochem*, *298*, 41–48.

37. Mortensen, O. V., and Amara, S. G. (2003). Dynamic regulation of the dopamine transporter. *Eur J Pharmacol*, *479*, 159–170.

38. Vaughan, R. A. (2004). Phosphorylation and regulation of psychostimulant-sensitive neurotransmitter transporters. *J Pharmacol Exp Ther*, *310*, 1–7.

39. Foster, J. D., Cervinski, M. A., Gorentla, B. K., and Vaughan, R. A. (2006). Regulation of the dopamine transporter by phosphorylation. *Handb Exp Pharmacol*, 197–214.

40. Melikian, H. E. (2004). Neurotransmitter transporter trafficking: endocytosis, recycling, and regulation. *Pharmacol Ther*, *104*, 17–27.

41. Kantor, L., and Gnegy, M. E. (1998). Protein kinase C inhibitors block amphetamine-mediated dopamine release in rat striatal slices. *J Pharmacol Exp Ther*, *284*, 592–598.

42. Cowell, R. M., Kantor, L., Hewlett, G. H., Frey, K. A., and Gnegy, M. E. (2000). Dopamine transporter antagonists block phorbol ester–induced dopamine release and dopamine transporter phosphorylation in striatal synaptosomes. *Eur J Pharmacol*, *389*, 59–65.

43. Huff, R. A., Vaughan, R. A., Kuhar, M. J., and Uhl, G. R. (1997). Phorbol esters increase dopamine transporter phosphorylation and decrease transport V_{max}. *J Neurochem, 68*, 225–232.

44. Vaughan, R. A., Huff, R. A., Uhl, G. R., and Kuhar, M. J. (1997). Protein kinase C–mediated phosphorylation and functional regulation of dopamine transporters in striatal synaptosomes. *J Biol Chem, 272*, 15541–15546.

45. Granas, C., Ferrer, J., Loland, C. J., Javitch, J. A., and Gether, U. (2003). N-Terminal truncation of the dopamine transporter abolishes phorbol ester- and substance P receptor–stimulated phosphorylation without impairing transporter internalization. *J Biol Chem, 278*, 4990–5000.

46. Foster, J. D., Pananusorn, B., and Vaughan, R. A. (2002). Dopamine transporters are phosphorylated on N-terminal serines in rat striatum. *J Biol Chem, 277*, 25178–25186.

47. Cervinski, M. A., Foster, J. D., and Vaughan, R. A. (2005). Psychoactive substrates stimulate dopamine transporter phosphorylation and down-regulation by cocaine-sensitive and protein kinase C–dependent mechanisms. *J Biol Chem, 280*, 40442–40449.

48. Gorentla, B. K., and Vaughan, R. A. (2005). Differential effects of dopamine and psychoactive drugs on dopamine transporter phosphorylation and regulation. *Neuropharmacology, 49*, 759–768.

49. Foster, J. D., Pananusorn, B., Cervinski, M. A., Holden, H. E., and Vaughan, R. A. (2003). Dopamine transporters are dephosphorylated in striatal homogenates and in vitro by protein phosphatase 1. *Brain Res Mol Brain Res, 110*, 100–108.

50. Fog, J. U., Khoshbouei, H., Holy, M., Owens, W. A., Vaegter, C. B., Sen, N., Nikandrova, Y., Bowton, E., McMahon, D. G., Colbran, R. J., Daws, L. C., Sitte, H. H., Javitch, J. A., Galli, A., and Gether, U. (2006). Calmodulin kinase II interacts with the dopamine transporter C terminus to regulate amphetamine-induced reverse transport. *Neuron, 51*, 417–429.

51. Lin, Z., Zhang, P. W., Zhu, X., Melgari, J. M., Huff, R., Spieldoch, R. L., and Uhl, G. R. (2003). Phosphatidylinositol 3-kinase, protein kinase C, and MEK1/2 kinase regulation of dopamine transporters (DAT) require N-terminal DAT phosphoacceptor sites. *J Biol Chem, 278*, 20162–20170.

52. Khoshbouei, H., Sen, N., Guptaroy, B., Johnson, L., Lund, D., Gnegy, M. E., Galli, A., and Javitch, J. A. (2004). N-Terminal phosphorylation of the dopamine transporter is required for amphetamine-induced efflux. *PLoS Biol, 2*, E78.

53. Cervinski, M. A., and Vaughan R. A. (2006). Syntaxin 1A influences the phosphorylation and regulation of the dopamine transporter. Program 35.10. 2006 *Abstract Viewer/itinerary Planner*. Society for Neuroscience, Atlanta, GA, Online.

54. Ciechanover, A. (2005). Proteolysis: from the lysosome to ubiquitin and the proteasome. *Nat Rev Mol Cell Biol, 6*, 79–87.

55. Glickman, M. H., and Ciechanover, A. (2002). The ubiquitin-proteasome proteolytic pathway: destruction for the sake of construction. *Physiol Rev, 82*, 373–428.

56. Hicke, L., and Dunn, R. (2003). Regulation of membrane protein transport by ubiquitin and ubiquitin-binding proteins. *Annu Rev Cell Dev Biol, 19*, 141–172.

57. Jiang, H., Jiang, Q., and Feng, J. (2004). Parkin increases dopamine uptake by enhancing the cell surface expression of dopamine transporter. *J Biol Chem, 279*, 54380–54386.

58. Oliveira, S. A., Scott, W. K., Martin, E. R., Nance, M. A., Watts, R. L., Hubble, J. P., Koller, W. C., Pahwa, R., Stern, M. B., Hiner, B. C., Ondo, W. G., Allen, F. H., Jr.,

Scott, B. L., Goetz, C. G., Small, G. W., Mastaglia, F., Stajich, J. M., Zhang, F., Booze, M. W., Winn, M. P., Middleton, L. T., Haines, J. L., Pericak-Vance, M. A., and Vance, J. M. (2003). Parkin mutations and susceptibility alleles in late-onset Parkinson's disease. *Ann Neurol*, *53*, 624–629.

59. Miranda, M., Wu, C. C., Sorkina, T., Korstjens, D. R., and Sorkin, A. (2005). Enhanced ubiquitylation and accelerated degradation of the dopamine transporter mediated by protein kinase C. *J Biol Chem*, *280*, 35617–35624.

60. Miranda, M., Dionne, K. R., Sorkina, T., and Sorkin, A. (2006). Three ubiquitin conjugation sites in the amino terminus of the dopamine transporter mediate protein kinase C–dependent endocytosis of the transporter. *Mol Biol Cell*, *18*, 313–323.

61. Sorkina, T., Miranda, M., Dionne, K. R., Hoover, B. R., Zahniser, N. R., and Sorkin, A. (2006). RNA interference screen reveals an essential role of Nedd4-2 in dopamine transporter ubiquitination and endocytosis. *J Neurosci*, *26*, 8195–8205.

62. Sitte, H. H., Farhan, H., and Javitch, J. A. (2004). Sodium-dependent neurotransmitter transporters: oligomerization as a determinant of transporter function and trafficking. *Mol Intervent*, *4*, 38–47.

63. Berger, S. P., Farrell, K., Conant, D., Kempner, E. S., and Paul, S. M. (1994). Radiation inactivation studies of the dopamine reuptake transporter protein. *Mol Pharmacol*, *46*, 726–731.

64. Milner, H. E., Beliveau, R., and Jarvis, S. M. (1994). The in situ size of the dopamine transporter is a tetramer as estimated by radiation inactivation. *Biochim Biophys Acta*, *1190*, 185–187.

65. Hastrup, H., Karlin, A., and Javitch, J. A. (2001). Symmetrical dimer of the human dopamine transporter revealed by cross-linking Cys-306 at the extracellular end of the sixth transmembrane segment. *Proc Natl Acad Sci U S A*, *98*, 10055–10060.

66. Hastrup, H., Sen, N., and Javitch, J. A. (2003). The human dopamine transporter forms a tetramer in the plasma membrane: cross-linking of a cysteine in the fourth transmembrane segment is sensitive to cocaine analogs. *J Biol Chem*, *278*, 45045–45048.

67. Sorkina, T., Doolen, S., Galperin, E., Zahniser, N. R., and Sorkin, A. (2003). Oligomerization of dopamine transporters visualized in living cells by fluorescence resonance energy transfer microscopy. *J Biol Chem*, *278*, 28274–28283.

68. Frederickson, C. J., Koh, J. Y., and Bush, A. I. (2005). The neurobiology of zinc in health and disease. *Nat Rev Neurosci*, *6*, 449–462.

69. Richfield, E. K. (1993). Zinc modulation of drug binding, cocaine affinity states, and dopamine uptake on the dopamine uptake complex. *Mol Pharmacol*, *43*, 100–108.

70. Bonnet, J. J., Benmansour, S., Amejdki-Chab, N., and Costentin, J. (1994). Effect of CH_3HgCl and several transition metals on the dopamine neuronal carrier; peculiar behaviour of Zn^{2+}. *Eur J Pharmacol*, *266*, 87–97.

71. Norregaard, L., Frederiksen, D., Nielsen, E. O., and Gether, U. (1998). Delineation of an endogenous zinc-binding site in the human dopamine transporter. *Embo J*, *17*, 4266–4273.

72. Loland, C. J., Norregaard, L., and Gether, U. (1999). Defining proximity relationships in the tertiary structure of the dopamine transporter: identification of a conserved glutamic acid as a third coordinate in the endogenous Zn^{2+}-binding site. *J Biol Chem*, *274*, 36928–36934.

73. Alberts, I. L., Nadassy, K., and Wodak, S. J. (1998). Analysis of zinc binding sites in protein crystal structures. *Protein Sci, 7*, 1700–1716.

74. Scholze, P., Norregaard, L., Singer, E. A., Freissmuth, M., Gether, U., and Sitte, H. H. (2002). The role of zinc ions in reverse transport mediated by monoamine transporters. *J Biol Chem, 277*, 21505–21513.

75. Pifl, C., Rebernik, P., Kattinger, A., and Reither, H. (2004). Zn^{2+} modulates currents generated by the dopamine transporter: parallel effects on amphetamine-induced charge transfer and release. *Neuropharmacology, 46*, 223–231.

76. Meinild, A. K., Sitte, H. H., and Gether, U. (2004). Zinc potentiates an uncoupled anion conductance associated with the dopamine transporter. *J Biol Chem, 279*, 49671–49679.

77. Norgaard-Nielsen, K., and Gether, U. (2006). Zn^{2+} modulation of neurotransmitter transporters. *Handb Exp Pharmacol, 175*, 1–22.

78. Norregaard, L., Loland, C. J., and Gether, U. (2003). Evidence for distinct sodium-, dopamine-, and cocaine-dependent conformational changes in transmembrane segments 7 and 8 of the dopamine transporter. *J Biol Chem, 278*, 30587–30596.

79. Norregaard, L., Visiers, I., Loland, C. J., Ballesteros, J., Weinstein, H., and Gether, U. (2000). Structural probing of a microdomain in the dopamine transporter by engineering of artificial Zn^{2+} binding sites. *Biochemistry, 39*, 15836–15846.

80. Loland, C. J., Norregaard, L., Litman, T., and Gether, U. (2002). Generation of an activating Zn^{2+} switch in the dopamine transporter: mutation of an intracellular tyrosine constitutively alters the conformational equilibrium of the transport cycle. *Proc Natl Acad Sci U S A, 99*, 1683–1688.

81. Loland, C. J., Norgaard-Nielsen, K., and Gether, U. (2003). Probing dopamine transporter structure and function by Zn^{2+}-site engineering. *Eur J Pharmacol, 479*, 187–197.

82. Torres, G. E. (2006). The dopamine transporter proteome. *J Neurochem, 97*(Suppl 1), 3–10.

83. Johnson, L. A., Guptaroy, B., Lund, D., Shamban, S., and Gnegy, M. E. (2005). Regulation of amphetamine-stimulated dopamine efflux by protein kinase C beta. *J Biol Chem, 280*, 10914–10919.

84. Bauman, A. L., Apparsundaram, S., Ramamoorthy, S., Wadzinski, B. E., Vaughan, R. A., and Blakely, R. D. (2000). Cocaine and antidepressant-sensitive biogenic amine transporters exist in regulated complexes with protein phosphatase 2A. *J Neurosci, 20*, 7571–7578.

85. Lee, K. H., Kim, M. Y., Kim, D. H., and Lee, Y. S. (2004). Syntaxin 1A and receptor for activated C kinase interact with the N-terminal region of human dopamine transporter. *Neurochem Res, 29*, 1405–1409.

86. Torres, G. E., Yao, W. D., Mohn, A. R., Quan, H., Kim, K. M., Levey, A. I., Staudinger, J., and Caron, M. G. (2001). Functional interaction between monoamine plasma membrane transporters and the synaptic PDZ domain-containing protein PICK1. *Neuron, 30*, 121–134.

87. Bjerggaard, C., Fog, J. U., Hastrup, H., Madsen, K., Loland, C. J., Javitch, J. A., and Gether, U. (2004). Surface targeting of the dopamine transporter involves discrete epitopes in the distal C terminus but does not require canonical PDZ domain interactions. *J Neurosci, 24*, 7024–7036.

88. Carneiro, A. M., Ingram, S. L., Beaulieu, J. M., Sweeney, A., Amara, S. G., Thomas, S. M., Caron, M. G., and Torres, G. E. (2002). The multiple LIM domain-containing adaptor

protein Hic-5 synaptically colocalizes and interacts with the dopamine transporter. *J Neurosci, 22,* 7045–7054.

89. Lee, F. J., Liu, F., Pristupa, Z. B., and Niznik, H. B. (2001). Direct binding and functional coupling of α-synuclein to the dopamine transporters accelerate dopamine-induced apoptosis. *FASEB J, 15,* 916–926.

90. Wersinger, C., and Sidhu, A. (2003). Attenuation of dopamine transporter activity by α-synuclein. *Neurosci Lett, 340,* 189–192.

91. Holton, K. L., Loder, M. K., and Melikian, H. E. (2005). Nonclassical, distinct endocytic signals dictate constitutive and PKC-regulated neurotransmitter transporter internalization. *Nat Neurosci, 8,* 881–888.

92. Maiya, R., and Mayfield, R. D. (2004). Dopamine transporter network and pathways. *Int Rev Neurobiol, 61,* 79–96.

93. Giros, B., Wang, Y. M., Suter, S., McLeskey, S. B., Pifl, C., and Caron, M. G. (1994). Delineation of discrete domains for substrate, cocaine, and tricyclic antidepressant interactions using chimeric dopamine–norepinephrine transporters. *J Biol Chem, 269,* 15985–15988.

94. Buck, K. J., and Amara, S. G. (1994). Chimeric dopamine–norepinephrine transporters delineate structural domains influencing selectivity for catecholamines and 1-methyl-4-phenylpyridinium. *Proc Natl Acad Sci U S A, 91,* 12584–12588.

95. Buck, K. J., and Amara, S. G. (1995). Structural domains of catecholamine transporter chimeras involved in selective inhibition by antidepressants and psychomotor stimulants. *Mol Pharmacol, 48,* 1030–1037.

96. Lee, S. H., Chang, M. Y., Jeon, D. J., Oh, D. Y., Son, H., Lee, C. H., and Lee, Y. S. (2002). The functional domains of dopamine transporter for cocaine analog, CFT binding. *Exp Mol Med, 34,* 90–94.

97. Lee, S. H., Kang, S. S., Son, H., and Lee, Y. S. (1998). The region of dopamine transporter encompassing the 3rd transmembrane domain is crucial for function. *Biochem Biophys Res Commun, 246,* 347–52.

98. Syringas, M., Janin, F., Mezghanni, S., Giros, B., Costentin, J., and Bonnet, J. J. (2000). Structural domains of chimeric dopamine–noradrenaline human transporters involved in the Na^+- and Cl^--dependence of dopamine transport. *Mol Pharmacol, 58,* 1404–1411.

99. Vaughan, R. A., and Kuhar, M. J. (1996). Dopamine transporter ligand binding domains: structural and functional properties revealed by limited proteolysis. *J Biol Chem, 271,* 21672–21680.

100. Vaughan, R. A., Agoston, G. E., Lever, J. R., and Newman, A. H. (1999). Differential binding of tropane-based photoaffinity ligands on the dopamine transporter. *J Neurosci, 19,* 630–636.

101. Vaughan, R. A., Gaffaney, J. D., Lever, J. R., Reith, M. E., and Dutta, A. K. (2001). Dual incorporation of photoaffinity ligands on dopamine transporters implicates proximity of labeled domains. *Mol Pharmacol, 59,* 1157–1164.

102. Vaughan, R. A., Parnas, M. L., Gaffaney, J. D., Lowe, M. J., Wirtz, S., Pham, A., Reed, B., Dutta, S. M., Murray, K. K., and Justice, J. B. (2005). Affinity labeling the dopamine transporter ligand binding site. *J Neurosci Methods, 143,* 33–40.

103. Vaughan, R. A., Sakrikar, D. J., Parnas, M. L., Adkins, S., Foster, J. D., Duval, R., Lever, J. R., Kulkarni, S., and Newman, A. H. (2007). Localization of cocaine analog [^{125}I]RTI 82 irreversible binding to transmembrane domain six of the dopamine transporter. *J Biol Chem, 282,* 8915–8925.

104. Parnas, M. L., Zou, M. F., Newman, A. H., Duval, R. A., Lever, J. R., and Vaughan, R. A. (2006). Localizing the binding site for cocaine on the dopamine transporter with [^{125}I]MFZ 2-24. Program 35.1. *2006 Abstract Viewer/Itinerary Planner*. Society for Neuroscience, Atlanta, GA, Online.

105. Surratt, C. K., Ukairo, O. T., and Ramanujapuram, S. (2005). Recognition of psychostimulants, antidepressants, and other inhibitors of synaptic neurotransmitter uptake by the plasma membrane monoamine transporters. *AAPS J, 7*, E739–E751.

106. Volz, T. J., and Schenk, J. O. (2005). A comprehensive atlas of the topography of functional groups of the dopamine transporter. *Synapse, 58*, 72–94.

107. Chen, N., Rickey, J., Berfield, J. L., and Reith, M. E. (2004). Aspartate 345 of the dopamine transporter is critical for conformational changes in substrate translocation and cocaine binding. *J Biol Chem, 279*, 5508–5519.

108. Chen, N., Zhen, J., and Reith, M. E. (2004). Mutation of Trp84 and Asp313 of the dopamine transporter reveals similar mode of binding interaction for GBR 12909 and benztropine as opposed to cocaine. *J Neurochem, 89*, 853–864.

109. Loland, C. J., Granas, C., Javitch, J. A., and Gether, U. (2004). Identification of intracellular residues in the dopamine transporter critical for regulation of transporter conformation and cocaine binding. *J Biol Chem, 279*, 3228–3238.

110. Kitayama, S., Shimada, S., Xu, H., Markham, L., Donovan, D. M., and Uhl, G. R. (1992). Dopamine transporter site-directed mutations differentially alter substrate transport and cocaine binding. *Proc Natl Acad Sci U S A, 89*, 7782–7785.

111. Barker, E. L., Perlman, M. A., Adkins, E. M., Houlihan, W. J., Pristupa, Z. B., Niznik, H. B., and Blakely, R. D. (1998). High affinity recognition of serotonin transporter antagonists defined by species-scanning mutagenesis: an aromatic residue in transmembrane domain I dictates species-selective recognition of citalopram and mazindol. *J Biol Chem, 273*, 19459–19468.

112. Barker, E. L., Moore, K. R., Rakhshan, F., and Blakely, R. D. (1999). Transmembrane domain I contributes to the permeation pathway for serotonin and ions in the serotonin transporter. *J Neurosci, 19*, 4705–4717.

113. Wang, W., Sonders, M. S., Ukairo, O. T., Scott, H., Kloetzel, M. K., and Surratt, C. K. (2003). Dissociation of high-affinity cocaine analog binding and dopamine uptake inhibition at the dopamine transporter. *Mol Pharmacol, 64*, 430–439.

114. Lin, Z., Wang, W., Kopajtic, T., Revay, R. S., and Uhl, G. R. (1999). Dopamine transporter: transmembrane phenylalanine mutations can selectively influence dopamine uptake and cocaine analog recognition. *Mol Pharmacol, 56*, 434–447.

115. Henry, L. K., Adkins, E. M., Han, Q., and Blakely, R. D. (2003). Serotonin and cocaine-sensitive inactivation of human serotonin transporters by methanethiosulfonates targeted to transmembrane domain I. *J Biol Chem, 278*, 37052–37063.

116. Henry, L. K., Field, J. R., Adkins, E. M., Parnas, M. L., Vaughan, R. A., Zou, M. F., Newman, A. H., and Blakely, R. D. (2006). Tyr-95 and Ile-172 in transmembrane segments 1 and 3 of human serotonin transporters interact to establish high affinity recognition of antidepressants. *J Biol Chem, 281*, 2012–2023.

117. Lee, S. H., Chang, M. Y., Lee, K. H., Park, B. S., Lee, Y. S., and Chin, H. R. (2000). Importance of valine at position 152 for the substrate transport and 2β-carbomethoxy-3β-(4-fluorophenyl)tropane binding of dopamine transporter. *Mol Pharmacol, 57*, 883–889.

118. Chen, R., Han, D. D., and Gu, H. H. (2005). A triple mutation in the second transmembrane domain of mouse dopamine transporter markedly decreases sensitivity to cocaine and methylphenidate. *J Neurochem, 94*, 352–359.

119. Wu, X., and Gu, H. H. (2003). Cocaine affinity decreased by mutations of aromatic residue phenylalanine 105 in the transmembrane domain 2 of dopamine transporter. *Mol Pharmacol, 63*, 653–658.

120. Sen, N., Shi, L., Beuming, T., Weinstein, H., and Javitch, J. A. (2005). A pincer-like configuration of TM2 in the human dopamine transporter is responsible for indirect effects on cocaine binding. *Neuropharmacology, 49*, 780–790.

121. Javitch, J. A. (1998). Probing structure of neurotransmitter transporters by substituted-cysteine accessibility method. *Methods Enzymol, 296*, 331–346.

122. Ferrer, J. V., and Javitch, J. A. (1998). Cocaine alters the accessibility of endogenous cysteines in putative extracellular and intracellular loops of the human dopamine transporter. *Proc Natl Acad Sci U S A, 95*, 9238–9243.

123. Chen, N., Ferrer, J. V., Javitch, J. A., and Justice, J. B., Jr. (2000). Transport-dependent accessibility of a cytoplasmic loop cysteine in the human dopamine transporter. *J Biol Chem, 275*, 1608–1614.

124. Gaffaney, J. D., Javitch, J. A., and Vaughan, R. A. (2004). Analysis of extracellular loop two of the dopamine transporter using the substituted cysteine accessibility method. Program 53.6. *2004 Abstract Viewer/Itinerary Planner*. Society for Neuroscience, Washington, DC. Online.

5

ELECTROCHEMICAL CHARACTERIZATION OF DOPAMINE TRANSPORTERS

EVGENY A. BUDYGIN AND SARA R. JONES

Department of Physiology and Pharmacology, Wake Forest University Health Sciences, Winston-Salem, North Carolina

5.1	Introduction	98
5.2	Advantages of Fast-Scan Cyclic Voltammetry to Measure Dopamine Uptake	98
5.3	Methodology for Fast-Scan Cyclic Voltammetry	98
	5.3.1 Electrode Fabrication and Calibration	99
	5.3.2 Voltammetry Data Acquisition	99
	5.3.3 Data Analysis and Uptake-Rate Determinations	100
5.4	Effects of Dopamine-Uptake Inhibitors and Releasers	100
	5.4.1 Cocaine	100
	5.4.2 Amphetamines	103
5.5	Cocaine Self-Administration and Dopamine Transporters	104
5.6	DAT Involvement in Psychostimulant Reinforcement	106
5.7	DAT Involvement in Variations in Regional Dopamine Dynamics	109
5.8	Fast Onset of DAT Inhibition in the Brain Following Intravenous Cocaine Infusion	110
5.9	Correlation of Behavioral Activation with DAT Inhibition	113
5.10	DAT and Phasic Dopamine Neurotransmission	115
5.11	Conclusions	117
	Acknowledgments	117
	References	118

5.1 INTRODUCTION

Dopamine transporters, which are responsible for clearing dopamine from the extracellular space of the brain following release from presynaptic terminals, are the primary target of psychostimulant drugs such as cocaine and amphetamine. These drugs inhibit the uptake of dopamine through the transporters, resulting in elevated extracellular dopamine levels and causing a variety of behavioral effects. Dopamine elevations have been linked to the acute reinforcing effects of many drugs of abuse. In addition, dopamine transporter inhibitors have been used for many different medical applications, such as the treatment of attention-deficit/hyperactivity disorders, Parkinson's disease, Alzheimers' disease, schizophrenia, Tourette's syndrome, Lesch–Nyhan disease, narcolepsy, depression, obesity, and stimulant abuse. Because of intense interest in the dopamine transporter as a target of pharmaceutical agents, many methods have been developed to measure different aspects of dopamine transporters. Density of binding sites for dopamine transporter ligands such as those discussed in other chapters has been used to measure the number of transporters in various regions of the brain. Measuring uptake of exogenously applied, radioactively labeled dopamine is typically accomplished using synaptosomal preparations. Voltammetry is one of the few methods used to assess functional uptake of endogenous dopamine.

5.2 ADVANTAGES OF FAST-SCAN CYCLIC VOLTAMMETRY TO MEASURE DOPAMINE UPTAKE

Fast-scan cyclic voltammetry offers several advantages for the characterization of dopamine transporters. This technique detects extracellular dopamine concentrations with subsecond resolution, allowing uptake to be measured in real time. It uses a micrometer-dimension probe that allows high spatial resolution with minimal tissue damage. The measured cyclic voltammograms identify the substance detected as dopamine. Voltammetry is able to resolve changes in dopamine uptake kinetically, separating changes in $V_{\max}$ and K_m parameters. Moreover, voltammetry recordings can be performed both in vitro and in vivo, including brain slices in both anesthetized and freely moving animals. This technique can effectively detect exogenously applied dopamine as well as electrically stimulated and naturally occurring release of endogenous dopamine.

5.3 METHODOLOGY FOR FAST-SCAN CYCLIC VOLTAMMETRY

Voltammetry takes advantage of the fact that dopamine is oxidized at fairly low voltages and can be identified based on voltage-dependent oxidation–reduction reactions. Briefly, every 100 ms, the voltage of the electrode is scanned first to oxidize and then to reduce any dopamine that is in the vicinity of the electrode tip. Following conversion of the current to concentration, it is possible to create concentration vs. time plots that reflect the fluctuations of dopamine in the extracellular space. In this way, the processes of dopamine release and uptake can be monitored in real time as they occur in the tissue.

5.3.1 Electrode Fabrication and Calibration

Voltammetry electrodes are made from carbon fibers typically 5 to 7 μm in diameter protruding from a pulled glass capillary. Carbon fibers are used in the construction of voltammetry electrodes because they have the advantage of excellent electrical conductance plus surface properties that make them resistant to adsorption of lipids and proteins from brain tissue. Voltammetry electrodes are made with carbon fibers (T-650, Thornell, Amoco Corp., Greenville, South Carolina) placed into glass capillaries (A-M Systems, Carlsborg, Washington), which are then heated at the center and pulled into two equal-sized electrodes using a vertical electrode puller (Narishige, Tokyo, Japan). Each electrode is then trimmed so that 50 to 200 μm of carbon fiber protrudes from the end of the glass. The longer the length of exposed fiber, the greater the sensitivity of the electrode. The usefulness of the long fiber length is, however, limited by fragility (high percentage of breakage), the dimensions of the sampling area desired, and increased electrical noise. Electrodes are soaked in isopropyl alcohol prior to use. Reference electrodes are made of chloridized silver wire (0.5-mm-diameter silver wire, Sigma-Aldrich, St. Louis, Missouri). All potentials are vs. Ag/AgCl.

For calibration of electrodes, a flow-injection system is used. This consists of a flow system wherein the electrode is placed in a stream of flowing buffer (approximately 1 to 2 mL/min) and the stream is rapidly switched to buffer containing a known concentration of dopamine (usually, 1 to 10 μM). The switching is accomplished with either a manual or a computer-controlled injector, such as those used in low-pressure liquid chromatography. The calibration buffer consists of (in mM) NaCl (126), KCl (2.5), NaH_2PO_4 (1.2), $CaCl_2$ (2.4), $MgCl_2$ (1.2), $NaHCO_3$ (25), and HEPES acid (20), pH 7.4 at 23°C. The speed and sensitivity of the electrode response to the injected dopamine are both recorded and used as criteria to select electrodes for experimental use. Only electrodes that respond quickly ($\leq$200 ms to peak) and with high sensitivity ($\geq$5 nA/μM dopamine) are used for experiments. Postcalibration of electrodes occurs immediately following the experiment, using the same protocol. The results of the postcalibration are used in calculating calibration factors for the Michaelis–Menten-based kinetic analyses of dopamine release and uptake information. Postcalibration values are more accurate than those measured during precalibration, because there is some slowing of electrode response times and reduction in sensitivity during an experiment due to adsorption of small amounts of lipid and protein from the brain tissue in contact with the electrode. In general, sensitivity is decreased by half and response times are doubled. For this reason, only the fastest and most sensitive electrodes are initially selected to be used in experiments.

5.3.2 Voltammetry Data Acquisition

Voltammetric recordings are made by applying a triangular waveform, scanning from -400 to 1200 mV at 300 V/s, every 100 ms. Electrode voltage is held at -400 mV between scans. The waveform is generated and the voltammetric signal is collected using Labview-based (National Instruments, Austin, Texas), locally written software, and a data acquisition board (PCI-6052E, National Instruments). A timing board (PCI-6711E, National Instruments) is used to synchronize all aspects of data collection and stimulus application.

5.3.3 Data Analysis and Uptake-Rate Determinations

Dopamine (DA) uptake can be modeled kinetically using the following Michaelis–Menten equation:

$$\frac{d[\text{DA}]}{dt} = f[\text{DA}]p - \frac{V_{\max}}{(K_m/[\text{DA}]) + 1}$$

where [DA] is the instantaneous concentration of DA in the extracellular space, f the frequency of stimulation, [DA]p the concentration of DA released per stimulus pulse, and $V_{\max}$ and K_m are Michaelis–Menten uptake parameters. For this type of curve fitting, three assumptions must be made: (1) each stimulus pulse releases a fixed quantity of dopamine from the presynaptic terminals, (2) uptake is a saturable process, and (3) uptake via the dopamine transporter is the predominant mechanism clearing dopamine from the extracellular space. This modeling also assumes a control K_m value (inversely related to the affinity of the transporter for dopamine) of 0.16 μM. An additional kinetic parameter, $V_{\max}$, which is related to the number and turnover rate of the dopamine transporter, is determined empirically. These assumptions are best suited for evaluating release and uptake in striatal regions where one-pulse or short-train stimulations are used and which have rapid uptake.

5.4 EFFECTS OF DOPAMINE-UPTAKE INHIBITORS AND RELEASERS

5.4.1 Cocaine

As shown in Figure 5-1 with cocaine, uptake inhibitors slow the uptake of dopamine, resulting in a decreased rate of return to baseline after electrically stimulated release. The rate of uptake is fit to a Michaelis–Menten-based kinetic model, which yields a value for $V_{\max}$, the maximal rate of uptake, and K_m, the affinity of dopamine for the transporter. The $V_{\max}$ value is determined before drug is applied, and for competitive uptake inhibitors it remains fixed for all measurements in the presence of the inhibitor. The K_m value appears to change to a lower affinity following application of an uptake inhibitor because of competition between the inhibitor and DA for a binding site. We use this apparent change in the affinity of DA for the transporter as a measure of uptake inhibition, and term the measured affinity *apparent K_m*. Uptake inhibition increases and apparent K_m decreases as a function of inhibitor concentration. This is shown graphically for cocaine in Figures 5-2 and 5-3.

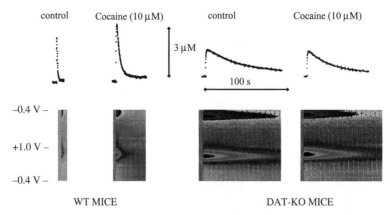

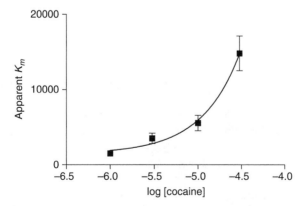

Figure 5-1 Effects of cocaine and genetic deletion of the dopamine transporter on dopamine clearance in core NAc. Single-pulse electrical stimulations were used to evoke dopamine release in slices containing the core NAc from control and DAT-KO mice. Dots are data points, collected every 100 ms. The overflow profile on the left shows a control stimulation, in the absence of drug, in a control mouse. Note that following stimulation, the extracellular level of dopamine returns to baseline levels within 1 second. The second profile shows that 10 μM cocaine markedly slows the return of dopamine to baseline levels, taking approximately 10 seconds to return. The third profile, recorded in a slice from a DAT-KO mouse, shows the dramatic slowing of dopamine clearance from the extracellular space following release. It takes approximately 100 seconds to clear the released dopamine. The fourth profile demonstrates that cocaine has no effect on dopamine clearance in DAT-KO mice. Insets: Below each profile, a pseudocolor plot gives information on voltage on the y-axis, time on the x-axis, and current due to dopamine oxidation and reduction on the z-axis in pseudocolor. Note the prolonged lifetime of dopamine in the extracellular space in the DAT-KO mice (green–blue colors). (*See insert for color representation of figure.*)

Figure 5-2 Apparent K_m changes with increasing concentrations of cocaine. This cocaine dose–response curve depicts the changes in apparent K_m for dopamine uptake in brain slices containing the core NAc. Increases in apparent K_m reflect competitive uptake inhibition of endogenous dopamine uptake through the dopamine transporter.

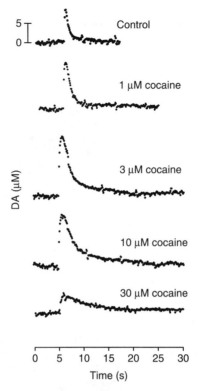

Figure 5-3 Dose–response curve for cocaine in slices of core NAc. Voltammetry recordings of dopamine release and uptake in the presence of increasing concentrations of cocaine applied to a brain slice containing the core NAc in a cumulative fashion, with a new concentration added every 30 minutes. It is clear that each higher concentration of cocaine inhibits the dopamine transporter more, although the clearance rate remains far faster than that measured in DAT-KO mice (see Fig. 5-1). At the highest doses, cocaine causes a decrease in dopamine release, which limits the concentration range over which recordings can be made.

Cocaine is a relatively weak dopamine uptake inhibitor. It produces a maximal apparent K_m value for DA of approximately 10 μM, while more potent and effica- cious uptake inhibitors such as nomifensine, GBR 12909, and PTT (2β-propanoyl- 3β-(4-tolyl)tropane [1]) can produce apparent K_m values nearing 100 μM at high concentrations. Cocaine causes decreases in dopamine release at high concentrations, which limit our ability to retain a sufficient signal size to quantify apparent K_m changes. Essentially, concentrations any higher than 30 μM cannot be used in slices because they abolish electrically stimulated release. Other drugs, however, do not produce such strong effects on release. For example, PTT is a high-affinity dopamine-uptake inhibitor created by Davies [2]. At a concentration of 2 μM, PTT appears to inhibit dopamine uptake completely without decreasing release. In fact, the peak height of the dopamine signal is increased markedly in the presence of PTT, probably due to diffusion of dopamine from distant sites. Since uptake is the primary determinant of how far dopamine can diffuse in the brain, inhibition of uptake allows greater diffusion and thereby detection of dopamine from a larger sampling area. This results in increased signal peak height. From previous work with dopamine transporter–knockout (DAT-KO) mice, we have determined the rate of dopamine clearance in the absence of uptake due to diffusion alone. Figure 5-1 shows a voltammetry recording of dopamine release and uptake in a slice containing the nucleus accumbens core from a DAT-KO mouse. Interestingly, in the presence of 2 μM PTT, the dopamine clearance rate was identical to that measured in DAT-KO mice. Such a magnitude of uptake inhibition was never seen with lower-affinity drugs such as cocaine or methylphenidate. We hypothesize that the differences between uptake inhibitors include not only affinity but also anesthetic and other drug properties, since we routinely see reductions in release with cocaine and methylphenidate but not with PTT or other drugs, such as GBR 12909.

5.4.2 Amphetamines

In contrast to the acute effects of uptake inhibitors, all releasers cause electrically stimulated dopamine release in brain slices to be reduced, and eventually abolished. This is due to the other well-known property of releasers such as amphetamine, which is to deplete synaptic vesicles of dopamine. The kinetics of dopamine uptake inhibition by amphetamines are complicated because there are two main effects, competitive uptake inhibition and transporter reversal. At low concentrations, the main effect of amphetamine is to compete with dopamine for the substrate site on the transporter, and the overflow curves can be fit in the same manner as pure uptake inhibitors, with an increase in the apparent K_m for dopamine uptake. At higher concentrations, amphetamine accumulates in the presynaptic terminal and causes reversal of the transporter (Fig. 5-4). This is due to several effects of amphetamine. First, amphetamine is a substrate for both the plasma membrane dopamine transporter and the vesicular monoamine transporter on synaptic vesicles, and so is taken into the presynaptic terminal and then into the vesicles. Once in the vesicles, amphetamine causes leakage of dopamine from the vesicles. One major theory about the mechanism of vesicular dopamine leakage is the weak base theory [3], which states that amphetamine acts as a weak base to disrupt the proton gradient across the vesicular membrane and thereby disrupt the functioning of the vesicular monoamine transporter. Without constant inward transport of dopamine into vesicles, dopamine leaks out into the cytoplasm of the presynaptic terminal. The accumulation of dopamine near the cytoplasmic side of the dopamine transporter sets the stage for reverse transport. As amphetamine continues to be taken up from the extracellular space, inward-facing transporters begin to bind and transport the cytoplasmic dopamine in the reverse direction into the extracellular space. This process is not specific to amphetamine, of course, and would occur in the presence of high extracellular concentrations of any dopamine transporter substrate, including dopamine itself. For this reason it is not advisable to apply exogenous dopamine to tissues in order to measure dopamine uptake. The kinetics of transporters that are dually functioning to take up substrate, as

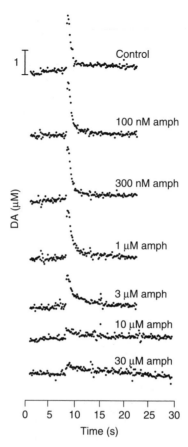

Figure 5-4 Dose–response curve for amphetamine in slices of core NAc. Similar to Figure 5-2, increasing concentrations of amphetamine applied cumulatively to a core NAc slice produce increasing amounts of uptake inhibition. At the higher doses of amphetamine, the voltammetric signal peak height is reduced, due to depletion of vesicular stores of releasable dopamine, and the signal does not return to baseline, indicating that a new equilibrium has been reached with the dopamine in the extracellular space.

well as release substrate, are complicated, and transporter kinetics under these conditions cannot be attributed to uptake alone (Fig. 5-5).

5.5   COCAINE SELF-ADMINISTRATION AND DOPAMINE TRANSPORTERS

In collaboration with David C. S. Roberts, we have examined the effects of high-dose, 24-hour-access cocaine self-administration in rats on dopamine uptake in the nucleus accumbens [4]. Rats had the opportunity to self-administer 1.5 mg/kg cocaine intravenously every 15 minutes, 24 hours a day, for 10 days. The average intake of cocaine

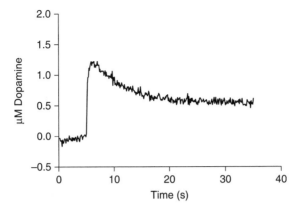

Figure 5-5 Representative dopamine uptake response to amphetamine. This dopamine over-flow curve was recorded in the presence of 3 μM amphetamine, a high dose that causes reversal of the transporter. Note that the signal does not return to baseline. This type of clearance curve cannot be explained purely in terms of uptake, since both uptake of extracellular DA and release of cytoplasmic DA are occurring.

per rat was approximately 800 mg/kg. DAT measurements were taken either 1 or 7 days after cessation of self-administration. Using a brain slice preparation, we electrically stimulated dopamine release in the core of the nucleus accumbens and measured the subsequent uptake of dopamine back into the presynaptic terminals and its inhibition by cocaine. As shown in Figure 5-6, we found that uptake rates were increased by approximately 50%, indicating either an increase in the number of dopamine transporters expressed on the surface of presynaptic membranes or an increase in the rate of uptake per transporter. Both phenomena have been documented, but chronic drug exposure has been shown multiple times to increase the expression of transporters [4–6]; therefore, it is most likely that the increased uptake is due to increased numbers of transporters. Rapidly increased V_{max} values for uptake have been shown with application of D2-type dopamine receptor agonists that activate dopamine autoreceptors [7–9] and during hyperpolarization of the membrane surrounding the transporters [10,11], reflecting the voltage dependency of the transport rate.

In addition to the increased rate of dopamine uptake, following cocaine self-administration it was found that cocaine showed a reduced ability to inhibit the dopamine transporter. These findings highlighted the utility of voltammetry to measure the effects of uptake inhibitors. It was found that the dose–response curves for cocaine to inhibit dopamine uptake were shifted to the right and showed reductions of 70% in the maximal inhibition achieved by cocaine. This is an unusual, if not unique finding and requires some explanation. In cocaine-exposed rats, cocaine was less potent (IC_{50} was shifted rightward) and less efficacious (maximal effect was reduced). This was not due to the documented change in rates of uptake in the absence of cocaine. The absolute rate of uptake is completely separated from the effects of an uptake inhibitor in these assays. Techniques such as in vivo microdialysis measure the effects of cocaine on extracellular dopamine levels, and

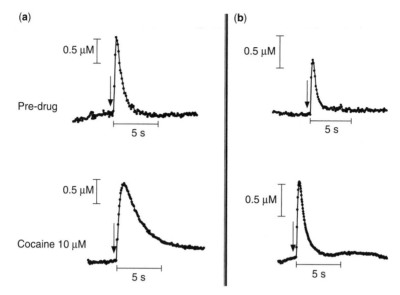

Figure 5-6 Effects of cocaine self-administration on dopamine uptake and sensitivity to cocaine. (a) Responses to single-pulse stimulation in brain slices containing the core NAc before drug is applied. Dopamine uptake is faster in naive rats than in rats that have self-administered cocaine (a 1-day-withdrawal animal is shown here). (b) Slowed clearance following cocaine, but the effect of 10 μM cocaine is less in rats that have self-administered cocaine. This demonstrates reduced sensitivity to cocaine at the transporter following chronic cocaine self-administration.

changes in the absolute rates of uptake as well as auto- and heteroreceptor activity and circuit-related alterations in neuronal activity are all factors in determining the effects of the drug. However, voltammetry allows the researcher to focus on cocaine effects on transport rate alone and to avoid confounding the inhibition effects with baseline rate. Thus, the novel finding of decreased cocaine activity indicates that the DAT is altered at the protein level to be less sensitive. Such alterations have been documented many times in G-protein and ion-channel receptors. Desensitization of receptors is a common occurrence but has not been documented before for DATs. The application of voltammetry techniques allowed the alteration in intrinsic sensitivity of the DAT to cocaine to be documented.

5.6 DAT INVOLVEMENT IN PSYCHOSTIMULANT REINFORCEMENT

Cocaine and amphetamine are believed to produce their stimulating and rewarding effects by enhancing extracellular dopamine concentrations in specific brain areas, including the caudate–putamen (CP) and nucleus accumbens (NAc). These drugs elevate dopamine in these regions by blocking the DAT. In fact, a high degree of correlation was found between the potency of cocainelike drugs as inhibitors of dopamine and their propensity to be self-administered [12]. Surprisingly, in knockout mice with a genetic deletion of the DAT, cocaine- and amphetamine-conditioned place preference [13–15] and self-administration of cocaine [16] were still found. At the same time, the stimulating effects of these drugs were completely eliminated [17]. These findings could argue against a primary role of dopamine in cocaine and amphetamine reinforcement. However, microdialysis studies from Carboni's group in Italy [18] have demonstrated that in the absence of the DAT, cocaine and amphetamine may still increase the levels of accumbal dopamine. These researchers have hypothesized that in the absence of the DAT, the norepinephrine transporter (NET) could provide an alternative uptake site for dopamine clearance. Although in normal mice, the NET does not take up dopamine in the NAc, uptake of dopamine via the NET may be a compensatory mechanism that takes place in the NAc of these mutants [18]. This hypothesis was supported by the finding that reboxetine, a NET inhibitor, increased dialysate dopamine levels in the NAc of these mice but not in wild-type control mice. However, extracellular dopamine is regulated by multiple factors, including release, uptake, and metabolism. Therefore, further studies were necessary to test this hypothesis. Since fast-scan cyclic voltammetry (FSCV) is uniquely suitable for evaluating dopamine uptake changes, this technique was used to further investigate this important question. In contrast to expectations, no evidence of alternative uptake was found when dopamine clearance was measured in NAc slices from DAT-KO mice [19]. In fact, the clearance rate of dopamine was indistinguishable between CP and the core and shell of NAc. More important, desipramine, a potent NET inhibitor, was not able to change dopamine clearance in the shell NAc of the KO mice, similar to findings in wild-type mice. Identical results were obtained with fluoxetine, a serotonin-transporter (SERT) inhibitor. Therefore, the voltammetric studies did not support the possibility that after genetic deletion of the DAT, the NET or SERT actively clear dopamine in the NAc. This is in agreement with the fact that cocaine, which inhibits transport at the DAT, NET, and SERT, was ineffective in inhibiting dopamine clearance in the shell and core of the NAc in DAT-KO mice (see Fig. 5-1) [19,20].

More recently, Carboni's group tried to obtain more evidence regarding the hypothesis that dopamine can be captured through NET sites in the NAc shell. To this end, they studied the effect of single and combined administration of DAT and NET blockers, GBR 12909 and reboxetine, respectively, on dopamine output in several brain regions of freely moving rats [21]. Interestingly, administration of reboxetine 20 minutes after the administration of GBR 12909 [10 mg/kg intraperitoneally (i.p.)], a selective blocker of the DAT, produced an increase in dopamine output in the NAc shell (+400% above basal) greater than that obtained by GBR 12909 alone (+300% above basal). On the contrary, reboxetine did not further increase the dopamine output produced by GBR 12909 in the NAc core or in the dorsal caudate, areas lacking concentrated norepinephrine innervations. A conclusion was made that in the situation of temporary dopamine transporter inactivation obtained by pharmacological blockade, dopamine can be cleared from the extracellular space by NET.

Therefore, compensatory dopamine reuptake by NET could take place in the NAc shell of the DAT-KO mice [18]. However, Carboni's conclusions are based on the assumption that GBR 12909 can completely block the striatal and accumbal DAT at a dose of 10 mg/kg, since a higher dose of GBR 12909 (20 mg/kg, i.p.) did not produce a further increase in extracellular dopamine concentration in these regions. In fact, it is unlikely that these GBR 12909 doses (10 and 20 mg/kg) can induce total DAT inhibition. Figure 5-7 shows the effect of GBR 12909 on dopamine uptake in rat NAc in comparison with the effect of complete genetic deletion of the DAT. Certainly, total dopamine-uptake inhibition in rat NAc is not observed following this GBR 12909 dose. Therefore, it is unclear why an additional increase in extracellular dopamine was found when GBR 12909 was combined with reboxetine [21].

At the present time, we cannot completely rule out the possibility that NET or SERT in the NAc shell may provide a minor clearance mechanism for dopamine that is masked by diffusion in the mutants. However, voltammetric data rule out the involvement of compensatory uptake of accumbal dopamine via other mono-amine transporters in the cocaine-induced dopamine increase in the NAc. Recently, it was demonstrated that modulation of the serotonergic system in the ventral tegmental area could explain the unexpected increase in accumbal dopamine following cocaine and amphetamine administration in mice with a genetic deletion of the DAT [13,14]. These findings suggest a role for the mesolimbic dopamine system in the reinforcing effects of cocaine and other psychostimulants, even in the situation where no DAT is present. One serious caveat, however, is that it is uncertain whether this drug-induced reinforcement in mutants who lack a DAT can be compared to that in normal animals. More important, cocaine, amphetamine, and fluoxetine actually lead to a marked calming effect in DAT-KO mice; and this effect has also been

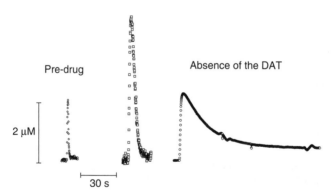

Figure 5-7 Changes in accumbal dopamine clearance following intraperitoneal adminis-tration of GBR 12909 and complete genetic deletion of the dopamine transporter. Dopamine efflux was measured following an experimenter-delivered stimulus train (1 s, 60 Hz, 300 μA) in the NAc of an anesthetized rat before and 15 minutes after GBR 12909 injection (first two signals), and in the NAc of an anesthetized dopamine transporter knockout mouse (third signal). The clearance rate of dopamine is markedly longer in the mutant mouse than the effect of GBR 12909 in the rat.

shown to be mediated by activation of the brain serotonergic system [17]. Therefore, it was speculated that perhaps these animals, suffering from chronic hyperactivity and restlessness, find SERT inhibition caused by these drugs pleasant, because of its soothing effect rather than experiencing further reward in a classical sense [22]. In line with this speculation, mice lacking both dopamine transporter and SERT no longer display cocaine place preference [23]. Moreover, generation of a mouse line carrying a mutant DAT (L104V/F105C/A109V) that is functional but insensitive to cocaine inhibition has been used to demonstrate that blockade of the DAT is required for cocaine reward in mice with a functional DAT [24]. Therefore, the dopamine transporter is the most important target for cocaine- and amphetamine-induced rewarding effects in normal animals.

5.7 DAT INVOLVEMENT IN VARIATIONS IN REGIONAL DOPAMINE DYNAMICS

The regulation of functional levels of dopamine neurotransmission in separate brain regions was the main focus for many research groups in the early 1990s. A good deal of biochemical, pharmacological, and electrophysiological evidence was obtained that subsets of the mesotelencephalic dopamine system display diverse characteristics. However, a detailed examination of the delicate equilibrium among dopamine release, uptake, metabolism, and diffusion became possible when voltammetry was applied [25–28]. In fact, meaningful evaluation of extracellular neurotransmitter dynamics requires the capability to make spatially and temporally resolved measurements. The temporal resolution of voltammetry and the spatial resolution afforded by carbon-fiber microelectrodes have allowed this goal to be achieved. Data that were obtained by voltammetric techniques have expanded and modified our understanding of the regional regulation of dopamine neurotransmission in the brain. In vivo, time-resolved measurements revealed that the instantaneous concentration of extracellular dopamine is a result of the balance between its release into the synaptic cleft and reuptake by the dopamine transporter. At the same time, regulation through metabolism takes place at a sufficiently slow rate that it does not affect the time course of extracellular dopamine measured by voltammetry following stimulation [29]. The regulation of extracellular dopamine concentrations was examined in vivo with voltammetry in the medial prefrontal cortex (MPFC), basal lateral amygdaloid nucleus (BAN), caudate–putamen (CP), and nucleus accumbens (NAc). Transient electrical stimulation of ascending dopamine fibers in a near-physiological range of frequencies (10 to 20 Hz) elicited similar levels of extracellular dopamine in the MPFC, BAN, CP, and NAc. This occurred despite the documented 90-fold disparity in dopamine tissue levels and terminal density. However, marked differences were observed in the dynamics and overall frequency dependence of the synaptic overflow of dopamine evoked. These differences are due to the significantly different rates of DA uptake and release found in each of the four regions. For example, rate constants for the release and uptake of dopamine were similar in the MPFC and BAN but approximately 8 and 50 times less, respectively, than those in the CP and NAc.

When the parameters were normalized to endogenous dopamine tissue content, a unique picture emerged: Compared to all other regions, relative release was 10-fold greater in the MPFC, while relative uptake was at least 10 times less in the BAN. These regional differences could have far-reaching consequences when the effects of drugs are considered. For example, inhibition of a DAT in a region where it tightly controls extracellular dopamine concentration, such as the striatum, would have a greater effect than in a brain region where the dopamine transporter does not regulate extracellular dopamine levels as closely, such as the BAN. Such effects have been observed for cocaine [30]. Surprisingly, in vivo studies demonstrated that cocaine increases extracellular dopamine to a greater extent in the limbic-related NAc than in the sensorimotor-related CP [31,32], which receives the densest dopaminergic innervation in the brain. The preferential effect of cocaine is additionally unexpected because the drug exhibits a similar potency for binding to the DAT and for inhibiting dopamine uptake in the two regions [33]. Whether the preferential effect of cocaine was mediated by regional differences in the presynaptic control of extracellular dopamine was examined by Wu and co-workers [32]. The combination of the lower rates for dopamine uptake and relatively large release was found to underlie the preferential increase in extracellular dopamine after cocaine in the NAc. This relationship explains the paradox that cocaine increases accumbal dopamine more effectively despite identical effects on the dopamine transporter in the two regions. This mechanism may also mediate the differential effects of psychostimulants in extrastriatal regions and other dopamine-uptake inhibitors in the striatum. These findings further differentiated the functional characteristics of the mesotelencephalic systems and revealed an essential impact of uptake via the dopamine transporter in the region-specific nature of dopamine neurotransmission in the brain.

5.8 FAST ONSET OF DAT INHIBITION IN THE BRAIN FOLLOWING INTRAVENOUS COCAINE INFUSION

The ability of cocaine to inhibit the DAT has been highlighted as the most important mechanism underlying its acute reinforcing effects [34]. DAT blockade appears to be responsible for the observed enhancement of extracellular dopamine that follows cocaine administration. However, it was unclear how quickly dopamine-uptake inhibition occurs following intravenous cocaine administration. In fact, increased locomotion, which may be correlated with increased accumbal dopamine, is observed in 6 to 10 s after intravenous (i.v.) cocaine administration [13,35]. However, estimates of the latency to a measurable neurochemical cocaine response vary widely. In vivo animal studies using microdialysis report that i.v. cocaine elevates dopamine levels beginning 2 to 5 minutes after infusion [36]. At the same time, electrophysiological studies have shown that dopamine cell inhibition occurs with a latency of tens of seconds [37]. This inhibition may result from rapid DAT blockade, elevation of dopamine levels, and consequent activation of autoreceptors. Therefore, there was disagreement between behavioral and neurochemical studies concerning the time course of cocaine effects. Because of the high temporal resolution of voltammetry, the processes of dopamine release and uptake could be temporally separated to make evaluation of cocaine effects on uptake easier to address. Also important, voltammetry is a technique that provides a rapid sampling rate (milliseconds) and therefore is suitable for evaluation of dopamine uptake changes over short time periods. These advantages allowed us to study the effects of i.v. cocaine (1.5 mg/kg) on dopamine dynamics in the NAc of anesthetized rats [13]. Anesthetized animals were chosen to avoid potentially confounding behavioral and conditioned influences, which are known to affect dopamine neurotransmission. We found that i.v. cocaine begins inhibiting dopamine uptake in rat NAc within 4 to 6 seconds after infusion (see Fig. 5-8). The peak of uptake inhibition was reached in approximately 1 minute (Fig. 5-9), which is consistent with the time when cocaine reaches peak brain levels [38]. In addition, uptake inhibition and cocaine levels are returned to normal in roughly 1 hour [38]. Importantly, changes in the peak height of dopamine overflow following stimulation were on a similar time scale to the alteration in apparent K_m for dopamine uptake (Fig. 5-10). This suggests that the effect of cocaine on dopamine peak height is a direct consequence of the decrease in dopamine-uptake rate.

There were several other voltammetric studies which demonstrated that cocaine exerts its effects on dopamine uptake within several seconds after its i.v. administration, and this effect peaked within 60 to 100 s [39,40]. However, some previous reports using voltammetry with a similar time resolution to the present results, combined with iontophoretically applied dopamine into the NAc of freely moving rats,

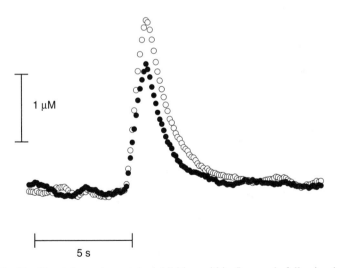

Figure 5-8 Significant dopamine-uptake inhibition within 5 seconds following intravenous cocaine administration. Electrically stimulated (1 s, 60 Hz, 300 μA) dopamine overflow measured by FSCV in the NAc of an anesthetized rat before (solid circles) and 5 seconds after cocaine (1.5 mg/kg) administration (open circles). The curves are arranged so that the descending portion of the curves align, to show that the clearance rate of dopamine is different after cocaine administration.

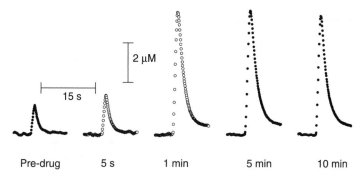

Figure 5-9 Maximal dopamine inhibition within 1 minute following intravenous cocaine administration. The peak of uptake inhibition in rat NAc was reached in 1 minute, which is consistent with the time when cocaine reaches peak brain levels.

found that dopamine uptake inhibition began at 2 minutes and peaked 6 to 7 minutes later, while behavioral activation occurred within seconds after i.v. cocaine (1 mg/kg) administration [35]. In contrast to our investigation, in the previous study, the clearance rate of exogenous dopamine, which was applied by iontophoresis at high concentrations directly in the NAc, was measured [35]. When high concentrations of exogenous dopamine are used, the ability of the transporter to function normally is compromised, as discussed in Section 5.4.2. In this situation the transporter takes up dopamine into the presynaptic terminal, but vesicular uptake and storage of dopamine quickly become saturated and the concentration of dopamine on the inside of the terminal increases dramatically. Under these conditions the DAT moves dopamine in both the forward and reverse directions. The kinetic

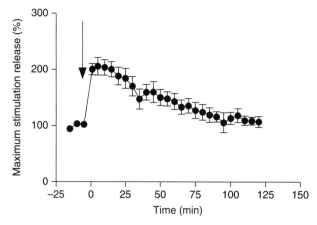

Figure 5-10 Time course of the effect of cocaine on dopamine peak height in rat NAc monitored by FSCV. Changes in dopamine peak height tightly follow a cocaine-induced dopamine uptake inhibiting effect. Data are means $\pm$ SEM ($n = 6$). Saline infusion did not change evoked dopamine height (data not shown).

analysis of uptake during this dual process is complicated and can be roughly equated with dopamine transport in the presence of high concentrations of amphetamine, another substrate for the DAT [41]. This represents a major difference between the former study and our investigations, where uptake of endogenously released dopamine, which was induced by electrical stimulation of dopamine cell bodies in the ventral tegmental area, was measured [13,40].

In summary, the higher temporal resolution of voltammetry allows the time course of DAT inhibition to be determined more precisely. This is particularly helpful when DAT inhibitors are injected intravenously and therefore reach the brain much faster than by other routes of administration. Here, we have shown that i.v.-administered cocaine begins inhibiting the DAT in the NAc within 4 to 6 seconds after infusion. This is at least 10-fold faster than neurochemical estimates published previously. Defining the time frame of the pharmacological effects of DAT inhibition may be a start toward understanding the variables that interact to produce the rewarding subjective effects of addictive drugs.

5.9 CORRELATION OF BEHAVIORAL ACTIVATION WITH DAT INHIBITION

Several lines of evidence suggest that inhibition of the DAT and consequent elevation in extracellular dopamine concentration are important for psychomotor activation such as a stereotypical behavior, and henceforth cocaine and other DAT inhibitors induce such behavioral enhancement. However, due to limited temporal resolution of available techniques, a clear relationship between the drug-induced changes in dopaminergic activity and behavior has been very difficult to establish. For example, a poor correlation was found between dopamine levels measured by microdialysis and behavioral changes produced by psychomotor stimulants [42]. A possible explanation for the discrepancies between drug evaluation by microdialysis and behavior is that dialysate dopamine does not precisely characterize time-dependent changes in extracellular dopamine. To test this hypothesis, we compared the effects of GBR 12909 (20 mg/kg, i.p.), a selective dopamine-uptake inhibitor, on dialysate dopamine and electrically evoked dopamine measured by FSCV. A close temporal association was observed between the increase of striatal dopamine detected by voltammetry and stereotypical behavior following administration of the dopamine inhibitor (Fig. 5-11). In contrast, dialysate dopamine correlated poorly, reaching a plateau approximately 40 minutes after that of the behavioral changes. It is important to note that our findings with GBR 12909 are consistent with previous microdialysis results obtained with similar doses, both in terms of potency and efficacy rate (maximum effect in 1 hour) [43,44]. Therefore, the results from microdialysis measurements notably lag behind the behavioral changes and voltammetric detection of striatal dopamine. This is not surprising when the processes that occur during microdialysis are considered. To be measured, dopamine has to diffuse through tissue to reach the probe, undergo partition across the dialysis membrane, and then undergo transit from the probe to the exit tubing. All of these processes

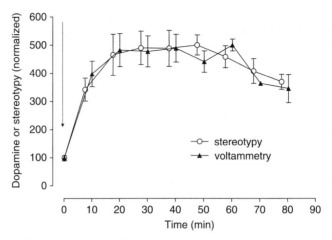

Figure 5-11 Changes in evoked dopamine overflow and stereotyped behavior in rats following intraperitoneal administration of GBR 12909 (20 mg/kg). Electrically stimulated (1 s, 60 Hz, 120 μA) dopamine overflow measured by FSCV in the striatum of freely moving rats closely paralleled the time course of the increase in stereotypy after the drug. Both voltammetric and behavioral assessments are normalized to basal level. Stereotypy was monitored using the scale described by Murray and Waddington [55]. Data are means ± SEM ($n = 6$).

require time. Moreover, the microdialysis probe causes tissue damage, distorting extracellular dopamine dynamics in the sampled area relative to intact tissue. In contrast, many of these time-consuming and tissue-disturbing processes are eliminated when using the voltammetry approach. Indeed, the small size of the carbon-fiber microelectrode (diameter 6 μm) allows its placement closer to release sites with minimal tissue damage compared with microdialysis probes, and the measurement occurs when dopamine encounters the electrode.

The relationship between drug-induced changes in DAT function and behavior was also evaluated following the systemic administration of nomifensine (7 mg/kg, i.p.), another dopamine-uptake inhibitor [45]. The positive correlation between the pattern of motor activation following nomifensine and extracellular dopamine was remarkable. It appears that increased basal levels of extracellular dopamine produced by nomifensine-induced blockade of DAT lead to dopamine receptor stimulation and ultimately, motor activation. The correlation between the evoked signal measured by voltammetry and behavior is therefore strong, because the presynaptic dopamine transporter activity is measured in real time. An important concern for this experiment was whether electrical stimulation, which must be applied to quantify altered DAT function, affects the drug-induced behavior under study. To determine potential artifacts of the electrical stimulation, drug-induced behavior with and without application of the pulse train was compared. Nomifensine-induced motor activation was almost identical in animals with simultaneous voltammetric measurements of dopamine and animals that were not surgically prepared for voltammetric detection [45]. Therefore, electrical stimulation has a negligible impact on the time course of drug-induced motor activation.

There was some evidence that the most reinforcing substance produces a rapid and short-lasting state of euphoria as well as psychomotor activation. However, whether the time course of DAT inhibition is essential for this difference was not clear. To address this issue, the effects of GBR 12909 (10 mg/kg, i.p.) and cocaine (10 mg/kg, i.p.) on DAT activity and behavior were compared using FSCV [46]. In fact, cocaine and GBR 12909 both increased evoked dopamine overflow, but on different time scales. Following cocaine administration, dopamine overflow increased to 330% of the pre-drug value within the first 4 minutes and returned to baseline values in 90 minutes. For GBR 12909, dopamine overflow increased to 340% at a more modest rate (19 minutes) and returned to the baseline value after about 300 minutes. Following administration of either drug, there was a marked increase in stereotypical behavior. Stereotypy scores increased from 0.6 to 2.4 after cocaine administration and from 0.5 to 2.5 for GBR 12909; however, the time course was again slower for GBR 12909. Throughout the duration of the experiment, the changes in stereotypy scores were perfectly temporally coincident with DAT inhibition. Indeed, when the change in stereotypic activity was plotted against the change in dopamine overflow, the correlation coefficients obtained were $r = 0.90$ and $r = 0.78$ for cocaine and GBR 12909, respectively. Therefore, following intra-peritoneal administration, cocaine has a faster time course than GBR 12909 on DAT inhibition and on behavior; and changes in these variables are temporally correlated on a minute time scale for each drug. The fact that GBR 12909 has a slower onset and longer action on DAT efficacy is in agreement with the lesser reinforcing properties of GBR 12909.

Taken together, the data emphasize that a dopamine transporter is an essential target for the manifestation of psychomotor actions. This finding was made possible by a voltammetric microprobe with millisecond temporal resolution and its use in the awake animal to assess dopamine uptake, an important mechanism of dopamine neurotransmission.

5.10   DAT AND PHASIC DOPAMINE NEUROTRANSMISSION

Extracellular dopamine is known to naturally reach high concentrations (transients) during several seconds with subsequent clearance via DAT. This is described as phasic dopamine activity or subsecond dopamine release [47–53], and its origin is not clearly understood. One hypothesis is that these dopamine transients arise from concerted burst firing of dopamine neurons, since such firing patterns are predicted to generate the brief, high dopamine concentrations (around 100 to 200 nM) that transients exhibit. The first FSCV recordings of such neurochemical events were performed by Rebec et al. [49]. These data demonstrated the regional heterogeneity of phasic dopamine response to novelty. In fact, results clearly distinguished accumbal core from shell and shore. The sheer magnitude of the dopamine response in the shell suggests that this area is extremely sensitive to entry into novelty. The dopamine response was limited to the brief time in which entry into novelty occurred, signifying that the DAT is an important player in novelty-related dopamine changes.

Later, dopamine transients were measured in the NAc core during sexual behavior in male rats by Robinson et al. [51]. Large (200 to 500 nmol/L), transient (<1 second) dopamine signals were associated with the introduction of a receptive female into the test cage as well as with subsequent sexual behaviors, such as approaching and sniffing the female. However, it is likely that additional transients were below the detection limits, since when dopamine uptake was inhibited, the number of transients detected was increased greatly. The dopamine signals associated with conspecific interactions were detected in the CP, NAc core and shell, and olfactory tubercle [50].

The increase in phasic dopamine activity was also observed in the NAc core during cocaine self-administration [48]. Dopamine transients were tightly associated with the lever-press response for cocaine as well as with the cues predicting cocaine infusion in cocaine-experienced rats. Notably, the initiation of cocaine-seeking behaviors could be induced by electrically evoking dopamine release in these experiments. Together, these findings are consistent with theories of the role of phasic dopamine neurotransmission as an alerting and/or switching signal that may be important in reinforcement and learning.

The functional significance of phasic dopamine and the role of the DAT in this transmission were explored further using pharmacological approaches. The findings that nomifensine and cocaine significantly enhance the amplitude and duration of dopamine transients [53] strongly support the important role of the DAT in the regulation of subsecond dopamine neurotransmission. The histograms of the distributions before and after nomifensine are consistent with the interpretation that the uptake blocker amplifies dopamine signals that were previously undetected. On the other hand, if the only effect of nomifensine was to amplify dopamine signals, similar increases in the frequency of dopamine transients would be expected in both the NAc and olfactory tubercle (OT), as its effect on electrically stimulated release was very similar between the two nuclei. Despite this, the number of dopamine transients increased after nomifensine by 570% in the NAc but only by 130% in the OT. One interpretation of this discrepancy is that nomifensine facilitates the occurrence of additional dopamine transients in the NAc by increasing the firing rate of the dopaminergic neurons innervating that nucleus or by an impulse-independent mechanism [53]. More important, locomotor activity after nomifensine correlated with the rate of dopamine transients only in the NAc. Since the frequency of dopamine transients following nomifensine administration was different between the two nuclei, it is possible that the additional dopaminergic activity in the NAc mediated the expression of stereotyped locomotion [53]. In any case, further research is required to verify this hypothesis.

The capacity of dopaminergic neurons to generate dopamine transients is dramatically impaired in anesthetized animals [54]. However, following inhibition of the dopamine transporter, dopamine transients appeared in the striatum in anesthetized preparations. For example, marked dopamine transients occurred shortly after intravenous cocaine administration in anesthetized rats (Fig. 5-12). Moreover, following combination of the dopamine uptake inhibition and blockade of dopamine receptors,

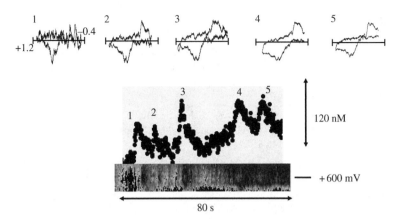

Figure 5-12 Cocaine-induced dopamine transients in the NAc of an anesthetized rat. Representative color plots (below), concentration–time plots (above), and cyclic voltammograms of dopamine (insets) 1 minute following cocaine infusion (1.5 mg/kg, i.v.). There was a significant increase in the frequency and amplitude of dopamine transients after cocaine compared to baseline. (*See insert for color representation of figure.*)

dopamine transients appeared in the striatum that qualitatively resembled those in freely moving animals following the blockade of dopamine uptake [53,54]. These data suggest that dramatically decreased dopamine phasic activity under anesthesia is due to an overactivation of D2 dopamine receptors and other mechanisms that inhibit cell firing. This is consistent with the current hypothesis that the dopamine transients are a product of burst firing of dopamine neurons and consequential uptake via the DAT.

5.11 CONCLUSIONS

Voltammetry is a quickly growing technique used in many different fields of science. It is particularly useful for evaluating the kinetics of dopamine uptake via the dopamine transporter, due to its high temporal resolution. The small size of the microelectrode probes used offers advantages for recording from small brain nuclei and in minimizing tissue damage during insertion in vivo. Sophisticated kinetic models are available for analyzing dopamine uptake data in the presence of inhibitors and following pharmacological treatment or self-administration. Overall, voltammetry is a highly useful technique for examining the details of dopamine transport.

ACKNOWLEDGMENTS

We would like to thank Dr. Marc G. Caron for the gift of DAT-KO mice. This work was supported by NIH grants AAO14091 and DA018815 to S.R.J. and DA021634 to E.A.B.

REFERENCES

1. Letchworth, S. R., Smith, H. R., Porrino, L. J., Bennett, B. A., Davies, H. M., Sexton T., and Childers, S. R. (2000). Characterization of a tropane radioligand, [(3)H]2β-propanoyl-3β-(4-tolyl)tropane ([(3)H]PTT), for dopamine transport sites in rat brain. *J Pharmacol Exp Ther*, *293*, 686–696.

2. Porrino, L. J., Migliarese, K., Davies, H. M., Saikali, E., and Childers, S. R. (1994). Behavioral effects of the novel tropane analog, 2β-propanoyl-3β-(4-toluyl)tropane (PTT). *Life Sci*, *54*, PL511–517.

3. Sulzer, D., Pothos, E., Sung, H. M., Maidment, N. T., Hoebel, B. G., and Rayport, S. (1992). Weak base model of amphetamine action. *Ann N Y Acad Sci*, *654*, 525–528.

4. Mateo, Y., Lack, C. M., Morgan, D., Roberts, D. C., and Jones, S. R. (2005). Reduced dopamine terminal function and insensitivity to cocaine following cocaine binge self-administration and deprivation. *Neuropsychopharmacology*, *30*, 1455–1463.

5. Letchworth, S. R., Nader, M. A., Smith, H. R., Friedman, D. P., and Porrino, L. J. (2001). Progression of changes in dopamine transporter binding site density as a result of cocaine self-administration in rhesus monkeys. *J Neurosci*, *21*, 2799–2807.

6. Mash, D. C., Pablo, J., Ouyang, Q., Hearn, W. L., and Izenwasser, S. (2002). Dopamine transport function is elevated in cocaine users. *J Neurochem*, *81*, 292–300.

7. Bolan, E. A., Kivell, B., Jaligam, V., Oz, M., Jayanthi, L. D., Han, Y., Sen, N., Urizar, E., Gomes, I., Devi, L. A., Ramamoorthy, S., Javitch, J. A., Zapata, A., and Shippenberg, T. S. (2007). D2 receptors regulate dopamine transporter function via an ERK 1/2–dependent and PI3 kinase–independent mechanism. *Mol Pharmacol*, *71*, 1222–1232.

8. Mayfield, R. D., and Zahniser, N. R. (2001). Dopamine D2 receptor regulation of the dopamine transporter expressed in *Xenopus laevis* oocytes is voltage-independent. *Mol Pharmacol*, *59*, 113–121.

9. Wu, Q., Reith, M. E., Walker, Q. D., Kuhn, C. M., Carroll, F. I., and Garris, P. A. (2002). Concurrent autoreceptor-mediated control of dopamine release and uptake during neurotransmission: an in vivo voltammetric study. *J Neurosci*, *22*, 6272–6281.

10. Ingram, S. L., Prasad, B. M., and Amara, S. G. (2002). Dopamine transporter–mediated conductances increase excitability of midbrain dopamine neurons. *Nat Neurosci*, *5*, 971–978.

11. Sonders, M. S., Zhu, S. J., Zahniser, N. R., Kavanaugh, M. P., and Amara, S. G. (1997). Multiple ionic conductances of the human dopamine transporter: the actions of dopamine and psychostimulants. *J Neurosci*, *17*, 960–974.

12. Ritz, M. C., Lamb, R. J., Goldberg, S. R., and Kuhar, M. J. (1987). Cocaine receptors on dopamine transporters are related to self-administration of cocaine. *Science*, *237*, 1219–1223.

13. Budygin, E. A., Brodie, M. S., Sotnikova, T. D., Mateo, Y., John, C. E., Cyr, M., Gainetdinov, R. R., and Jones, S. R. (2004). Dissociation of rewarding and dopamine transporter-mediated properties of amphetamine. *Proc Natl Acad Sci U S A*, *101*, 7781–7786.

14. Mateo, Y., Budygin, E. A., John, C. E., and Jones, S. R. (2004). Role of serotonin in cocaine effects in mice with reduced dopamine transporter function. *Proc Natl Acad Sci U S A*, *101*, 372–377.

15. Sora, I., Wichems, C., Takahashi, N., Li, X. F., Zeng, Z., Revay, R., Lesch, K. P., Murphy, D. L., and Uhl, G. R. (1998). Cocaine reward models: conditioned place preference can be established in dopamine- and in serotonin-transporter knockout mice. *Proc Natl Acad Sci U S A*, *95*, 7699–7704.

16. Rocha, B. A., Fumagalli, F., Gainetdinov, R. R., Jones, S. R., Ator, R., Giros, B., Miller, G. W., and Caron, M. G. (1998). Cocaine self-administration in dopamine-transporter knockout mice. *Nat Neurosci*, *1*, 132–137.

17. Gainetdinov, R. R., Wetsel, W. C., Jones, S. R., Levin, E. D., Jaber, M., and Caron, M. G. (1999). Role of serotonin in the paradoxical calming effect of psychostimulants on hyperactivity. *Science*, *283*, 397–401.

18. Carboni, E., Spielewoy, C., Vacca, C., Nosten-Bertrand, M., Giros, B., and Di Chiara, G. (2001). Cocaine and amphetamine increase extracellular dopamine in the nucleus accumbens of mice lacking the dopamine transporter gene. *J Neurosci*, *21*(RC141), 141–144.

19. Budygin, E. A., John, C. E., Mateo, Y., and Jones, S. R. (2002). Lack of cocaine effect on dopamine clearance in the core and shell of the nucleus accumbens of dopamine transporter knock-out mice. *J Neurosci*, *22*(RC222).

20. Mateo, Y., Budygin, E. A., Morgan, D., Roberts, D. C., and Jones, S. R. (2004). Fast onset of dopamine uptake inhibition by intravenous cocaine. *Eur J Neurosci*, *20*, 2838–2842.

21. Carboni, E., Silvagni, A., Vacca, C., and Di Chiara, G. (2006). Cumulative effect of norepinephrine and dopamine carrier blockade on extracellular dopamine increase in the nucleus accumbens shell, bed nucleus of stria terminalis and prefrontal cortex. *J Neurochem*, *96*, 473–481.

22. Laakso, A., Mohn, A. R., Gainetdinov, R. R., and Caron, M. G. (2002). Experimental genetic approaches to addiction. *Neuron*, *36*, 213–228.

23. Sora, I., Hall, F. S., Andrews, A. M., Itokawa, M., Li, X. F., Wei, H. B., Wichems, C., Lesch, K. P., Murphy, D. L., and Uhl, G. R. (2001). Molecular mechanisms of cocaine reward: combined dopamine and serotonin transporter knockouts eliminate cocaine place preference. *Proc Natl Acad Sci U S A*, *98*, 5300–5305.

24. Chen, R., Tilley, M. R., Wei, H., Zhou, F., Zhou, F. M., Ching, S., Quan, N., Stephens, R. L., Hill, E. R., Nottoli, T., Han, D. D., and Gu, H. H. (2006). Abolished cocaine reward in mice with a cocaine-insensitive dopamine transporter. *Proc Natl Acad Sci U S A*, *103*, 9333–9338.

25. Garris, P. A., Ciolkowski, E. L., and Wightman, R. M. (1994). Heterogeneity of evoked dopamine overflow within the striatal and striatoamygdaloid regions. *Neuroscience*, *59*, 417–427.

26. Garris, P. A., and Wightman, R. M. (1994). In vivo voltammetric measurement of evoked extracellular dopamine in the rat basolateral amygdaloid nucleus. *J Physiol*, *478*(Pt 2), 239–249.

27. Garris, P. A., and Wightman, R. M. (1994). Different kinetics govern dopaminergic transmission in the amygdala, prefrontal cortex, and striatum: an in vivo voltammetric study. *J Neurosci*, *14*, 442–450.

28. Garris, P. A., and Wightman, R. M. (1995). Distinct pharmacological regulation of evoked dopamine efflux in the amygdala and striatum of the rat in vivo. *Synapse*, *20*, 269–279.

29. Wightman, R. M., Amatore, C., Engstrom, R. C., Hale, P. D., Kristensen, E. W., Kuhr, W. G., and May, L. J. (1988). Real-time characterization of dopamine overflow and uptake in the rat striatum. *Neuroscience*, *25*, 513–523.

30. Pontieri, F. E., Tanda, G., and Di Chiara, G. (1995). Intravenous cocaine, morphine, and amphetamine preferentially increase extracellular dopamine in the "shell" as compared with the "core" of the rat nucleus accumbens. *Proc Natl Acad Sci U S A, 92*, 12304–12308.

31. Carboni, E., Imperato, A., Perezzani, L., and Di Chiara, G. (1989). Amphetamine, cocaine, phencyclidine and nomifensine increase extracellular dopamine concentrations preferentially in the nucleus accumbens of freely moving rats. *Neuroscience, 28*, 653–661.

32. Wu, Q., Reith, M. E., Kuhar, M. J., Carroll, F. I., and Garris, P. A. (2001). Preferential increases in nucleus accumbens dopamine after systemic cocaine administration are caused by unique characteristics of dopamine neurotransmission. *J Neurosci, 21*, 6338–6347.

33. Jones, S. R., Garris, P. A., and Wightman, R. M. (1995). Different effects of cocaine and nomifensine on dopamine uptake in the caudate–putamen and nucleus accumbens. *J Pharmacol Exp Ther, 274*, 396–403.

34. Koob, G. F., and Bloom, F. E. (1988). Cellular and molecular mechanisms of drug dependence. *Science, 242*, 715–723.

35. Kiyatkin, E. A., Kiyatkin, D. E., and Rebec, G. V. (2000). Phasic inhibition of dopamine uptake in nucleus accumbens induced by intravenous cocaine in freely behaving rats. *Neuroscience, 98*, 729–741.

36. Ahmed, S. H., Lin, D., Koob, G. F., and Parsons, L. H. (2003). Escalation of cocaine self-administration does not depend on altered cocaine-induced nucleus accumbens dopamine levels. *J Neurochem, 86*, 102–113.

37. Hinerth, M. A., Collins, H. A., Baniecki, M., Hanson, R. N., and Waszczak, B. L. (2000). Novel in vivo electrophysiological assay for the effects of cocaine and putative "cocaine antagonists" on dopamine transporter activity of substantia nigra and ventral tegmental area dopamine neurons. *Synapse, 38*, 305–312.

38. Fowler, J. S., Volkow, N. D., Logan, J., Gatley, S. J., Pappas, N., King, P., Ding, Y. S., and Wang, G. J. (1998). Measuring dopamine transporter occupancy by cocaine in vivo: radiotracer considerations. *Synapse, 28*, 111–116.

39. Heien, M. L., Khan, A. S., Ariansen, J. L., Cheer, J. F., Phillips, P. E., Wassum, K. M., and Wightman, R. M. (2005). Real-time measurement of dopamine fluctuations after cocaine in the brain of behaving rats. *Proc Natl Acad Sci U S A, 102*, 10023–10028.

40. Samaha, A. N., Mallet, N., Ferguson, S. M., Gonon, F., and Robinson, T. E. (2004). The rate of cocaine administration alters gene regulation and behavioral plasticity: implications for addiction. *J Neurosci, 24*, 6362–6370.

41. Jones, S. R., Joseph, J. D., Barak, L. S., Caron, M. G., and Wightman, R. M. (1999). Dopamine neuronal transport kinetics and effects of amphetamine. *J Neurochem, 73*, 2406–2414.

42. Budygin, E. A., Kilpatrick, M. R., Gainetdinov, R. R., and Wightman, R. M. (2000). Correlation between behavior and extracellular dopamine levels in rat striatum: comparison of microdialysis and fast-scan cyclic voltammetry. *Neurosci Lett, 281*, 9–12.

43. Rothman, R. B., Mele, A., Reid, A. A., Akunne, H. C., Greig, N., Thurkauf, A., de Costa, B. R., Rice, K. C., and Pert, A. (1991). GBR 12909 antagonizes the ability of cocaine to elevate extracellular levels of dopamine. *Pharmacol Biochem Behav, 40*, 387–397.

44. Westerink, B. H., Damsma, G., De Vries, J. B., and Koning, H. (1987). Dopamine re-uptake inhibitors show inconsistent effects on the in vivo release of dopamine as measured by intracerebral dialysis in the rat. *Eur J Pharmacol, 135*, 123–128.

45. Garris, P. A., Budygin, E. A., Phillips, P. E., Venton, B. J., Robinson, D. L., Bergstrom, B. P., Rebec, G. V., and Wightman, R. M. (2003). A role for presynaptic mechanisms in the actions of nomifensine and haloperidol. *Neuroscience, 118*, 819–829.

46. Phillips, P. E., Budygin, E. A., Robinson, D. L., and Wightman, R. M. (2001). Correlation of stereotypy with evoked dopamine overflow following administration of uptake blockers. 270–271.

47. Phillips, P. E., Robinson, D. L., Stuber, G. D., Carelli, R. M., and Wightman, R. M. (2003). Real-time measurements of phasic changes in extracellular dopamine concentration in freely moving rats by fast-scan cyclic voltammetry. *Methods Mol Med, 79*, 443–464.

48. Phillips, P. E., Stuber, G. D., Heien, M. L., Wightman, R. M., and Carelli, R. M. (2003). Subsecond dopamine release promotes cocaine seeking. *Nature, 422*, 614–618.

49. Rebec, G. V., Christensen, J. R., Guerra, C., and Bardo, M. T. (1997). Regional and temporal differences in real-time dopamine efflux in the nucleus accumbens during free-choice novelty. *Brain Res, 776*, 61–67.

50. Robinson, D. L., Heien, M. L., and Wightman, R. M. (2002). Frequency of dopamine concentration transients increases in dorsal and ventral striatum of male rats during introduction of conspecifics. *J Neurosci, 22*, 10477–10486.

51. Robinson, D. L., Phillips, P. E., Budygin, E. A., Trafton, B. J., Garris, P. A., and Wightman, R. M. (2001). Sub-second changes in accumbal dopamine during sexual behavior in male rats. *Neuroreport, 12*, 2549–2552.

52. Robinson, D. L., Venton, B. J., Heien, M. L., and Wightman, R. M. (2003). Detecting subsecond dopamine release with fast-scan cyclic voltammetry in vivo. *Clin Chem, 49*, 1763–1773.

53. Robinson, D. L., and Wightman, R. M. (2004). Nomifensine amplifies subsecond dopamine signals in the ventral striatum of freely-moving rats. *J Neurochem, 90*, 894–903.

54. Venton, B. J., and Wightman, R. M. (2007). Pharmacologically induced, subsecond dopamine transients in the caudate–putamen of the anesthetized rat. *Synapse, 61*, 37–39.

PART II

MEDICINAL CHEMISTRY

6

TROPANE-BASED DOPAMINE TRANSPORTER–UPTAKE INHIBITORS

Scott P. Runyon and F. Ivy Carroll

Center for Organic and Medicinal Chemistry, Research Triangle Institute, Research Triangle Park, North Carolina

6.1 Introduction 125

6.2 Structure–Activity Studies 127
    6.2.1 N-Substitution 138
    6.2.2 Tropane Ring Modifications 140

6.3 Radioligands 142

6.4 Irreversible Binding and Fluorescent Ligands 147

6.5 Animal Behavioral and Clinical Studies 150

6.6 Conclusions 151

References 151

References for Table 6-4 164

6.1 INTRODUCTION

Cocaine (1) is one of the most commonly recognized tropane alkaloids, due in part to its significant abuse and dependence liability. Thus, substituted derivatives of cocaine have received much attention as potential pharmacotherapies for psychostimulant abuse. In 1973, Clark and co-workers set out to separate the stimulant and depressant

Dopamine Transporters: Chemistry, Biology, and Pharmacology. Edited by Mark L. Trudell and Sari Izenwasser
Copyright © 2008 John Wiley & Sons, Inc.

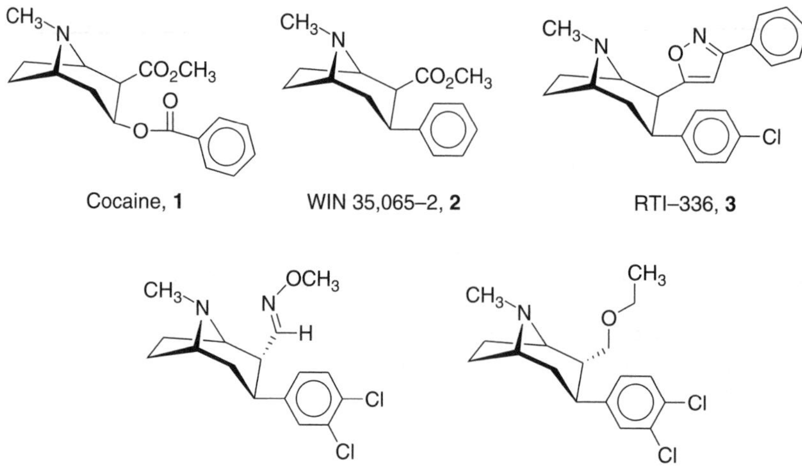

Cocaine, **1** WIN 35,065–2, **2** RTI–336, **3**

Brasofensine, **4** Tesofensine, **5**

Figure 6-1 Structures for 3-phenyltropane analogs **1**–**5**.

actions of cocaine from its toxicity and dependence liability [1,2]. The initial studies by Clark and colleagues resulted in the development of WIN 35,065-2 (**2**) (Fig. 6-1) [1]. WIN 35,065-2 ($IC_{50} = 23$ nM) is an analog of cocaine in which the benzoyl ester has been replaced by a phenyl group linked directly to the tropane ring and is fourfold more potent than cocaine ($IC_{50} = 89$ nM) at the dopamine transporter (DAT) [3]. Discovery of WIN 35,065-2 provided a new molecular template on which to further investigate the pharmacological properties of compounds related to cocaine.

The *Chemical Abstracts* name for WIN 35,065-2 is methyl [1*R*(exo,exo)]-8-methyl-3-phenyl-8-azabicyclo[3.2.1]octane-2-carboxylate. However, the 3β-phenyltropane-2β-carboxylic acid methyl ester nomenclature introduced initially by Clark et al. [2] is used in this chapter, and the compound class is referred to as 3-phenyltropanes. WIN 35,065-2 was derived originally from natural (−)-cocaine. Unless stated otherwise, all 3-phenyltropanes presented herein will possess the same absolute stereochemistry as that of (−)-cocaine and WIN 35,065-2.

Studies in 1992 strongly suggested that interaction of cocaine with the DAT was one of the molecular mechanisms responsible for the reinforcing effects of cocaine [4]. Since that time, studies with DAT-knock out mice [5] and positron emission tomography (PET) [6] further clarified the linkage between cocaine's physiological effects and DAT occupancy.

Once the molecular target for cocaine had been identified, numerous tropane-based ligands were synthesized for the purpose of selectively modulating the DAT and ultimately providing pharmacotherapies for psychostimulant abuse [4], depression, attention-deficit hyperactivity disorder (ADHD) [7,8], and Parkinson's disease [9–11]. Although no effective pharmacotherapy for cocaine abuse has been discovered to date, the development of tropane-based ligands has led to some of the most promising leads for treatment of substance abuse (RTI-336, **3**) [12], Parkinson's disease (brasofensine, **4**) [13], and obesity (tesofensine, **5**) [1] (Fig. 6-1).

Due to their increasing promise as pharmacotherapies, 3-phenyltropanes have been the focus of numerous structure–activity relationship (SAR) studies and drug discovery efforts. Most important, structural modification of the 2- and 3-positions has provided compounds with excellent potency at the DAT and selectivity over the serotonin transporter (5-HTT) and the norepinephrine transporter (NET). In addition, modification of the 6- and 7-positions, as well as altering the N-substituent, has led to the formulation of a comprehensive pharmacophore for 3-phenyltropanes and has facilitated a more comprehensive study of the DAT and its role in human disease. The subject of 3-phenyltropanes has been reviewed several times during the last 30 years [15–19]. In this chapter we summarize these prior studies and present additional SAR data since the last review.

6.2 STRUCTURE–ACTIVITY STUDIES

As with cocaine, WIN 35,065-2 has four sites of asymmetry and eight possible isomers. Prior studies with cocaine [20] suggested that stereochemistry at the 1-, 2-, and 3-positions was important for modulating the pharmacological properties of phenyltropanes. A series of substituted (1*R*)2-carbomethoxy-3-(4-substituted phenyl)tropanes encompassing all four 2- and 3-position stereoisomers were synthesized and evaluated for DAT binding, locomotor activity, and drug discrimination properties [21]. Interestingly, the relative order of affinity at the DAT for 3-phenyltropane isomers does not exactly parallel the rank order of affinity demonstrated for cocaine. In the cocaine series 2β,3β > 2β,3α > 2α,3β > 2α,3α [20], whereas the order of affinity for 3-phenyltropanes at the DAT is 2β,3β ≈ 2β,3α > 2α,3α > 2α,3β [21] (Fig. 6-2). 3-Phenyltropanes show minor differences in affinity between the 2β,3β and 2β,3α isomers, whereas in the cocaine series, affinity at the DAT varied 60-fold between the 2β,3β and 2β,3α isomers and as much as 600-fold between the 2β,3β, 2α,3β, and 2α,3α isomers. In both the cocaine and the 3-phenyltropane series, the 1*R* isomer is significantly more potent than the 1*S* isomer. These studies demonstrate that the relative stereochemistry at the 2- and

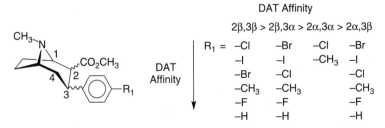

Figure 6-2 Structures for isomers of (1*R*)-2-carbomethoxy-3-(4-substituted phenyl)tropanes.

3-positions and the absolute configuration of 3-phenyltropanes are important contributors to DAT affinity.

Early SAR work [3,22–26] on 3-phenyltropanes incorporated a diverse set of $3'$- or $4'$-substituents on the aryl ring. Affinity for the DAT relative to WIN 35,065-2 was examined to gain information concerning the pharmacophore for the cocaine-binding site on the DAT. Substitution at the $4'$-position has received the most attention due in part to the critical role it plays in enhancing DAT affinity. The addition of $4'$-methyl, $4'$-fluoro, $4'$-chloro, $4'$-iodo, $4'$-ethylene, and $4'$-acetylene to WIN 35,065-2 ($IC_{50} = 23$ nM) provided a series of compounds with affinities of less than 2 nM at the DAT [16] and references cited therein (Table 6-1). Recently, Tamagnan and co-workers further explored the $4'$-position of 3β-phenyltropanes with various heterocyles [27]. This work identified **6l** and **6m** as potent inhibitors of the DAT with K_i of 6.4 and 4.4 nM, respectively. The most interesting aspect of this study was the selectivity of $4'$-heterocycle substituents for the 5-HTT. The $4'$-thiophene **6n** had K_i values of 0.017 nM at the 5-HTT and was 710- and 11,118-fold selective for the 5-HTT over the DAT and NET, respectively. It is also interesting that $4'$-substituted 3-phenyltropanes having very diverse steric and electronic parameters bind with moderate affinity at the DAT. Derivatives such as $4'$-nitro (**6h**), $4'$-hydroxy (**6j**), $4'$-(5-bromo-3-thiophene) (**6l**), $4'$-methoxy (**6k**), and $4'$-amino (**6i**) have IC_{50} values between 2 and 12 nM at the DAT [16] and references cited therein (Table 6-1).

The addition of $3'$-substituents to WIN 35,065-2 does not affect DAT binding affinity as significantly as does monosubstitution at the $4'$-position. For example, addition of halogens such as $3'$-bromo (**7b**) [23], $3'$-chloro (**7c**) [3], $3'$-fluoro (**7d**) [3], and $3'$-iodo (**7f**) [23] provided compounds with DAT IC_{50} values between 6 and 29 nM (Table 6-2). More recently, the tolerance for $3'$-position substitution on 3β-phenyltropanes has been examined by Bois and colleagues [28]. Introduction of $3'$-(3-thiophene) (**7e**), $3'$-(2-furanyl) (**7g**), and $3'$-phenyl (**7a**) provided compounds roughly four- to ninefold more potent than cocaine (Table 6-2). Although monosubstitution at the $3'$-position can enhance DAT affinity moderately in relation to WIN 35,065-2, $3',4'$-disubstitution has provided some of the most potent 3-phenyltropanes to date [25]. Substitution of the 3β-phenyl group with combinations of $3'$- and $4'$-chloro, -bromo, -iodo, or -methyl (**8a–n**) groups provides compounds with subnanomolar affinity at the DAT [29] (Table 6-3). The most potent derivative (**8h**) was nonselective among transporters, having an IC_{50} values of 0.12 nM at the DAT, and had K_i values of 0.23 and 0.65 nM at the 5-HTT and NET, respectively. Although $3'$-halogens such as chloro, iodo, and bromo are tolerated at the DAT, there appears to be a limited tolerance for $3'$-hydroxy substituents. Meltzer and colleagues prepared catechol derivative **9** to determine if the aryl group of 3-phenyltropanes could potentially interact with the DAT in a manner similar to the 3,4-dihydroxyphenyl group of dopamine **10** [30] (Fig. 6-3). The $4'$-hydroxy-substituted 3-phenyltropane **6j** had an IC_{50} value of 12.1 nM [3], while the $3',4'$-dihydroxy analog **9** was 113-fold less potent at the DAT. This suggests that the endogenous neurotransmitter dopamine does not occupy the same binding site on the DAT as that of 3β-phenyltropanes.

TABLE 6-1 DAT-Binding Affinity of 3β-(4′-Substituted Phenyl)tropane-2β-Carboxylic Acid Methyl Esters

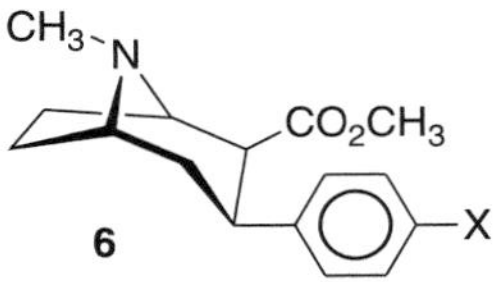

IC$_{50}$ Values

Compound	<2 nM, X =	Compound	2 to 12 nM, X =	Compound	12 to 30 nM, X =	Compound	30 to 200 nM, X =
a	CH$_3$	g	N$_3$	t	CF$_3$	aa	NHCOCH$_3$
b	Cl	h	NO$_2$	u	C(=CH$_2$)CH$_3$	bb	NHCOC$_2$H$_5$
c	Br	i	NH$_2$	v	cis-CH=CHCH$_3$	cc	NHCO$_2$C$_2$H$_5$
d	I	j	OH	w[a]	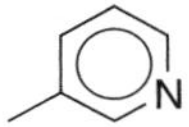 (pyridine, 3-methyl)	dd	Sn(CH$_3$)$_3$
e	—CH=CH$_2$	k	OCH$_3$	x[a]	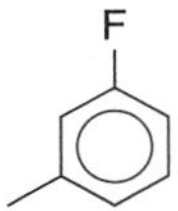 (3-fluoro-5-methylphenyl)	ee	CH$_2$CH$_3$
f	—C≡CH	l[a]	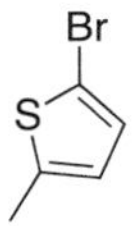 (5-bromo-2-methylthiophene)	y[a]	 (4-methylfuran)	ff	C$_3$H$_7$

(Continued)

TABLE 6.1 *Continued*

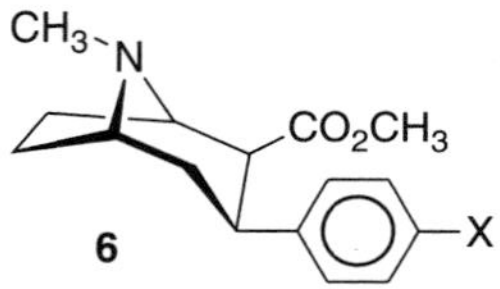

CH₃–N structure (compound **6**) with CO₂CH₃ and X-substituted phenyl group:

$$\text{6: } CH_3\text{-N tropane with } CO_2CH_3 \text{ and } p\text{-X-phenyl}$$

IC₅₀ Values

Compound	<2 nM, X =	Compound	2 to 12 nM, X =	Compound	12 to 30 nM, X =	Compound	30 to 200 nM, X =
		m[a]	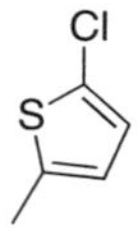 (5-chloro-2-methylthiophene)	**z**[a]	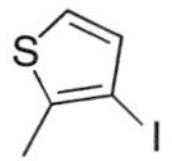 (3-iodo-2-methylthiophene)	**gg**	$CH(CH_3)_2$
		n[a]	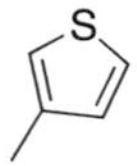 (3-methylthiophene)			**hh**	$CH_2CH=CH_2$
		o	F			**ii**[a]	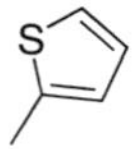 (2-methylthiophene)

p^a

q *trans-*
$CH_2=CHCH_3$

r $-C\equiv CCH_3$

s C_6H_5

jj^a

kk^a

ll^a

mm^a

[a] Data represents K_i values.

TABLE 6-2   DAT-Binding Affinity of 3β-(3'-Substituted Phenyl)tropane-2β-Carboxylic Acid Methyl Esters

	IC$_{50}$ Values				
	2 to 12 nM,		12 to 30 nM,		30 to 200 nM,
Compound	X =	Compound	X =	Compound	X =
a	C$_6$H$_5$	d	F	g^a	
b	Br	e^a			
c	Cl	f	I		

^{a}Data represent K_i values.

Considering the diversity of 4'-substituents that bind with moderate to high affinity at the DAT, several attempts have been made to develop quantitative structure–activity relationships (QSARs) and predictive molecular models to better describe ligand–receptor interactions at the DAT. A generalized three-point pharmacophore model was generated by Wang and colleagues for cocaine and 3-phenyltropane analogs that depicts the distances between the basic nitrogen and the centroid of the aryl ring (5.0 to 7.0 Å), the basic nitrogen and the 2-position carbonyl oxygen (2.2 to 4.5 Å), and the 2-position carbonyl oxygen and the centroid of the aromatic ring (3.4 to 6.1 Å). These distances were derived from all energetically accessible conformers of both cocaine and WIN 35,065-2 [31,32]. Models such as this may prove useful for future three-dimensional database searches. Molecular modeling investigations have also been used to identify potential correlations between substituents on the phenyl group of 3-phenyltropanes and their DAT affinity. CoMFA models were developed for a 25-compound set in which the substituents on the 3-phenyl ring were altered [3]. These models indicate a region of bulk tolerance extends from the 4'-position for a limited distance, while molecules with excessively large 4'-substituents had reduced potency. Electrostatic forces contributed marginally to binding (27%), although such contributions may be more important than described by this model.

Since the 2α-isomer of WIN 35,065-2 is significantly less potent at the DAT than the 2β-isomer, structural modifications have focused primarily on various 2β-derivatives [21]. The broad tolerance of the 2β-position to structural modification is evident through the diversity of analogs that have equal or greater affinity than

TABLE 6-3 DAT Binding Affinity of 3β-(3′,4′-Disubstituted Phenyl)tropane-2β-Carboxylic Acid Methyl Esters

IC$_{50}$ Values

Compound	<2 nM, X,Y =	Compound	<2 nM, X,Y =	Compound	<2 nM, X, Y =	Compound	2 to 12 nM, X, Y =
a	NH$_2$, I	f	Cl, Br	j	Br, I	o	N$_3$, I
b	Cl, Cl	g	Cl, I	k	I, Cl	p	F, CH$_3$
c	Cl, CH$_3$	h	Br, Cl	l	I, Br	q	NH$_2$, Br
d		i	Br, Br	m	I, I		
e	Cl, Cl			n	CH$_3$,CH$_3$		

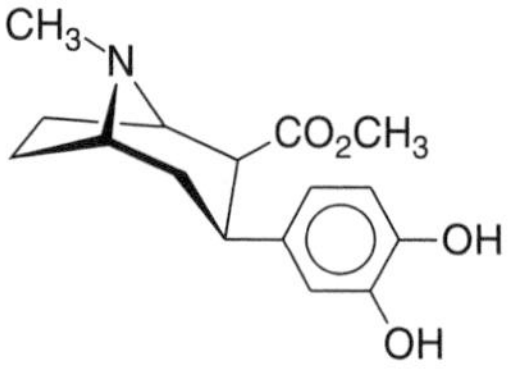

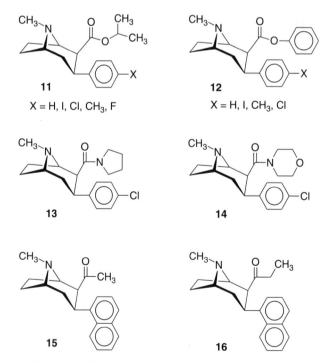

9 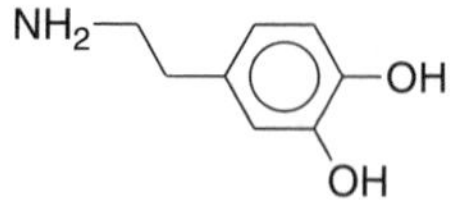   Dopamine, **10**

Figure 6-3 Structures of 3-phenyltropane analog **9** and dopamine **10**.

cocaine and WIN 35,065-2 at the DAT. Isopropyl (**11**) and phenyl esters (**12**) of various 4′-substituted 3-phenyltropanes are highly potent and tend to be more selective for the DAT relative to the NET and 5-HTT [33,34] (Fig. 6-4).

An SAR study of a diverse set of amides showed that this type of substituent is tolerated at the 2β-position of 3-phenyltropanes. In studies by Carroll and colleagues, 2β-tertiary amides were in general more potent than 2β-secondary amides [33,35]. Two notable examples are the 2β-*N*-morpholinocarboxamide (**13**) and the 2β-*N*-pyrrolidinocarboxamide (**14**), both of which are highly potent and selective at the DAT (Fig. 6-4).

Several 2β-ketones with potent DAT binding affinity have also been reported by Davies [36–38] and Bennett and colleagues [39]. 3β-4′-(2-Naphthylphenyl)-2β-acetyl (**15**) and 2β-propanoyltropanes (**16**) were the most potent in the series,

11

X = H, I, Cl, CH₃, F

12

X = H, I, CH₃, Cl

13

14

15

16

Figure 6-4 Structures for 3-phenyltropane analogs **11–16**.

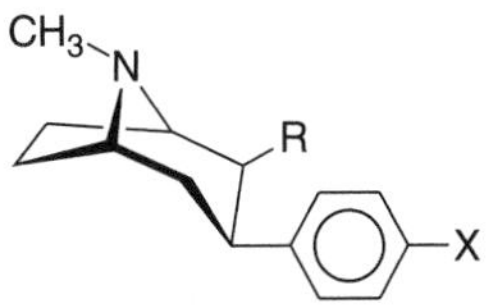

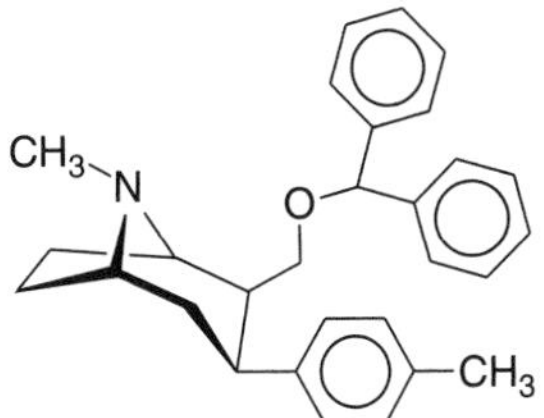

17a, R = –(CH₂)₂CH₃, X = H
  b, R = –(CH₂)₂CH₃, X = Cl
  c, R = –CH=CH₂, X = Cl
  d, R = –CH=CH–Cl, X = Cl
  e, R = –CH=CH–C₆H₅, X = Cl
  f, R = –(CH₂)₂C₆H₅, X = Cl
  g, R = –CH=CH–CO₂CH₃, X = H
  h, R = –CH=CH–CH₂–OH, X = H
  i, R = –CH₂CH₂CH₂–OH, X = H

Figure 6-5 Structures for 3-phenyltropane analogs **17a–i** and **18**.

having IC$_{50}$ values of 0.28 and 0.15 nM, respectively (Fig. 6-4). 2β-Alkyl and 2β-alkenyl 3-phenyltropanes have also been synthesized and shown to be potent inhibitors of the DAT [40–42] (Fig. 6-5). 2β-Propyl (**17a,b**), 2β-vinyl (**17c**), 2β-chlorovinyl (**17d**), 2β-phenylvinyl (**17e**), 2β-phenylethyl (**17f**), 2β-(2-carbomethoxyvinyl) (**17g**), 2β-(3-hydroxymethylvinyl) (**17h**), and 2β-(3-hydroxypropyl) (**17i**) provide compounds having potent DAT affinity when combined with the appropriately substituted 3-phenyl group.

   To better define the structural requirements for high affinity at the 2-position, Xu and colleagues [43] synthesized a series of 2-diarylmethoxy tropanes. In general, the 2β-isomers displayed higher affinity (K_is = 34 to 112 nM) than the 2α-isomers (K_i = 60 to 690 nM), and the unsubstituted diarylmethoxy analog **18** had the highest affinity at the DAT (K_i = 34 nM) (Fig. 6-5). Modification of the 2-position through replacement of the 2-carbomethoxy group with substituted aromatic rings has provided compounds with high DAT affinity and selectivity. Koturri and others have prepared and evaluated a series of 2β,3β- and 2α,3α-diaryltropanes for binding to monoamine transporters. In all cases, the 2β,3β derivatives had higher affinity at the DAT relative to the 2α,3α analogs [44]. Compound **19a**, with a phenyl and 4-methylphenyl group at the 2- and 3-positions, respectively, had an IC$_{50}$ value of 1.96 nM at the DAT and IC$_{50}$ values of 480 and 1100 nM at the NET and 5-HTT, respectively (Fig. 6-6). Addition of a 4′-CH₃ (**19b**) or a 4′-OCH₃ (**19c**) group to the 2-phenyl ring did not alter DAT affinity or selectivity significantly; however, a compound with a 2-(3,4-dimethoxyphenyl) group (**19d**) decreased affinity at the DAT 10-fold but enhanced selectivity over the 5-HTT and NET.

   Even though the DAT allows a broad tolerance for 2-position substitution, the exact nature of the 2-position substituent can impart selectivity for the DAT over the NET and the 5-HTT [16]. RTI-177 (**20a**) was prepared through bioisosteric replacement of the 2β-carbomethoxy group with a 2β-isoxazole group [45] (Fig. 6-7).

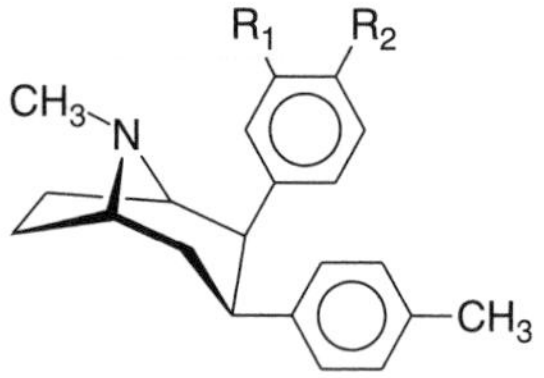

19a, R_1 = H, R_2 = H
b, R_1 = H, R_2 = CH_3
c, R_1 = H, R_2 = OCH_3
d, R_1 = OCH_3, R_2 = OCH_3

Figure 6-6 Structures for 3-phenyltropane analogs **19a–d**.

Compound **20a** had an IC_{50} value of 1.28 nM at the DAT and prominent DAT selectivity vs. the 5-HTT (1890-fold). Further exploration of this structural class led to numerous 2β-(1,2-isoxazoles) with high affinity and selectivity for the DAT [46]. Substitution on the 1,2-isoxazole ring with methyl (**21a–c**) or 4-substituted phenyl groups (**20b,c, 3**) provided a series of selective compounds with IC_{50} values ranging from 0.50 to 26 nM at the DAT. In all cases, compounds having a 3β-(4-chlorophenyl) group as opposed to a 3β-(4-methylphenyl) group were more potent at the DAT. In addition to 2β-(1,2-isoxazoles), several other 3β-(4-chlorophenyl)-2β-heterocycles have high affinity at the DAT. Most notably, oxazole (**22a**), thiazole (**22b**), oxadiazoles (**22c** and **22d**), thiadiazole (**22e**), and benzthiazole (**22f**) all display affinity between 1 and 20 nM at the DAT (Fig. 6-7) [45,47].

Although 2β-(1,2-isoxazole) compounds in general were potent and selective inhibitors of the DAT, **20b** (RTI-370) and **20c** (RTI-371, see Fig. 6-7) displayed

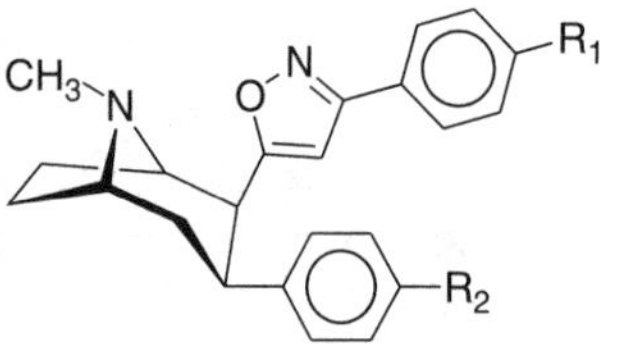

20a, R_1 = H, R_2 = Cl
RTI-370, b, R_1 = CH_3, R_2 = CH_3
RTI-371, c, R_1 = Cl, R_2 = CH_3
RTI-336, 3, R_1 = CH_3, R_2 = Cl

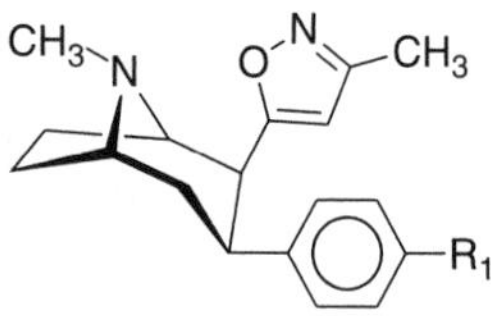

21a, R_1 = Cl
b, R_1 = CH_3
c, R_1 = I

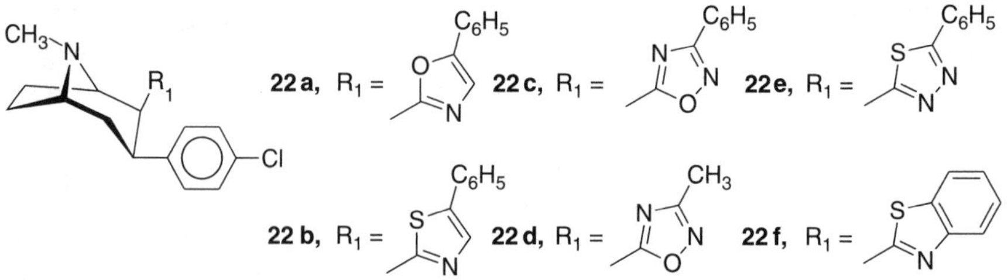

22a, R_1 =   **22c,** R_1 =   **22e,** R_1 =

22b, R_1 =   **22d,** R_1 =   **22f,** R_1 =

Figure 6-7 Structures for 3-phenyltropane analogs **3**, **20a–c**, **21a–c**, and **22a–f**.

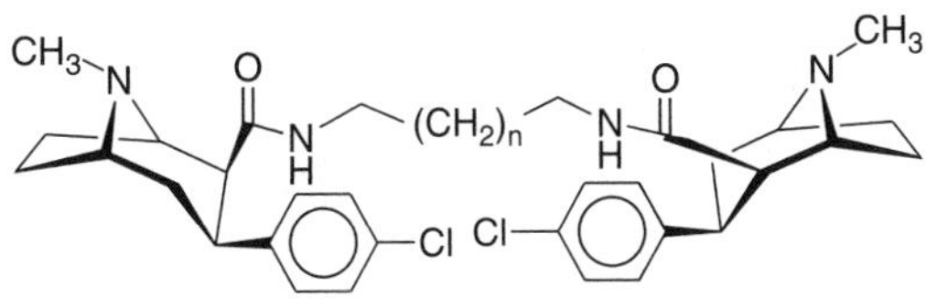

23a, n = 1
 b, n = 2
 c, n = 4
 d, n = 6

Figure 6-8   Structures for 3-phenyltropane analogs **23a–d**.

unusual properties in vivo [46]. Both were potent and extremely selective ligands, with IC_{50} values of 8.74 and 13 nM at the DAT and had IC_{50} values of 6980 and 100,000 nM at the NET and IC_{50} values of 100,000 nM at the 5-HTT (Fig. 6-7). These compounds appear to have a unique pharmacological profile because neither compound produced locomotor stimulation in mice. In addition, **20b** generalized weakly to cocaine, whereas **20c** did not generalize at all. This phenomenon has also been observed for other classes of DAT-uptake inhibitors [48]. Clearly, compounds that lack locomotor stimulation but possess potent and selective DAT affinity require further investigation.

The preparation of bivalent ligands with varied linkers has led to the development of a number of compounds with interesting biological activity. Taking advantage of the two-position tolerance for substitution, Fandrick and colleagues [49] synthesized a series of bivalent analogs of RTI-31 having amide linkers separated by spacers of three to eight methylene units (**23a–d**) (Fig. 6-8). The order of potency increased with increasing spacer lengths. The most potent compound, **23d**, had a K_i value of 6.7 nM at the DAT with an IC_{50} value of 871 nM for DA uptake. Although DA uptake did not correlate with potency, the apparent affinity of **23d** demonstrates a unique mode of binding for bivalent tropane analogs.

The synthesis of conformationally constrained analogs can provide valuable information regarding the potential modes of ligand–receptor interaction. A series of tropane macrocycles in which the 2-position substituent was cyclized to the 4'-position through linkers having varied size, hydrophobicity, and electronic properties were prepared and analyzed by Carroll and colleagues [50]. The most potent derivatives at the DAT (**24**, $IC_{50} = 3.8$ nM) and 5-HTT (**25**, $K_i = 1.9$ nM) utilized pimelic and sebacic acid linkers, respectively (Fig. 6-9). Molecular models comparing x-ray

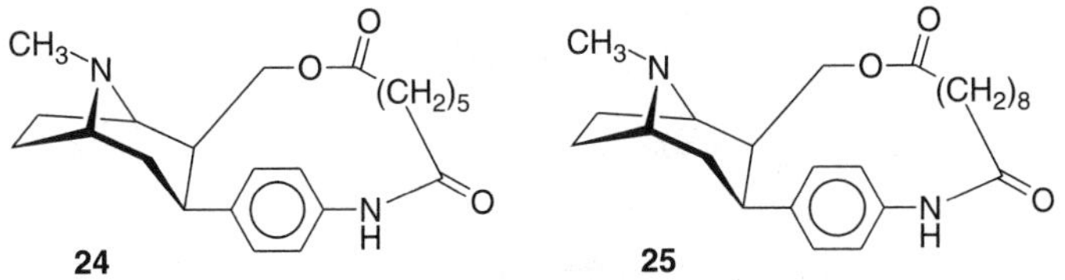

Figure 6-9   Structures for 3-phenyltropane analogs **24** and **25**.

structures with computer-generated minimum-energy conformations suggest that the differences in the geometry of the methylene linkers and amide oxygen may account for the selectivity and potency of these derivatives.

6.2.1 N-Substitution

Removal of the N-CH$_3$ group from 3-phenyltropanes has little effect on DAT potency but generally enhances 5-HTT and NET potency [23,26,38,51–53]. Substituting N-propyl (**26a**), N-allyl (**26b**), N-phenylethyl (**26c**), N-phenylpropyl (**26d**), and N-iodopropenyl (**26e**) for N-CH$_3$ on WIN 35,428 (**60**) also has little effect on DAT-binding affinity (Fig. 6-10). N-Substituted derivatives of **RTI-55** (**6d**), where R$_1$ = ethyl (**26f**), propyl (**26g**), butyl (**26h**), fluoroethyl (**26i**), and fluoropropyl (**26j**), all retain high affinity at the DAT [51,52,54]. In addition, the large tolerance for bulky N-substituents at the DAT is observed through the high-affinity phthalimido analogs **26k** and **26l** [55,56]. N-Desmethyl 3-phenyltropanes having 2β ketones instead of 2β esters have also been shown to be more potent than their N-methyl counterparts. Davies and colleagues [38] demonstrated that compound **27** (Fig. 6-10) was more potent at the DAT than the N-CH$_3$ parent, **16** (Fig. 6-4). Combining the synthetic ease of obtaining various N-substituted 3-phenyltropanes with the limited effects on DAT potency has led to the development of a number of tools to study real-time monoamine transporter occupancy and function. In particular, [^{11}C] and [^{18}F] N-substituted derivatives where R$_1$ = ^{11}CH$_3$ and CH$_2$CH$_2$CH$_2$ ^{18}F have been developed as positron emission tomography (PET) ligands. A more detailed description of the SAR for such radioligands is presented in Section 6.2.

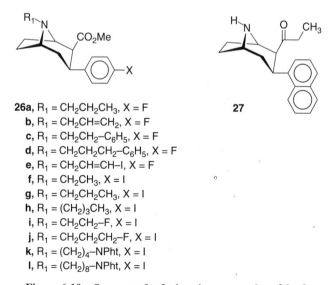

26a, R$_1$ = CH$_2$CH$_2$CH$_3$, X = F
  b, R$_1$ = CH$_2$CH=CH$_2$, X = F
  c, R$_1$ = CH$_2$CH$_2$–C$_6$H$_5$, X = F
  d, R$_1$ = CH$_2$CH$_2$CH$_2$–C$_6$H$_5$, X = F
  e, R$_1$ = CH$_2$CH=CH–I, X = F
  f, R$_1$ = CH$_2$CH$_3$, X = I
  g, R$_1$ = CH$_2$CH$_2$CH$_3$, X = I
  h, R$_1$ = (CH$_2$)$_3$CH$_3$, X = I
  i, R$_1$ = CH$_2$CH$_2$–F, X = I
  j, R$_1$ = CH$_2$CH$_2$CH$_2$–F, X = I
  k, R$_1$ = (CH$_2$)$_4$–NPht, X = I
  l, R$_1$ = (CH$_2$)$_8$–NPht, X = I

27

Figure 6-10 Structures for 3-phenyltropane analogs **26a–l**.

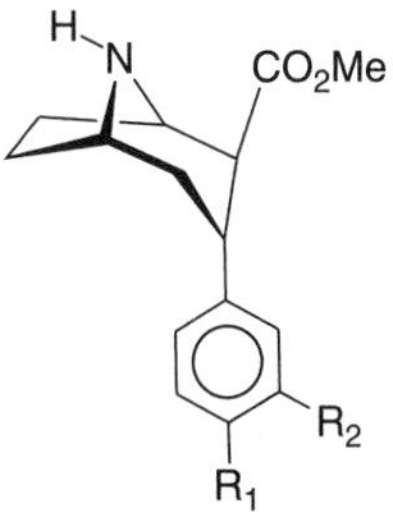

28a, $R_1 = CH_3$, $R_2 = H$
b, $R_1 = Cl$, $R_2 = H$
c, $R_1 = F$, $R_2 = H$
d, $R_1 = CH_3$, $R_2 = F$

Figure 6-11   Structures for 3-phenyltropane analogs **28a–d**.

Tropane ring conformations can be influenced to favor a boat or chair by altering the stereochemistry of the 2- and 3-position substituents. 2β,3β-Substitution favors the chair conformer, and 2β,3α substitution favors a flattened boat conformer [57]. Carroll and co-workers have synthesized a series of 2β,3α-nortropanes to determine the structural features necessary for affinity and selectivity for the NET [58]. Substitution of the 4′-position on 2β,3α-aryl nortropanes with methyl (**28a**), fluoro (**28c**), or chloro (**28d**) substituents provides relatively nonselective compounds with IC$_{50}$ values ranging between 5 and 9 nM at the NET, 3 to 34 nM at the DAT, and 53 to 500 nM at the 5-HTT (Fig. 6-11). However, the 2β,3α derivative (**28d**) has the greatest affinity and selectivity for the NET. It incorporates a 3-fluoro-4-methylphenyl group, had a K_i value of 0.43 nM at the NET, and was 21-fold selective vs. the DAT and 55-fold selective vs. the 5-HTT.

Meltzer and colleagues have determined that replacement of the 8-aza group on 3-phenyltropanes with 8-carba (**29a**) [59], 8-oxa (**29b**) [59], or 8-thia (**29c**) [60] groups did not significantly alter DAT potency (Fig. 6-12). Flattening the tropane

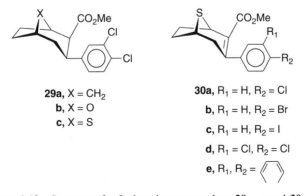

29a, $X = CH_2$
b, $X = O$
c, $X = S$

30a, $R_1 = H$, $R_2 = Cl$
b, $R_1 = H$, $R_2 = Br$
c, $R_1 = H$, $R_2 = I$
d, $R_1 = Cl$, $R_2 = Cl$
e, R_1, $R_2 =$

Figure 6-12   Structures for 3-phenyltropane analogs **29a–c** and **30a–e**.

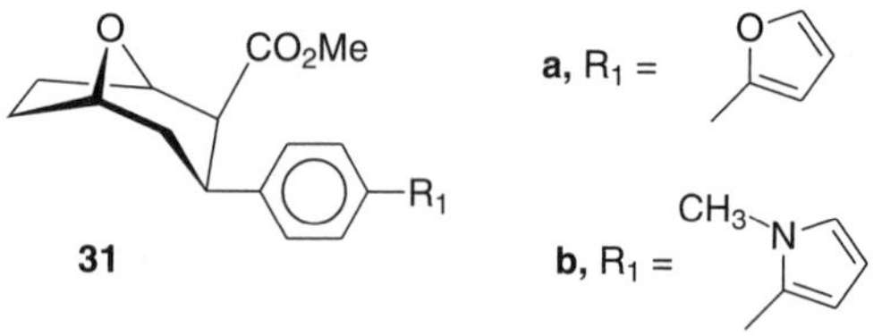

Figure 6-13   Structures for 3-phenyltropane analogs **31a**–**b**.

ring through introduction of a 2,3-double bond was shown to enhance DAT selectivity over the 5-HTT. Meltzer and co-workers [60] have extended their work to include 8-thiabicyclo[3.2.1]oct-2-enes. Affinity for the DAT was dependent on the size of the 4-substituent on the 3-aryl group. Replacement of 4-chloro (**30a**, IC_{50} = 13 nM) with a 4-bromo (**30b**, IC_{50} = 9.1 nM), 4-iodo (**30c**, IC_{50} = 6.7 nM), 3,4-dichloro (**30d**, IC_{50} = 4.5 nM), and 3-(2-naphthyl) group (**30e**, IC_{50} = 8.0 nM) enhanced affinity at the DAT.

Torun and others synthesized a series of 2β,3α- and 2β,3β-3-aryl-8-oxabicyclo [3.2.1]octanes with various 4′-heterocycles [61]. The 2β,3β-4′-(2-furanyl) derivative **31a** was the most potent, with an IC_{50} value of 64 nM at the DAT and 30 nM at the 5-HTT. The 2β,3β-4′-N-methylpyrrole derivative **31b** was the most selective for the 5-HTT, having an IC_{50} value of 566 nM and at the DAT of 12,400 nM (Fig. 6-13).

6.2.2   Tropane Ring Modifications

A number of modifications have been made to the basic tropane framework that resulted in compounds with only moderate or reduced affinity at the DAT. Both expansion [62] (**32**) and contraction [63] (**33**) of the tropane ring by one methylene unit resulted in compounds with significantly reduced affinity (Fig. 6-14). Moving the tropane two-carbon bridge to form 3-azabicyclo[3.2.1]octane resulted in compound **34** having an IC_{50} value of 81 nM at the DAT [64]. Altering the position of the nitrogen to form 6-azabicyclo[3.2.1]octanes resulted in compound **35** with comparable DAT uptake inhibition as cocaine [65]. Heteroatom-rich 3-phenyltropanes with a 3-(3′,4′-dichlorophenyl) substituent (**36**) showed moderate affinity at the DAT (IC_{50} = 21 nM) [66]. The addition of hydroxyl or methyl substituents to either the 6- (**37a,b**) or 7-position (**38a,b**) results in compounds with reduced potency at the DAT [67–69].

A number of ring-constrained 3-phenyltropane derivatives have been examined in which the direction of the nitrogen lone pair is fixed. Forming a two-carbon bridge between the basic nitrogen and the 6- (**39**) or 7-position (**40**) resulted in compounds with moderately reduced potency at the DAT compared to the parent [70] (Fig. 6-15). Linking the basic nitrogen with the 2-position via a two-(**41**) or three-carbon (**42**) bridge results in compounds with reduced affinity at the DAT [71,72]. Joining the basic nitrogen with the 3 position (**43**) provides compounds with reduced DAT potency, however, this series is more potent and selective for the NET [73–76].

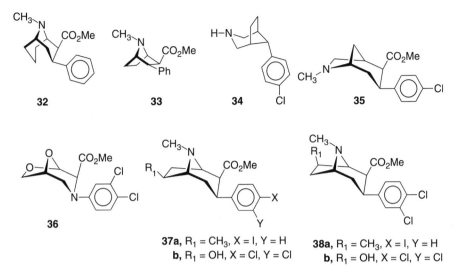

32
33
34
35
36
37a, R_1 = CH_3, X = I, Y = H
b, R_1 = OH, X = Cl, Y = Cl
38a, R_1 = CH_3, X = I, Y = H
b, R_1 = OH, X = Cl, Y = Cl

Figure 6-14 Structures for 3-phenyltropane analogs **32–36**, **37a–b**, and **38a–b**.

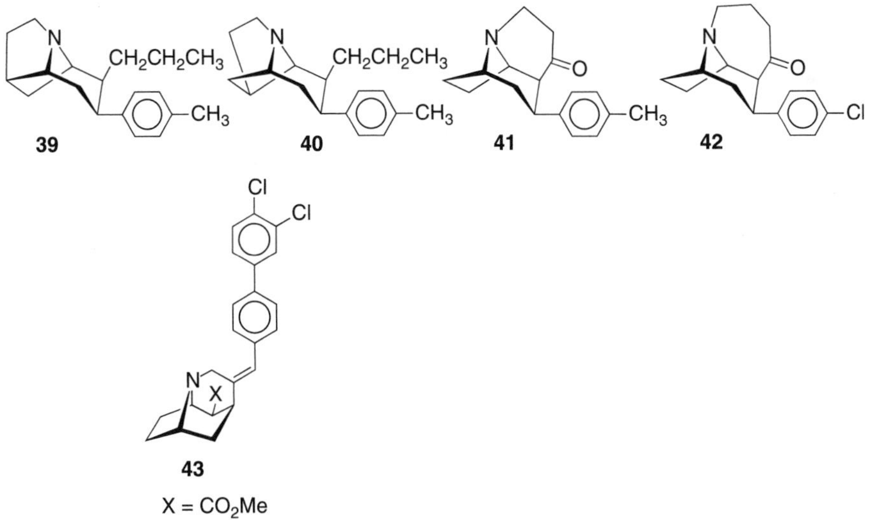

39
40
41
42
43
X = CO_2Me

Figure 6-15 Structures for 3-phenyltropane analogs **39–43**.

6.3   RADIOLIGANDS

The first tritium-labeled 3-phenyltropane analog reported was [³H]WIN 35,428 (**44**) [52], followed shortly thereafter by [³H]WIN 35,065-2 (**45**) [4,77,78] and [³H]RTI-31 (**46**) [79] (Fig. 6-16). A very early in vivo labeling study comparing [³H]WIN 35,065-2 and [³H]WIN 35,428 to [³H]cocaine (**47**) showed that both 3-phenyltropane analogs provided greater and longer-lasting specific binding in vivo than did cocaine

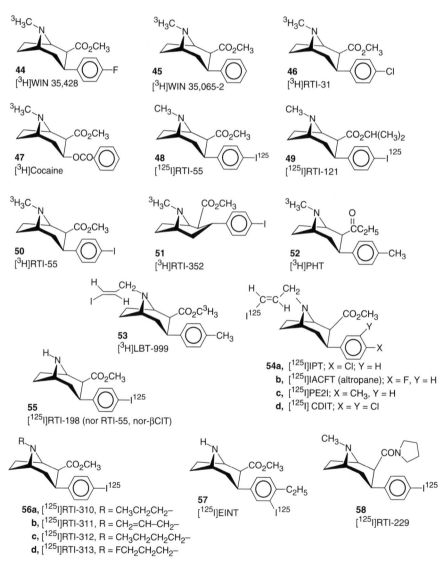

Figure 6-16   Structures for 3-phenyltropane analogs **44–55**, **56a–d**, **57**, and **58**.

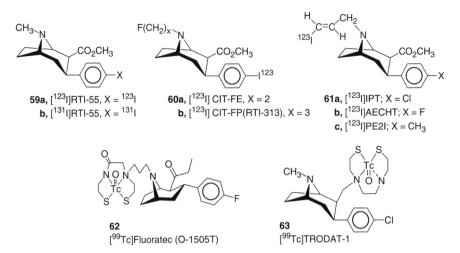

Figure 6-17   Structures for 3-phenyltropane analogs **59a–b**, **60a–b**, **61a–c**, **62**, and **63**.

[80]. Importantly, this showed that the 3-phenyltropane class of compounds had potential as useful PET ligands.

The first [125I]-labeled 3-phenyltropane analogs developed were [125I]RTI-55 (**48**), also called [125I]β-CIT, and [125I]RTI-121 (**49**) [79,81–87]. [3H]WIN 35,428, [125I]RTI-55, and [125I]RTI-121 have been the most widely used radioligands for

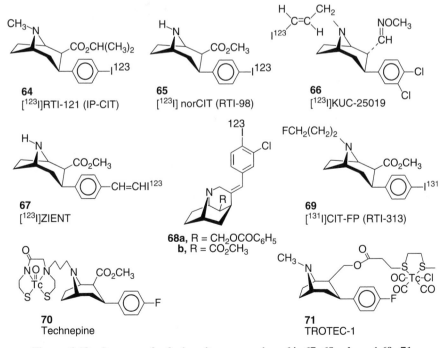

Figure 6-18   Structures for 3-phenyltropane analogs **64–67**, **68a–b**, and **69–71**.

TABLE 6-4   Chemical Structures of 3-Phenyltropane PET Ligands

Code	R_1	R_2	R_3	R_4	R_5	Refs.[a,b]
β-CFT (WIN 35,428)	[11C]CH$_3$	[11C]CH$_3$	[18F]F	H	H	1–16
β-CIT (RTI-55)	[11C]CH$_3$	[11C]CH$_3$	I	H	H	17–26
β-CCT (RTI-31)	CH$_3$	[11C]CH$_3$	Cl	H	H	27–29
β-CMT (RTI-32)	CH$_3$	[11C]CH$_3$	CH$_3$	H	H	27,28
FECNT	CH$_2$CH$_2$[18F]F	CH$_3$	Cl	H	H	30–38
β-CCT-FP (β-FPCT)	(CH$_2$)$_2$CH$_2$[18F] F	CH$_3$	Cl	H	H	30,39
β-CBT	CH$_3$	CH$_3$	[76Br]Br	H	H	40
β-FECT (β-FE-CCT)	CH$_3$	CH$_2$CH$_2$[18F]F	Cl	H	H	28,41
β-FETT (β-FE-CMT)	CH$_3$	CH$_2$CH$_2$[18F]F	CH$_3$	H	H	28,41
β-FIPCT (β-FiP-CCT)	CH$_3$	CH(CH$_3$)CH$_2$[18F]F	Cl	H	H	42,43
β-CFT-FE	CH$_2$CH$_2$[18F]F	CH$_3$	F	H	H	44,45
β-CFT-FP	(CH$_2$)$_2$CH$_2$[18F] F	CH$_3$	F	H	H	44
β-CBT-FE	CH$_2$CH$_2$F	CH$_3$	[76Br]Br	H	H	46

β-CIT-FP (FP-CIT)	$(CH_2)_2CH_2[^{18}F]$ F	$[^{11}C]CH_3$	I	H	H	47–51
β-CMT-FP (FP-CMT)	$(CH_2)_2CH_2[^{18}F]$ F	CH_3	CH_3	H	H	52
β-CBT-FP (FP-CBT)	$(CH_2)_2CH_2[^{18}F]$ F	CH_3	$[^{76}Br]Br$	H	H	46,53
β-CDCT	CH_3	$[^{11}C]CH_3$	Cl	Cl	H	29
β-IP-CIT (RTI-121)	$[^{11}C]CH_3$	$CH(CH_3)_2$	I	H	H	22
NS-2214 (BMS-204756)	$[^{11}C]CH_3$	b	Cl	Cl	H	54
β-CIT-FE	CH_2CH_2F	$[^{11}C]CH_3$	I	H	H	14,55,56
β-CpFMT	CH_3	CH_3	$CH_2[^{18}F]F$	H	H	57,58
β-CmFMT	CH_3	CH_3	H	$CH_2[^{18}F]F$	H	57
β-CoFMT (o-FWIN)	CH_3	CH_3	H	H	$CH_2[^{18}F]F$	58
β-CPPIT	$[^{11}C]CH_3$	c	Cl	H	H	59,60
FCT	$4-[^{18}F]FBn$	CH_2CH_3	Cl	H	H	61
E-IACFT (altropane)	$CH_2CH{=}CHI$	$[^{11}C]CH_3$	F	H	H	62
MCL301	$CH_2CH_2[^{18}F]F$	CH_3	I	H	H	63,64
MCL322	$CH_2CH_2[^{18}F]F$	CH_3	Br	H	H	63,64
FEβCIT	$CH_2CH_2[^{18}F]F$	CH_3	I	H	H	65

Source: Modified from Dollé et al. [91].

[a] These references are listed separately following the chapter references and are referred to as References for Table 6-4.

[b] 2α-(N–OCH₃, —C(=O)–H).

[c] 2β-(isoxazole O–N ring with C₆H₅).

145

studying monoamine transporters. However, a number of other radioligands have been developed (Fig. 6-16). The tritium-labeled 3-phenyltropane analogs include [³H]RTI-55 (**50**) [88], [³H]RTI-352 (**51**) [89], [³H]PHT (**52**) [90], and [³H]LBT-999 (**53**) [91]. The additional 3-phenyltropane analogs possessing an iodine-125 label include [¹²⁵I]IPT (**54a**) [92]; [¹²⁵I]IACFT (**54b**) [56]; [¹²⁵I]PE2I (**54c**) [93]; [¹²⁵I] β-CDIT (**54d**) [94]; [¹²⁵I]RTI-98 (nor-RTI-55 and nor-β-CIT) (**55**) [95]; [¹²⁵I]RTI-310, -311, -312, and -313 (**56a–d**) [96]; [¹²⁵I]EINT (**57**) [97,98]; and [¹²⁵I]RTI-229 (**58**) [97].

Since the dopamine transporter (DAT) is a marker for dopamine (DA) neurons, there is interest in studying regulatory or drug-induced changes in DAT levels as well as the loss of DA neurons caused by neurodegenerative diseases such as Parkinson's. PET and SPECT (single-photon emission computed tomography) could measure striatal (DA) terminal function in vivo as reflected by DAT binding. In vivo preimaging studies with RTI-55 showed that this 3-phenyltropane analog possessed a greater than 15 : 1 striatal/cerebellum binding ratio, a measure of specific binding. This was followed by the synthesis and development of [¹²³I]RTI-55 (**59**) as the first SPECT ligand for imaging DAT-uptake sites in primate brain [99,100] (Fig. 6-17). These and other studies led to the development of [¹²³I]RTI-55 (Dopascan, Iometopane) as a diagnostic ligand for Parkinson's disease [101].

These early successful studies with [¹²³I]RTI-55 led to the syntheses and evaluation of other 3-phenyltropane analogs as potential SPECT imaging agents. The structures of compounds studied in humans are shown in Figure 6-17. (See [101–111] and the references cited therein.) Of all the 3-phenyltropane SPECT ligands developed, [¹²³I]RTI-55 has been the ligand studied most extensively in humans. Thousands of patients have been studied worldwide. Other 3-phenyltropane SPECT ligands [64–71] that have been developed but have not reached clinical evaluation are listed in Fig. 6-18 [106,112–117].

A wealth of 3-phenyltropane PET ligands has been developed. These have been summarized in a recent article and are listed in Table 6-4. Those that have been studied in humans are listed in Fig. 6-19. (See [91,118–121] and the references cited therein.)

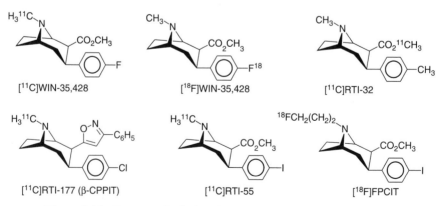

Figure 6-19   Structures for 3-phenyltropane PET ligands used in humans.

6.4 IRREVERSIBLE BINDING AND FLUORESCENT LIGANDS

Affinity ligands have been used for the pharmacological and structural characterization of numerous macromolecules (enzymes and receptors). In many cases these affinity ligands have a reactive electrophilic moiety (or photoactivated group) as part of their structure, and thus have potential for covalently binding to or near the recognition site of a receptor. The 3-phenyltropane affinity ligands **72** (RTI-76), **73** (RTI-86), and **74**, as well as the photoaffinity ligands **75** (RTI-73), **76** (RTI-78), **77** (RTI-82), and **78** (see Fig. 6-20) were developed for studying the DAT, 5-HTT, or both [122,123]. The affinity ligands RTI-76, RTI-86, and **74** and the photoaffinity ligands RTI-73 and RTI-82 (after photoactivation) all showed irreversible binding to the DAT in various preparations [122–125].

RTI-76 has been the most studied affinity ligand. RTI-76 was used to show that the in vitro turnover half-life of the DAT protein in rat DA LLC-PK1 cells was 23 hours [124]. In vivo studies showed that the half-life of recovery for the DAT protein in the striatum and the nucleus accumbens after intracerebroventricular (i.c.v.) injection in rats with RTI-76 was 2 days in both [126]. In a similar study, RTI-76 was used to

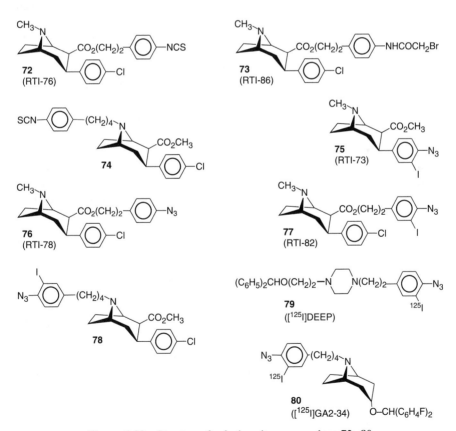

Figure 6-20 Structures for 3-phenyltropane analogs **72**–**80**.

show that the half-life of recovery of the serotonin transporter (5-HTT) protein in the hippocampus was 3.4 days in the rat [127]. RTI-76 was used to show that stimulation and blockade of the D1 and D2 dopamine receptor families affected the degradation rate constant and the production rate of the DAT, and thus its half-life in the rat striatum and nucleus accumbens [128]. However, the physiological significance of the D1 and D2 receptor-mediated influences is not clear since there was no significant change in the DAT density following treatment with any of the dopaminergic ligands. RTI-76 increased locomotor activity in rats only after 50% of the DAT protein was occupied [129].

Several studies have shown that repeated cocaine administration followed by withdrawal alters DAT levels in rats [130–135]. RTI-76 was used to show that cocaine pretreatment (10 days) and withdrawal (10 days) in rats reduced the half-life of the DAT from 2.1 to 0.94 days in the striatum, but did not alter the half-life in the nucleus accumbens [136]. The half-life of the DAT has been associated with sensitization to the locomotor effects of cocaine in rats [137–139]. Studies with RTI-76 show that the effects of cocaine are long lasting, and these changes may be partially responsible for the withdrawal symptoms observed in cocaine addicts. The mechanism(s) of these effects remain to be elucidated.

[^{125}I]RTI-82 was prepared [140] and used to show that it labeled a DAT protein having the same molecular weight and possessed a similar pharmacological profile and similar sensitivity to a neuramidase as that labeled by **79** ([^{125}I]DEEP), a GBR analog [141]. [^{125}I]RTI-82 was also used as a probe for aiding in the identification of the binding domain on the DAT. Epitope-specific immunoprecipitation of protolytic fragments generated from SDS-solubilized DATs showed that [^{125}I]RTI-82 was localized to the C-terminal half of the protein near the transmembrane domain (TM) 4–7 [142,143]. Later studies using a chemical cleavage and site-directed approach identified its incorporation in a sequence containing TM6 [144]. Similar studies with the GBR photoaffinity DAT ligand [^{125}I]DEEP and the benztropine photoaffinity ligand **80** ([^{125}I]GA2-34) showed that both of these ligands were incorporated near the TM 1–2 domain [145,146]. This suggests that the DAT might be folded into a structure where the TM 1–2 and TM 4–7 domains are in proximity.

Fluorescently labeled high-affinity ligands can be used to characterize the biophysical microenvironment of a binding site as well as its accessibility to the aqueous milieu [147]. Compound **81** (RTI-233), **82** (RTI-587), and **83** (RTI-592) (Fig. 6-21), which contain the environmentally sensitive 7-nitrobenz-2-oxa-1,3-diazol-4-yl (NBD) group, possess high affinity and specific binding to the 5-HTT. RTI-233 was used as a molecular reporter to characterize the microenvironment of the cocaine-binding site of the 5-HTT expressed in Sf-9 insect cells [148]. Steady-state fluorescence anisotropy measurement and collisional quenching experiments showed constrained mobility of bound RTI-233 relative to RTI-233 free in solution. The studies were consistent with a highly hydrophobic microenvironment in the binding pocket for cocaine like uptake inhibitors. However, the bound RTI-233 was still accessible for aqueous quenching and, thus, partially exposed to solvent. Fluorescence resonance energy transfer (FRET) studies using RTI-233 as the

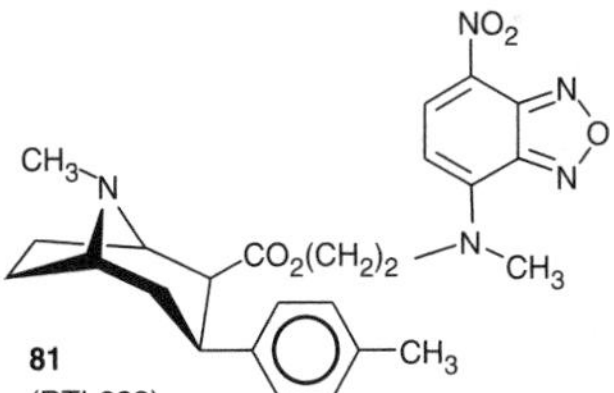

81
(RTI-233)

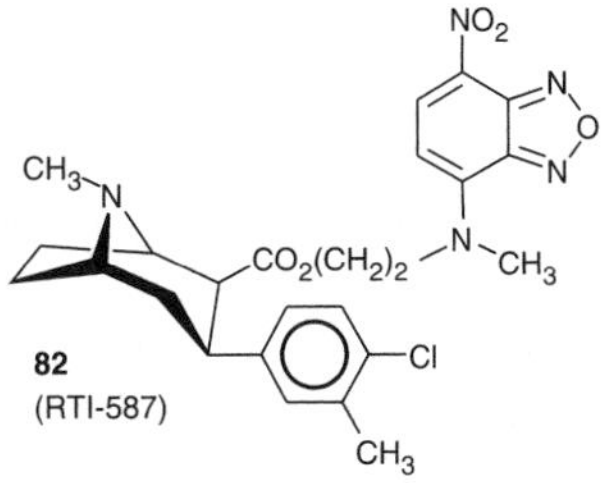

82
(RTI-587)

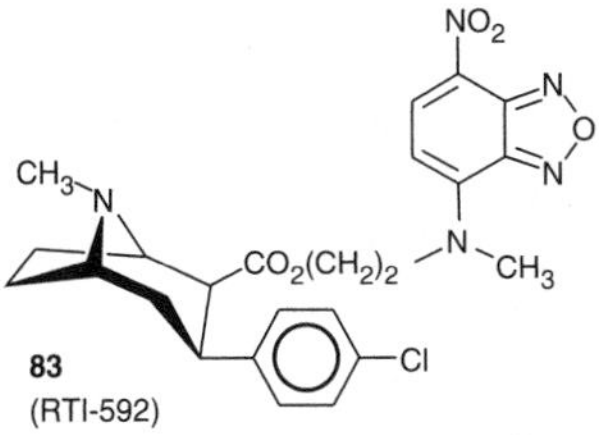

83
(RTI-592)

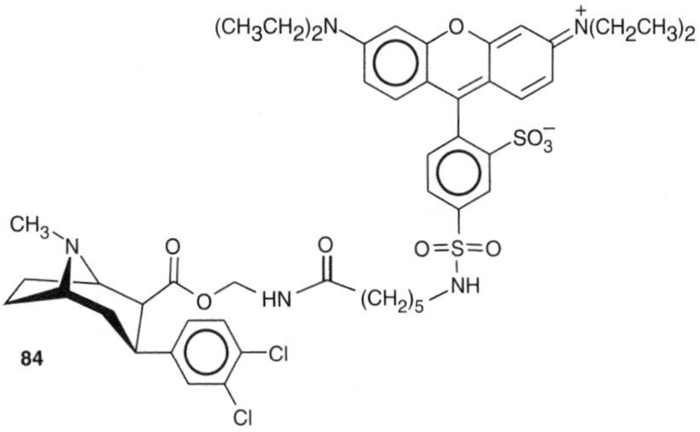

84

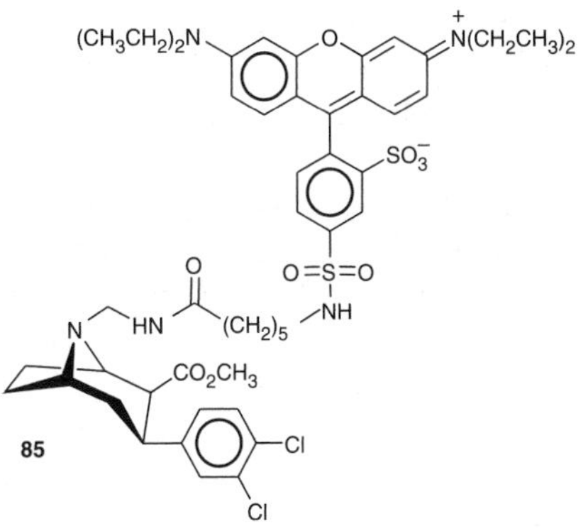

85

Figure 6-21 Structures for 3-phenyltropane analogs **81–85**.

acceptor fluorophore and Texas Red covalently attached to endogenous cysteine in the 5-HTT showed that the distances from the cocaine-binding site to the labeled endogenous cysteines might be estimated [147].

Confocal fluorescence correlation spectroscopy (confocal FCS) is a powerful single-molecule fluorescence technology that can be used to explore protein structure and protein function in vitro as well as in living cells [147]. FCS studies were used to show that in the presence of unlabeled 5-HTT, RTI-233 displayed a substantial increase in the diffusion time of the same order of magnitude as determined by other methods [147]. Recently, the rhodamine-labeled 3-phenyltropanes **84** and **85** were found useful for visualizing the DAT in HEK293 cells expressing DAT by confocal laser scanning microscopy [149].

6.5 ANIMAL BEHAVIORAL AND CLINICAL STUDIES

3-Phenyltropane analogs have been highly useful neurochemical tools for exploring the behavioral effects of monoamine uptake inhibitors as well as the mechanisms underlying these behaviors. The first 3-phenyltropanes discovered, WIN 35,065-2, WIN 35,428, and the N-demethylated analog of the former compound WIN 35,981, were evaluated for locomotor activity and were reported to be more active than cocaine [2]. Subsequent studies with WIN 35,065-2 and WIN 35,428 found similar results with locomotor activity and other drug-induced behaviors in rodents. In addition to being more potent in inducing locomotor activity [150,151], both compounds were also more potent than cocaine in epilateral rotation in 6-hydroxydopamine-lesioned animals [152] and sniffing behavior [153]. In drug discrimination studies with pigeons, both WIN 35,065-2 and WIN 35,428 were fully generalized to cocaine and were more potent [154].

WIN 35,065-2 and WIN 35,428 were three to 10 times more potent than cocaine when evaluated for their effects on schedule-controlled behavior using squirrel monkeys responding under a multiple-fixed-ratio, fixed-interval schedule of termination of stimulus associated with electric shock, or a similar schedule of food presentation [155–157]. WIN 35,065-3, the enantiomer of WIN 35,065-2, was much less potent. WIN 35,065-2 and WIN 35,981 were three to 10 times more potent than cocaine in maintaining self-administration in squirrel monkeys responding under a fixed-interval schedule of drug self-administration [158]. These reports prompted the interest of medicinal chemists, and a number of 3-phenyltropane analogs were synthesized and evaluated for locomotor activity and generalization to cocaine in drug-discrimination tests. Although there are exceptions, in general, the 3-phenyltropane analogs tended to have potencies greater than cocaine that paralleled their affinities for the DAT.

A few of the 3-phenyltropane analogs possessed properties in the locomotor activity and drug discrimination tests that suggested they might have potential as indirect dopamine agonist pharmacotherapies to treat cocaine abuse and dependence [159–162]. These compounds were evaluated in monkey self-administration tests. The most studied DAT-selective 3-phenyltropane analog is RTI-336 (**3**). RTI-336 provided a dose-dependent reduction in cocaine self-administration in rhesus monkeys and was equally effective at 0.1 and 0.3 mg/kg per injection of cocaine [163]. The ED_{50} dose of RTI-336 for reducing cocaine self-administration resulted in approximately 90% DAT occupancy. RTI-113, another DAT-selective 3-phenyl-tropane analog, also reduced cocaine self-administration in rhesus monkeys [164]. In the case of RTI-113, a DAT occupancy of 72 to 78% was required for the pretreatment dose of RTI-113 in order to reduce responding maintained by cocaine. RTI-177, a 3-phenyltropane analog with good affinity for both the DAT and NET, also reduced cocaine self-administration in rhesus monkeys under a second-order schedule with a 73% DAT occupancy at doses that reduced cocaine self-administration by 50% [165]. RTI-112, a 3-phenyltropane with roughly equal reuptake efficacy at the DAT, 5-HTT, and NET, did not exhibit levels of DAT occupancy above the detection threshold at doses that produced significant decrease in cocaine self-administration in rhesus monkeys [165]. However, the same dose of RTI-112 displayed high levels of 5-HTT occupancy.

Interestingly, coadministration of the ED_{50} dose of RTI-336, used in the rhesus monkey studies above in combination with ineffective doses of the SERT-selective uptake inhibitor fluoxetine or citalopram, completely suppressed cocaine self-administration without affecting DAT occupancy. The nonselective monoamine uptake inhibitor PTT [2β-propanoyl-3β-(4-tolyl)tropane] has also been shown to reduce cocaine self-administration in rhesus monkeys under a fixed-interval schedule [166]. Two 3-phenyltropane analogs, brasofensine (**4**) and tesofensine (**5**), have undergone clinical evaluation [13,167,168]. Brasofensine advanced through phase II clinical trials as a potential drug to treat Parkinson's disease. Tesofensine recently entered phase II clinical trials for obesity.

6.6   CONCLUSIONS

Using WIN 35,065-2 as the lead compound, a large number of 3-phenyltropane analogs have been synthesized and evaluated for binding affinity at the DAT, 5-HTT, and NET. In addition to having high affinity for all three transporters, analogs selective for each of the transporters have been developed. Many of the analogs have been evaluated in locomotor activity and cocaine discrimination studies. Although there are exceptions, for the most part, the potency in these assays correlate with their affinity at the DAT. Similar to other classes of DAT-uptake inhibitors, DAT-selective as well as monoamine transporter nonselective analogs show reduction of cocaine self-administration in both rat and monkey tests. Analogs have been developed that are in clinical or advanced preclinical studies as potential pharmacotherapies for treating cocaine addiction, Parkinson's disease, and obesity.

Numerous 3-phenyltropane radioligands have been synthesized and used for various biochemical and pharmacological studies. Many SPECT and PET imaging 3-phenyltropanes have been developed largely as diagnostic agents for Parkinson's disease. However, these imaging agents have been used for basic studies of drug addiction as well as other diseases. The 3-phenyltropane class of monoamine transporter inhibitors has also provided a number of irreversible binding and fluorescent ligands that have been used as biochemical and pharmacological tools to characterize the cocaine binding sites on monoamine transporters and to help study the pharmacological effects of the DAT.

REFERENCES

1. Clarke, R. L. (1977). The tropane alkaloids. In *The Alkaloids*, Manske, R. H. F., Ed. Academic Press, New York, Vol. 16, Chap. 2.

2. Clarke, R. L., Daum, S. J., Gambino, A. J., Aceto, M. D., Pearl, J., Levitt, M., Cumiskey, W. R., and Bogado, E. F. (1973). Compounds affecting the central nervous system. 3β-Phenyltropane-2-carboxylic esters and analogs. *J Med Chem*, *16*, 1260–1267.

3. Carroll, F. I., Mascarella, S. W., Kuzemko, M. A., Gao, Y., Abraham, P., Lewin, A. H., Boja, J. W., and Kuhar, M. J. (1994). Synthesis, ligand binding, and QSAR (CoMFA and classical) study of 3β-(3′-substituted phenyl)-, 3β-(4′-substituted phenyl)-, and 3β-(3′,4′-disubstituted phenyl)tropane-2β-carboxylic acid methyl esters. *J Med Chem, 37,* 2865–2873.

4. Carroll, F. I., Lewin, A. H., Boja, J. W., and Kuhar, M. J. (1992). Cocaine receptor: Biochemical characterization and structure–activity relationships for the dopamine transporter. *J Med Chem, 35,* 969–981.

5. Rocha, B. A., Fumagalli, F., Gainetdinov, R. R., Jones, S. R., Ator, R., Giros, B., Miller, G. W., and Caron, M. G. (1998). Cocaine self-administration in dopamine-transporter knockout mice. *Nat Neurosci, 1,* 132–137.

6. Volkow, N. D., Wang, G.-J., Fischman, M. W., Foltin, R. W., Fowler, J. S., Abumrad, N. N., Vitkun, S., Logan, J., Gatley, S. J., Pappas, N., Hitzemann, R., and Shea, C. E. (1997). Relationship between subjective effects of cocaine and dopamine transporter occupancy. *Nature, 386,* 827–833.

7. DiMaio, S., Grizenko, N., and Joober, R. (2003). Dopamine genes and attention-deficit hyperactivity disorder: a review. *J Psychiatry Neurosci, 28,* 27–38.

8. Glase, S. A. and Dooley, D. J. (2004). Attention deficit hyperactivity disorder: pathophysiology and design of new treatments. In *Annual Reports in Medicinal Chemistry*, Vol. 39, Doherty, A.M., Ed. Academic Press, San Diego, CA.

9. Uhl, G. R. (1998). Hypothesis: the role of dopaminergic transporters in selective vulnerability of cells in Parkinson's disease. *Ann Neurol, 43,* 555–560.

10. Uhl, G. R. (2003). Dopamine transporter: basic science and human variation of a key molecule for dopaminergic function, locomotion, and parkinsonism. *Mov Disord, 18,* S71–S80.

11. Johnston, T. H., and Brotchie, J. M. (2004). Drugs in development for Parkinson's disease. *Curr Opin Investig Drugs, 5,* 720–726.

12. Carroll, F. I., Howard, J. L., Howell, L. L., Fox, B. S., and Kuhar, M. J. (2006). Development of the dopamine transporter selective RTI-336 as a pharmacotherapy for cocaine abuse. *AAPS, 8*(Art 24), E196–E203.

13. Moldt, P., Wåtjen, F., and Scheel-Krüger, J. (1995). Tropane-2-aldoxine derivatives as neurotransmitter reuptake inhibitors. Neurosearch A/S, Denmark.

14. Neurosearch (2004). Press release.

15. Dutta, A. K., Zhang, S., Kolhatkar, R., and Reith, M. E. (2003). Dopamine transporter as target for drug development of cocaine dependence medications. *Eur J Pharmacol, 479,* 93–106.

16. Carroll, F. I., Lewin, A. H., and Mascarella, S. W. (2001). Dopamine transporter uptake blockers: structure–activity relationships. In *Neurotransmitter Transporters: Structure, Function, and Regulation*, 2nd ed., Reith, M. E. A., Ed. Humana Press. Totowa, NJ, pp. 381–432.

17. Reith, M. E. A. (2002). *Neurotransmitter Transporters: Structure, Function, and Regulation*, Humana Press, Totowa, NJ.

18. Singh, S. (2000). Chemistry, design, and structure–activity relationship of cocaine antagonists. *Chem Rev, 100,* 925–1024.

19. Runyon, S. P., and Carroll, F. I. (2006). Dopamine transporter ligands: recent developments and therapeutic potential. *Curr Top Med Chem, 6,* 1825–1843.

20. Carroll, F. I., Lewin, A. H., Abraham, P., Parham, K., Boja, J. W., and Kuhar, M. J. (1991). Synthesis and ligand binding of cocaine isomers at the cocaine receptor. *J Med Chem*, *34*, 883–886.

21. Carroll, F. I., Runyon, S. P., Abraham, P., Navarro, H., Kuhar, M. J., Pollard, G. T., and Howard, J. L. (2004). Monoamine transporter binding, locomotor activity, and drug discrimination properties of 3-(4-substituted-phenyl)tropane-2-carboxylic acid methyl ester isomers. *J Med Chem*, *47*, 6401–6409.

22. Carroll, F. I., Gao, Y., Rahman, M. A., Abraham, P., Parham, K., Lewin, A. H., Boja, J. W., and Kuhar, M. J. (1991). Synthesis, ligand binding, QSAR, and CoMFA study of 3β-(*p*-substituted phenyl)tropane-2β-carboxylic acid methyl esters. *J Med Chem*, *34*, 2719–2925.

23. Meltzer, P. C., Liang, A. Y., Brownell, A.-L., Elmaleh, D. R., and Madras, B. K. (1993). Substituted 3-phenyltropane analogs of cocaine: synthesis, inhibition of binding at cocaine recognition sites, and positron emission tomography imaging. *J Med Chem*, *36*, 855–862.

24. Kline, R. H., Jr., Wright, J., Fox, K. M., and Eldefrawi, M. E. (1990). Synthesis of 3-arylecgonine analogues as inhibitors of cocaine binding and dopamine uptake. *J Med Chem*, *33*, 2024–2027.

25. Carroll, F. I., Kuzemko, M. A., Gao, Y., Abraham, P., Lewin, A. H., Boja, J. W., and Kuhar, M. J. (1992). Synthesis and ligand binding of 3β-(3-substituted phenyl)- and 3β-(3,4-disubstituted phenyl)tropane-2β-carboxylic acid methyl esters. *Med Chem Res*, *1*, 382–387.

26. Blough, B. E., Abraham, P., Lewin, A. H., Kuhar, M. J., Boja, J. W., and Carroll, F. I. (1996). Synthesis and transporter binding properties of 3β-(4′-alkyl-, 4′-alkenyl-, and 4′-alkynylphenyl)nortropane-2β-carboxylic acid methyl esters: serotonin transporter selective analogs. *J Med Chem*, *39*, 4027–4035.

27. Tamagnan, G., Alagille, D., Fu, X., Kula, N. S., Baldessarini, R. J., Innis, R. B., and Baldwin, R. M. (2005). Synthesis and monoamine transporter affinity of new 2β-carbomethoxy-3β-[4-(substituted thiophenyl)]phenyltropanes: discovery of a selective SERT antagonist with picomolar potency. *Bioorg Med Chem Lett*, *15*, 1131–1133.

28. Bois, F., Baldwin, R. M., Kula, N. S., Baldessarini, R. J., Innis, R. B., and Tamagnan, G. (2004). Synthesis and monoamine transporter affinity of 3′-analogs of 2-β-carbomethoxy-3-β-(4′-iodophenyl)tropane (β-CIT). *Bioorg Med Chem Lett*, *14*, 2117–2120.

29. Carroll, F. I., Blough, B. E., Nie, Z., Kuhar, M. J., Howell, L. L., and Navarro, H. A. (2005). Synthesis and monoamine transporter binding properties of 3β-(3′,4′-disubstituted phenyl)tropane-2β-carboxylic acid methyl esters. *J Med Chem*, *48*, 2767–2771.

30. Meltzer, P. C., McPhee, M., and Madras, B. K. (2003). Synthesis and biological activity of 2-carbomethoxy-3-catechol-8-azabicyclo[3.2.1]octanes. *Bioorg Med Chem Lett*, *13*, 4133–4137.

31. Enyedy, I. J., Sakamuri, S., Zaman, W. A., Johnson, K. M., and Wang, S. (2003). Pharmacophore-based discovery of substituted pyridines as novel dopamine transporter inhibitors. *Bioorg Med Chem Lett*, *13*, 513–517.

32. Wang, S., Sakamuri, S., Enyedy, I. J., Kozikowski, A. P., Deschaux, O., Bandyopadhyay, B. C., Tella, S. R., Zaman, W. A., and Johnson, K. M. (2000). Discovery of a novel dopamine transporter inhibitor, 4-hydroxy-1-methyl-4-(4-methylphenyl)-3-piperidyl 4-methylphenyl ketone, as a potential cocaine antagonist through 3D-database

pharmacophore searching: molecular modeling, structure–activity relationships, and behavioral pharmacological studies. *J Med Chem*, *43*, 351–360.

33. Carroll, F. I., Kotian, P., Dehghani, A., Gray, J. L., Kuzemko, M. A., Parham, K. A., Abraham, P., Lewin, A. H., Boja, J. W., and Kuhar, M. J. (1995). Cocaine and 3β-(4′-substituted phenyl)tropane-2β-carboxylic acid ester and amide analogues. New high-affinity and selective compounds for the dopamine transporter. *J Med Chem*, *38*, 379–388.

34. Carroll, F. I., Abraham, P., Lewin, A. H., Parham, K. A., Boja, J. W., and Kuhar, M. J. (1992). Isopropyl and phenyl esters of 3β-(4-substituted phenyl)tropan-2β-carboxylic acids. Potent and selective compounds for the dopamine transporter. *J Med Chem*, *35*, 2497–2500.

35. Carroll, F. I., Kotian, P., Gray, J. L., Abraham, P., Kuzemko, M. A., Lewin, A. H., Boja, J. W., and Kuhar, M. J. (1993). 3β-(4′-Chlorophenyl)tropan-2β-carboxamides and cocaine amide analogues: new high affinity and selective compounds for the dopamine transporter. *Med Chem Res*, *3*, 468–472.

36. Davies, H. M. L., Saikali, E., Huby, N. J. S., Gilliat, V. J., Matasi, J. J., Sexton, T., and Childers, S. R. (1994). Synthesis of 2β-acyl-3β-aryl-8-azabicyclo[3.2.1]octanes and their binding affinities at dopamine and serotonin transport sites in rat striatum and frontal cortex. *J Med Chem*, *37*, 1262–1268.

37. Davies, H. M. L., Saikali, E., Sexton, T., and Childers, S. R. (1993). Novel 2-substituted cocaine analogs: binding properties at dopamine transport sites in rat striatum. *Eur J Pharmacol Mol Pharmacol Sect-1*, *244*, 93–97.

38. Davies, H. M. L., Kuhn, L. A., Thornley, C., Matasi, J. J., Sexton, T., and Childers, S. R. C. (1996). Synthesis of 3β-aryl-8-azabicyclo[3.2.1]octanes with high binding affinities and selectivities for the serotonin transporter site. *J Med Chem*, *39*, 2554–2558.

39. Bennett, B. A., Wichems, C. H., Hollingsworth, C. K., Davies, H. M. L., Thornley, C., Sexton, T., and Childers, S. R. (1995). Novel 2-substituted cocaine analogs: uptake and ligand binding studies at dopamine, serotonin and norepinephrine transport sites in the rat brain. *J Pharmacol Exp Ther*, *272*, 1176–1186.

40. Kozikowski, A. P., Roberti, M., Xiang, L., Bergmann, J. S., Callahan, P. M., Cunningham, K. A., and Johnson, K. M. (1992). Structure–activity relationship studies of cocaine: replacement of the C-2 ester group by vinyl argues against H-bonding and provides an esterase-resistant, high-affinity cocaine analogue. *J Med Chem*, *35*, 4764–4766.

41. Kozikowski, A. P., Saiah, M. K. E., Johnson, K. M., and Bergmann, J. S. (1995). Chemistry and biology of the 2β-alkyl-3β-phenyl analogues of cocaine: subnanomolar affinity ligands that suggest a new pharmacophore model at the C-2 position. *J Med Chem*, *38*, 3086–3093.

42. Kelkar, S. V., Izenwasser, S., Katz, J. L., Klein, C. L., Zhu, N., and Trudell, M. L. (1994). Synthesis, cocaine receptor affinity, and dopamine uptake inhibition of several new 2β-substituted 3β-phenyltropanes. *J Med Chem*, *37*, 3875–3877.

43. Xu, L., Kulkarni, S. S., Izenwasser, S., Katz, J. L., Kopajtic, T., Lomenzo, S. A., Newman, A. H., and Trudell, M. L. (2004). Synthesis and monoamine transporter binding of 2-(diarylmethoxymethyl)-3β-aryltropane derivatives. *J Med Chem*, *47*, 1676–1682.

44. Kotturi, S. V., Jiang, S., Chang, A. C., Abraham, P., Navarro, H. A., Kuhar, M. J., and Carroll, F. I. (2005). Synthesis and monoamine transporter binding properties of 2,3-diaryltropanes. *J Med Chem*, *48*, 7437–7444.

45. Kotian, P., Mascarella, S. W., Abraham, P., Lewin, A. H., Boja, J. W., Kuhar, M. J., and Carroll, F. I. (1996). Synthesis, ligand binding, and quantitative structure–activity relationship study of 3β-(4′-substituted phenyl)-2β-(heterocyclic)tropanes: evidence for an electrostatic interaction at the 2β-position. *J Med Chem*, *39*, 2753–2763.

46. Carroll, F. I., Pawlush, N., Kuhar, M. J., Pollard, G. T., and Howard, J. L. (2004). Synthesis, monoamine transporter binding properties, and behavioral pharmacology of a series of 3β-(substituted phenyl)-2β-(3′-substituted isoxazol-5-yl)tropanes. *J Med Chem*, *47*, 296–302.

47. Kotian, P., Abraham, P., Lewin, A. H., Mascarella, S. W., Boja, J. W., Kuhar, M. J., and Carroll, F. I. (1995). Synthesis and ligand binding study of 3β-(4′-substituted phenyl)-2β-(heterocyclic)tropanes. *J Med Chem*, *38*, 3451–3453.

48. Newman, A. H., Allen, A. C., Izenwasser, S., and Katz, J. L. (1994). Novel 3α-(diphenylmethoxy)tropane analogs: potent dopamine uptake inhibitors without cocaine-like behavioral profiles. *J Med Chem*, *37*, 2258–2261.

49. Fandrick, K., Feng, X., Janowsky, A., Johnson, R., and Cashman, J. R. (2003). Bivalent biogenic amine reuptake inhibitors. *Bioorg Med Chem Lett*, *13*, 2151–2154.

50. Carroll, F. I., Blough, B. E., Huang, X., Nie, Z., Mascarella, S. W., Deschamps, J., and Navarro, H. A. (2006). Synthesis and monoamine transporter binding properties of 2,3-cyclo analogues of 3β-(4′-aminophenyl)-2β-tropanemethanol. *J Med Chem*, *49*, 4589–4594.

51. Milius, R. A., Saha, J. K., Madras, B. K., and Neumeyer, J. L. (1991). Synthesis and receptor binding of N-substituted tropane derivatives: high-affinity ligands for the cocaine receptor. *J Med Chem*, *34*, 1728–1731.

52. Madras, B. K., Kamien, J. B., Fahey, M. A., Canfield, D. R., Milius, R. A., Saha, J. K., Neumeyer, J. L., and Spealman, R. D. (1990). N-Modified fluorophenyltropane analogs of cocaine with high affinity for cocaine receptors. *Pharmacol Biochem Behav*, *35*, 949–953.

53. Boja, J. W., Kuhar, M. J., Kopajtic, T., Yang, E., Abraham, P., Lewin, A. H., and Carroll, F. I. (1994). Secondary amine analogues of 3β-(4′-substituted phenyl)tropane-2β-carboxylic acid esters and *N*-norcocaine exhibit enhanced affinity for serotonin and norepinephrine transporters. *J Med Chem*, *37*, 1220–1223.

54. Neumeyer, J. L., Wang, S., Gao, Y., Milius, R. A., Kula, N. S., Campbell, A., Baldessarini, R. J., Zea-Ponce, Y., Baldwin, R. M., and Innis, R. B. (1994). *N*-ω-Fluoroalkyl analogs of (1R)-2β-carbomethoxy-3β-(4-iodophenyl)tropane (β-CIT): radiotracers for positron emission tomography and single photon emission computed tomography imaging of dopamine transporters. *J Med Chem*, *37*, 1558–1561.

55. Neumeyer, J. L., Tamagnan, G., Wang, S., Gao, Y., Milius, R. A., Kula, N. S., and Baldessarini, R. J. (1996). N-Substituted analogs of 2β-carbomethoxy-3β-(4′-iodophenyl) tropane (β-CIT) with selective affinity to dopamine or serotonin transporters in rat forebrain. *J Med Chem*, *39*, 543–548.

56. Elmaleh, D. R., Fischman, A. J., Shoup, T. M., Byon, C., Hanson, R. N., Liang, A. Y., Meltzer, P. C., and Madras, B. K. (1996). Preparation and biological evaluation of

iodine-125-IACFT: a selective SPECT agent for imaging dopamine transporter sites. *J Nucl Med*, *37*, 1197–1202.

57. Holmquist, C. R., Keverline-Frantz, K. I., Abraham, P., Boja, J. W., Kuhar, M. J. K., and Carroll, F. I. (1996). 3α-(4′-Substituted phenyl)tropane-2β-carboxylic acid methyl esters: novel ligands with high affinity and selectivity at the dopamine transporter. *J Med Chem*, *39*, 4139–4141.

58. Carroll, F. I., Tyagi, S., Blough, B. E., Kuhar, M. J., and Navarro, H. A. (2005). Synthesis and monoamine transporter binding properties of 3β-(substituted phenyl)nortropane-2β-carboxylic acid methyl esters: norepinephrine transporter selective compounds. *J Med Chem*, *48*, 3852–3857.

59. Meltzer, P. C., Blundell, P., Yong, Y. F., Chen, Z., George, C., Gonzalez, M. D., and Madras, B. K. (2000). 2-Carbomethoxy-3-aryl-8-bicyclo[3.2.1]octanes: potent non-nitrogen inhibitors of monoamine transporters. *J Med Chem*, *43*, 2982–2991.

60. Meltzer, P. C., Pham-Huu, D. P., and Madras, B. K. (2004). Synthesis of 8-thiabicyclo [3.2.1]oct-2-enes and their binding affinity for the dopamine and serotonin transporters. *Bioorg Med Chem Lett*, *14*, 6007–6010.

61. Torun, L., Liu, S., Madras, B. K., and Meltzer, P. C. (2006). Synthesis of 3-(4-hetero-arylphenyl)-8-oxabicyclo[3.2.1]octane-2-carboxylic acid methyl esters. *Tetrahedron Lett*, *47*, 599–603.

62. Chen, Z., Izenwasser, S., Katz, J. L., Zhu, N., Klein, C. L., and Trudell, M. L. (1996). Synthesis and dopamine transporter affinity of 2-(methoxycarbonyl)-9-methyl-3-phenyl-9-azabicyclo[3.3.1]nonane derivatives. *J Med Chem*, *39*, 4744–4749.

63. Zhang, C., Izenwasser, S., Katz, J. L., Terry, P. D., and Trudell, M. L. (1998). Synthesis and dopamine transporter affinity of the four stereoisomers of (+/−)-2-(methoxycarbo-nyl)-7-methyl-3-phenyl-7-azabicyclo[2.2.1]heptane. *J Med Chem*, *41*, 2430–2435.

64. Kim, D. I., Schweri, M. M., and Deutsch, H. M. (2003). Synthesis and pharmacology of site specific cocaine abuse treatment agents: 8-substituted isotropane (3-azabicyclo [3.2.1]octane) dopamine uptake inhibitors. *J Med Chem*, *46*, 1456–1464.

65. Quirante, J., Vila, X., Bonjoch, J., Kozikowski, A. P., and Johnson, K. M. (2004). 2,3-Disubstituted 6-azabicyclo[3.2.1]octanes as novel dopamine transporter inhibitors. *Bioorg Med Chem*, *12*, 1383–1391.

66. Cini, N., Danieli, E., Menchi, G., Trabocchi, A., Bottoncetti, A., Raspanti, S., Pupi, A., and Guarna, A. (2006). 3-Aza-6,8-dioxabicyclo[3.2.1]octanes as new enantiopure heteroatom-rich tropane-like ligands of human dopamine transporter. *Bioorg Med Chem*, *14*, 5110–5120.

67. Zhao, L., Johnson, K. M., Zhang, M., Flippen-Anderson, J., and Kozikowski, A. P. (2000). Chemical synthesis and pharmacology of 6- and 7-hydroxylated 2-carbo-methoxy-3-(*p*-tolyl)tropanes: antagonism of cocaine's locomotor stimulant effects. *J Med Chem*, *43*, 3283–3294.

68. Meltzer, P. C., Wang, B., Chen, Z., Blundell, P., Jayaraman, M., Gonzalez, M. D., George, C., and Madras, B. K. (2001). Synthesis of 6- and 7- hydroxy-8-azabicyclo [3.2.1]octanes and their binding affinity for the dopamine and serotonin transporters. *J Med Chem*, *44*, 2619–2635.

69. Airaksinen, A. J., Huotari, M., Shvetsov, A., Vainiotalo, P., Mannisto, P. T., Tuomisto, L., Bergstrom, K. A., and Vepsalainen, J. (2005). Synthesis and biological evaluation

of 6/7-exo-methyl-3β-(4-iodo)phenyltropane-2β-carboxylic acid methyl esters. *Eur J Med Chem, 40,* 299–304.

70. Smith, M. P., Johnson, K. M., Zhang, M., Flippen-Anderson, J. L., and Kozikowski, A. P. (1998). Tuning the selectivity of monoamine transporter Inhibitors by the stereochemistry of the nitrogen lone pair. *J Am Chem Soc, 120,* 9072–9073.

71. Tamiz, A. P., Smith, M. P., and Kozikowski, A. P. (2000). Design, synthesis and biological evaluation of 7-azatricyclodecanes: analogues of cocaine. *Bioorg Med Chem Lett, 10,* 297–300.

72. Carroll, F. I., et al. (unpublished).

73. Zhou, J., Zhang, A., Klass, T., Johnson, K. M., Wang, C. Z., Ye, Y. P., and Kozikowski, A. P. (2003). Biaryl analogues of conformationally constrained tricyclic tropanes as potent and selective norepinephrine reuptake inhibitors: synthesis and evaluation of their uptake inhibition at monoamine transporter sites. *J Med Chem, 46,* 1997–2007.

74. Zhou, J., Klass, T., Johnson, K. M., Giberson, K. M., and Kozikowski, A. P. (2005). Discovery of novel conformationally constrained tropane-based biaryl and arylacetylene ligands as potent and selective norepinephrine transporter inhibitors and potential antidepressants. *Bioorg Med Chem Lett, 15,* 2461–2465.

75. Zhou, J., Klass, T., Zhang, A., Johnson, K. M., Wang, C. Z., Ye, Y., and Kozikowski, A. P. (2003). Synthesis and pharmacological evaluation of (Z)-9-(heteroarylmethylene)-7-azatricyclo[4.3.1.0(3,7)]decanes: thiophene analogues as potent norepinephrine transporter inhibitors. *Bioorg Med Chem Lett, 13,* 3565–3569.

76. Hoepping, A., Johnson, K. M., George, C., Flippen-Anderson, J., and Kozikowski, A. P. (2000). Novel conformationally constrained tropane analogues by 6-endo-trig radical cyclization and stille coupling: switch of activity toward the serotonin and/or norepinephrine transporter. *J Med Chem, 43,* 2064–2071.

77. Naseree, T. M., Abraham, P., Kepler, J. A., Carroll, F. I., Lewin, A. H., and Kuhar, M. J. (1990). Synthesis of [³H]WIN 35,065–2: a new radioligand for cocaine receptors. *J Label Compounds Radiopharm, XXVIII,* 1011–1016.

78. Ritz, M. C., Cone, E. J., and Kuhar, M. J. (1990). Cocaine inhibition of ligand binding at dopamine, norepinephrine and serotonin transporters: a structure–activity study. *Life Sci, 46,* 635–645.

79. Boja, J. W., Cline, E. J., Carroll, F. I., Lewin, A. H., Philip, A., Dannals, R., Wong, D., Scheffel, U., and Kuhar, M. J. (1992). High potency cocaine analogs: neurochemical, imaging, and behavioral studies. In *Annals of the New York Academy of Sciences,* Vol. 654, *The Neurobiology of Drug and Alcohol Addiction,,* pp. 282–291.

80. Scheffel, U., Boja, J. W., and Kuhar, M. J. (1989). Cocaine receptors: in vivo labeling with ³H-(−)-cocaine, ³H-WIN 35,065-2 and ³H-WIN 35,428. *Synapse, 4,* 390–392.

81. Kuhar, M. J., Carroll, F. I., Lewin, A. H., Boja, J. W., Scheffel, U., and Wong, D. F. (1997). Imaging transporters for dopamine and other neurotransmitters in brain. In *Neurotransmitter Transporters: Structure and Function,* Reith, M. E. A., Ed. Humana Press, Totowa, NJ, pp. 297–313.

82. Boja, J. W., Mitchell, W. M., Patel, A., Kopajtic, T. A., Carroll, F. I., Lewin, A. H., Abraham, P., and Kuhar, M. J. (1992). High affinity binding of [¹²⁵I]RTI-55 to dopamine and serotonin transporters in rat brain. *Synapse, 12,* 27–36.

83. Cline, E. J., Scheffel, U., Boja, J. W., Mitchell, W. M., Carroll, F. I., Abraham, P., Lewin, A. H., and Kuhar, M. J. (1992). In vivo binding of [^{125}I]RTI-55 to dopamine transporters: pharmacology and regional distribution with autoradiography. *Synapse, 12,* 37–46.

84. Lever, J. R., Scheffel, U., Stathis, M., Seltzman, H. H., Wyrick, C. D., Abraham, P., Parham, K., Thomas, B. F., Boja, J. W., Kuhar, M. J., and Carroll, F. I. (1996). Synthesis and in vivo studies of a selective ligand for the dopamine transporter: 3β-(4-[^{125}I]iodophenyl)tropan-2β-carboxylic acid isopropyl ester ([^{125}I]RTI-121). *Nucl Med Biol, 23,* 277–284.

85. Boja, J. W., Cadet, J. L., Kopajtic, T. A., Lever, J., Seltzman, H. H., Wyrick, C. D., Lewin, A. H., Abraham, P., and Carroll, F. I. (1995). Selective labeling of the dopamine transporter by the high affinity ligand [^{125}I]3β-(iodophenyl)tropan-2β-carboxylic acid isopropyl ester. *Mol Pharmacol, 47,* 779–786.

86. Boja, J. W., Carroll, F. I., Vaughan, R. A., Kopajtic, T., and Kuhar, M. J. (1998). Multiple binding sites for [^{125}I]RTI-121 and other cocaine analogs in rat frontal cerebral cortex. *Synapse, 30,* 9–17.

87. Musachio, J. L., Keverline, K. I., Carroll, F. I., and Dannals, R. F. (1996). 3β-(*p*-Trimethylsilylphenyl)tropane-2β-carboxylic acid methyl ester: a new precursor for the preparation of [^{123}I]RTI-55. *Appl Radiat Isot, 47,* 79–81.

88. Kula, N. S., Baldessarini, R. J., Tarazi, F. I., Fisser, R., Wang, S., Trometer, J., and Neumeyer, J. L. (1999). [^{3}H]β-CIT: a radioligand for dopamine transporters in rat brain tissue. *Eur J Pharmacol, 385,* 291–294.

89. Zhan, Y., Saindane, A. M., Scheffel, U., Carroll, F. I., Holmquist C. R., Kepler, J. A., Taylor, G. F., and Kuhar, M. J. (1997). RTI-352: A 3α analogue of RTI-55 as an in vivo dopamine transporter binding ligand. *Synapse, 25,* 389–392.

90. Letchworth, S. R., Smith, H. R., Porrino, L. J., Bennett, B. A., Davies, H. M., Sexton, T., and Childers, S. R. (2000). Characterization of a tropane radioligand, [^{3}H]2β-propanoyl-3β-(4-tolyl)tropane ([^{3}H]PTT), for dopamine transport sites in rat brain. *J Pharmacol Exp Ther, 293,* 686–696.

91. Dollé, F., Emond, P., Mavel, S., Demphel, S., Hinnen, F., Mincheva, Z., Saba, W., Valette, H., Chalon, S., Halldin, C., Helfenbein, J., Legaillard, J., Madelmont, J. C., Deloye, J. B., Bottlaender, M., and Guilloteau, D. (2006). Synthesis, radiosynthesis and in vivo preliminary evaluation of [^{11}C]LBT-999, a selective radioligand for the visualisation of the dopamine transporter with, PET. *Bioorg Med Chem, 14,* 1115–1125.

92. Kung, M. P., Essman, W. D., Frederick, D., Meegalla, S., Goodman, M., Mu, M., Lucki, I., and Kung, H. F. (1995). IPT: a novel iodinated ligand for the CNS dopamine transporter. *Synapse, 20,* 316–324.

93. Chalon, S., Garreau, L., Emond, P., Zimmer, L., Vilar, M. P., Besnard, J. C., and Guilloteau, D. (1999). Pharmacological characterization of (*E*)-*N*-(3-Iodoprop-2-enyl)-2β-carbomethoxy-3β-(4′-methylphenyl)nortropane as a selective and potent inhibitor of the neuronal dopamine transporter. *J Pharmacol Exp Ther, 291,* 648–654.

94. Garreau, L., Emond, P., Belzung, C., Guilloteau, D., Frangin, Y., Besnard, J. C., and Chalon, S. (1997). *N*-(3-Iodoprop-2*E*-enyl)-2β-carbomethoxy-3β-(3′,4′-dichloro phenyl) nortropane (β-CDIT), a tropane derivative: pharmacological characterization as a specific ligand for the dopamine transporter in the rodent brain. *J Pharmacol Exp Ther, 282,* 467–474.

95. Bergstroem, K. A., Halldin, C., Hall, H., Lundkvist, C., Ginovart, N., Swahn, C. G., and Farde, L. (1997). In vitro and in vivo characterisation of nor-β-CIT: a potential

radioligand for visualisation of the serotonin transporter in the brain. *Eur J Nucl Med*, *24*, 596–601.

96. Scheffel, U., Lever, J. R., Abraham, P., Parham, K. R., Mathews, W. B., Kopajtic, T., Carroll, F. I., and Kuhar, M. J. (1997). N-Substituted phenyltropanes as in vivo binding ligands for rapid imaging studies of the dopamine transporter. *Synapse*, *25*, 345–349.

97. Zhong, D., Kotian, P., Wyrick, C. D., Seltzman, H. H., Kepler, J. A., Boja, J. W., Kuhar, M. J., and Carroll, F. I. (1999). Synthesis of 3β-(4-[^{125}I]iodophenyl)tropane-2β-pyrrolidine carboxamide ([^{125}I]RTI-229). *J Label Compounds Radiopharm*, *42*, 281–286.

98. Navarro, H. A., Xu, H., Zhong, D., Blough, B. E., Ross, W. P., Kuhar, M. J., and Carroll, F. I. (2001). [^{125}I]3β-(4-Ethyl-3-iodophenyl)nortropane-2β-carboxylic acid methyl ester ([^{125}I]EINT): a potent and selective radioligand for the brain serotonin transporter. *Synapse*, *41*, 241–247.

99. Carroll, F. I., Rahman, M. A., Abraham, P., Parham, K., Lewin, A. H., Dannals, R. F., Shaya, E., Scheffel, U., Wong, D. F., Boja, J. W., and Kuhar, M. J. (1991). [^{123}I]3β-(4-Iodophenyl)tropan-2β-carboxylic acid methyl ester (RTI-55), a unique cocaine receptor ligand for imaging the dopamine and serotonin transporters in vivo. *Med Chem Res*, *1*, 289–294.

100. Shaya, E. K., Scheffel, U., Dannals, R. F., Ricaurte, G. A., Carroll, F. I., Wagner, H. N., Jr., Kuhar, M. J., and Wong, D. F. (1992). In vivo imaging of dopamine reuptake sites in the primate brain using single photon emission computer tomography (SPECT) and iodine-123 labeled RTI-55. *Synapse*, *10*, 169–172.

101. Morgan, G. F., and Nowotnik, D. P. (1999). Development of dopamine transporter imaging agents for diagnosis of Parkinson's disease. *Drug News Perspect*, *12* 137.

102. Brooks, D. J., and Piccini, P. (2006). Imaging in Parkinson's disease: the role of mono-amines in behavior. *Biol Psychiatry*, *59*, 908–918.

103. Filippi, L., Manni, C., Pierantozzi, M., Brusa, L., Danieli, R., Stanzione, P., and Schillaci, O. (2005). ^{123}I-FP-CIT semi-quantitative SPECT detects preclinical bilateral dopaminergic deficit in early Parkinson's disease with unilateral symptoms. *Nucl Med Commun*, *26*, 421–426.

104. Van Laere, K., De Ceuninck, L., Dom, R., Van den Eynden, J., Vanbilloen, H., Cleynhens, J., Dupont, P., Bormans, G., Verbruggen, A., and Mortelmans, L. (2004). Dopamine transporter SPECT using fast kinetic ligands: ^{123}I-FP-β-CIT versus ^{99m}Tc-TRODAT-1. *Eur J Nucl Med Mol Imaging*, *31*, 1119–1127.

105. Cheon, K. A., Ryu, Y. H., Namkoong, K., Kim, C. H., Kim, J. J., and Lee, J. D. (2004). Dopamine transporter density of the basal ganglia assessed with [^{123}I]IPT SPECT in drug-naive children with Tourette's disorder. *Psychiatry Res*, *130*, 85–95.

106. Meltzer, P. C., Blundell, P., Zona, T., Yang, L., Huang, H., Bonab, A. A., Livni, E., Fischman, A., and Madras, B. K. (2003). A second-generation 99m technetium single photon emission computed tomography agent that provides in vivo images of the dopamine transporter in primate brain. *J Med Chem*, *46*, 3483–3496.

107. Brooks, D. J., Frey, K. A., Marek, K. L., Oakes, D., Paty, D., Prentice, R., Shults, C. W., and Stoessl, A. J. (2003). Assessment of neuroimaging techniques as biomarkers of the progression of Parkinson's disease. *Exp Neurol*, *184*(Suppl 1), S68–S79.

108. Dhawan, V., and Eidelberg, D. (2001). SPECT imaging in Parkinson's disease. *Adv Neurol*, *86*, 205–213.

109. Booij, J., Reneman, L., Lavalaye, J., Knol, R. J. J., de Bruin, K., Speelman, J. D., and Janssen, A. G. M. (2001). Dopamine transporter imaging as an in vivo marker of dopaminergic neurons. *Drugs Future*, *26*, 271–279.

110. Mozley, P. D., Stubbs, J. B., Plossl, K., Dresel, S. H., Barraclough, E. D., Alavi, A., Araujo, L. I., and Kung, H. F. (1998). Biodistribution and dosimetry of TRODAT-1: a technetium-99 m tropane for imaging dopamine transporters. *J Nucl Med*, *39*, 2069–2076.

111. Ye, B., Kuang, A., et al. (2002). β-CIT labeled with ^{131}I and its preliminary clinical practice. *Tongweisu*, *15*, 141–144.

112. al-Tikriti, M. S., Zea-Ponce, Y., Baldwin, R. M., Zoghbi, S. S., Laruelle, M., Seibyl, J. P., Giddings, S. S., Scanley, B. E., Charney, D. S., Hoffer, P. B., et al. (1995). Characterization of the dopamine transporter in nonhuman primate brain: homogenate binding, whole body imaging, and ex vivo autoradiography using [^{125}I] and [^{123}I]IPCIT. *Nucl Med Biol*, *22*, 649–658.

113. Reneman, L., Booij, J., Lavalaye, J., De Bruin, K., De Wolff, F. A., Koopmans, R. P., Stoof, J. C., and Den Heeten, G. J. (1999). Comparative in vivo study of iodine-123-labeled β-CIT and nor-β-CIT binding to serotonin transporters in rat brain. *Synapse*, *34*, 77–80.

114. Goodman, M. M., Chen, P., Plisson, C., Martarello, L., Galt, J., Votaw, J. R., Kilts, C. D., Malveaux, G., Camp, V. M., Shi, B., Ely, T. D., Howell, L., McConathy, J., and Nemeroff, C. B. (2003). Synthesis and characterization of iodine-123 labeled 2β-carbomethoxy-3β-(4'-((Z)-2-iodoethenyl)phenyl)nortropane: a ligand for in vivo imaging of serotonin transporters by single-photon-emission tomography. *J Med Chem*, *46*, 925–935.

115. Plisson, C., McConathy, J., Martarello, L., Malveaux, E. J., Camp, V. M., Williams, L., Votaw, J. R., and Goodman, M. M. (2004). Synthesis, radiosynthesis, and biological evaluation of carbon-11 and iodine-123 labeled 2β-carbomethoxy-3β-[4'-((Z)-2-haloethenyl)phenyl]tropanes: candidate radioligands for in vivo imaging of the serotonin transporter. *J Med Chem*, *47*, 1122–1135.

116. Quinlivan, M., Mattner, F., Papazian, V., Zhou, J., Katsifis, A., Emond, P., Chalon, S., Kozikowski, A., Guilloteau, D., Kassiou, M. (2003). Synthesis and evaluation of iodine-123 labelled tricyclic tropanes as radioligands for the serotonin transporter. *Nucl Med Biol*, *30*, 741–746.

117. Chen, Z., Zhou, X., et al. (2003). Preclinical pharmacological studies of ^{131}I-FP-β-CIT. *Hejishu*, *26*, 858–862.

118. Elsinga, P. H., Hatano, K., and Ishiwata, K. (2006). PET tracers for imaging of the dopaminergic system. *Curr Med Chem*, *13*, 2139–2153.

119. Schönbächler, R. D., Gucker, P. M., Arigoni, M., Kneifel, S., Vollenweider, F. X., Buck, A., Burger, C., Berthold, T., Bruhlmeier, M., Schubiger, P. A., and Ametamey, S. M. (2002). PET imaging of dopamine transporters in the human brain using [^{11}C]-β-CPPIT, a cocaine derivative lacking the 2β-ester function. *Nucl Med Biol*, *29*, 19–27.

120. Ma, Y., Dhawan, V., Mentis, M., Chaly, T., Spetsieris, P. G., and Eidelberg, D. (2002). Parametric mapping of [^{18}F]FPCIT binding in early stage Parkinson's disease: a PET study. *Synapse*, *45*, 125–133.

121. Laihinen, A. O., Rinne, J. O., Nagren, K. A., Lehikoinen, P. K., Oikonen, V. J., Ruotsalainen, U. H., Ruottinen, H. M., and Rinne, U. K. (1995). PET studies on brain monoamine transporters with carbon-11β-CIT in Parkinson's disease. *J Nucl Med*, *36*, 1263–1267.

122. Carroll, F. I., Gao, Y., Abraham, P., Lewin, A. H., Lew, R., Patel, A., Boja, J. W., and Kuhar, M. J. (1992). Probes for the cocaine receptor: potentially irreversible ligands for the dopamine transporter. *J Med Chem*, *35*, 1813–1817.

123. Zou, M. F., Kopajtic, T., Katz, J. L., Wirtz, S., Justice, J. B., Jr., and Newman, A. H. (2001). Novel tropane-based irreversible ligands for the dopamine transporter. *J Med Chem*, *44*, 4453–4461.

124. Patel, A. P., Carroll, F. I., and Kuhar, M. J. (1997). Turnover of rat dopamine transporter protein in rDAT-LLC-PK1 cells. In *Neurotransmitter Release and Uptake*, Pogun, S. Ed. *NATO ASI Series*, Springer-Verlag. Berlin Vol. H 100.

125. Wang, L. C., Berfield, J. L., Kuhar, M. J., Carroll, F. I., and Reith, M. E. A. (2000). RTI-76, an isothiocyanate derivative of a phenyltropane cocaine analog, as a tool for irreversibly inactivating dopamine transporter function in vitro. *Naunyn Schmiedebergs Arch Pharmacol*, *362*, 238–247.

126. Kimmel, H. L., Carroll, F. I., and Kuhar, M. J. (2000). Dopamine transporter synthesis and degradation rate in rat striatum and nucleus accumbens using RTI-76. *Neuropharmacology*, *39*, 578–585.

127. Vicentic, A., Battaglia, G., Carroll, F. I., and Kuhar, M. J. (1999). Serotonin transporter production and degradation rates: studies with RTI-76. *Brain Res*, *841*, 1–10.

128. Kimmel, H. L., Joyce, A. R., Carroll, F. I., and Kuhar, M. J. (2001). Dopamine D1 and D2 receptors influence dopamine transporter synthesis and degradation in the rat. *J Pharmacol Exp Ther*, *298*, 129–140.

129. Kimmel, H. L., Carroll, F. I., and Kuhar, M. J. (2001). RTI-76, an irreversible inhibitor of dopamine transporter binding, increases locomotor activity in the rat at high doses. *Brain Res*, *897*, 157–163.

130. Arroyo, M., Baker, W. A., and Everitt, B. J. (2000). Cocaine self-administration in rats differentially alters mRNA levels of the monoamine transporters and striatal neuropeptides. *Brain Res Mol Brain Res*, *83*, 107–120.

131. Boulay, D., Duterte-Boucher, D., Leroux-Nicollet, I., Naudon, L., and Costentin, J. (1996). Locomotor sensitization and decrease in [³H]mazindol binding to the dopamine transporter in the nucleus accumbens are delayed after chronic treatments by GBR12783 or cocaine. *J Pharmacol Exp Ther*, *278*, 330–337.

132. Cerruti, C., Pilotte, N. S., Uhl, G., and Kuhar, M. J. (1994). Reduction in dopamine transporter mRNA after cessation of repeated cocaine administration. *Mol Brain Res*, *22*, 132–138.

133. Claye, L. H., Akunne, H. C., Davis, M. D., DeMattos, S., and Soliman, K. F. (1995). Behavioral and neurochemical changes in the dopaminergic system after repeated cocaine administration. *Mol Neurobiol*, *11*, 55–66.

134. Maggos, C. E., Spangler, R., Zhou, Y., Schlussman, S. D., Ho, A., and Kreek, M. J. (1997). Quantitation of dopamine transporter mRNA in the rat brain: mapping, effects of "binge" cocaine administration and withdrawal. *Synapse*, *26*, 55–61.

135. Xia, Y., Goebel, D. J., Kapatos, G., and Bannon, M. J. (1992). Quantitation of rat dopamine transporter mRNA: effects of cocaine treatment and withdrawal. *J Neurochem, 59,* 1179–1182.

136. Kimmel, H. L., Carroll, F. I., and Kuhar, M. J. (2003). Withdrawal from repeated cocaine alters dopamine transporter protein turnover in the rat striatum. *J Pharmacol Exp Ther, 304,* 15–21.

137. Filip, M., and Siwanowicz, J. (2001). Implication of the nucleus accumbens shell, but not core, in the acute and sensitizing effects of cocaine in rats. *Pol J Pharmacol, 53,* 459–466.

138. Pierce, R. C., and Kalivas, P. W. (1995). Amphetamine produces sensitized increases in locomotion and extracellular dopamine preferentially in the nucleus accumbens shell of rats administered repeated cocaine. *J Pharmacol Exp Ther, 275,* 1019–1029.

139. Todtenkopf, M. S., Carreiras, T., Melloni, R. H., and Stellar, J. R. (2002). The dorsomedial shell of the nucleus accumbens facilitates cocaine-induced locomotor activity during the induction of behavioral sensitization. *Behav Brain Res, 131,* 9–16.

140. Lever, J. R., Carroll, F. I., Patel, A., Abraham, P., Boja, J., Lewin, A., and Lew, R. (1993). Radiosynthesis of a photoaffinity probe for the cocaine receptor of the dopamine transporter: 3β-(*p*-Chlorophenyl)tropan-2β-carboxylic acid *m*-([^{125}i]-iodo)-*p*-azido-phenethyl ester ([^{125}i]-RTI-82). *J Label Compounds Radiopharm, 33,* 1131–1137.

141. Patel, A., Boja, J. W., Lever, J., Lew, R., Simantov, R., Carroll, F. I., Lewin, A. H., Abraham, P., Gao, Y., and Kuhar, M. J. (1992). A cocaine analog and a GBR analog label the same protein in rat striatal membranes. *Brain Res, 576,* 173–174.

142. Vaughan, R. A. (1995). Photoaffinity-labeled ligand binding domains on dopamine transporters identified by peptide mapping. *Mol Pharmacol, 47,* 956–964.

143. Vaughan, R. A., and Kuhar, M. J. (1996). Dopamine transporter ligand binding domains. *J Biol Chem, 271,* 21672–21680.

144. Vaughan, R. A., Sakrikar, D. S., Parnas, M. L., Adkins, S., Foster, J. D., Duval, R. A., Lever, J. R., Kulkarni, S. S., and Hauck-Newman, A. (2007). Localization of cocaine analog [^{125}I]RTI 82 irreversible binding to transmembrane domain 6 of the dopamine transporter. *J Biol Chem, 282,* 8915–8925.

145. Newman, A. H., and Kulkarni, S. (2002). Probes for the dopamine transporter: new leads toward a cocaine-abuse therapeutic–a focus on analogues of benztropine and rimcazole. *Med Res Rev, 22,* 429–464.

146. Vaughan, R. A., Agoston, G. E., Lever, J. R., and Newman, A. H. (1999). Differential binding of tropane-based photoaffinity ligands on the dopamine transporter. *J Neurosci, 19,* 630–636.

147. Rasmussen, S. G., Adkins, E. M., Carroll, F. I., Maresch, M. J., and Gether, U. (2003). Structural and functional probing of the biogenic amine transporters by fluorescence spectroscopy. *Eur J Pharmacol, 479,* 13–22.

148. Rasmussen, S. G. F., Carroll, F. I., Maresch, M. J., Jensen, A. D., Tate, C. G., and Gether, U. (2001). Biophysical characterization of the cocaine binding pocket in the serotonin transporter using a fluorescent cocaine-analogue as a molecular reporter. *J Biol Chem, 276,* 4717–4723.

149. Cha, J. H., Zou, M. F., Adkins, E. M., Rasmussen, S. G., Loland, C. J., Schoenenberger, B., Gether, U., and Newman, A. H. (2005). Rhodamine-labeled

2β-carbomethoxy-3β-(3,4-dichlorophenyl)tropane analogues as high-affinity fluorescent probes for the dopamine transporter. *J Med Chem*, *48*, 7513–7516.

150. Heikkila, R. E., Manzino, L., and Cabbat, F. S. (1981). Stereospecific effects of cocaine derivatives on ^{3}H-dopamine uptake: correlations with behavioral effects. *Subst Alcohol Actions/Misuse*, *2*, 115–121.

151. Reith, M. E. A., Meisler, B. E., and Lajtha, A. (1985). Locomotor effects of cocaine, cocaine congeners, and local anesthetics in mice. *Pharmacol Biochem Behav*, *23*, 831–836.

152. Heikkila, R. E., Cabbat, F. S., Manzino, L., and Duvoisin, R. C. (1979). Rotational behavior induced by cocaine analogs in rats with unilateral 6-hydroxydopamine lesions of the substantia nigra: dependence upon dopamine uptake inhibition. *J Pharmacol Exp Ther*, *211*, 189–194.

153. Reith, M. E. A., Meisler, B. E., Sershen, H., and Lajtha, A. (1986). Structural requirements for cocaine congeners to interact with dopamine and serotonin uptake sites in mouse brain and to induce stereotyped behavior. *Biochem Pharmacol*, *35*, 1123–1129.

154. Jarbe, T. U. C. (1981). Cocaine cue in pigeons: time course studies and generalization to structurally related compounds (norcocaine, WIN 35,428 and 335,065-2) and (+)-amphetamine. *Br J Pharmacol*, *73*, 843–852.

155. Spealman, R. D., Goldberg, S. R., Kelleher, R. T., Goldberg, D. M., and Charlton, J. P. (1977). Some effects of cocaine and two cocaine analogs on schedule-controlled behavior of squirrel monkeys. *J Pharmacol Exp Ther*, *202*, 500–509.

156. Spealman, R. D., Goldberg, S. R., Kelleher, R. T., Morse, W. H., Goldberg, D. M., and Hakansson, C. G. (1979). Effects of norcocaine and some norcocaine derivatives on schedule-controlled behavior of pigeons and squirrel monkeys. *J Pharmacol Exp Ther*, *210*, 196–205.

157. Spealman, R. D., Kelleher, R. T., and Goldberg, S. R. (1983). Stereoselective behavioral effects of cocaine and a phenyltropane analog. *J Pharmacol Exp Ther*, *225*, 509–514.

158. Spealman, R. D., and Kelleher, R. T. (1981). Self-administration of cocaine derivatives by squirrel monkeys. *J Pharmacol Exp Ther*, *216*, 532–536.

159. Carroll, F. I., Howell, L. L., and Kuhar, M. J. (1999). Pharmacotherapies for treatment of cocaine abuse: preclinical aspects. *J Med Chem*, *42*, 2721–2736.

160. Carroll, F. I. (2003). 2002 Medicinal Chemistry Division Award address: Monoamine transporters and opioid receptors—targets for addiction therapy. *J Med Chem*, *46*, 1775–1794.

161. Howell, L. L., and Wilcox, K. M. (2001). The dopamine transporter and cocaine medication development: drug self-administration in nonhuman primates. *J Pharmacol Exp Ther*, *298*, 1–6.

162. Mello, N. K., and Negus, S. S. (1996). Preclinical evaluation of pharmacotherapies for treatment of cocaine and opioid abuse using drug self-administration procedures. *Neuropsychopharmacology*, *14*, 375–424.

163. Howell, L. L., Carroll, F. I., Votaw, J. R., Goodman, M. M., and Kimmel, H. L. (2007). Effects of combined dopamine and serotonin transporter inhibitors on cocaine self-administration in rhesus monkeys. *J Pharmacol Exp Ther*, *320*, 757–765.

164. Wilcox, K. M., Lindsey, K. P., Votaw, J. R., Goodman, M. M., Martarello, L., Carroll, F. I., and Howell, L. L. (2002). Self-administration of cocaine and the cocaine analog RTI-113: relationship to dopamine transporter occupancy determined by PET neuroimaging in rhesus monkeys. *Synapse, 43*, 78–85.

165. Lindsey, K. P., Wilcox, K. M., Votaw, J. R., Goodman, M. M., Plisson, C., Carroll, F. I., Rice, K. C., and Howell, L. L. (2004). Effects of dopamine transporter inhibitors on cocaine self-administration in rhesus monkeys: relationship to transporter occupancy determined by positron emission tomography neuroimaging. *J Pharmacol Exp Ther, 309*, 959–969.

166. Nader, M. A., Grant, K. A., Davies, H. M., Mach, R. H., and Childers, S. R. (1997). The reinforcing and discriminative stimulus effects of the novel cocaine analog 2β-propanoyl-3β-(4-tolyl)-tropane in rhesus monkeys. *J Pharmacol Exp Ther, 280*, 541–550.

167. Frackiewicz, E. J., Jhee, S. S., Shiovitz, T. M., Webster, J., Topham, C., Dockens, R. C., Whigan, D., Salazar, D. E., and Cutler, N. R. (2002). Brasofensine treatment for Parkinson's disease in combination with levodopa/carbidopa. *Ann Pharmacother, 36*, 225–230.

168. Yu, P. (2000). Brasofensine neurosearch. *Curr Opin Investig Drugs, 1*, 504–507.

REFERENCES FOR TABLE 6-4

1. Hantraye, P., Brownell, A. L., Elmaleh, D., Spealman, R. D., Wullner, U., Brownell, G. L., Madras, B. K., and Isacson, O. (1992). Dopamine fiber detection by [^{11}C]-CFT and PET in a primate model of parkinsonism. *Neuroreport, 3*, 265–268.

2. Wong, D. W., Yung, B., Dannals, R. F., Shaya, E. S., Ravert, H. T., Chen, C. A., Chan, B., Folio, T., Scheffel, U., Ricaurte, G. A., Neumeyer, J. L., Wagner, H. N., Jr., and Kuhar, M. J. (1993). In vivo imaging of baboon and human dopamine transporters by positron emission tomography using [^{11}C]WIN 35,428. *Synapse, 15*, 130–142.

3. Dannals, R. F., Neumeyer, J. L., Milius, R. A., Ravert, H. T., Wilson, A. A., and Wagner, H. N., Jr. (1993). Synthesis of a radiotracer for studying dopamine uptake sites in vivo using PET: 2β-carbomethoxy-3β-(4-fluorophenyl) [N-^{11}C-methyl]tropane ([^{11}C]CFT or [^{11}C]WIN-35,428). *J Label Compounds Radiopharm, 33*, 147–153.

4. Meltzer, P. C., Liang, A. Y., Brownell, A. L., Elmaleh, D. R., and Madras, B. K. (1993). Substituted 3-phenyltropane analogs of cocaine: synthesis, inhibition of binding at cocaine recognition sites, and positron emission tomography imaging. *J Med Chem, 36*, 855–862.

5. Frost, J. J., Rosier, A. J., Reich, S. G., Smith, J. S., Ehlers, M. D., Snyder, S. H., Ravert, H. T., and Dannals, R. F. (1993). Positron emission tomographic imaging of the dopamine transporter with ^{11}C-WIN 35,428 reveals marked declines in mild Parkinson's disease. *Ann Neurol, 34*, 423–431.

6. Haaparanta, M., Bergman, J., Laakso, A., Hietala, J., and Solin, O. (1996). [^{18}F]CFT ([^{18}F]WIN 35,428), a radioligand to study the dopamine transporter with PET: biodistribution in rats. *Synapse, 23*, 321–327.

7. Bergman, J., and Solin, O. (1997). Fluorine-18-labeled fluorine gas for synthesis of tracer molecules. *Nucl Med Biol, 24*, 677–683.

8. Villemagne, V., Yuan, J., Wong, D. F., Dannals, R. F., Hatzidimitriou, G., Mathews, W. B., Ravert, H. T., Musachio, J., McCann, U. D., and Ricaurte, G. A. (1998). Brain

dopamine neurotoxicity in baboons treated with doses of methamphetamine comparable to those recreationally abused by humans: evidence from [^{11}C]WIN-35,428 positron emission tomography studies and direct in vitro determinations. *J Neurosci, 18,* 419–427.

9. Laakso, A., Bergman, J., Haaparanta, M., Vilkman, H., Solin, O., and Hietala, J. (1998). [^{18}F]CFT [(^{18}F)WIN 35,428], a radioligand to study the dopamine transporter with PET: characterization in human subjects. *Synapse, 28,* 244–250.

10. Ouchi, Y., Yoshikawa, E., Okada, H., Futatsubashi, M., Sekine, Y., Iyo, M., and Sakamoto, M. (1999). Alterations in binding site density of dopamine transporter in the striatum, orbitofrontal cortex, and amygdala in early Parkinson's disease: compartment analysis for β-CFT binding with positron emission tomography. *Ann Neurol, 45,* 601–610.

11. Nurmi, E., Bergman, J., Eskola, O., Solin, O., Hinkka, S. M., Sonninen, P., and Rinne, J. O. (2000). Reproducibility and effect of levodopa on dopamine transporter function measurements: a [^{18}F]CFT PET study. *J Cereb Blood Flow Metab, 20,* 1604–1609.

12. Ouchi, Y., Kanno, T., Okada, H., Yoshikawa, E., Futatsubashi, M., Nobezawa, S., Torizuka, T., and Tanaka, K. (2001). Changes in dopamine availability in the nigrostriatal and mesocortical dopaminergic systems by gait in Parkinson's disease. *Brain, 124,* 784–792.

13. Sekine, Y., Iyo, M., Ouchi, Y., Matsunaga, T., Tsukada, H., Okada, H., Yoshikawa, E., Futatsubashi, M., Takei, N., and Mori, N. (2001). Methamphetamine-related psychiatric symptoms and reduced brain dopamine transporters studied with, PET. *Am J Psychiatry, 158,* 1206–1214.

14. Tsukada, H., Nishiyama, S., Kakiuchi, T., Ohba, H., Sato, K., and Harada, N. (2001). Ketamine alters the availability of striatal dopamine transporter as measured by [^{11}C]β-CFT and [^{11}C]β-CIT-FE in the monkey brain. *Synapse, 42,* 273–280.

15. Rinne, O. J., Nurmi, E., Ruottinen, H. M., Bergman, J., Eskola, O., and Solin, O. (2001). [^{18}F]FDOPA and [^{18}F]CFT are both sensitive PET markers to detect presynaptic dopaminergic hypofunction in early Parkinson's disease. *Synapse, 40,* 193–200.

16. Nurmi, E., Bergman, J., Eskola, O., Solin, O., Vahlberg, T., Sonninen, P., and Rinne, J. O. (2003). Progression of dopaminergic hypofunction in striatal subregions in Parkinson's disease using [^{18}F]CFT, PET. *Synapse, 48,* 109–115.

17. Muller, L., Halldin, C., Farde, L., Karlsson, P., Hall, H., Swahn, C. G., Neumeyer, J., Gao, Y., and Milius, R. (1993). [^{11}C]β-CIT, a cocaine analogue: preparation, autoradiography and preliminary PET investigations. *Nucl Med Biol, 20,* 249–255.

18. Farde, L., Halldin, C., Muller, L., Suhara, T., Karlsson, P., and Hall, H. (1994). PET study of [^{11}C]β-CIT binding to monoamine transporters in the monkey and human brain. *Synapse, 16,* 93–103.

19. Laruelle, M., Wallace, E., Seibyl, J. P., Baldwin, R. M., Zea-Ponce, Y., Zoghbi, S. S., Neumeyer, J. L., Charney, D. S., Hoffer, P. B., and Innis, R. B. (1994). Graphical, kinetic, and equilibrium analyses of in vivo [^{123}I]β-CIT binding to dopamine transporters in healthy human subjects. *J Cereb Blood Flow Metab, 14,* 982–994.

20. Seibyl, J. P., Marek, K. L., Quinlan, D., Sheff, K., Zoghbi, S., Zea-Ponce, Y., Baldwin, R. M., Fussell, B., Smith, E. O., and Charney, D. S., et al. (1995). Decreased single-photon emission computed tomographic [^{123}I]β-CIT striatal uptake correlates with symptom severity in Parkinson's disease. *Ann Neurol, 38,* 589–598.

21. Nagren, K., Muller, L., Halldin, C., Swahn, C. G., and Lehikoinen, P. (1995). Improved synthesis of some commonly used PET radioligands by the use of [^{11}C]methyl triflate. *Nucl Med Biol, 22*, 235–239.

22. Hume, S. P., Luthra, S. K., Brown, D. J., Opacka-Juffry, J., Osman, S., Ashworth, S., Myers, R., Brady, F., Carroll, F. I., Kuhar, M. J., and Brooks, D. J. (1996). Evaluation of [^{11}C]RTI-121 as a selective radioligand for PET studies of the dopamine transporter. *Nucl Med Biol, 23*, 377–384.

23. Zheng, Q. H., and Mulholland, G. K. (1996). Improved synthesis of β-CIT and [^{11}C]β-CIT labeled at nitrogen or oxygen positions. *Nucl Med Biol, 23*, 981–986.

24. Ginovart, N., Lundin, A., Farde, L., Halldin, C., Backman, L., Swahn, C. G., Pauli, S., and Sedvall, G. (1997). PET study of the pre- and post-synaptic dopaminergic markers for the neurodegenerative process in Huntington's disease. *Brain, 120*(Pt 3), 503–514.

25. Lundkvist, C., Halldin, C., Swahn, C. G., Ginovart, N., and Farde, L. (1999). Different brain radioactivity curves in a PET study with [^{11}C]β-CIT labelled in two different positions. *Nucl Med Biol, 26*, 343–350.

26. Cicchetti, F., Brownell, A. L., Williams, K., Chen, Y. I., Livni, E., and Isacson, O. (2002). Neuroinflammation of the nigrostriatal pathway during progressive 6-OHDA dopamine degeneration in rats monitored by immunohistochemistry and PET imaging. *Eur J Neurosci, 15*, 991–998.

27. Wilson, A. A., DaSilva, J. N., and Houle, S. (1994). Facile radiolabelling and purification of 2β-[O-^{11}CH$_3$]-carbomethoxy-3β-aryltropanes: radiotracers for the dopamine transporter. *J Label Compounds Radiopharm, 34*, 759–765.

28. Wilson, A. A., DaSilva, J. N., and Houle, S. (1996). In vivo evaluation of [^{11}C]- and [^{18}F]-labelled cocaine analogues as potential dopamine transporter ligands for positron emission tomography. *Nucl Med Biol, 23*, 141–146.

29. Brownell, A. L., Elmaleh, D. R., Meltzer, P. C., Shoup, T. M., Brownell, G. L., Fischman, A. J., and Madras, B. K. (1996). Cocaine congeners as PET imaging probes for dopamine terminals. *J Nucl Med, 37*, 1186–1192.

30. Goodman, M. M., Kabalka, G. W., Kung, M. P., Kung, H. F., and Meyer, M. A. (1994). Synthesis of *N*-3-[^{18}F]fluoropropyl-2β-carbomethoxy-3β-(4-chlorophenyl)tropane: a high affinity neuroligand to map dopamine reuptake sites by PET. *J Label Compounds Radiopharm, 35*, 488–490.

31. Goodman, M. M., Kilts, C. D., Keil, R., Shi, B., Martarello, L., Xing, D., Votaw, J., Ely, T. D., Lambert, P., Owens, M. J., Camp, V. M., Malveaux, E., and Hoffman, J. M. (2000). ^{18}F-Labeled FECNT: a selective radioligand for PET imaging of brain dopamine transporters. *Nucl Med Biol, 27*, 1–12.

32. Deterding, T. A., Votaw, J. R., Wang, C. K., Eshima, D., Eshima, L., Keil, R., Malveaux, E., Kilts, C. D., Goodman, M. M., and Hoffman, J. M. (2001). Biodistribution and radiation dosimetry of the dopamine transporter ligand. *J Nucl Med, 42*, 376–381.

33. Votaw, J. R., Howell, L. L., Martarello, L., Hoffman, J. M., Kilts, C. D., Lindsey, K. P., and Goodman, M. M. (2002). Measurement of dopamine transporter occupancy for multiple injections of cocaine using a single injection of [F-18]FECNT. *Synapse, 44*, 203–210.

34. Wilcox, K. M., Lindsey, K. P., Votaw, J. R., Goodman, M. M., Martarello, L., Carroll, F. I., and Howell, L. L. (2002). Self-administration of cocaine and the cocaine analog RTI-113: relationship to dopamine transporter occupancy determined by PET neuroimaging in rhesus monkeys. *Synapse, 43*, 78–85.

35. Davis, M. R., Votaw, J. R., Bremner, J. D., Byas-Smith, M. G., Faber, T. L., Voll, R. J., Hoffman, J. M., Grafton, S. T., Kilts, C. D., and Goodman, M. M. (2003). Initial human PET imaging studies with the dopamine transporter ligand ^{18}F-FECNT. *J Nucl Med, 44,* 855–861.

36. Votaw, J., Byas-Smith, M., Hua, J., Voll, R., Martarello, L., Levey, A. I., Bowman, F. D., and Goodman, M. (2003). Interaction of isoflurane with the dopamine transporter. *Anesthesiology, 98,* 404–411.

37. Lindsey, K. P., Wilcox, K. M., Votaw, J. R., Goodman, M. M., Plisson, C., Carroll, F. I., Rice, K. C., and Howell, L. L. (2004). Effects of dopamine transporter inhibitors on cocaine self-administration in rhesus monkeys: relationship to transporter occupancy determined by positron emission tomography neuroimaging. *J Pharmacol Exp Ther, 309,* 959–969.

38. Votaw, J. R., Byas-Smith, M. G., Voll, R., Halkar, R., and Goodman, M. M. (2004). Isoflurane alters the amount of dopamine transporter expressed on the plasma membrane in humans. *Anesthesiology, 101,* 1128–1135.

39. Goodman, M. M., Keil, R., Shoup, T. M., Eshima, D., Eshima, L., Kilts, C., Votaw, J., Camp, V. M., Votaw, D., Smith, E., Kung, M. P., Malveaux, E., Watts, R., Huerkamp, M., Wu, D., Garcia, E., and Hoffman, J. M. (1997). Fluorine-18-FPCT: a PET radiotracer for imaging dopamine transporters. *J Nucl Med, 38,* 119–126.

40. Loc'h, C., Müller, L., Ottaviani, M., Halldin, C., Farde, L., and Mazière, B. (1995). Synthesis of 2β-carbomethoxy-3β-(4-[76Br]bromophenyl)tropane ([76Br]β-CBT), a pet tracer for in vivo imaging of the dopamine uptake sites. *J Label Compounds Radiopharm, 36,* 385–392.

41. Wilson, A. A., Dasilva, J. N., and Houle, S. (1995). Synthesis of two radiofluorinated cocaine analogues using distilled 2-[18F]fluoroethyl bromide. *Appl Radiat Isot, 46,* 765–770.

42. Goodman, M. M., Shi, B., Keil, R., Hoffman, J., Kilts, C., Camp, V., Eshima, D., Shattuck, L., and Colla, M. (1995). Abstract of the 11th International Symposium on Radiopharmaceutical Chemistry, Vancouver, BC, Canada, August 13–17. *J Label Compounds Radiopharm, 37,* 58–60.

43. Xing, D., Chen, P., Keil, R., Kilts, C. D., Shi, B., Camp, V. M., Malveaux, G., Ely, T., Owens, M. J., Votaw, J., Davis, M., Hoffman, J. M., BaKay, R. A., Subramanian, T., Watts, R. L., and Goodman, M. M. (2000). Synthesis, biodistribution, and primate imaging of fluorine-18 labeled 2β-carbo-1'-fluoro-2-propoxy-3β-(4-chlorophenyl)tropanes: ligands for the imaging of dopamine transporters by positron emission tomography. *J Med Chem, 43,* 639–648.

44. Firnau, G., Chen, J. J., and Murthy, D. (1995). Abstract of the 11th International Symposium on Radiopharmaceutical Chemistry, Vancouver, BC, Canada, August 13–17, 1995. *J Label Compounds Radiopharm, 37,* 55–57.

45. Harada, N., Ohba, H., Fukumoto, D., Kakiuchi, T., and Tsukada, H. (2004). Potential of [18F]β-CFT-FE (2β-carbomethoxy-3β-(4-fluorophenyl)-8-(2-[18F]fluoroethyl)nortropane) as a dopamine transporter ligand: a PET study in the conscious monkey brain. *Synapse, 54,* 37–45.

46. Loc'h, C., Halldin, C., Hantraye, P., Swahn, C. G., Lundqvist, C., Patt, J., Mazière, M., Farde, L., and Mazière, B. (1995). Abstract of the 11th International Symposium on Radiopharmaceutical Chemistry, Vancouver, BC, Canada, August, 13–17. *J Label Compounds Radiopharm, 37,* 64–65.

47. Chaly, T., Dhawan, V., Kazumata, K., Antonini, A., Margouleff, C., Dahl, J. R., Belakhlef, A., Margouleff, D., Yee, A., Wang, S., Tamagnan, G., Neumeyer, J. L., and Eidelberg, D. (1996). Radiosynthesis of [^{18}F] N-3-fluoropropyl-2β-carbomethoxy-3β-(4-iodophenyl)-nortropane and the first human study with positron emission tomography. *Nucl Med Biol*, 23, 999–1004.

48. Kazumata, K., Dhawan, V., Chaly, T., Antonini, A., Margouleff, C., Belakhlef, A., Neumeyer, J., and Eidelberg, D. (1998). Dopamine transporter imaging with fluorine-18-FPCIT and, PET. *J Nucl Med*, 39, 1521–1530.

49. Lundkvist, C., Halldin, C., Swahn, C. G., Hall, H., Karlsson, P., Nakashima, Y., Wang, S., Milius, R. A., Neumeyer, J. L., and Farde, L. (1995). [O-Methyl-^{11}C]β-CIT-FP, a potential radioligand for quantitation of the dopamine transporter: preparation, autoradiography, metabolite studies, and positron emission tomography examinations. *Nucl Med Biol*, 22, 905–913.

50. Lundkvist, C., Halldin, C., Ginovart, N., Swahn, C. G., and Farde, L. (1997). [^{18}F]β-CIT-FP is superior to [^{11}C]β-CIT-FP for quantitation of the dopamine transporter. *Nucl Med Biol*, 24, 621–627.

51. Ma, Y., Dhawan, V., Mentis, M., Chaly, T., Spetsieris, P. G., and Eidelberg, D. (2002). Parametric mapping of [^{18}F]FPCIT binding in early stage Parkinson's disease: a PET study. *Synapse*, 45, 125–133.

52. Chaly, T., Jr., Matacchieri, R., Dahl, R., Dhawan, V., and Eidelberg, D. (1999). Radiosynthesis of [^{18}F] N-3-fluoropropyl-2β-carbomethoxy-3β-(4'methylphenyl) nortropane (FPCMT). *Appl Radiat Isot*, 51, 299–305.

53. Chaly, T., Baldwin, R. M., Neumeyer, J. L., Hellman, M. J., Dhawan, V., Garg, P. K., Tamagnan, G., Staley, J. K., Al-Tikriti, M. S., Hou, Y., Zoghbi, S. S., Gu, X. H., Zong, R., and Eidelberg, D. (2004). Radiosynthesis of [^{18}F] N-(3-Fluoropropyl)-2β-carbomethoxy-3β-(4-bromophenyl)nortropane and the regional brain uptake in non human primate using, PET. *Nucl Med Biol*, 31, 125–131.

54. Gee, A. D., Smith, D. F., and Gjedde, A. (1997). The synthesis of O-methyl-[^{11}C]venlafaxine: a non-classical, fast-acting antidepressant. *J Label Compounds Radiopharm*, 39, 89–95.

55. Halldin, C., Farde, L., Lundkvist, C., Ginovart, N., Nakashima, Y., Karlsson, P., and Swahn, C. G. (1996). [^{11}C]β-CIT-FE, a radioligand for quantitation of the dopamine transporter in the living brain using positron emission tomography. *Synapse*, 22, 386–390.

56. Lundkvist, C., Sandell, J., Nagren, K., Pike, V. W., and Halldin, C. (1998). Improved syntheses of the PET radioligands, [^{11}C]FLB 457, [^{11}C]MDL 100907 and [^{11}C]β-CIT-FE, by the use of [^{11}C]methyl triflate. *J Label Compounds Radiopharm*, 41, 545–556.

57. Petric, A., Barrio, J. R., Namavari, M., Huang, S. C., and Satyamurthy, N. (1999). Synthesis of 3β-(4-[^{18}F]fluoromethylphenyl)- and 3β-(2-[^{18}F] fluoromethylphenyl)-tropane-2β-carboxylic acid methyl esters: new ligands for mapping brain dopamine transporter with positron emission tomography. *Nucl Med Biol*, 26, 529–535.

58. Stout, D., Petric, A., Satyamurthy, N., Nguyen, Q., Huang, S. C., Namavari, M., and Barrio, J. R. 1999). 2β-Carbomethoxy-3β-(4- and 2-[^{18}F]fluoromethylphenyl)tropanes: specific probes for in vivo quantification of central dopamine transporter sites. *Nucl Med Biol*, 26, 897–903.

59. Schönbächler, R. D., Gucker, P. M., Arigoni, M., Kneifel, S., Vollenweider, F. X., Buck, A., Burger, C., Berthold, T., Bruhlmeier, M., Schubiger, P. A., and Ametamey, S. M.

(2002). PET imaging of dopamine transporters in the human brain using [^{11}C]-β-CPPIT, a cocaine derivative lacking the 2β-ester function. *Nucl Med Biol, 29*, 19–27.

60. Schönbächler, R., Ametamey, S. M., and Schubiger, P. A. (1999). Synthesis and ^{11}C-radiolabelling of a tropane derivative lacking the 2 ester group: a potential PET-tracer for the dopamine transporter. *J Label Compounds Radiopharm, 42*, 447–456.

61. Mach, R. H., Nader, M. A., Ehrenkaufer, R. L., Gage, H. D., Childers, S. R., Hodges, L. M., Hodges, M. M., and Davies, H. M. (2000). Fluorine-18-labeled tropane analogs for PET imaging studies of the dopamine transporter. *Synapse, 37*, 109–117.

62. Fischman, A. J., Bonab, A. A., Babich, J. W., Livni, E., Alpert, N. M., Meltzer, P. C., and Madras, B. K. (2001). [^{11}C,^{127}I]Altropane: a highly selective ligand for PET imaging of dopamine transporter sites. *Synapse, 39*, 332–342.

63. Baldwin, R. M., Cosgrove, K., Staley, J. K., Vogel, R. S., Amici, L., Brenner, E., Tian, H., Peng, X., Zhang, A., Neumeyer, J. L., and Tamagnan, G.-D. (2004). Abstract CS15, presented at the 5th International Symposium on Radiohalogens, Vancouver, BC, Canada.

64. Peng, X., Zhang, A., Kula, N. S., Baldessarini, R. J., and Neumeyer, J. L. (2004). Synthesis and amine transporter affinities of novel phenyltropane derivatives as potential positron emission tomography (PET) imaging agents. *Bioorg Med Chem Lett, 14*, 5635–5639.

65. Mitterhauser, M., Wadsak, W., Mien, L. K., Hoepping, A., Viernstein, H., Dudczak, R., and Kletter, K. (2005). Synthesis and biodistribution of [^{18}F]FE-β-CIT, a new potential tracer for the dopamine transporter. *Synapse, 55*, 73–79.

7

THE BENZTROPINES: ATYPICAL DOPAMINE-UPTAKE INHIBITORS THAT PROVIDE CLUES ABOUT COCAINE'S MECHANISM AT THE DOPAMINE TRANSPORTER

AMY HAUCK NEWMAN AND JONATHAN L. KATZ

Medications Discovery Research Branch, National Institute on Drug Abuse Intramural Research Program, National Institutes of Health, Baltimore, Maryland

7.1	Introduction	172
7.2	Design, Synthesis, and Structure–Activity Relationships of the Benztropines	173
	7.2.1 3-Substituted Benztropines	173
	7.2.2 N-Substituted Benztropines	178
	7.2.3 6/7-Substituted Benztropines	184
	7.2.4 Benztropinamines	184
	7.2.5 N- and 2-Substituted Benztropines	189
7.3	In Vivo Findings	192
	7.3.1 Stimulation of Locomotor Activity	192
	7.3.2 Cocainelike Subjective Effects	194
	7.3.3 Benztropine Self-Administration	195
	7.3.4 Place Conditioning	196
	7.3.5 Summary of In Vivo Findings	196
7.4	Mechanisms for Differences from Cocaine	196
	7.4.1 Histamine Antagonist Effects	197
	7.4.2 Muscarinic Receptor Antagonist Effects	197
7.5	Differences between In Vivo and In Vitro Actions	199
7.6	Molecular Effects of Benztropine Analogs	203
	Acknowledgments	204
	References	204

Dopamine Transporters: Chemistry, Biology, and Pharmacology. Edited by Mark L. Trudell and Sari Izenwasser
Copyright © 2008 John Wiley & Sons, Inc

7.1   INTRODUCTION

The starting point for this line of research is the dopamine transporter (DAT) hypothesis of cocaine and its behavioral effects. Ritz et al. [47] showed a significant and positive correlation of binding affinities at the DAT and the potency for self-administration of a variety of monoamine-uptake inhibitors. That correlation was better than the correlations for these same compounds in binding to either the norepinephrine or serotonin transporters. Thus, the DAT was considered the primary biological target relevant to the effects of cocaine that contribute to its abuse liability.

Over the years a significant number of studies have supported the hypothesis, and much of the data is summarized in other chapters in this book. However, there are some limitations to the DAT hypothesis. For example, studies on DAT-knockout mice have shown place preference and self-administration of cocaine [49,52]. Although the effects of cocaine in DAT-knockout mice are not clearly understood at this time, it is clear that the reinforcing effects of cocaine can be obtained in animals lacking what is thought to be the primary biological substrate for cocaine's actions. One of the limitations, discussed in this chapter, was suggested initially in papers by Rothman et al. [50] and Vaugeois et al. [60]. In the paper by Rothman and colleagues, the locomotor-stimulating effects of dopamine (DA)-reuptake blockers were determined and the DAT occupancy produced by behaviorally equivalent doses was assessed. There were differences in the apparent DAT occupancy produced by doses of GBR 12909, nomifensine, and WIN 35-065-2, which were comparable in their behavioral effects to cocaine, suggesting that it takes different amounts of DAT occupancy by these drugs to produce behavioral effects equivalent to those produced by cocaine. In addition there is the observation that some dopamine-uptake inhibitors that are used clinically are not abused, and other dopamine-uptake inhibitors have pharmacological effects that differ from those of cocaine in animal models.

As can be seen in Figure 7-1, benztropine (BZT) shares structural features with cocaine and with GBR 12909, which will be elaborated on in the next section. GBR 12909 is a selective dopamine-uptake inhibitor, and therefore from a structural perspective, BZT and its analogs were of interest. In addition, BZT is in clinical use and is not subject to any significant abuse. In an early study, Colpaert et al. [13] showed that BZT did not fully substitute in rats trained to discriminate cocaine from saline injections. A lack of cocainelike effects of BZT suggested that it may be of some interest beyond its structural features. It is also possible that BZT is a typical dopamine-uptake inhibitor that also has other actions that interfere with its cocainelike effects.

Before proceeding with the BZT story, it is important to emphasize that although BZT analogs have pharmacological effects that differ from those of cocaine, this group of drugs may not be the only chemical class of DAT inhibitors in which this disparity exists. For example, some analogs of the sigma receptor ligand rimcazole bind to the DAT but do not produce cocainelike effects [9,20,25]. Further, there has been a recent preliminary report [36] of a cocaine analog with high affinity for the DAT, low efficacy in stimulating locomotor

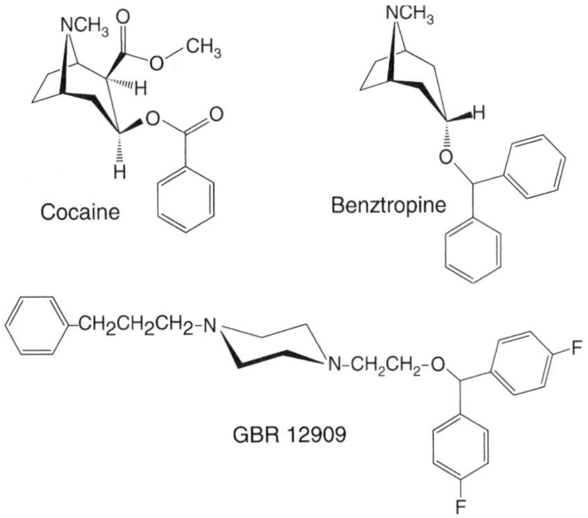

Figure 7-1 Chemical structures of cocaine, benztropine (BZT), and GBR 12909.

activity, and antagonism of the stimulation of locomotor activity produced by cocaine. Taken together, these studies are consistent with the idea that binding to the DAT and its resulting inhibition of dopamine uptake does not uniformly produce cocainelike effects.

7.2 DESIGN, SYNTHESIS, AND STRUCTURE–ACTIVITY RELATIONSHIPS OF THE BENZTROPINES

As mentioned above, BZT (3α-diphenylmethoxytropane) is a molecule with the shared features of a tropane ring from cocaine and a diphenyl ether of the phenylpiperazines (e.g., GBR 12909), which when combined create a unique group of dopamine-uptake inhibitors. In Figure 7-2, the chemical modifications that have been explored in our laboratory on the parent molecule BZT are illustrated. Initially, we noted that the diphenyl ether at the 3-position of the tropane ring was in the α-configuration and that this opposed most of the cocaine analogs that had been reported to have high affinity for the DAT (see [10] for a review). Furthermore, replacement of the β-benzoyl group of cocaine with various substituted phenyl rings significantly affected binding affinities at the DAT and generally improved potency for inhibition of dopamine uptake, and hence exploration of both stereochemistry and optimal substitution on these phenyl rings was an initial avenue of pursuit.

7.2.1 3-Substituted Benztropines

The synthesis of the 3α-BZTs generally followed the strategy depicted in Scheme 7-1. Various substituted benzhydrols were either commercially available or

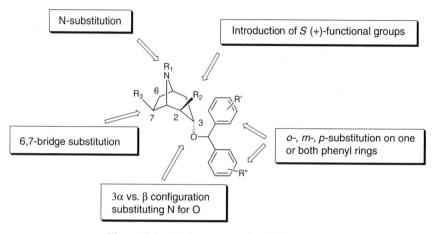

Figure 7-2   Design strategy for BZT analogs.

could be readily obtained from the benzophenones via $NaBH_4$ reduction. Conversion to the benzylchlorides with $SOCl_2$ provided the necessary synthons to combine with tropine under melt reaction conditions to give the desired BZT analogs [29,37,38]. Only one 3β-analog (AHN 1-063) was synthesized using a similar route but with pseudotropine (not shown). This compound had a lower affinity for the DAT than did its stereoisomer (Table 7-1), demonstrating stereoselectivity of the effects of these BZT analogs.

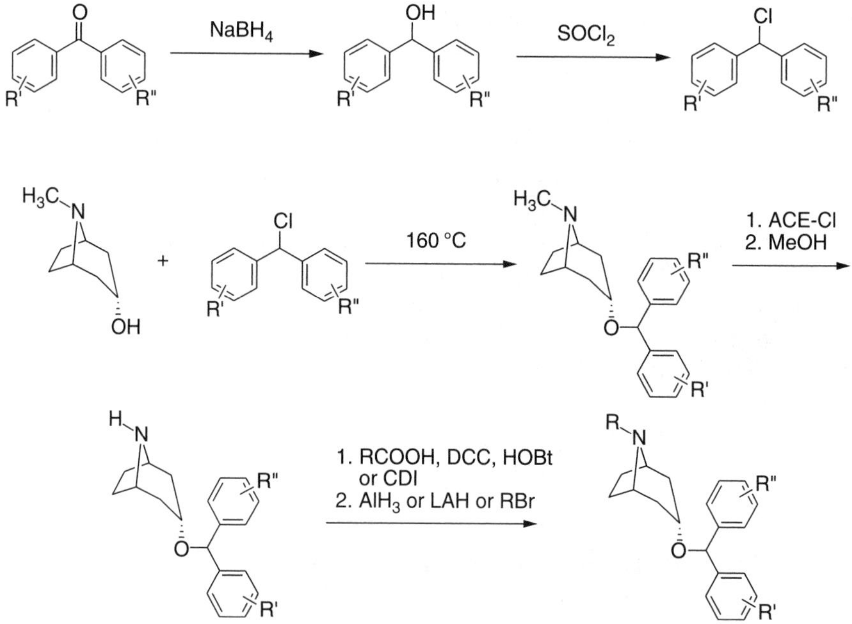

Scheme 7-1   General synthesis of N- and 3-substituted BZTs.

TABLE 7-1 Binding Data for the 3-Substituted Benztropines at DAT, SERT, NET, and M1 Receptors and Inhibition of Dopamine Uptake[1]

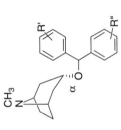

Compound	R', R''	DAT	SERT	NET[2]	M1	H1	DAUI[2,3]
AHN 1-055	4',4''-diF	11.8 ± 1.3[a]	3260 ± 110[k]	610 ± 80.5[k] 1 hour 844 ± 57 3 hours	11.6 ± 0.930[g]	19.7 ± 1.32[n]	71 ± 12[a,h] 13.8 ± 1.71[m]
RIK 11	4'-F, 4''-Br	15.2 ± 2.9[f]	—	—	22.4 ± 2.0[h]	64.8 ± 3.97[n]	27.2 ± 14.7[a,h]
AHN 2-018	3',4'-diCl, 4'-F	18.9 ± 2.6[a]	344 ± 31.4	655 ± 61.1 3 hours	38.9 ± 3.5[h]	215 ± 31.7[n]	23.9 ± 14.7[a,h]
4',4''-DiCl BZT	4',4''-diCl	20.0 ± 2.8[a]	1640 ± 236[i]	2980 ± 182[i] 1 hour	40.6 ± 3.2[h] 47.9 ± 5.18[i]	122 ± 4.55[n]	75 ± 24[a,h] 23.4 ± 3.00[i]
AHN 2-017	3',4'-diCl	21.1 ± 4.0[a]	—	413 ± 53.2 3 hours	14.7 ± 1.6[h]	147 ± 9.89[n]	46.7 ± 14.5[a,h]
RIK 7	3'-Cl	21.6 ± 1.5[b]	259 ± 19.3[i]	451 ± 62.5[i] 1 hour	0.98 ± 0.01[b]	—	228 ± 77.1[b] 12.5 ± 0.906[i]
RIK 15	3',4'-diF	23.3 ± 1.9[f]	4930 ± 255	694 ± 45.7 1 hour	5.87 ± 0.52[h]	29.1 ± 0.983[n]	139 ± 23.8[h]

(Continued)

TABLE 7.1 *Continued*

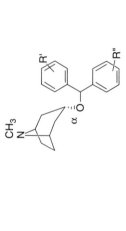

Compound	R', R''	DAT	SERT	NET[2]	M1	H1	DAUI[2,3]
RIK 17	3'-Cl,4''-F	23.4 ± 4.2^f	—	—	9.86 ± 1.02	—	123 ± 23^f
RIK 18	3'-Br	27.0 ± 3.2^f	—	—	—	—	104 ± 27^f
RIK 14	3',4'-diF	27.9 ± 3.1^f	—	—	3.79 ± 0.34^h	32.7 ± 4.58^n	181 ± 45.6^h
4'-Cl-BZT	4'-Cl	30.0 ± 3.6^a	5120 ± 395^i	1470 ± 180^i 1 hour	1.48 ± 0.02^b 7.90 ± 0.85^j	39.9 ± 1.57^n	115 ± 27.6^a 23.1 ± 1.80^i
RIK 19	3'-CH$_3$, 4''-F	30.9 ± 2.8^f	—	—	—	—	108 ± 23^f
AHN 2-009	4'-F	32.2 ± 3.2^a	$>10000^a$	1470 ± 87.0 1 hour	5.2 ± 0.4^h	17.7 ± 1.55^n	—
GA 1-89	3',4''-diCl	32.5 ± 4.88^i	3870 ± 303^i	1660 ± 239^i	21.5 ± 2.63^i	—	12.3 ± 1.23^i
AN012	4'-Br	37.9 ± 2.7^a	$>10000^a$	$>10000^a$	6.24 ± 6.2^h	64.7 ± 6.64^n	$28.6 \pm 15.5^{a,h}$
RIK 16	3'-Br, 4''-F	38.2 ± 5.3^f	—	—	—	—	179 ± 51^f
RIK 13	3',3''-diF	47.4 ± 5.2^b	$>10000^b$	$>10000^b$	0.85 ± 0.01^b	—	407 ± 63.9^b
AHN 2-016	2'-F	50.0 ± 6.0^b	$>10000^b$	$>10000^b$	0.43 ± 0.02^b	—	140 ± 17.2^b
RIK 12	3'-F	68.5 ± 8.2^b	$>10000^b$	$>10000^b$	0.60 ± 0.05^b	—	250 ± 64.7^b
RIK 20	3'-CH$_3$	78.0 ± 5.5^f	—	—	—	—	226 ± 70^b
AHN 1-056	4'-OCH$_3$	78.4 ± 6.3^a	$>10000^a$	$>10000^a$	$12.2 \pm 1.0^{e,h}$	28.6 ± 1.88^n	$468 \pm 114^{a,h}$

RIK 21	3'-CF$_3$, 4''-F	88.1 ± 8.8[f]	—	—	—	—	182 ± 36[f]
AN 011	4',4''-diBr	91.6 ± 11.9[a]	>10000[a]	>10000[a]	85.1 ± 8.5[e,h]	320 ± 22.7[n]	34.3 ± 13.7[a,h]
Benztropine	H,H	118 ± 10.6[a]	>10000[a]	1390 ± 134 (1 hour)	0.59 ± 0.01[b]; 2.1 ± 0.29[i]	15.7 ± 2.13[n]	403 ± 115[a,h]
RIK 04	4'-CH$_3$	187 ± 9.4[a]	>10000[a]	>10000[a]	11.6 ± 1.0[h]	23.1 ± 2.02[n]	512 ± 121[a,h]
RIK 09	3'-CF$_3$	187 ± 9.5[b]	>10000[b]	>10000[b]	2.52 ± 0.28[b]	—	457 ± 72.0[b]
AHN 2-020	4'-CN	196 ± 17.6[a]	>10000[a]	>10000[a]	22.0 ± 1.3[e,h]	27.7 ± 1.34[n]	222 ± 55.2[a,h]
AHN 2-021	4'-NO$_2$	197 ± 15.8[a]	>10000[a]	>10000[b]	—	—	—
RIK 10	2'-Cl	228 ± 20.5[b]	>10000[b]	>10000[b]	0.41 ± 0.01[b]	—	977 ± 109[b]
RIK 06	4'-OH	297 ± 38.6[a]	>10000[a]	>10000[a]	8.54 ± 0.6[h]	37.8 ± 4.74[n]	677 ± 164[a,h]
RIK 08	2'-CH$_3$	309 ± 18.5[b]	>10000[b]	>10000[b]	0.50 ± 0.01[b]	—	1200 ± 164[b]
RIK 03	4',4''-diCH$_3$	420 ± 29.4[a]	>10000[a]	>10000[a]	—	—	—
RIK 05	4'-CH$_2$CH$_3$	520 ± 41.6[a]	>10000[a]	>10000[a]	27.6 ± 2.2[h]	114 ± 14.7[n]	2155 ± 710[a,h]
RIK 01	4'-CF$_3$	635 ± 64[a]	>10000[a]	>10000[a]	18.9 ± 1.7[h]	2200 ± 373[n]	—
AHN 1-063	3β-4'-Cl	854 ± 59.8[a]	>10000[a]	2120 ± 276 (1 hour)	—	—	—
2'-Amino-BZT	2'-NH$_2$	1840 ± 147[b]	—	—	1.12 ± 0.04[b]	—	373 ± 117[b]
RIK 02	4'-C(CH$_3$)$_3$	1918 ± 134.3[a]	>10000[a]	>10000[a]	—	157 ± 2.09[n]	2880 ± 956[a,h]
AHN 1-057	4',4''-diOCH$_3$	2000 ± 140[a]	>10000[a]	3300 ± 170[n]; 3210 ± 149 (1 hour)	120 ± 6.0[h]	1050 ± 43.0[n]	236 ± 20.5[i]
Cocaine	—	187 ± 18.7[i]	172 ± 15[n]	2120 ± 314 (3 hours)	61400 ± 10900[o]	—	304 ± 29.6[o]

[1] All binding data are recorded in K_i ± SEM and are published in the references cited. [a]Newman et al. [38]; [b]Kline et al. [29]; [c]Agoston et al. [1]; [d]Agoston et al. [2]; [e]Newman and Agoston [70]; [f]Husbands et al. [20]; [g]Robarge et al. [48]; [h]Katz et al. [24]; [i]Newman et al. [39]; [j]Newman et al. [39]; [k]Katz et al. [71]; [k]Zou et al. [67]; [m]Kulkarni et al. [30]; [n]Campbell et al. [8]; [o]Katz et al. [26]. If no reference is cited, the data are previously unpublished but the methods used are identical to those used for the values referenced.

[2] Where there are two values, different radiolabeled ligands or analysis programs or incubation times were used, the details of which can be found in the primary references.

[3] The inhibition of dopamine uptake (DAUI) data are recorded as IC_{50} ± SEM and were obtained in rat brain chopped tissue or synaptosomes according to methods detailed in the primary references.

More than 40 3α-diphenyl ether analogs of BZT were prepared and evaluated for binding at the DAT, serotonin (SERT), and norepinephrine (NET) transporters and at muscarinic M1 and histamine H1 receptors, as well as for inhibition of dopamine uptake. These data are displayed in Table 7-1. Throughout the rest of this chapter, in vitro data will be displayed similarly in the tables, starting with the compound with the highest DAT affinity of the series. The 3α-stereochemistry provided optimally active compounds, as did small substituents such as F or Cl in the *para*- or *meta*-positions, with 4′,4″-diF giving the highest affinity analog in this series (AHN 1-055). However, it must be noted that small halogens in these positions uniformly gave high-affinity analogs ($K_i = 11$ to 30 nM), whereas increasing steric bulk or substitution in the 2′-position caused decreases in DAT affinity [29]. Note that compared to BZT and cocaine, many of these halogenated analogs had a higher affinity at the DAT. In addition, muscarinic receptor affinities were generally lower than that for BZT, so that instead of compounds with 60- to 100-fold selectivity for muscarinic receptors, these analogs showed similar affinities at DAT and M1 receptors. Although none of these analogs demonstrated high affinity for SERT or NET, the high-to-moderate muscarinic receptor binding could contribute to their central nervous system (CNS) effects. Like the parent compound, it should also be noted that several of these analogs demonstrated high to moderate affinity for histamine H1 receptors [8,31]. Because these additional actions may affect the behavioral profile of these agents significantly, their contributions were investigated (see below). In addition, it was desirable to find structural features that would discriminate between them, and we thus initiated studies to alter muscarinic M1 receptor affinity.

7.2.2 N-Substituted Benztropines

GBR 12909 has a propylphenyl substituent appended to its piperazine N-terminus and does not bind appreciably to muscarinic M1 receptors. As BZT can be considered a rigid analog of GBR 12909, a series of N-substituted analogs of AHN 1-055 and several other small halogenated diphenyl ether analogs were designed. As seen in Scheme 7-1, simple N-demethylation with α-chloroethylchloroformate (ACE-Cl) followed by methanolysis gave the N-nor analogs. Conversion to amides using either Schotten–Baumann reaction conditions with acid chlorides or peptide synthesis conditions followed by reduction (LiAlH$_4$ or AlH$_3$), or via direct N-alkylation, gave the desired *N*-alkylamines [2,30,39,48]. N-Substitution turned out to be a fruitful area of investigation, and more than 60 analogs have been prepared and evaluated in vitro (Table 7-2).

In general, N-substituted analogs with the 4′,4″-diF substitution on the 3α-diphenyl ether resulted in high-affinity binding at the DAT, with several extended alkyl and alkylaryl substituents being well tolerated [DAT (HEPES buffer), $K_i = 11$ to 30 nM]. However, there was an optimal length of the N-substituents which, if exceeded, resulted in low-affinity compounds at the DAT (e.g., SSK 1-033, SSK 2-005). Furthermore, the tropane N must be a secondary or tertiary amine, as the amides were inactive at the DAT (e.g., AHN 2-032, JHW 021) [2]. When the

TABLE 7-2 Binding Data for the N- and 3-Substituted Benztropines at DAT, SERT, NET, and M1 and H1 Receptors and Inhibition of Dopamine Uptake[1]

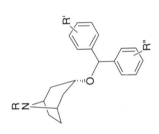

Compound	R	R', R''	DAT	SERT	NET[2]	M1	H1	DAUI[2,3]
GA 103	4''-Phenyl-n-butyl	4',4''-diF	8.51 ± 1.2^c	376 ± 51.8^m	2210 ± 240^m 1 hour	576 ± 10.7^g	141 ± 6.72^n	—
AHN 2-003	H	4',4''-diF	11.2 ± 1.2^c	922 ± 87.2^o	902 ± 60.9 1 hour 457 ± 69.8 1 hour	203 ± 16.5^g	65.4 ± 7.28^n	15.1 ± 1.27
SSK 2-032	CH_2CH_2NHCO-4-pyr	4',4''-diF	12.6 ± 0.073^m	572 ± 76.4^m	652 ± 61.9^m 1 hour	394 ± 15.9^m	—	14.2 ± 1.68^m
SSK 2-013	$CH_2CH_2NHCOPh$	4',4''-diF	13.1 ± 1.47^m	597 ± 13.8^m	637 ± 48.7^m 1 hour	556 ± 49.4^m	—	6.5 ± 0.530^m
GA 2-99	2''-Aminoethyl	4',4''-diF	13.9 ± 1.7^g	4600 ± 680^m	1420 ± 130^m 1 hour	1250 ± 138^g	240 ± 32.6^n	26.0 ± 2.88^m
GA 1-106	4''-(4'''-nitrophenyl)-n-butyl	4',4''-diF	20.2 ± 2.2^c	—	—	299 ± 19.7^g	—	—

(*Continued*)

TABLE 7.2 *Continued*

Compound	R	R', R''	DAT	SERT	NET[2]	M1	H1	DAUI[2,3]
JHW 007	n-butyl	4',4''-diF	24.6 ± 2.0[c]	1350 ± 151[o]	1490 ± 190 3 hours 1670 ± 232[o] 1 hour	251 ± 12.5[c] 399 ± 28.3[o]	—	—
PG1088	$CH_2CH_2CH_2CH_2NH_2$	4',4''-diF	26.0 ± 1.6[m]	555 ± 69[m]	—	2370 ± 170[m]	—	—
JHW 084	H	3'-Cl	26.2 ± 2.1[j]	2100 ± 285[j]	508 ± 70.0[j] 1 hour	91.7 ± 7.65[j]	—	—
SSK 2-046	$CH_2CH_2CONHPh$	—	27.6 ± 1.92[m]	1490 ± 81.7[m] 1 hour	1420 ± 165[m] 1 hour	3280 ± 420[m]	140 ± 17.3[n]	4.99 ± 0.011[m]
GA 2-95	S-2''-Amino-3''-methyl-n-butyl	4',4''-diF	29.5 ± 3.5[g]	—	—	2180 ± 85.8	—	—
GA 2-32	4-(4'-Aminophenyl)-n-butyl	4',4''-diF	29.7 ± 3.6[d]			134 ± 16[g]		
AHN 2-005	Allyl	4',4''-diF	29.9 ± 3.0[c]	2850 ± 62.5[o]	1570 ± 162 1 hour 1740 ± 242 3 hours	126 ± 7.6[c] 177 ± 21[g]	24.9 ± 1.16[n]	19.7 ± 0.568
GA 2-8	3''-Butene	4',4''-diF	31.9 ± 2.8[g]	—	—	278 ± 18.5[g]	—	—

JHW 013	Cyclopropylmethyl	4',4''-diF	32.4 ± 2.9^c	1420 ± 116	1640 ± 153 1 hour	136 ± 11^c 257 ± 28.9^g	—	—
IMH 023	Allyl	3'-Cl	33.6 ± 3.40^j	1200 ± 154^j	1230 ± 71.0^j 1 hour	69.7 ± 7.13^j	24.9 ± 1.16	—
SSK 2-020	$CH_2CH_2NHCH_2Ph$	4',4''-diF	36.0 ± 2.89^m	1090 ± 81.3^m	1130 ± 117^m 1 hour	1230 ± 150^m	—	8.12 ± 1.2^m
SLW 011	H	4'-Cl	36.8 ± 3.30^j	1320 ± 194^j	1010 ± 116^j 1 hour	95.5 ± 12.7^j	—	—
JHW 010	3''-Phenyl-n-propyl	4',4''-diF	41.9 ± 4.6^c	—	—	136 ± 12^c 312 ± 10.6^g	—	—
SSK 2-025	$CH_2CH_2OCH_2CH_2OH$	4',4''-diF	42.6 ± 2.21^m	14300 ± 634^m	3650 ± 520^m 1 hour	336 ± 46.5^m	—	48.9 ± 6.59^m
GA 1-69	Indol-3''-ethyl	4',4''-diF	44.6 ± 4.9^c	490 ± 56.4	7350 ± 934	3280 ± 220^g	333 ± 22.6^n	—
PG2-032	$CH_2CH_2CH_2N(CH_3)_2$	4',4''-diF	47.7 ± 5.48^m	499 ± 64.2^m	731 ± 11.5 3 hours	368 ± 49^m	167 ± 24.4	—
SSK 1-026	$CH_2CH_2OCH_2Ph$	—	48.5 ± 6.72^m	1300 ± 75.1^m	3060 ± 346^m 1 hour	390 ± 50.6^m	—	22.3 ± 2.61^m
SLW 019	Butylphenyl	4'-Cl	53.3 ± 38.8^j	737 ± 38.8^j	2320 ± 276^j 1 hour	1430 ± 202^j	—	—
IMH 022	Allyl	4',4''-diCl	54.6 ± 7.10^j	568 ± 32.0^j	4140 ± 438^j 1 hour	127 ± 12.2^j	—	—
GA 2-50	R-2''-Amino-3''-methyl-n-butyl	4',4''-diF	56.4 ± 9.6^g	3870 ± 135	2130 ± 160 1 hour	4020 ± 592	218 ± 15.5^n	20.7 ± 2.85
GA 1-104	2'-{[(4''-Nitrophenyl)phenyl]methoxy}ethyl	4',4''-diF	57.0 ± 9.7^c	—	—	460 ± 51.4^g	—	—
GA 1-120	3''-(4''-Fluorophenyl)-n-Propyl	4',4''-diF	60.7 ± 7.3^c	—	—	686 ± 85.6^g	—	—
IMH 011	H	4',4''-diCl	61.7 ± 6.20^j	727 ± 47.3^j	6040 ± 693^j 1 hour	649 ± 61.2^j	—	—
JHW 005	Benzyl	4',4''-diF	82.2 ± 12.3^c	2090 ± 125°	4180 ± 657 1 hour	1030 ± 150^g	116 ± 8.41^n	61.0 ± 5.69
SSK 1-050	CH-CH_2OCH_2-2-thiophene	—	65.9 ± 4.97^m	2090 ± 220^m	3450 ± 438^m 1 hour	613 ± 89.5^m	—	18.7 ± 2.0^m

(*Continued*)

181

TABLE 7.2 *Continued*

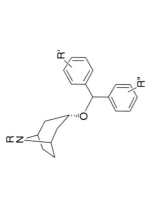

Compound	R	R', R''	DAT	SERT	NET[2]	M1	H1	DAUI[2,3]
SLW 017	Allyl	4'-Cl	84.2 ± 7.60[j]	2330 ± 166[j]	4720 ± 690[j] 1 hour	127 ± 18.7[j]	—	—
IMH 028	Butylphenyl	4',4''-diCl	85.0 ± 8.50[j]	659 ± 42.0[j]	4950 ± 324[j] 1 hour	3730 ± 493[j]	—	—
AHN 2-006	Cinnamyl	4',4''-diF	86.4 ± 10.4[c]	—	—	401 ± 21.8[g]	—	—
JHW 009	4''-Fluorobenzyl	4',4''-diF	95.6 ± 9.6[c]	—	>10000[c]	1540 ± 120[g]	—	—
MJR 1-21	Propylphenyl	3'-Cl	101 ± 11.1[j]	429 ± 53.9[j]	1810 ± 192[j] 1 hour	337 ± 14.0[j]	—	—
JHW 025	Dimethyl quaternary	4',4''-diF	108 ± 13.0[c]	1220 ± 166[o]	3350 ± 534 1 hour	29.1 ± 3.5[c]; 11.8 ± 0.590[o]	—	—
SLW 020	Propylphenyl	4'-Cl	148 ± 16.3[j]	302 ± 15.3[j]	10300 ± 660[j] 1 hour	397 ± 21.9[j]	—	—
MJR 1-25	Butylphenyl	3'-Cl	153 ± 19.9[j]	591 ± 73.3[j]	1470 ± 76.4[j] 1 hour	564 ± 48.5[j]	—	—
SLW 021	Propylphenyl	4',4''-diCl	158 ± 20.5[j]	585 ± 60.1[j]	19800 ± 3090[j] 1 hour	3050 ± 329[j]	—	—

SSK 2-082	CH$_2$CH$_2$CONHPh	4',4''-diF	181 ± 5.64[m]	558 ± 81.0[m]	7780 ± 784 3 hours	4140 ± 340[m]	509 ± 65.5	—
MJR 1-17	Benzyl	3'-Cl	191 ± 15.3[j]	1680 ± 235[j]	6460 ± 882[j] 1 hour	674 ± 71.2[j]	—	—
GA 2-61	S-2''-Phenyl-n-butyl	4',4''-diF	228 ± 21[g]	—	—	2360 ± 296[g]	—	—
SLW 007	Benzyl	4'-Cl	365 ± 25.5[j]	3530 ± 403[j]	13000 ± 629[j] 1 hour	2560 ± 365[j]	—	—
GA 2-53	Naphthylmethyl	4',4''-diF	367 ± 37[g]	—	—	1790 ± 49.0[g]	—	—
GA 2-52	3''',4'''-diCl-benzyl	4',4''-diF	392 ± 35[g]	—	—	3880 ± 367[g]	—	—
SLW 018	Benzyl	4',4''-diCl	399 ± 27.9[j]	3610 ± 214[j]	7660 ± 1,240[j] 1 hour	8630 ± 1300[j]	—	—
GA 2-38	R-2''-phenyl-n-propyl	4',4''-diF	477 ± 53[g]	—	—	392 ± 15.4[g]	—	—
SSK 1-033	CH$_2$CH$_2$CH$_2$OPh	4',4''-diF	703 ± 100[m]	7730 ± 735[m]	12100 ± 1,580[m] 1 hour	9280 ± 1040[m]	—	227 ± 29.2[m]
SSK 2-005	CH$_2$CH$_2$OCH$_2$CH$_2$-N-morpholine	4',4''-diF	1620 ± 31[m]	23000 ± 2250[m]	28100 ± 2760[m] 1 hour	15800 ± 2260[m]	—	659 ± 96.1[m]
AHN 2-032	Formyl	4',4''-diF	2020 ± 263[c]	>10000[c]	>10000[c]	12300 ± 1480[c]	—	—
JHW 021	Acetyl	4',4''-diF	2340 ± 117[c]	>10000[c]	>10000[c]	—	—	—
GBR12909	—		11.9 ± 1.9	105 ± 11.4	497 ± 17.0	—	—	2.30 ± 0.144

[1] All binding data are recorded in K_i ± SEM; these data and the methods used to obtain them are published in the references cited. [a]Newman et al. [38]; [b]Kline et al. [29]; [c]Agoston et al. [1]; [d]Agoston et al. [2]; [g]Robarge et al. [2]; [i]Katz et al. [48]; [j]Newman et al. [39]; [k]Zou et al. [67]; [m]Kulkarni et al. [30]; [n]Campbell et al. [8]; [o]Katz et al. [26]. If no reference is cited, the data are previously unpublished but the methods used are identical to those used for the values referenced in this column.

[2] Where there are two values, different radiolabeled ligands or analysis programs or incubation times were used, the details of which can be found in the primary references.

[3] The inhibition of dopamine uptake (DAUI) data are recorded as IC$_{50}$ ± SEM and were obtained in rat brain chopped tissue or brain synaptosomes according to methods detailed in the primary references.

183

diphenyl ether substituents increased in steric bulk (e.g., 4′,4″-diCl), N-substitution resulted in further decreases in DAT affinity. However, a notable separation of DAT from muscarinic M1 receptor binding was achieved with these N-substitutions, with several analogs in this series having >100-fold selectivity for the DAT. This is in remarkable contrast to the parent BZT, which is 60- to 100-fold selective for muscarinic M1 receptors over DAT. Stereoselectivity was originally reported for DAT binding with GA 2-50 and GA 2-95: however, subsequent testing (unpublished results) of these analogs showed no stereoselectivity at DAT and only a twofold difference in affinity at muscarinic M1 receptors. Several of the analogs in this series have been evaluated in vivo and will be described in detail below. Further, as many N-substituents were well tolerated at the DAT, this position has been used to design and synthesize molecular tools such as the photoaffinity ligand GA 2-34 for structure–function studies of the DAT [1,61].

7.2.3 6/7-Substituted Benztropines

Once identification of the optimal substitutions on the 3α-diphenyl ether and the tropane N8 were identified, additional modifications at the 6/7 bridge were investigated. These compounds were synthesized according to the strategy outlined in Scheme 7-2. The 6-acetyl tropine was prepared using a literature method [45] and the melt reaction described previously was used to obtain the 6β-O-acetyl BZT. Deacetylation to the 6β-alcohol using LiAlH$_4$ was followed by reacetylation with various acetyl chlorides or anhydrides to allow the exploration of SAR at this position [18]. One N-demethylated analog was also prepared. In addition, the 6-β-OH BZT was oxidized and then stereoselectively reduced to the 6α-alcohol. Upon testing (Table 7-3) it was discovered that none of the analogs showed high affinity for the SERT or NET, and most were poorly active at the DAT, with the exception of those analogs with very small substituents in the 6/7-bridge. Notably, both alcohols (β-: PG01053, α-: PG02045) and the carboxylate analog (PG01086) were well tolerated at DAT, but the additional steric bulk of the other analogs lowered DAT affinities significantly. Also, the 6α-OH analog was equiactive at DAT and muscarinic M1 receptors, whereas a small degree of selectivity (<10-fold) was exhibited for the other two active analogs, and most of the other compounds were too poorly active to be of interest. Although originally we had plans to make N-substituted analogs of some of the 6/7-substituted compounds, based on the SAR, it appeared that this would not be especially profitable.

7.2.4 Benztropinamines

While preparing the BZT analogs, it was noted that under acidic conditions, the diphenyl ethers would sometimes cleave from the tropane ring. We therefore decided to explore a series of compounds wherein the ether linkage was replaced by a secondary or tertiary amine. This substitution provided some additional

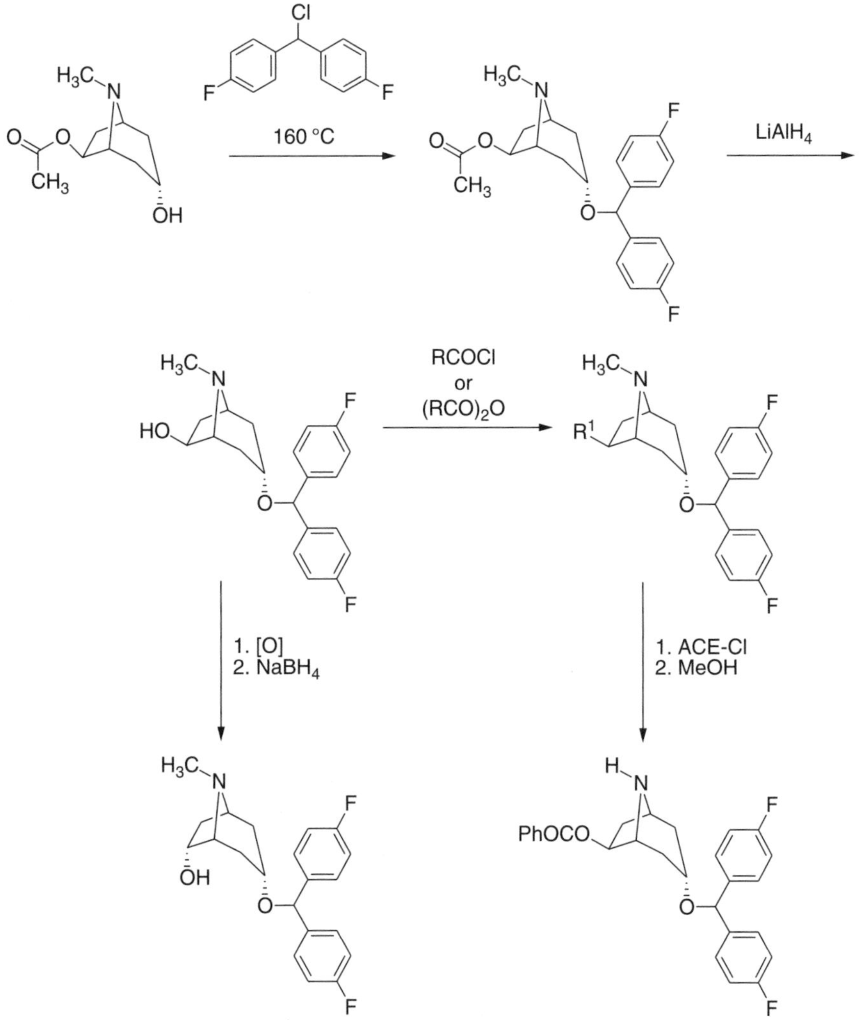

Scheme 7-2 General synthesis of 6/7-substituted BZTs.

flexibility for SAR and also had the potential to improve the water solubility of these compounds, due to the possibility of forming a more-water soluble salt. The synthesis of these benztropinamines, as we refer to them, is depicted in Scheme 7-3. The 3α-aminotropane was prepared by a literature procedure [5] and then reacted with the benzhydrylchloride in refluxing acetonitrile [19]. In this way, a number of diphenyl ether analogs could be obtained. N-Demethylation was performed as described above and was followed by similar methods for introducing N-alkyl and arylalkyl substituents on the tropane N8. These substituents were selected from the substituents

TABLE 7-3 Binding Data for the 6/7-Substituted Benztropines at DAT, SERT, NET, and M1 Receptors[a]

Compound	R_1	R_2	DAT	SERT	NET	M1
PG01086	β-HCOO	CH_3	5.81 ± 0.310	8290 ± 279	1380 ± 190	46.8 ± 6.56
PG01053	β-HO	CH_3	6.28 ± 0.451	9090 ± 615	1300 ± 142	48.3 ± 6.86
PG02045	α-HO	CH_3	31.3 ± 2.27	11700 ± 977	6390 ± 486	27.0 ± 3.43
PG01050	β-CH_3COO	CH_3	57.1 ± 5.56	16500 ± 1610	6910 ± 573	432 ± 47.8
PG01084	β-CH_3CH_2COO	CH_3	75.4 ± 3.01	14000 ± 1130	3720 ± 457	142 ± 21.1
PG02031	β-$MeSO_3$	CH_3	81.0 ± 3.27	13800 ± 380	6880 ± 805	617 ± 66.5
PG01091	β-PhCOO	CH_3	107 ± 10.0	3730 ± 533	1370 ± 82.3	552 ± 52.9
PG02029	β-$PhCH_2COO$	CH_3	119 ± 14.9	5850 ± 682	6206 ± 394	2270 ± 307
PG02009	β-$(CH_3)_2CHCH_2COO$	CH_3	148 ± 23.3	7960 ± 1010	6610 ± 754	723 ± 66.5
PG02025	β-$(CH_3)_2CHCOO$	CH_3	151 ± 9.42	14800 ± 1850	10200 ± 948	784 ± 86.4
PG01089	β-$PhCH_2CH_2COO$	CH_3	171 ± 11.7	7960 ± 671	5690 ± 276	3480 ± 409
PG02036	β-PhCOO	H	175 ± 17.5	4440 ± 148	3170 ± 408	8430 ± 461

Source: Data from Grundt et al. [18].

[a]Each K_i value represents data from at least three independent experiments, each performed in triplicate. K_i values were analyzed by PRISM.

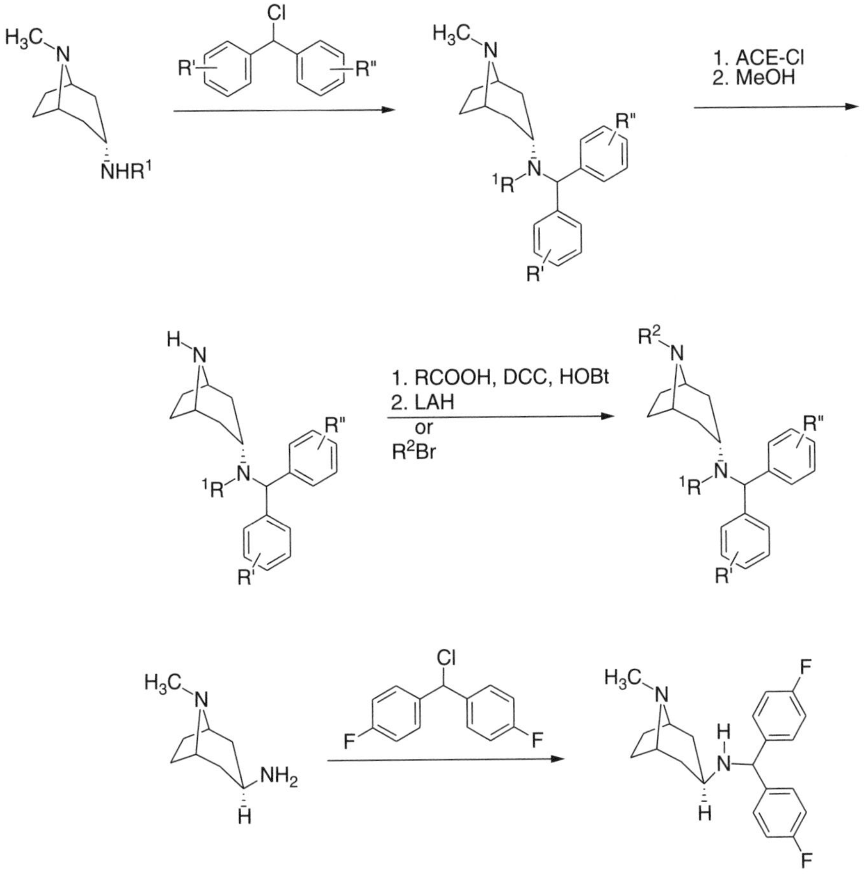

Scheme 7-3   General synthesis of benztropinamines.

that gave optimal binding in the BZT series for direct comparison of SAR. One 3β-analog was prepared as shown.

In this series, the 3-amino substitution readily replaced the ether linkage and gave a series of highly active compounds at the DAT that were comparable to the similarly substituted BZT analogs (Table 7-4). As with the 3β-BZT, AHN 1-063, the least active analog in the benztropinamines was the similarly substituted 3β-analog PG 02048. 3-N-Methylation also served to reduce DAT-binding affinity (PG 439 in Table 7-4). Once again, none of the compounds showed high affinity for the SERT or NET. Most of the compounds were similarly selective for the DAT over muscarinic M1 receptors as their ether counterparts. These compounds appeared to be more stable to acidic conditions, and in vivo testing is underway.

TABLE 7-4   Binding Data for the Benztropinamines at DAT, SERT, NET, and M1 Receptors[a]

Compound	R'	R''	R$_1$	R$_2$	DAT[b]	SERT	NET	M1
PG 459	4-F	4-F	H	[3-(N-Phenyl)-propionamido]-	4.61 ± 0.537	608 ± 41.0	2470 ± 83	2540 ± 124
PG 568	3,4-Cl	3,4-Cl	H	CH$_3$	5.35 ± 0.247	3010 ± 332	259 ± 33.6	17.8 ± 2.27
PG 455	4-F	4-F	H	H	8.45 ± 0.228	4150 ± 368	997 ± 107	154 ± 19.6
PG02046	4-F	4-F	H	CH$_3$	11.3 ± 1.61	8690 ± 436	1810 ± 269	7.81 ± 1.17
PG 554	4-F	4-F	H	4-Phenylbutyl	11.7 ± 0.385	502 ± 68.1	1630 ± 115	438 ± 56.2
PG 556	4-F	4-F	H	2-Ethylamino	12.5 ± 1.73	10900 ± 1090	3550 ± 222	2110 ± 109
PG 466	4-F	4-F	H	n-Butyl	21.5 ± 2.31	2640 ± 27.6	2920 ± 209	454 ± 36.7
PG 458	4-F	4-F	H	Allyl	26.8 ± 3.43	3920 ± 581	5580 ± 821	130 ± 10.5
PG 549	4-Cl	H	H	CH$_3$	36.5 ± 0.662	13600 ± 1040	3170 ± 233	9.30 ± 0.271
PG 436	4-Cl	4-Cl	H	CH$_3$	38.1 ± 5.35	4180 ± 623	7580 ± 325	50.0 ± 7.10
PG 566	4-F	4-F	H	[2-(1H-indol-3-yl)-ethyl]-	64.5 ± 4.32	347 ± 41.3	8250 ± 939	413 ± 6.55
PG 439	4-F	4-F	CH$_3$	CH$_3$	123 ± 15.5	17200 ± 2460	8410 ± 724	18.6 ± 2.48
PG02048[c]	4-F	4-F	H	CH$_3$	661 ± 34.9	48500 ± 6370	152000 ± 12300	59.4 ± 7.32

Source: Data are from Grundt et al. [19].

[a]Each K_i value represents data from at least three independent experiments, each performed in triplicate. K_i values were analyzed by GraphPad Prism.
[b]Sucrose buffer.
[c]The configuration at C-3 is beta.

7.2.5 N- and 2-Substituted Benztropines

The 2-substituted BZT analogs were first prepared by Meltzer and colleagues, who made all eight stereoisomers and found that only the S-(+)-2-COOCH$_3$-substituted analog of 4′,4″-diF BZT (difluoropine, MFZ 1-76 in Table 7-5) showed any activity at the DAT [34] This observation was notable from the standpoint that although cocaine and its 3-phenyl analogs all need a substituent at the 2-position, it must also be in the R-(−)-stereochemistry. The BZTs clearly do not need a substituent in this position, as hundreds of analogs from our lab and others have shown, but if this substituent is on the tropane ring, it must be the equal and opposite enantiomer to cocaine [34,35,67]. These findings supported our original hypothesis that the cocainelike and BZT analogs demonstrated very different SAR and probably interacted with different binding domains on the DAT protein (see [40] for a review). Thus, we began to devise a stereoselective synthesis for the S-(+)-substituted BZTs using chiral amine technology and published the first series of 2-substituted compounds using this method [67], as depicted in Scheme 7-4. The synthetic chiralamine NDPPA was prepared and used to give the 2-carbomethoxytropinone in ~85% ee. Subsequent treatment with D-tartaric acid gave the key intermediate in >99% ee. Stereoselective reduction using hydrogenation and PtO$_2$ as the catalyst followed by isomerization to the desired carboxylate was followed by ether formation. At this stage numerous ethers could be prepared depending on the alcohol chosen. The 2-substituted analogs were not stable under the melt reaction conditions used previously, so pTsOH-catalyzed ether formation with a Dean Stark trap was utilized to give the various substituted 3α-BZTs. These esters could then be reduced to the alcohol and oxidized to the key intermediate aldehyde, from which various olefinic analogs were prepared via Wittig reaction conditions. Conversely, N-demethylation opened the door to a variety of N-substituted esters. This strategy has allowed variation at all three positions (N, 2, and 3) on the tropane ring, which has allowed for rich deduction of SAR in this series and direct comparison to cocaine analogs that are substituted in these three positions [68,69].

This complex series of compounds has provided the most DAT-selective agents of all the BZTs, with several compounds showing >1000-fold DAT selectivity over muscarinic M1 receptors (Table 7-5). Furthermore, both SERT/DAT and NET/DAT selectivity ratios were discovered to be quite high. Nearly all of the compounds in this series prepared so far have demonstrated high affinity for DAT ($K_i < 40$ nM), and these same compounds have shown remarkably complex pharmacology in vivo. In addition to the opposite enantiomeric requirement at the 2-position, the substituents that gave very high DAT affinity [e.g., CH$_2$OH (MFZ 6-21 in Table 7-5)] are not the same substituents that give high DAT affinity to cocaine, further demonstrating that these dopamine-uptake inhibitors do not access the same DAT-binding domain. This hypothesis is further strengthened by the unique behavioral profile that these compounds have shown in animal models of cocaine abuse [69].

TABLE 7-5 Binding Data for the 2-Substituted Benztropines at DAT, SERT, NET, M1 and Inhibition of Dopamine Uptake[1]

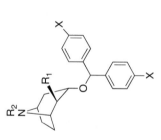

Compound[2]	R₁, R₂	X	DAT[3]	SERT	NET	M1	DAUI[4]
MFZ 6-35	CH=CH₂, CH₃	F	1.81 ± 0.21[a]	1790 ± 115[a]	473 ± 64[a]	163 ± 23.1[a]	3.99 ± 0.520
MFZ 1-76	COOCH₃, CH₃	F	12.9 ± 1.80[b]	690 ± 58.4[b]	269 ± 38.9[b]	133 ± 4.16[b]	1.50 ± 0.020[b]
			2.94 ± 0.360[a]				
MFZ 6-21	CH₂OH, CH₃	F	3.40 ± 0.321[a]	910 ± 64.5[a]	983 ± 97.0[a]	109 ± 15.1[a]	13.8 ± 1.55
MFZ 6-30	CH₂CH₂CO₂CH₃, CH₃	F	3.74 ± 0.070[a]	1070 ± 118[a]	454 ± 43.2[a]	3110 ± 435[a]	3.66 ± 0.526
MFZ 6-26	CH=CHCO₂CH₃, CH₃	F	4.69 ± 0.601[a]	572 ± 51.0[a]	269 ± 34.8[a]	1380 ± 81[a]	1.97 ± 0.090
MFZ 6-95	CO₂CH₂CH₃, butyl	F	8.18 ± 0.171[a]	2500 ± 327[a]	580 ± 15.9[a]	3200 ± 298[a]	9.96 ± 1.03
MFZ 6-93	CO₂CH₂CH₃, H	F	8.87 ± 1.02[a]	2150 ± 221[a]	563 ± 73.9[a]	17100 ± 2480[a]	4.71 ± 0.667
MFZ 6-96	CO₂CH₂CH₃, allyl	F	11.0 ± 0.941[a]	2280 ± 125[a]	935 ± 113[a]	11400 ± 1650[a]	8.84 ± 0.183
MFZ 4-87	CO₂CH₃, CH₃	Cl	12.6 ± 0.399[a]	528 ± 39.4[a]	2150 ± 325[a]	382 ± 36.5[a]	2.46 ± 0.201[a]
MFZ 4-86	CO₂CH₂CH₃, CH₃	Cl	14.6 ± 0.388[a]	1560 ± 90.6[a]	3350 ± 154[a]	3060 ± 147[a]	1.52 ± 0.215[a]
MFZ 2-71	COOCH₂CH₃, CH₃	F	16.8 ± 2.00[b]	1850 ± 270[b]	629 ± 31.0[b]	1890 ± 13[b]	1.84 ± 0.159[b]
			6.87 ± 0.333[a]				
(±)MFZ 1-76R	COOCH₃, CH₃	F	21.3 ± 3.62[b]	1750 ± 243[b]	474 ± 65.9[b]	302 ± 42.8[b]	2.94 ± 0.397[b]

Compound	R groups	X	K_i				
(±)MFZ 4-86R	$CO_2CH_2CH_3$, CH_3	Cl	22.0 ± 0.844[a]	—	—	—	2.79 ± 0.027
MFZ 2-74	COO-2-propyl, CH_3	F	23.3 ± 3.26[b]	12000 ± 1280[b]	642 ± 13.4[b]	2680 ± 139[b]	6.24 ± 0.248[b]
(±)MFZ 4-87R	CO_2CH_3, CH_3	Cl	23.4 ± 1.54[a]	—	—	382.4 ± 36.5	3.95 ± 0.510
(±)MFZ 2-71R	$COOCH_2CH_3$, CH_3	F	26.2 ± 3.40[b]	3740 ± 485[b]	1020 ± 117[b]	1860 ± 189[b]	3.61 ± 0.404[b]
MFZ 6-97	$CH_2OCO(CH_2)_3C_6H_4$-4'NH_2, CH_3	F	27.6 ± 0.85[a]	1390 ± 150[a]	440 ± 18[a]	342 ± 34[a]	
MFZ 3-68	$COOCH_2CH_2Ph$	F	36.7 ± 4.16[c]	3160 ± 461[c]	1030 ± 84.8[c]	354 ± 52.4[c]	2.57 ± 0.180[c]
MFZ 6-83	$CH_2OCO(CH_2)_3C_6H_4$-4'-NO_2, CH_3	F	39.9 ± 4.31[a]	3780 ± 390[a]	2130 ± 160[a]	708 ± 100[a]	—
MFZ 2-82	$COOCH_2Ph$, CH_3	F	40.2 ± 9.25[b]	2040 ± 303[b]	2230 ± 200[b]	4380 ± 528[b]	2.20 ± 0.340[b]
MFZ 4-38	$COOCH_2CH_2Ph$-3'-1,4'NH_2,H CH_3	F	42.3 ± 3.97[c]	—	—	—	—
(±)MFZ 2-74R	COO-2-propyl, CH_3	F	49.1 ± 6.39[b]	13800 ± 679[b]	1990 ± 380[b]	7640 ± 104[b]	6.75 ± 0.180[b]
(±)MFZ 2-82R	$COOCH_2Ph$, CH_3	F	90.3 ± 18.1[b]	3320 ± 290[b]	4110 ± 72.7[b]	5100 ± 310[b]	3.47 ± 0.202[b]
MFZ 4-35	$COOCH_2CH_2Ph$-4'NO_2, CH_3	F	97.7 ± 13.8[c]				3.30 ± 0.299[c]

[1] All binding data are recorded in $K_i \pm$ SEM and are published in the references cited. [a]Zou et al. [69]; [b]Zou et al. [67]; [c]Zou et al. [68]. If reference no cited, the data are previously unpublished but the methods used are identical to those used for the values referenced.

[2] All compounds are optically active and the S-(+)-enantiomer is cited unless indicated otherwise.

[3] Where there are two values, different radiolabeled ligands or analysis programs or incubation times were used, the details of which can be found in the primary references.

[4] The inhibition of dopamine uptake (DAUI) data are recorded as $IC_{50} \pm$ SEM and were obtained in rat brain synaptosomes according to methods detailed in the primary references.

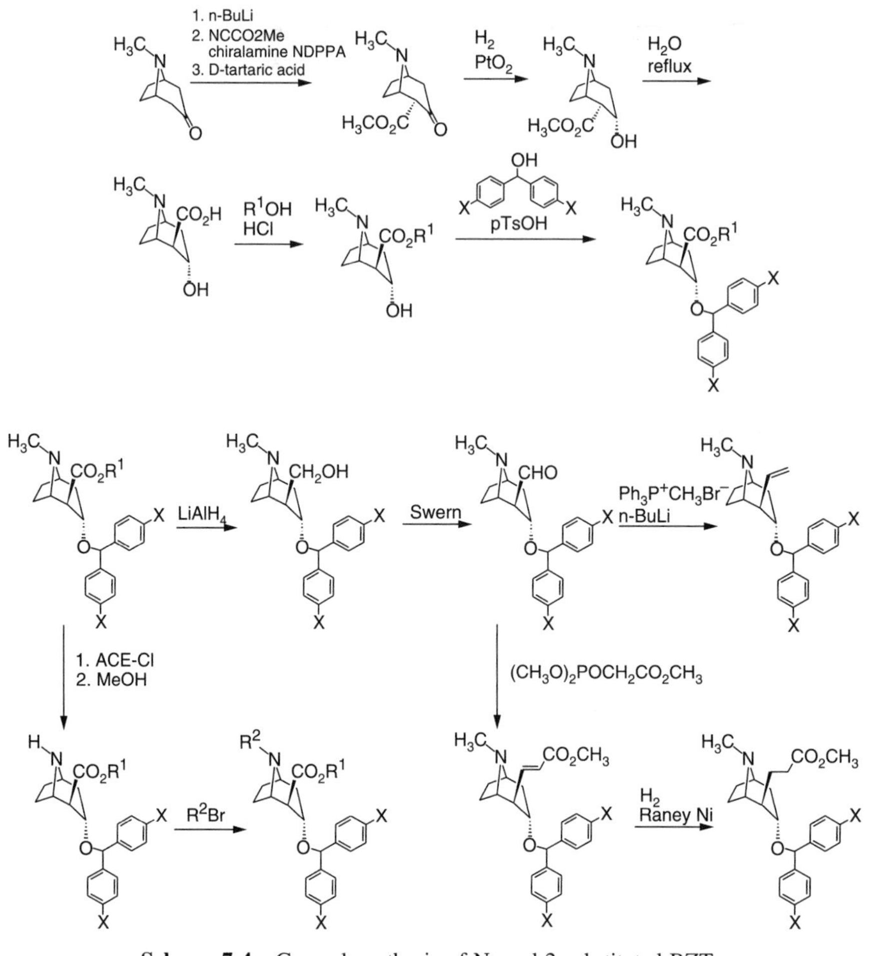

Scheme 7-4   General synthesis of N- and 2-substituted BZTs.

7.3   IN VIVO FINDINGS

As detailed above, BZT and its analogs bind to the DAT with affinities that are related to variations in structure. In addition, most of the BZT analogs are selective for the DAT compared to the other monoamine transporters. According to the DAT hypothesis of cocaine's actions, these actions should confer upon these drugs cocainelike behavioral effects. However, when studied in various laboratory models of cocaine action and cocaine abuse, the BZT analogs were typically less effective than cocaine.

7.3.1   Stimulation of Locomotor Activity

The stimulation of locomotor activity is a benchmark effect of psychomotor stimulant drugs [27]. Most dopamine-uptake inhibitors increase activity at low to intermediate

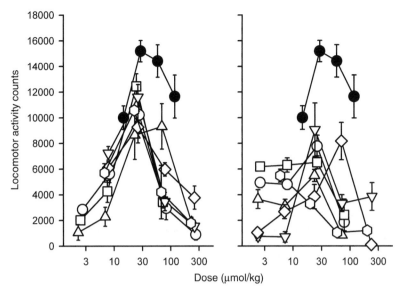

Figure 7-3 Dose-dependent effects of 3β-diphenylmethoxytropane analogs on locomotor activity in mice: horizontal activity counts after drug administration vs. dose of drug. Each point represents the average effect determined in eight mice. The data are from the 30-minute period during the first 60 minutes after drug administration, in which the greatest stimulant effects were obtained. Note that the fluoro-substituted compounds (left panel) were generally more efficacious than the other compounds (right panel). Left panel: Solid circles, cocaine; open circles, 4′-F-BZT; squares 4′,4′-diF-BZT; triangles, 3′,4′-diCl,4′-F-BZT; downward triangles, 3′,4′-diF-BZT; diamonds, 3′,4′-diF-BZT; hexagons, 4′-Br,4″-F-BZT. Right panel: Solid circles, cocaine; open circles, 4′-Cl-BZT; squares, 4′-Cl-BZT (ß); triangles, 4′,4″-diCl-BZT; downward triangles, 3′,4′-diCl-BZT; diamonds, 4′-Br-BZT; hexagons 4′,4′-diBr-BZT. (Adapted from Katz et al. [24].)

doses. At the highest doses, activity is often increased to a lesser extent than it is at intermediate doses. In a comparison of various standard dopamine uptake inhibitors, we found that the maximal effects were generally comparable, if relatively restricted time points for measurement are selected, eliminating duration of action as an influence on the measurement of maximal effects [21]. In contrast, BZT analogs showed variations in effectiveness in stimulating locomotor activity [24]. Figure 7-3 shows results of initial studies with the first generation of BZT analogs. In those studies cocaine produced a typical bell-shaped dose response for stimulation of locomotor activity. Several of the BZT analogs had maximal effects that were substantially less than those of cocaine. Compounds with a fluoro-substitution in the *para*-position on either of the phenyl rings had an efficacy that approached that of cocaine (Fig. 7-3, left panel). However, several other structural variants, in particular those with chloro-substitutions, retained relatively high affinity for the DAT in binding assays but did not stimulate activity to the same extent as did cocaine (Fig. 7-3, right panel).

7.3.2   Cocainelike Subjective Effects

Most monoamine-uptake inhibitors with affinity for the DAT produced dose-related substitution in subjects trained to discriminate cocaine from saline injections [3,28]. The potency differences among these dopamine-uptake inhibitors are generally directly related to their affinity for the DAT. In addition, monoamine-uptake inhibitors with affinity primarily for either the serotonin or norepinephrine transporters generally do not fully substitute for the cocaine discriminative stimulus [3]. In contrast to those findings with known dopamine-uptake inhibitors, BZT analogs did not fully substitute for cocaine in rats trained to discriminate cocaine from saline [24]. As with the stimulation of locomotor activity, there were differences among the analogs with regard to their effectiveness that were related to their chemical structure. As shown in Figure 7-4, *para*-fluoro-substituted analogs (left panels) tended to be among the most effective, whereas BZT analogs with *para*-chloro substitutions, despite binding affinities comparable to those of the fluoro-substituted compounds (Table 7-1), were clearly less effective (right panels).

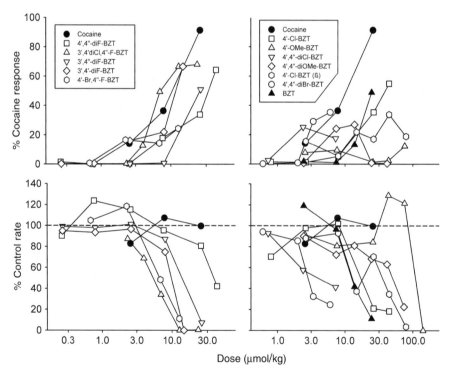

Figure 7-4   Effects of 3α-diphenylmethoxytropane analogs in rats trained to discriminate injections of cocaine from saline: percentage of responses on the cocaine-appropriate key (top panels) or rates at which responses were emitted (as a percentage of response rate after saline administration; lower panels) vs. drug dose. Each point represents the effect in four or six rats. The percentage of responses emitted on the cocaine-appropriate key was considered unreliable and not plotted if fewer than half of the subjects responded at that dose. Note that the fluoro-substituted compounds (left panels) were generally more effective in substituting for cocaine than the other compounds. (Adapted from Katz et al. [24].)

7.3.3 Benztropine Self-Administration

Cocaine is well known for its reinforcing effects (e.g., [27]), and indeed, there is a wealth of published findings documenting those effects (see [22] for a review). Typically, as dose per injection is increased to intermediate values, rates of responding maintained by cocaine increase. At the highest doses, however, rates of responding are typically maintained at lower than maximal rates. We examined the reinforcing effects of BZT analogs in two studies and compared them directly to those of cocaine. In the first study [64] the reinforcing effects of cocaine were compared to those of 3'-Cl- and 4'-Cl-BZT in rhesus monkeys trained to self-administer intravenous (i.v.) cocaine under a fixed-ratio 10 (FR10) schedule of reinforcement. Figure 7-5 shows that self-administration was maintained with cocaine, showing a maximum of approximately 60 injections per hour with cocaine at 0.01 mg/kg per injection. Self administration was also maintained above vehicle levels with both 3'-Cl-BZT and 4'-Cl-BZT, but not to the same extent as by cocaine. In addition, BZT itself did not maintain responding significantly above vehicle levels. In a second study, 3'-Cl-, 4'-Cl-, and 3',4''-diCl-BZT were further compared to cocaine and GBR 12909 using both fixed- and progressive-ratio schedules in rhesus monkeys trained to self-administer i.v. cocaine [65]. As in the previous study, some responding above that maintained by the vehicle was obtained with 3'-Cl-BZT and 4'-Cl-BZT, but not with 3',4''-diCl-BZT. The effects were most pronounced in one of the four subjects, and were generally weaker than those obtained with either cocaine or GBR 12909. Results with the progressive-ratio schedule suggested that the rank order of these compounds for their effectiveness was

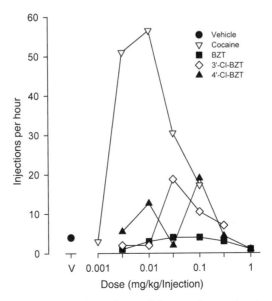

Figure 7-5 Self-administration of cocaine and benztropine analog by rhesus monkeys. Symbols represent the mean of the last three sessions of drug availability averaged across subjects. Points above "V" are those obtained with vehicle. (Adapted from Woolverton et al. [64].)

cocaine >GBR 12909 > 3'-Cl-BZT = 4''-Cl-BZT ≫3',4''-diCl-BZT. Although mechanisms accounting for the variations in reinforcing effectiveness of these drugs were not identified, it was clear that there were differences in the effectiveness of cocaine and the BZT analogs.

7.3.4 Place Conditioning

Some BZT analogs have a slower onset of action than that of cocaine, and it has been suggested that the slow onset may contribute to a decreased reinforcing effectiveness, as delays in reinforcement are known to decrease the effectiveness of reinforcers generally (e.g., [4,17,53]. Delays in onset of action may be readily accommodated in place-conditioning procedures by modifying the time between injection and placement of the subject in the conditioning chamber [14], whereas they are not so readily accommodated in self-administration procedures. The assessment of effects of a drug with delayed onset of effect in place-conditioning procedures may involve selection of an appropriate time between injection and conditioning session.

Li et al. [32] examined place conditioning with BZT analogs administered at various times before conditioning trials. The *N*-methyl-substituted BZT analog AHN 1-055 was without significant effects at doses that ranged from 0.3 to 3.0 mg/kg and when administered up to 90 minutes before conditioning trials. In contrast, effects of indivdual doses of AHN 2-005 and JHW 007 were significant when administered 45 minutes but not immediately or 90 minutes before trials. The results support and extend the previous self-administration results, showing reinforcing effects of the BZT analogs that were less than those obtained with cocaine, and further suggest that these differences from cocaine are not entirely accounted for by a slower onset of action.

7.3.5 Summary of In Vivo Findings

In summary, BZT analogs combine structural features of the class of dopamine-uptake inhibitors cocaine and GBR 12909. In addition, these compounds selectively bind the DAT among the monoamine transporters and inhibit dopamine uptake. They generally do not, however, fully reproduce a cocainelike behavioral profile. The reasons for this disparity between biochemical and behavioral effects have been the subject of various lines of investigation.

7.4 MECHANISMS FOR DIFFERENCES FROM COCAINE

The parent compound, BZT, has actions as an antihistaminic and as an antimuscarinic agent [33,46]. The antimuscarinic effects of BZT analogs were discussed briefly above with regard to their potential to interact and dampen the cocainelike effects of the BZT analogs. We discuss both antimuscarinic and antihistaminic effects below as potential mechanisms contributing to the differences between BZT analogs and cocainelike dopamine-uptake inhibitors.

7.4.1 Histamine Antagonist Effects

If the histamine antagonist effects of the BZT analogs interfere with the expression of their cocainelike actions, histamine antagonists should attenuate the effects of cocaine [8]. Rats trained to discriminate injections of 10 mg/kg of cocaine from saline were tested with different doses of cocaine either alone or after pretreatment with histamine antagonists. Increasing doses of promethazine, an H1 histamine antagonist, did not alter the effects of cocaine. Similarly, other H1 antagonists, including triprolidine, chlorpheniramine, and mepyramine, were not effective in blocking the effects of cocaine. Chlorpheniramine and mepyramine, in contrast to the others, shifted the cocaine dose–response curve to the left. This leftward shift is probably due to the actions of these two latter compounds at the DAT [58].

Campbell et al. also examined whether affinity for histamine H1 receptors relative to affinity for the DAT predicted outcome for the locomotor stimulant effects of BZT analogs [8]. Various assessments of these numbers (e.g., Fisher's exact test, correlations) indicated that the ratios of H1 to DAT affinities were not significantly related to whether the drugs produced a significant stimulation of activity. Because the affinities of the BZT analogs for H2 and H3 receptors were uniformly lower than affinities for the DAT, we did not conduct the same analysis for the other histamine receptors.

In summary, it does not appear that histamine agonist or antagonist actions can interfere substantially with either the discriminative-stimulus or locomotor stimulant effects of cocaine. In addition, studies in the literature suggest that H1 antagonists have some behavioral effects in common with psychomotor stimulant drugs (e.g., [7]). Thus, it does not appear that actions at histamine receptors interfere in any substantial way with the effects of cocaine.

7.4.2 Muscarinic Receptor Antagonist Effects

As with histaminic effects, if muscarinic antagonist effects of the BZT analogs interfere with the expression of cocainelike effects, a muscarinic receptor antagonist should attenuate the effects of cocaine. Instead, both atropine and scopolamine shifted the dose–effect curve to the left for the discriminative-stimulus effects of cocaine [24]. In addition, atropine and other muscarinic antagonists are known to potentiate various effects of stimulant drugs, including cocaine [51]. The BZT analogs have preferential affinity for muscarinic M1 receptors, whereas atropine and scopolamine are nonselective [55]. Therefore, we investigated the effects of the preferential M1 antagonists telenzepine and trihexyphenidyl on the discriminative-stimulus effects of cocaine. Both of these drugs produced a small leftward shift in the cocaine dose–effect curve for the discriminative-stimulus effects [56]. However, trihexyphenidyl potentiated and telenzepine attenuated the effect of cocaine on locomotor activity [55]. In sum, there was no consistent attenuation of a wide range of effects of cocaine by either telenzepine or trihexyphenidyl, suggesting that the differences between BZT analogs and those of cocaine are not due to preferential M_1 antagonist effects.

In addition to the pharmacological studies, Agoston et al. [2], Robarge et al. [48], and Kulkarni et al. [30] synthesized BZT analogs with reduced affinity at muscarinic M_1 receptors. Substitutions on the tropane nitrogen decreased ($\geq$100-fold) M_1 muscarinic receptor affinity significantly. In addition, these substitutions did not decrease the high affinity of the BZT analogs for the DAT, and selectivity among monoamine transporters was retained. The N-methyl-4′,4″diF analog (AHN 1-055) was the most effective of the BZT analogs and produced a stimulation of locomotor activity that approached that of cocaine. In contrast to that compound, various other N-substituted analogs had substantially lower efficacy as either locomotor stimulants or in substituting in rats trained to discriminate cocaine from saline injections (e.g., AHN 2-005 and JHW 007 [26]).

The antagonist activity at muscarinic receptors may also contribute to the differences in the reinforcing effectiveness of BZT analogs and cocaine. In one study [43], rhesus monkeys were trained to self-administer cocaine under fixed- and progressive-ratio schedules. In most cases, combinations of cocaine and scopolamine maintained less self-administration than did cocaine alone. The authors concluded that anticholinergic actions contribute to the diminished self-administration of BZT analogs relative to cocaine and suggested that the mechanism involves either antagonism of the reinforcing effect of cocaine or punishment of the cocaine self-administration by the anticholinergic effect.

Wilson and Schuster [63] previously found atropine to increase rather than decrease rates of responding maintained by cocaine, a result that is not consistent with punishment by the anticholinergic agent. Further, the atropine-induced increases in response rates were similar to the effect of lowering the dose of cocaine. Because atropine was administered independent of responding before the experimental session, it was not functioning as a punishing stimulus. All of these considerations suggest a pharmacological noncompetitive antagonism of the reinforcing effects of cocaine by the anticholinergic agents.

Li et al. [32] investigated the effects of atropine in the place-conditioning procedure. Several trends that failed to reach significance in that study mitigate firm conclusions regarding the basis for the interaction of anticholinergics with cocaine. Atropine alone produced a trend toward conditioned-place avoidance and a trend toward decreases in effectiveness of cocaine, suggesting a behavioral basis to the interaction. Thus, selected aspects of the results suggest the potential of an anticholinergic punishing effect.

The hypothesis that the anticholinergic effect of the BZT analogs punishes behavior relates specifically to self-administration procedures using the BZT analogs. However, as noted above, there is a wider range of behavioral differences between the BZT analogs and cocainelike DA uptake inhibitors, as evidenced by the drug discrimination and locomotor activity results. Hypotheses accounting for those effects need to be broader than those that apply specifically to reinforcing effects. In that regard, it should be noted that the compounds with reduced M_1 affinity (e.g., AHN 2-005 and JHW 007) have less cocainelike activity than compounds with greater M1 affinity (e.g., AHN 1-055, RIK 7). Examining a wide range of compounds and various behavioral endpoints provides little evidence that actions at muscarinic receptors interfere with the expression of cocainelike effects of BZT analogs.

7.5   DIFFERENCES BETWEEN IN VIVO AND IN VITRO ACTIONS

That the BZT analogs have effects in vitro that predict an effect that generally is not observed in vivo, has suggested relatively poor CNS penetration of BZT analogs. A study of the pharmacokinetics of selected BZT analogs in rats found high concentrations in the brain within minutes after injection, and that the elimination rates of BZT analogs were much slower than that for cocaine [42]. All of the BZT analogs were detected at initial concentrations approximating 4 to 15 μg per gram of tissue, approximating 3 to 27 μM concentrations, depending on the drug. These molar concentrations are well above the K_i values for these drugs, which range from 11 to 30 nM ([42], see also [41]). In addition, the parent compound is known to exhibit reliable CNS activity, and a previous study showed increases in extracellular DA levels assessed by in vivo microdialysis after systemic injection [12]. Thus, the BZT analogs are available in the brain at concentrations well above those necessary for accessing the DAT within a short time after injection.

We compared the effects of 4′-Cl-BZT with those of cocaine on extracellular DA concentrations in the nucleus accumbens [57] and in several specific brain regions, including the nucleus accumbens shell and core, prefrontal cortex, and dorsal caudate [54]. Cocaine and 4′-Cl-BZT produced dose-related increases in the concentration of DA in all regions, and at comparable doses, the effects of 4′-Cl-BZT on DA levels in all brain areas except the PFCX were generally reduced compared with those of cocaine. Possibly, the most notable difference between cocaine and 4′-Cl-BZT was that the latter compound had a long duration of action with DA levels substantially elevated at the higher doses up to 5 hours after injection. These long-lasting effects are consistent with the slow elimination of BZT analogs noted in pharmacokinetic studies. Possibly more important, however, was that in addition to the long duration of action of 4′-Cl-BZT, the rate of increase in extracellular DA levels was slower than that for cocaine.

The slower onset of effects on extracellular DA concentrations seemingly conflicts with the results of pharmacokinetic studies, which indicated that the BZT analogs were in brain within minutes after injection at concentrations well above their K_i values. We therefore conducted studies of the binding of the analogs in vivo using intravenous administration of [125I]RTI-121 in mice to label the DAT and compared these effects to those of cocaine [15,16]. Figure 7-6 shows the effects of cocaine and several N-substituted BZT analogs. As has been shown in the literature, cocaine displaced [125I]RTI-121 in a dose- and time-dependent manner. Maximal displacement of RTI-121 by cocaine was obtained at 30 minutes after injection. The BZT analogs also displaced RTI-121 binding in a dose- and time-related manner; however, each of the compounds had an apparent association rate that was substantially slower than that for cocaine. Both AHN 1-055 and AHN 2-005 reached their maximum displacement of RTI-121 at 150 minutes after injection compared to 30 minutes for cocaine.

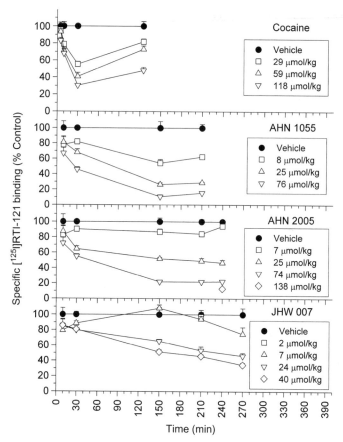

Figure 7-6 Displacement of specific [^{125}I]RTI-121 accumulation in striatum at various times following intraperitoneal injection of cocaine, AHN 1-055, AHN 2-005, and JHW 007. For each point the number of replicates was from 5 to 10 or 13. Note that maximal displacement of [^{125}I]RTI-121 was obtained with cocaine at 30 minutes after injection, and at later times with the other compounds. (Adapted from Desai et al. [15,16].)

Consistent with the slow elimination seen in the pharmacokinetic studies, there was little evidence of the dissociation of any of the BZT analogs up to 4 to 4.5 hours after injection. There was evidence of a plateau in the in vivo displacement of RTI-121 around 150 minutes after injection for AHN 1-055 and AHN 2-005; however, for JHW 007 a plateau was not reached up to 4.5 hours after injection. When the apparent association rate of JHW 007 was calculated as the percent displacement of RTI-121 per minute over the linear portions of the curves shown in Figure 7-5, it was found that the association of JHW 007 was more than 10-fold lower than that for cocaine.

Concurrently with the time course for the in vivo binding to the DAT, we also examined the behavioral effects of the BZT analogs, and compared them to those of cocaine over an 8-hour period. The onset of effects was relatively rapid, and cocaine produced its maximal effects at 30 minutes after injection, as it did in the binding study. The effects of cocaine decreased substantially over the next 2 hours and were essentially absent after that point. In contrast, the maximal effects of the BZT analogs AHN 1-055 and AHN 2-005 were delayed substantially. Because the "baseline" level of activity in mice changes with time in the chamber (habituation), it is difficult to determine exactly at what point(s) the BZT analogs produced their maximal effects. However, it appears that maximal effects occurred at about 90 to 120 minutes after injection and were sustained throughout the 8-hour observation period. Regardless of time, each of the drugs, while stimulating locomotor activity in mice, was less effective than cocaine. In contrast to the effects of AHN 1-055 and AHN 2-005, JHW 007 did not produce significant stimulation of locomotor activity at any dose and at any time.

We examined the relationship between the DAT occupancy and stimulation of locomotor activity produced by cocaine by correlating occupancy and stimulation at the various doses and time points [15]. There was a significant correlation of DAT occupancy and stimulant effects (Fig. 7-7). However, the correlation was not nearly as strong as expected, particularly when the studies by Ritz et al. [47] and others (e.g., [6]) were considered. One important potential reason for the differences may

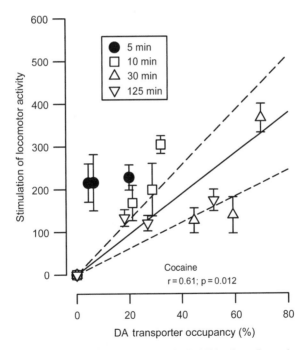

Figure 7-7 Relationship between occupancy of the DAT in the striatum by cocaine and its locomotor stimulant effects: amount of stimulation of horizontal locomotor activity expressed as counts after drug administration minus counts after vehicle administration; vs. DAT occupancy produced by cocaine (the inverse of specific [^{125}I]RTI-121 binding) expressed as a percentage of vehicle controls. The solid line represents the regression of stimulation of locomotor activity on the occupancy of the DAT with the line forced through the origin. The dashed lines represent the 95% confidence limits for the regression line. The error bars on points represent ±1 SEM. (Adapted from Desai et al. [15].)

be that previous studies related the in vivo effects of the various compounds to binding constants obtained in vitro with various buffer systems at equilibrium.

Interestingly, with cocaine at times soon after injection there was a disproportionately greater effect on locomotor activity than was predicted by DAT occupancy. That the divergence from the regression was observed for data obtained soon after injection is consistent with suggestions that association rate plays an important role in the stimulant effects of cocaine (e.g., [62]). As mentioned above, the apparent association rate of JHW 007 was approximately 10-fold lower than that for cocaine, and this drug had the lowest cocainelike efficacy in stimulating locomotor activity. Moreover, JHW 007 never reached an apparent equilibrium in the in vivo binding studies (Fig. 7-6). However, 4.5 hours after injection, the displacement of [^{125}]RTI-121 produced by JHW 007 at the higher doses was comparable to that produced by maximal stimulating doses of cocaine.

Because JHW 007 lacked robust cocainelike stimulant effects over that time period, we conducted studies of its potential antagonist effects against cocaine. Saline or JHW 007 was administered 4.5 hours before cocaine, after which locomotor activity was assessed. Following a saline injection, cocaine produced dose-related increases in locomotor activity that reached a maximum of 40 mg/kg; however, after

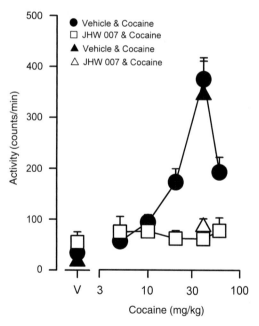

Figure 7-8 Interactions of cocaine and JHW 007: locomotor activity counts vs. treatment condition for vehicle (V) or dose of cocaine. Each point represents the average effect determined in eight mice, except $n = 6$ for the combination of 10 mg/kg of JHW 007 with 60 mg/kg cocaine. The unconnected points above 40 mg/kg cocaine show replicates determined several months later in a separate shipment of mice. The error bars represent ± 1 SEM. (Adapted from Desai et al. [15].)

pretreatment with JHW 007, the effects of cocaine were blocked (Fig. 7-8). Other drugs that bind to the DAT when administered in combination with cocaine produce dose-dependent leftward shifts in the cocaine dose–effect curve (e.g., [25]). Nonetheless, there are reports, as mentioned in Section 7.1, of DA-uptake inhibitors with unusual effects, some resembling those of JHW 007. Thus, a BZT structure might not be necessary for high-affinity DAT binding and cocaine antagonist effects.

7.6 MOLECULAR EFFECTS OF BENZTROPINE ANALOGS

There are several lines of evidence that suggest that the BZTs bind differently than cocaine to the DAT and that those differences in interaction, at the molecular level, may be related to the actions of BZT analogs that differ from those of cocaine [23,40]. First, SAR between the BZTs and the 3-phenyltropane class of dopamine uptake inhibitors at the DAT are very different and were described in detail above. Specifically, the 3β-diphenyl ethers with small halogen substituents in the *para-* or *meta*-positions of the phenyl rings provide optimal binding affinity at the DAT for BZTs, whereas various substituted 3β-phenyl rings in the cocaine class of molecules is generally optimal for high-affinity DAT binding. Tropane-*N*-alkyl and arylalkyl substitutions on BZT analogs are generally well tolerated at the DAT, and this substitution provides a point of departure from high-affinity binding at muscarinic receptors. In the case of the 3-phenyl tropanes, significant exploration of tropane N-substitution has not yielded higher DAT affinities or selectivities across other transporters and receptors to which the parent molecule binds. The 3β-phenyl tropanes require substitution in the 2-position with *R*-(−)-2β-substituents being optimal at the DAT. In contrast, only *S*-(+)-2β-substitution in the BZTs results in high-affinity binding at the DAT, and the BZTs without substitution at this position have essentially equal DAT affinities, suggesting that this substituent is not necessary for high-affinity binding.

Early studies employing site-directed mutagenesis of the DAT revealed differences in binding modes between cocaine and BZT. Reith et al. [44] compared the abilities of, among other drugs, cocaine and BZT to affect the reaction of a methane-thiosulfonate (MTS) to various cysteine residues within the DA transporter. In contrast to cocaine, BZT had no effect on the reaction of Cys-90 with an MTS reagent. In addition, the decrease in the reaction of Cys-135 with an MTS reagent produced by cocaine and other DA-uptake inhibitors was not obtained with BZT [44]. Several other studies have demonstrated differences in the actions of BZT or its analogs with wild-type and mutant DA transporters (e.g., [11]). A more recent study with our BZT analogs, using site-directed mutagenesis of aspartate 79 in TM1 of DAT, demonstrated that these compounds bind the DAT differently than cocaine, its analogs, and other structurally diverse DAT inhibitors [59]. Studies using photoaffinity labeling techniques have also provided evidence that structurally different dopamine uptake inhibitors covalently attach to different TMs in the DAT [1,61,66]. These results together support the idea that different DA-uptake inhibitors bind to the DAT in different ways and suggest that there may be particular conformational changes of the DAT induced by cocaine and cocainelike drugs that contribute to its substantial abuse liability. Additional studies are ongoing combining the molecular techniques that are available with the arsenal of BZT and 3β-phenyltropane analogs that we have prepared over the years to ultimately identify and characterize the DAT-binding domains of these BZT analogs and compare those to cocaine in hopes of better understanding what might contribute to the slow association rate that is observed for the BZT analogs, as described above. As we have clearly demonstrated, these compounds are bioavailable, readily penetrate the blood–brain barrier, and gain high levels of brain/plasma ratios within minutes of injection. Nevertheless, DAT occupancy and subsequent increases in dopamine, measured by microdialysis, are delayed significantly compared to cocaine. Furthermore, DAT occupancy and maintained levels of dopamine are also long lasting, and the combination of these effects is probably responsible for the differing pharmacological properties of the compounds and may be exploited for development of a medication to treat cocaine addiction. Our studies dismiss potential confounds with muscarinic or histaminic receptor binding, as these actions do not uniformly attenuate the behavioral effects of cocaine, nor do they seem to contribute substantial untoward side effects. Nevertheless, intensive efforts to design and synthesize compounds with both high affinities and selectivities at DAT have been rewarded in recent years [30,69]. Future investigations will continue to elucidate the interactions of these drugs compared to cocaine at the DAT, and to reveal how those differences affect other neural systems and circuitry, so as to provide a better understanding of cocaine's mechanism of action and to feed this information into our drug design. Further, as the BZTs do not appear to have significant abuse liability, they may also have therapeutic utility in other disorders, such as attention-deficit/hyperactivity disorder, and drug development in this area is another line of current pursuit.

ACKNOWLEDGMENTS

We acknowledge the dedicated members of our laboratories, past and present, who have diligently taken on the arduous tasks of synthesizing, characterizing, and biologically evaluating all of our novel BZTs. We are particularly indebted to Ms. Theresa Kopajtic, who performed the vast majority of binding and uptake studies on these compounds and helped us keep these data straight. We also thank Ms. J. Cao for aiding in the construction of the data tables, and Dr. Noel Paul for his editorial comments on an earlier version of this review. The work was supported by the National Institute on Drug Abuse Intramural Research Program.

REFERENCES

1. Agoston, G. E., Vaughan, R., Lever, J. R., Izenwasser, S., Terry, P. D., and Newman, A. H. (1997). A novel photoaffinity label for the dopamine transporter based on N-substituted-4′,4″-difluoro-3β-(diphenylmethoxy)tropane. *Bioorg Med Chem Lett, 7*, 3027–3032.

2. Agoston, G. E., Wu, J. H., Izenwasser, S., George, C., Katz, J., Kline, R. H., and Newman, A. H. (1997). Novel N-substituted 4′,4″-difluoro-3β-(diphenylmethoxy)tropane analogs: selective ligands for the dopamine transporter. *J Med Chem, 40*, 4329–4339.

3. Baker, L. E., Riddle, E. E., Saunders, R. B., and Appel, J. B. (1993). The role of monoamine uptake in the discriminative stimulus effects of cocaine and related compounds. *Behav Pharmacol, 1*, 69–79.

4. Beardsley, P. M., and Balster, R. L. (1993). The effects of delay of reinforcement and dose on the self-administration of cocaine and procaine in rhesus monkeys. *Drug Alcohol Depend, 34*, 37–43.

5. Berdini, V., Cesta, M. C., Curti, R., D'annibale, G., Di Bello, N., Nano, G., Nicolini, L., Topai, A., and Allegretti, M. (2002). A modified palladium catalysed reductive amination procedure. *Tetrahedron, 58*, 5669–5674.

6. Bergman, J., Madras, B. K., Johnson, S. E., and Spealman, R. D. (1989). Effects of cocaine and related drugs in nonhuman primates: III. Self-administration by squirrel monkeys. *J Pharmacol Exp Ther, 251*, 150–155.

7. Bergman, J., and Spealman, R. D. (1986). Some behavioral effects of histamine H1 antagonists in squirrel monkeys. *J Pharmacol Exp Ther, 239*, 104–110.

8. Campbell, V. C., Kopajtic, T. A., Newman, A. H., and Katz, J. L. (2005). Assessment of the influence of histaminergic actions on cocaine-like effects of 3β-diphenylmethoxy-tropane analogues. *J Pharmacol Exp Ther, 315*, 631–640.

9. Cao, J., Kulkarni, S. S., Husbands, S. M., Bowen, W. D., Williams, W., Kopajtic, T., Katz, J. L., George, C., and Newman, A. H. (2003). Dual probes for the dopamine transporter and σ1 receptors: novel piperazinyl alkyl-bis-(4′-fluorophenyl)amine analogues as potential cocaine abuse therapeutic agents. *J Med Chem, 46*, 2589–2598.

10. Carroll, F. I. (2003). Monoamine transporters and opioid receptors: targets for addiction therapy. *J Med Chem, 46*, 1775–1794.

11. Chen, N., Zhen, J., and Reith, M. E. A. (2004). Mutation of Trp84 and Asp313 of the dopamine transporter reveals similar mode of binding interaction for GBR 12909 and benztropine as opposed to cocaine. *J Neurochem, 89*, 853–864.

12. Church, W. H., Justice, J. B., Jr, and Byrd, L. D. (1987). Extracellular dopamine in rat striatum following uptake inhibition by cocaine, nomifensine and benztropine. *Eur J Pharmacol, 139*, 345–348.

13. Colpaert, F. C., Niemegeers, C. J., and Janssen, P. A. (1979). Discriminative stimulus properties of cocaine: neuropharmacological characteristics as derived from stimulus generalization experiments. *Pharmacol Biochem Behav, 10*, 535–546.

14. De Beun, R., Jansen, E., Geerts, N. E., Slangen, J. L., and Van De Poll, N. E. (1992). Temporal characteristics of appetitive stimulus effects of luteinizing hormone-releasing hormone in male rats. *Pharmacol Biochem Behav, 42*, 445–450.

15. Desai, R., Kopajtic, T., Koffarnus, M., and Newman, A. H., and Katz, J. L. (2005). Identification of a dopamine transporter ligands that functions as a cocaine antagonist. *J Neurosci, 25*, 1889–1893.

16. Desai, R., Kopajtic, T., French, D., Newman, A. H., and Katz, J. L. (2005). Relationship between in vivo occupancy at the dopamine transporter and behavioral effects of cocaine, GBR 12909 and benztropine analogues. *J Pharmacol Exp Ther, 315*, 397–404.

17. Gollub, L., and Yanagita, T. (1974). Delayed cocaine reinforcement on lever-pressing in rhesus monkeys. *Psychonom Soc, 4*, 263.

18. Grundt, P., Kopajtic, T. A., Katz, J. L., and Newman, A. H. (2004). The effect of 6-substituted-4′,4″-difluorobenztropines on monoamine transporters and the muscarinic M1 receptor. *Bioorg Med Chem Lett, 14*, 3295–3298.

19. Grundt, P., Kopajtic, T., Katz, J. L., and Newman, A. H. (2005). N-8-Substituted-benztropinamine analogs as selective dopamine transporter ligands. *Bioorg Med Chem Lett, 15*, 5419–5423.

20. Husbands, S. H., Kopajtic, T., Izenwasser, S., Bowen, W. D., Vilner, B. J., Katz, J. L., and Newman, A. H. (1999). Structure–activity relationships at the monoamine transporters and σ receptors for a novel series of 9-[3-*cis*-3,5-dimethyl-1-piperazinyl)propyl] carbazole (Rimcazole) Analogs. *J Med Chem, 42*, 4446–4455.

21. Izenwasser, S., Terry, P., Heller, B., Witkin, J. M., and Katz, J. L. (1994). Differential relationships among dopamine transporter affinities and stimulant potencies of various uptake inhibitors. *Eur J Pharmacol, 263*, 277–283.

22. Johanson, C. E., and Fischman, M. W. (1989). The pharmacology of cocaine related to its abuse. *Pharmacol Rev, 41*, 3–52.

23. Katz, J. L., Izenwasser, S., and Newman, A. H. (1997). Relations between heterogeneity of dopamine transporter binding and punction and the behavioral pharmacology of cocaine. *Pharmacol Biochem Behav, 57*, 505–512.

24. Katz, J. L., Izenwasser, S., Kline, R. H., Allen, A. C., and Newman, A. H. (1999). Novel 3α-diphenylmethoxytropane analogs: selective dopamine uptake inhibitors with behavioral effects distinct from those of cocaine. *J Pharmacol Exp Ther, 288*, 302–315.

25. Katz, J. L., Libby, T. A., Kopajtic, T., Husbands, S. M., and Newman, A. H. (2003). Behavioral effects of rimcazole analogs alone and in combination with cocaine. *Eur J Pharmacol, 468*, 109–119.

26. Katz, J. L., Kopajtic, T., Agoston, G. E., and Newman, A. H. (2004). Effects of N-substituted analogues of benztropine: diminished cocaine-like effects in dopamine transporter ligands. *J Pharmacol Exp Ther, 288*, 302–315.

27. Kelleher, R. T. (1977). Psychomotor stimulants. In *Drug Abuse: The Clinical and Basic Aspects*. C. V. Mosby St. Louis, MO, pp. 116–147.

28. Kleven, M. S., Anthony, E. W., and Woolverton, W. L. (1990). Pharmacological characterization of the discriminative stimulus effects of cocaine in rhesus monkeys. *J Pharmacol Exp Ther, 254*, 312–317.

29. Kline, R. H., Izenwasser, S., Katz, J. L., and Newman, A. H. (1997). 3′-Chloro-3β-(diphenylmethoxy)tropane but not 4′-chloro-3β-(diphenylmethoxy)tropane produces a cocaine like behavioral profile. *J Med Chem, 40*, 851–857.

30. Kulkarni, S. S., Grundt, P., Kopajtic, T., Katz, J. L., and Newman, A. H. (2004). Structure–activity relationships at monoamine transporters for a series of N-substituted-3α-(bis[4-fluorophenyl]methoxy)tropanes: comparative molecular field analysis, synthesis and pharmacological evaluation. *J Med Chem, 47*, 3388–3398.

31. Kulkarni, S. S., Kopajtic, T., Katz, J. L., and Newman, A. H. (2006). Comparative structure–activity relationships of benztropine analogues at the dopamine transporter and histamine H1 receptors. *Bioorg Med Chem, 14*, 3625–3634.

32. Li, S.-M., Newman, A. H., and Katz, J. L. (2005). Place conditioning and locomotor effects of N-substituted, 4′,4″-difluorobenztropine analogues in rats. *J Pharmacol Exp Ther, 313*, 1223–1230.

33. McKearney, J. W. (1982). Stimulant actions of histamine H1 antagonists on operant behavior in the squirrel monkey. *Psychopharmacology*, 77, 156–158.

34. Meltzer, P. C., Liang, A. Y., and Madras, B. K. (1994). The discovery of an unusually selective and novel cocaine analog: difluoropine, synthesis and inhibition of binding at cocaine recognition sites. *J Med Chem*, 37, 2001–2110.

35. Meltzer, P. C., Liang, A. Y., Blundell, P., Gonzales, M. D., Chen, Z., George, C., and Madras, B. K. (1997). 2-Carbomethoxy-3-aryl-8-oxabicyclo[3.2.1]octanes: potent and non-nitrogen inhibitors of monoamine transporters. *J Med Chem*, 40, 2662–2673.

36. Navarro, H. A., Howard, J. L., Pollard, G. T., and Carroll, F. I. (2005). The DAT-selective 3-phenyltropane RTI-371 antagonizes the in vivo effect of cocaine. Presented at the 67th Annual Meeting of the College on Problems of Drug Dependence.

37. Newman, A. H., Allen, A. C., Izenwasser, S., and Katz, J. L. (1994). Novel 3α-diphenylmethoxytropane analogs are potent dopamine uptake inhibitors without cocaine-like behavioral profiles. *J Med Chem*, 37, 2258–2261.

38. Newman, A. H., Kline, R. H., Allen, A. C., Izenwasser, S., George, C., and Katz, J. L. (1995). Novel 4′- and 4′,4″-substituted-3α-(diphenylmethoxy)tropane analogs are potent and selective dopamine uptake inhibitors. *J Med Chem*, 38, 3933–3940.

39. Newman, A. H., Robarge, M. J., Howard, I. M., Wittkopp, S. L., Kopajtic, T., Izenwasser, S., and Katz, J. L. (2001). Structure–activity relationships at the monoamine transporters and muscarinic receptors for N-substituted-3α-[3′-Cl-, 4′-Cl-, 4′,4″-diCl-substituted) diphenyl]methoxytropanes. *J Med Chem*, 44, 633–640.

40. Newman, A. H. and Kulkarni, S. S. (2002). Probes for the dopamine transporter: new leads toward a cocaine-abuse therapeutic: a focus on analogues of benztropine and rimcazole. *Med Res Rev*, 22, 1–36.

41. Othman, A. A., Syed, S. A., Newman, A. H., and Eddington, N. D. (2007). In vitro transport, human and rat CYP phenotyping and in vivo population pharmacokinetics and brain distribution of the chloro benztropine (BZT) analogs, a class of compounds extensively evaluated in animal models of drug abuse. *J Pharmacol Exp Ther*, 320, 344–353.

42. Raje, S., Cao, J., Newman, A. H., Gao, H., and Eddington, N. (2003). Evaluation of the blood brain barrier transport, population pharmacokinetics and brain distribution of benztropine analogs and cocaine using in vitro and in vivo techniques. *J Pharmacol Exp Ther*, 307, 801–808.

43. Ranaldi, R., and Woolverton, W. L. (2002). Self-administration of cocaine: scopolamine combinations by rhesus monkeys. *Psychopharmacology (Berl)*, 161, 442–448.

44. Reith, M. E. A., Berfield, J. L., Wang, L. C., Ferrer, J. V., and Javitch, J. A. (2001). The uptake inhibitors cocaine and benztropine differentially alter the conformation of the human dopamine transporter. *J Biol Chem*, 276, 29012–29018.

45. Renz, J., and Lindenmann, A. (1960). Alkaloidsynthesen: XIII. Synthesen von Scopin-benzhydryläthern. *Z Physiol Chem*, 321, 148–160.

46. Richelson, E. (1981). Tricyclic antidepressants: interactions with histamine and muscarinic acetylcholine receptors. In *Antidepressants: Neurochemical, Behavioral and Clinical Perspectives*. Raven Press, New York, pp. 53–73.

47. Ritz, M. C., Lamb, R. J., Goldberg, S. R., and Kuhar, M. J. (1987). Cocaine receptors on dopamine transporters are related to self-administration of cocaine. *Science*, 237, 1219–1223.

48. Robarge, M. J., Agoston, G. E., Izenwasser, S., Kopajtic, T., George, C., Katz, J. L., and Newman, A. H. (2000). Highly selective chiral N-substituted 3α-[bis(4'-fluorophenyl)-methoxy]tropane analogues for the dopamine transporter: synthesis and comparative molecular field analysis. *J Med Chem, 43*, 1085–1093.

49. Rocha, B. A., Fumagalli, F., Gainetdinov, R. R., Jones, S. R., Ator, R., Giros, B., Miller, G. W., and Caron, M. G. (1998). Cocaine self-administration in dopamine-transporter knockout mice. *Nat Neurosci, 1*, 132–137.

50. Rothman, R. B., Greig, N., Kim, A., De Costa, B. R., Rice, K. C., Carroll, F. I., and Pert, A. (1992). Cocaine and GBR12909 produce equivalent motoric responses at different occupancy of the dopamine transporter. *Pharmacol Biochem Behav, 43*, 1135–1142.

51. Scheckel, C. L., and Boff, E. (1964). Behavioral effects of interacting imipramine and other drugs with *d*-amphetamine, cocaine, and tetrabenazine. *Psychopharmacologia, 5*, 198–208.

52. Sora, I., Wichems, C., Takahashi, N., Li, X. F., Zeng, Z., Revay, R., Lesch, K. P., Murphy, D. L., and Uhl, G. R. (1998). Cocaine reward models: conditioned place preference can be established in dopamine- and in serotonin-transporter knockout mice. *Proc Natl Acad Sci U S A, 95*, 7699–7704.

53. Stretch, R., Gerber, G. J., and Lane, E. (1976). Cocaine self-injection behaviour under schedules of delayed reinforcement in monkeys. *Can J Physiol Pharmacol, 54*, 632–638.

54. Tanda, G., Ebbs, A., and Newman, A. H., and Katz, J. L. (2005). Effects of 4-Cl-BZT on mesostriatal, mesocortical and mesolimbic dopamine transmission: comparison with effects of cocaine. *J Pharmacol Exp Ther, 313*, 613–620.

55. Tanda, G., Ebbs, A., Kopajtic, T., Campbell, B., Newman, A. H., and Katz, J. L. (2007). Effects of muscarinic M1 receptor blockade on cocaine-induced elevations of brain dopamine levels and locomotor behavior in rats. *J Pharmacol Exp Ther, 321*, 332–344.

56. Tanda, G., and Katz, J. L. (2007). Muscarinic preferential M1 receptor antagonists enhance the discriminative-stimulus effects of cocaine in rats. *Pharmacol Biochem Behav, 87*, 400–404.

57. Tolliver, B. K., Ho, L. B., Newman, A. H., Fox, L. M., Katz, J. L., and Berger, S. P. (1999). Behavioral and neurochemical effects of dopamine transporter ligands alone and in combination with cocaine: characterization of 4-chlorobenztropine in vivo. *J Pharmacol Exp Ther, 103*, 110–122.

58. Tuomisto, J., and Tuomisto, L. (1980). Effects of histamine and histamine antagonists on the uptake and release of catecholamines and 5-HT in brain synaptosomes. *Med Biol, 58*, 33–37.

59. Ukairo, O. T., Bondi, C. D., Newman, A. H., Kulkarni, S. S., Kosikowski, A. P., Pan, S., and Surratt, C. K. (2005). Recognition of benztropine by the dopamine transporter (DAT) differs from that of the classical dopamine-uptake inhibitors cocaine, methylphenidate and mazindol as a function of a DAT transmembrane 1 aspartic acid residue. *J Pharmacol Exp Ther, 314*, 575–583.

60. Vaugeois, J. M., Bonnet, J. J., Duterte-Boucher, D., Costentin, J. (1993). In vivo occupancy of the striatal dopamine uptake complex by various inhibitors does not predict their effects on locomotion. *Eur J Pharmacol, 230*, 195–201.

61. Vaughan, R. A., Agoston, G. E., Lever, J. R., and Newman, A. H. (1999). Differential binding sites of tropane-based photoaffinity ligands on the dopamine transporter. *J Neurosci, 19*, 630–636.

62. Volkow, N. D., Fowler, J. S., and Wang, G. J. (2002). Role of dopamine in drug reinforcement and addiction in humans: results from imaging studies. *Behav Pharmacol*, *13*, 355–366.

63. Wilson, M. C., and Schuster, C. R. (1973). Cholinergic influence on intravenous cocaine self-administration by rhesus monkeys. *Pharmacol Biochem Behav*, *1*, 643–649.

64. Woolverton, W. L., Rowlett, J. K., Wilcox, K. M., Paul, I. A., Kline, R. H., Newman, A. H., and Katz, J. L. (2000). 3′- and 4″-Chloro-substituted analogs of benztropine: intravenous self-administration and in vitro radioligand binding studies in rhesus monkeys. *Psychopharmacology*, *147*, 426–435.

65. Woolverton, W. L., Hecht, G. S., Agoston, G. E., Katz, J. L., and Newman, A. H. (2001). Further studies of the reinforcing effects of benztropine analogs in rhesus monkeys. *Psychopharmacology*, *154*, 375–382.

66. Zou, M., Kopajtic, T., Katz, J. L., Wirtz, S., Justice, J. J.Jr., and Newman, A. H. (2001). Novel tropane-based irreversible ligands for the dopamine transporter. *J Med Chem*, *44*, 4453–4461.

67. Zou, M.-F., Agoston, G. E., Belov, Y., Kopajtic, T., Katz, J. L., and Newman, A. H. (2002). Enantioselective synthesis of S-(+)-2β-carboalkoxy-3α-[bis(4-fluorophenyl)-methoxy]tropanes as novel probes for the dopamine transporter. *Bioorg Med Chem Lett*, *12*, 1249–1252.

68. Zou, M.-F., Kopajtic, T., Katz, J. L., and Newman, A. H. (2003). Structure–activity relationship comparison of (*S*)-2β-substituted-3α-(bis[4-fluorophenyl]methoxy)tropanes and (R)-2β-substituted 3β-(3,4-dichlorophenyl)tropanes at the dopamine transporter. *J Med Chem*, *46*, 2908–2916.

69. Zou, M.-F., Cao, J., Kopajtic, T., Desai, R. I., Katz, J. L., and Newman, A. H. (2006). Structure–activity relationship studies on a novel series of (S)-2β-Substituted 3α-[bis(4-fluoro or 4-chloro-phenyl)methoxy]tropane analogs for in vivo investigation. *J Med Chem*, *49*, 6391–6399.

70. Newman, A. H., Agoston, G. E. (1998). Novel benztropine [3α-(diphenylmethoxy)-tropane] analogs as probes for the dopamine transporter. *Curr Med Chem*, *5*, 301–315.

71. Katz, J. L., Agoston, G. E., Alling, K. L., Kline, R. H., Forster, M. J., Woolverton, W. L., Kopajtic, T. A., and Newman, A. H. (2001). Dopamine transporter binding without cocaine-like behavioral effects: synthesis and evaluation of benztropine analogs alone and in combination with cocaine in rodents. *Psychopharmacol*, *154*, 362–374.

8

STRUCTURE–ACTIVITY RELATIONSHIP OF GBR 12909 LIGANDS

THOMAS E. PRISINZANO

Department of Medicinal Chemistry, The University of Kansas, Lawrence, Kansas

KENNER C. RICE

Chemical Biology Research Branch, National Institute on Drug Abuse, National Institutes of Health, Bethesda, Maryland

8.1	Introduction	211
8.2	Structure–Activity Relationships of GBR 12909 Analogs	212
	8.2.1 Role of the 2-Benzhydryloxyethyl Group	212
	8.2.2 Role of the Linker	213
	8.2.3 Variation of the N-Substituent	215
8.3	Piperazine Analogs	216
8.4	Piperidine Analogs	219
8.5	Conformationally Restricted Analogs	221
8.6	QSAR and Modeling Studies of GBR 12909 Analogs	223
8.7	Conclusions	224
	References	225

8.1 INTRODUCTION

Among the first agents to be characterized as high-affinity and selective inhibitors of dopamine reuptake were the aryl-1,4-dialkylpiperazines GBR 12909 (**1**) and GBR

Dopamine Transporters: Chemistry, Biology, and Pharmacology. Edited by Mark L. Trudell and Sari Izenwasser

211

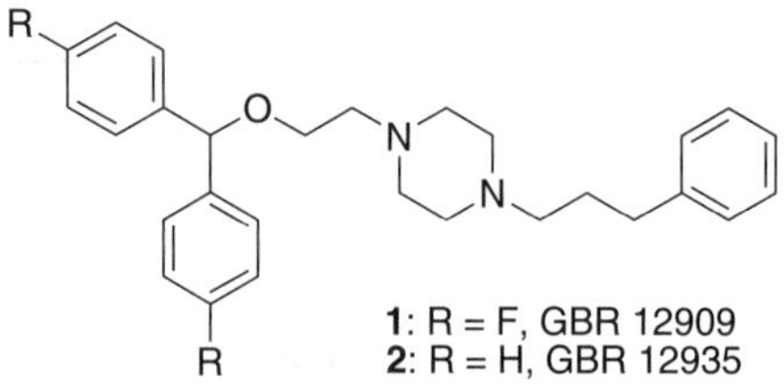

Figure 8-1 Structures of GBR 12909 (**1**) and GBR 12935 (**2**).

12935 (**2**) [1] (Fig. 8-1). This series of compounds was originally developed and tested in Europe as potential antidepressants [2]. In particular, **1** and **2** were identified as potent and selective ligands for the dopamine transporter (DAT) [1,3]. Further work identified these compounds as moderately selective inhibitors of dopamine uptake over serotonin and norepinephrine uptake [2,4–6].

Early studies with **1** noted that it differed from cocaine in several key ways [7]. It has a slower onset of action and longer duration of action than that of cocaine [8–11]. GBR 12909 possesses a much higher affinity for, and slower dissociation rate from, the DAT than cocaine [5,9,12]. It functionally antagonizes cocaine's ability to elevate extracellular levels of dopamine [9,13] and has a nonstimulantlike profile in normal human volunteers following oral administration [14]. Based on the pharmacological data above, efforts were begun to further explore **1** and **2** as potential stimulant abuse therapeutics. In particular, studies focused mainly on enhancing selectivity for the DAT over other monoamine transporters. Additional efforts were aimed at better understanding the effect of structure on the in vivo properties of GBR-based ligands.

8.2 STRUCTURE–ACTIVITY RELATIONSHIPS OF GBR 12909 ANALOGS

Initial analogs of **1** were prepared in an effort to identify its pharmacophore necessary for high affinity and selectivity for the DAT. To help elucidate the potential bioactive conformation of **1**, two x-ray crystallographic studies have examined its molecular structure and that of a derivative [15,16]. The structure of **1** can be divided into three distinct fragments: (1) the 2-benzhydryloxyethyl moiety, (2) the linker, and (3) the N-alkyl chain [17]. The role of each of these particular fragments has been probed systematically and the structure–activity relationships (SARs) have been summarized previously [18,19]. In this chapter we provide an overview as well as sections on efforts to develop conformationally restricted analogs and QSAR studies involving GBR 12909 analogs.

8.2.1 Role of the 2-Benzhydryloxyethyl Group

Most derivatives of **1** have contained either a 2-[bis-(4-fluorophenyl)methoxy]ethyl (**1**-like) or a 2-benzhydryloxyethyl- (**2**-like) substituent. Generally, the presence of

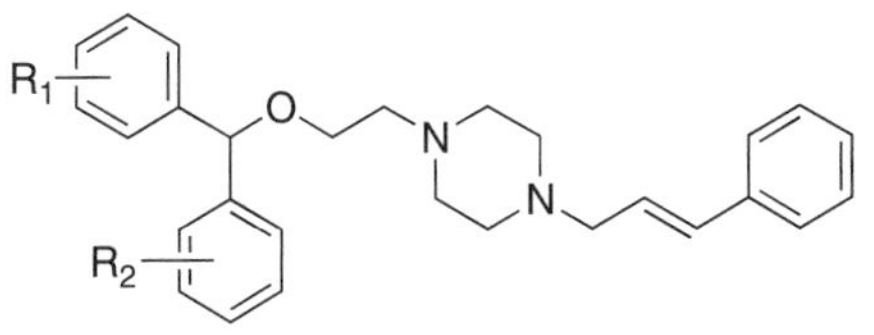

3a: $R_1 = R_2 = H$
3b: $R_1 = R_2 = 4\text{-Cl}$
3c: $R_1 = R_2 = 4\text{-F}$
3d: $R_1 = H; R_2 = 4\text{-CH}_3$
3e: $R_1 = H; R_2 = 4\text{-OCH}_3$
3f: $R_1 = 4\text{-F}; R_2 = 2,4\text{-diCl}$
3g: $R_1 = 3\text{-Br}; R_2 = 3\text{-OCH}_3$
3h: $R1 = H; R_2 = 3\text{-NH}_2$
3i: $R_1 = H; R_2 = 3\text{-NO}_2$

Figure 8-2 Structures of **3a–3i**.

a 2-[bis-(4-fluorophenyl)methoxy]ethyl moeity affords higher affinity for the DAT compared to the corresponding 2-benzhydryloxyethyl group [20–24]. However, the 2-benzhydryloxyethyl group usually has enhanced selectivity for the DAT over the serotonin transporter (SERT) [20,21,23].

The effect of other substituents in the 2-benzhydryloxyethyl fragment of GBR 12783 (**3a**) has been studied (Fig. 8-2) [1]. The addition of a 4-Cl group to both of the phenyl rings (**3b**) was well tolerated (DAT $IC_{50} = 20$ nM). However, this was not as well tolerated as the addition of a 4-F group to both of the phenyl rings (**3c**) ($IC_{50} = 1.9$ nM). Other substitution patterns explored were the combinations of a 4′-H and 4′-CH$_3$ (**3d**), 4′-H and 4′-OCH$_3$ (**3e**), 4′-F and 2′,4′-diCl (**3f**), and 3′-Br and 3′-OCH$_3$ (**3g**). This study [1] did not, however, take into account the asymmetric center created when only one ring was modified. A quantitative SAR (QSAR) study based on these substitution patterns concluded that substituents with a large inductive effect and small size provide the most potent compound. Later work by Deutsch et al. introduced —NO$_2$, —NH$_2$, —NCS, and —maleimido groups into the 3′- and 4′-positions of one of the phenyl rings of **3a** [25]. Generally, these substitutions were better tolerated in the 3′-position over the 4′-position in the approximate order of —NO$_2 \approx$ —NH$_2$ > —NCS > —maleimido. Interestingly, there was little difference in affinity between a 3′-NH$_2$ (**3h**) and 3′-NO$_2$ (**3i**) derivatives, despite the large difference in inductive effects when [^{3}H]methylphenidate was used as a radioligand. The effect of all of the above-mentioned modifications on selectivity for the DAT over the SERT and NET was not examined.

8.2.2 Role of the Linker

The piperazine ring in **1** has been replaced by several different functionalities (Fig. 8-3). Among the first alterations was the addition of methyl groups onto the piperazine ring in **1** [26]. It was shown that if the piperazine ring was replaced with a trans-(2S,5R)-dimethylpiperazine ring (**4a**), affinity and selectivity for the DAT were nearly identical to **1** [26]. Interestingly, **4a** had 30-fold higher activity than its enantiomer for inhibiting dopamine. Another modification to the piperazine ring was the addition of two keto groups to **1** (**4b**) [27]. This change, although lowering the lipophilicity of **1**, was not well tolerated, leading to a 60-fold loss of affinity for the DAT.

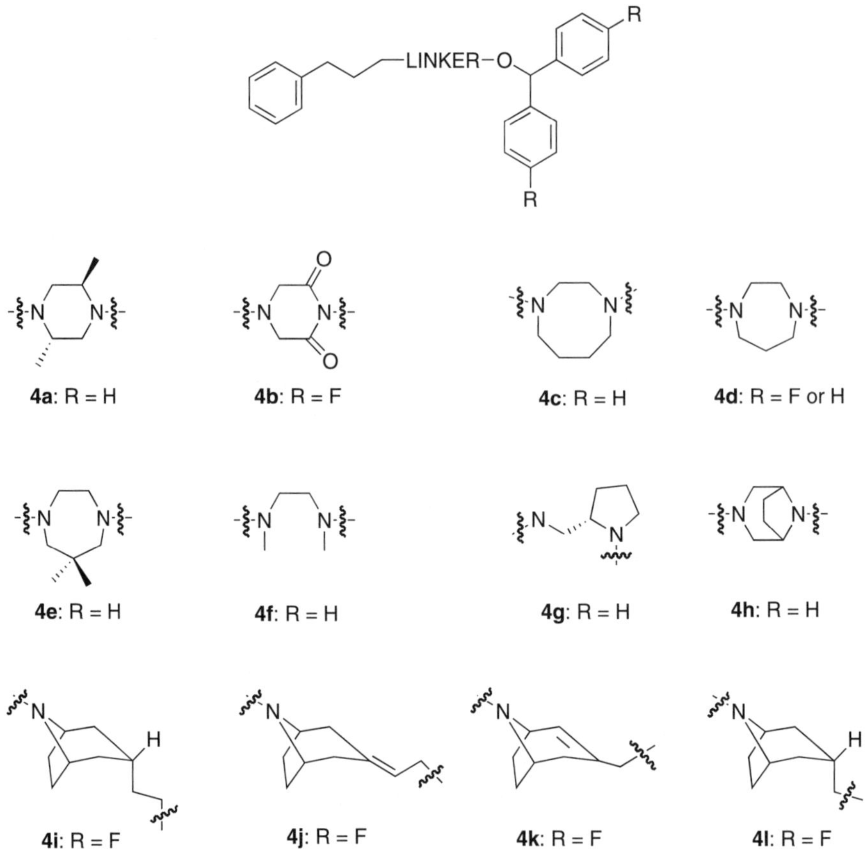

Figure 8-3   Structures of **4a**–**4l**.

Other studies have probed the size and shape of the piperazine ring in **1** [28]. Extending the piperazine ring into a homopiperazine ring (**4c**) resulted in a modest increase in the selectivity of inhibiting dopamine uptake over serotonin uptake. However, if the fluorine atoms were removed, the compound became more than 4000-fold selective for inhibiting dopamine uptake over serotonin uptake [28]. If the seven-membered ring was extended to an eight-membered ring (**4d**), affinity and selectivity for inhibiting dopamine over serotonin were decreased compared to **4c** [20]. Similarly, if a 6,6-dimethylhomopiperazine ring (**4e**) replaced the homo-piperazine ring, this also resulted in reduced affinity and selectivity for dopamine compared to **4c**. The piperazine moieties were further modified by opening the rings to create several different ring-opened congeners [20]. None of these ring-opened analogs, such as **4f**, had greater activity or selectivity for dopamine than did **4c**. An additional series of ring-opened analogs were prepared and found to have higher affinity and be more selective in inhibiting dopamine over serotonin than cocaine, but they showed less affinity and selectivity than **1** [29].

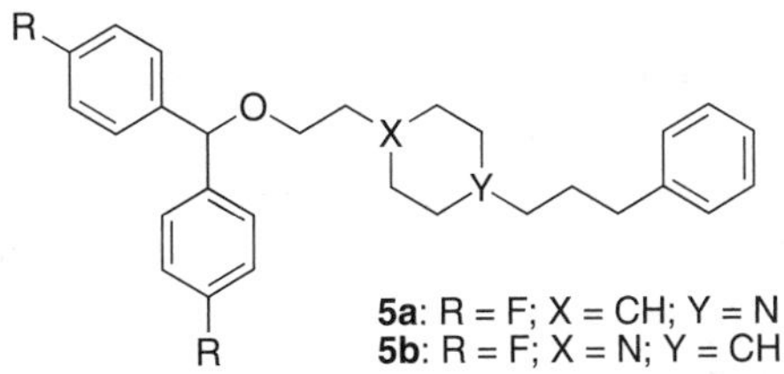

5a: R = F; X = CH; Y = N
5b: R = F; X = N; Y = CH

Figure 8-4   Structures of **5a** and **5b**.

The piperazine ring in **1** has also been replaced by several different diamines, including several bridged analogs [20,30–32]. These particular modifications showed that a wide degree of structural variation is tolerated [33]. Among the modified diamines, compound **4g** was identified as being the highest-affinity dopamine-reuptake inhibitor ($IC_{50} = 0.7$ nM) as well as the most selective over serotonin reuptake (1400-fold) [20]. However, other modifications do not result in increased selectivity for dopamine-uptake inhibition relative to **4c**. A 2,6-ethanobridged analog (**4h**) was prepared as well [32]. Several tropane ring–containing analogs have been prepared [34]. Tropane analog **4i** was found to be more potent than benztropine at inhibiting dopamine reuptake but was not as selective over serotonin reuptake. Tropane **4i** was also found to have greater than 220-fold more selectivity than benztropine over muscarinic receptors [34]. The most DAT-selective ligand in this series was alkene **4j** (143-fold); however, **1** was eightfold more potent in inhibiting dopamine uptake. Additional tropanelike GBR 12909 analogs have been reported [35,36]. Tropene **4k** was found to have similar affinity for the DAT as **1** [36]. Tropane **4l** was found to have higher affinity than **1** for the DAT ($K_i = 5.1$ nM vs. $K_i = 12$ nM), roughly the same as its one-carbon homolog **4i** ($K_i = 5.1$ nM vs. $K_i = 7.4$ nM) [36].

A final modification to the central piperazine ring in **1** studied was its conversion to a piperidine ring. This modification, however, initially created a question as to which of the two nitrogen atoms in **1** was essential for activity. If one of the nitrogens in **1** is replaced with a carbon atom, either **5a** or **5b** is formed (Fig. 8-4). Dutta et al. showed that compound **5a** had higher affinity than **5b** for the DAT, indicating that the distal nitrogen atom relative to the benzhydryloxy group was more essential [37]. In addition, **5a** was shown to have reduced affinity compared to **1** for a "piperazine acceptor site" [38]. However, if the fluorine atoms in **5a** were replaced with —Cl, —Br, or —OCH₃, this resulted in compounds with reduced affinity and selectivity for the DAT over the SERT [39]. If the fluorine atoms were replaced with H, this modification increased selectivity over the SERT.

8.2.3   Variation of the N-Substituent

The most widely studied change to the core structure of **1** regarding the variation in the N-substituent to the piperazine ring in **1** was its conversion to a piperidine ring [1,21–24,40,41]. Among the piperazine analogs of GBR 12909, the optimal chain length for inhibition of dopamine reuptake between the nitrogen and the phenyl ring was three carbons [1]. However, the optimal chain length for affinity to the DAT in the GBR 12909 piperidine analogs was first reported to be one carbon [39]. Given this discrepancy, a study was undertaken to compare the mode of binding of piperazine and piperidine analogs of **1** [42]. A lack of parallel effects given parallel changes was observed and indicates that the two series of ligands are not binding in an identical manner at the DAT [42,43]. Therefore, the role of the N-substituent will be discussed separately for piperazine and piperidine analogs.

8.3 PIPERAZINE ANALOGS

One of the first modifications in the *N*-phenylpropyl chain of **1** was incorporation of a *trans*-alkene (**6a**) [1] (Fig. 8-5). Among piperazine derivatives, this modification has been shown to increase affinity and selectivity for the DAT [21]. Additionally, a QSAR study has indicated that optimal DAT inhibitory activity is obtained if a small-volume electron-withdrawing group such as a halogen is placed in the phenyl ring of the *N*-phenylpropyl substituent [1]. Also, it has been shown that the phenyl ring in the *N*-phenylpropyl substituent in **1** and **6a** may be replaced by other heterocycles, such as a 2-thienyl group (**6b**) or a 3-pyridyl group (**6c**) [21]. These changes result in analogs that retain high affinity for the DAT and selectivity over the SERT.

Further variation of the *N*-alkyl group in **1** was used in the preparation of photo-affinity labels [44,45]. Tritiated analog **7** (Fig. 8-6) was shown to bind reversibly to and successfully label the DAT, enabling its purification [45]. In a similar manner, other photoaffinity labels based on GBR 12935, such as the piperazine [^{125}I]DEEP (**8**) and the piperdine [^{125}I]AD-96-129, have also been developed [46–48]. Importantly, these studies indicated that GBR-based ligands and phenyltropanes may bind to different domains of the DAT [48].

Another approach to varying the *N*-alkyl group was to develop potentially long-acting formulations [22,23,40,41]. The idea here was to find a polar hydroxyl-containing derivative that could easily be converted into a medium- or

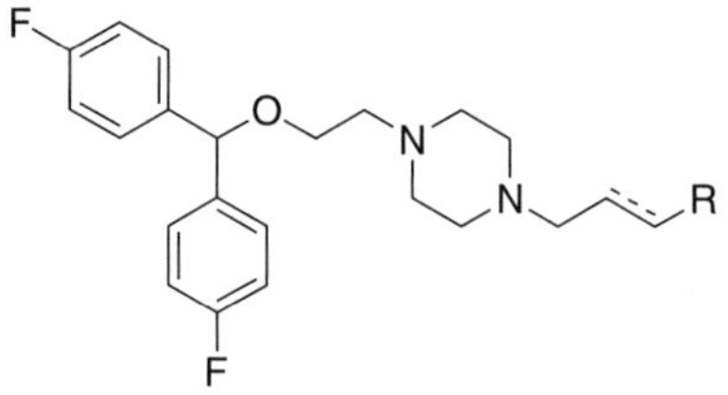

6a: R = C$_6$H$_5$; Bond = Double
6b: R = 2-C$_4$H$_3$S; Bond = Single or Double
6c: R = 3-C$_5$H$_4$N; Bond = Single or Double
6d: R = 3-OMeC$_6$H$_5$; Bond = Double

Figure 8-5   Structures of **6a–6d**.

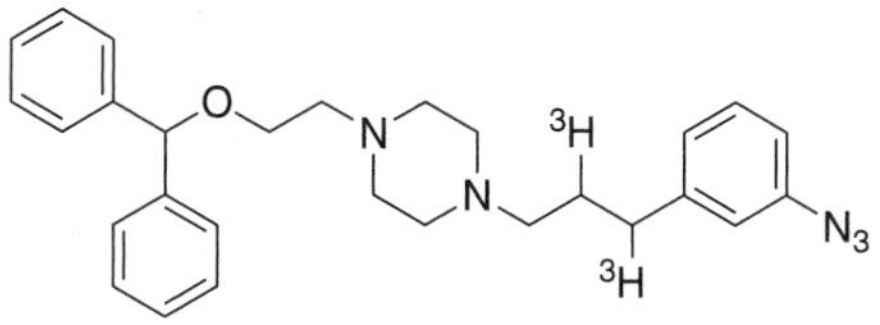

Figure 8-6 Structures of **7** and **8**.

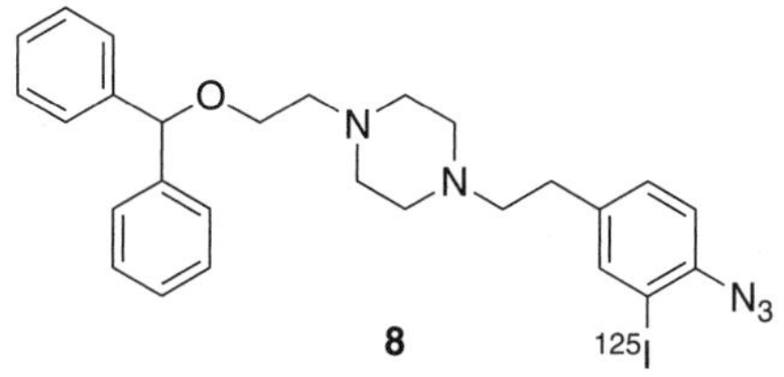

long-chain ester [40]. The rationale behind this type of preparation was that it would potentially extend the duration of **1**. Initially, racemic alcohol **9a** was shown to have affinity and activity similar at the DAT to that of **1** and could be converted to the corresponding decanoate ester (**9b**) for use as a long-acting formulation (Fig. 8-7). A single dose of **9b** largely suppressed cocaine self-adminstration in nonhuman primate for nearly a month [40]. Binding studies also showed that **9b** was capable of potently inhibiting binding. The study did not, however, distinguish whether **9a** or **9b** were responsible for the in vivo activity seen [40]. Additional work showed that both enantiomers of **9a** were potent and selective inhibitors of dopamine reuptake [22]. Interestingly, this study also showed that alkene **6d** had affinity similar to that of **1** but was more than 300-fold selective in inhibiting dopamine over serotonin reuptake [22]. This result was in contrast to a previous QSAR study which indicated that activity is optimal when a small electron-withdrawing group is placed in the phenyl ring of **6a** [1]. The presence of a hydroxyl group in the 2-position of the *N*-phenylpropyl group in **1** was also explored [23]. In contrast to the benzylic position, where little enantioselectivity in DAT affinity was seen, (*S*)-**9c** had

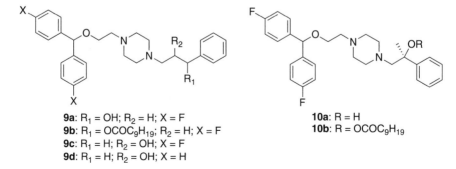

9a: R$_1$ = OH; R$_2$ = H; X = F
9b: R$_1$ = OCOC$_9$H$_{19}$; R$_2$ = H; X = F
9c: R$_1$ = H; R$_2$ = OH; X = F
9d: R$_1$ = H; R$_2$ = OH; X = H

10a: R = H
10b: R = OCOC$_9$H$_{19}$

Figure 8-7 Structures of **9a–9d** and **10a,b**.

higher DAT affinity (15-fold) and was more selective (20-fold) over the SERT than (*R*)-**9c**. As expected, removal of the fluorine atoms **9d** gave enhanced selectivity over the SERT (900-fold). (*S*)-**9d** was found to have high DAT affinity (2.3 nM) and the most selective over the SERT of any piperazine analog described to date [23]. Given the high affinity and selectivity of (*S*)-**9c**, additional studies were initiated to better understand its mode of binding [41]. If the phenyl ring in (*S*)-**9c** is removed, a 20-fold loss in affinity for the DAT is seen, indicating that the phenyl substituent is crucial to high affinity and selectivity [41]. These studies also identified **10a** as having similar affinity for the DAT as **1** ($K_i = 5.6$ nM vs. $K_i = 3.7$ nM). This compound has a methyl group that was added to potentially slow metabolism by esterases of a medium- or long-chain ester derivative such as **10b**, yielding a potential ultralong-acting formulation [16].

Further work has probed the effects of oxygenation in the *N*-phenylpropyl side chain as well as conformation restriction of the side chain [24]. The most selective compound identified in this series, **11** (Fig. 8-8), was more than 200-fold selective in binding to the DAT over the SERT. This study also identified compound **12**, a locked conformer of **6a**, as having lower affinity and SERT selectivity than **6a**. This indicated that the greater flexibility of **6a** adds to its high DAT affinity (0.9 nM) and SERT selectivity (158-fold) [24].

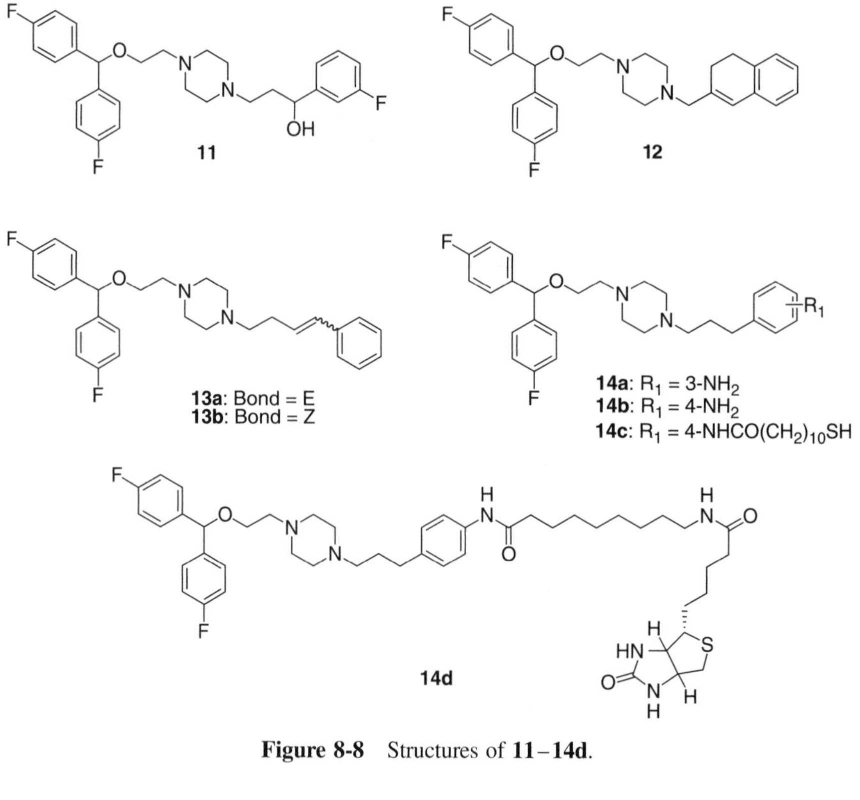

Figure 8-8   Structures of **11**–**14d**.

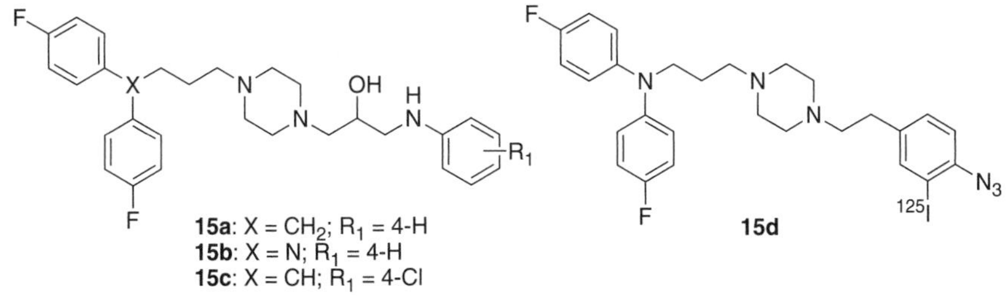

Figure 8-9 Structures of **15a–15d**.

Other N-substituent modifications have also been explored [49–51]. Alkenes **13a** and **13b** were found to have affinity for the DAT, but both compounds had lower affinity than **1** [49]. Interestingly, *Z*-alkene **13b** had higher affinity and selectivity over the SERT and NET than did *E*-alkene **13a**. Introduction of an amino group to the 3- or 4-position of the phenyl ring, **14a** and **14b**, had little effect on affinity or selectivity for the DAT compared to **1**. Furthermore, **14b** was converted into thiol **14c**, which allows conjugation to highly fluorescent nanocrystals [50]. However, when **14c** is bound to the quantum dots, the fluorescence of the dots is diminished [51]. In an effort to circumvent this problem, **14d** was prepared. Biotin conjugate **14d** retains high affinity for the DAT relative to quantum dot concentration [51].

In the course of trying to find a calcium antagonist with antioxidant activity, diphenyl piperazine **15a** (Fig. 8-9) was prepared [52]. In vivo experiments showed that **15a** produced a noticeable increase in locomotor activity. Given that activity is seen with DAT inhibition, **15a** and various analogs were evaluated for DAT affinity. Compound **15b** was found to have similar affinity for the DAT as **1**. An additional study probed the role of substitution in the phenyl ring of **15a** [53]. Generally, substitution in the 4-position was well tolerated, and **15c** was among the analogs found to have similar affinity for the DAT as **1**. Furthermore, radioiodinated analog **15d** was prepared and used to photolabel human DAT-transfected HEK-293 cells [54].

8.4 PIPERIDINE ANALOGS

Investigation of **5a** sought to identify the optimal chain length in the N-substituent [39]. In contrast to previous SAR seen with GBR 12909 piperazines, *N*-benzyl analog **16a** (Fig. 8-10) had similar affinity to **1**. This indicated that these new piperidine analogs were not binding in a manner identical to that of other GBR derivatives. The most selective analog over the SERT (49-fold) of the series was desfluoro analog **16b** [39]. Further research has shown that when the phenyl ring in **16a** is replaced by other aromatic rings, reduced affinity and selectivity for the DAT results [55]; however, its removal was not studied. An additional study showed that selectivity for the DAT over the SERT could be increased by the addition of different 4-position substituents (i.e., **16c** to **16f**) [43,55,56]. Iodo analog **16f** was shown to possess high affinity for the DAT ($IC_{50} = 0.96$ nM) and good selectivity over the SERT

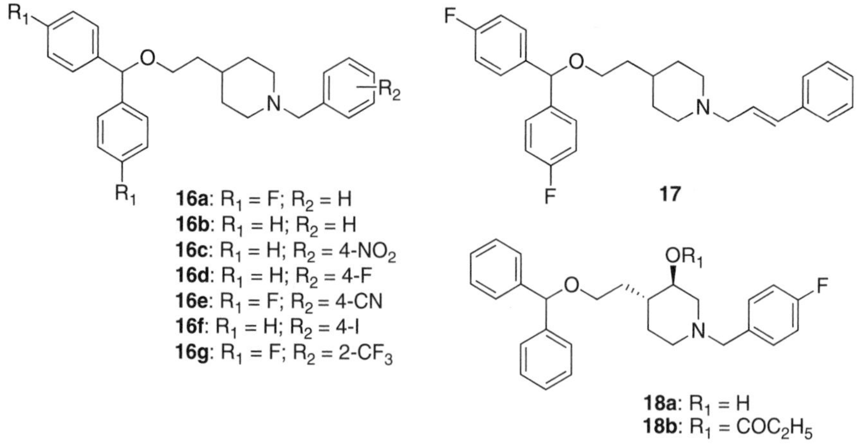

16a: R_1 = F; R_2 = H
16b: R_1 = H; R_2 = H
16c: R_1 = H; R_2 = 4-NO$_2$
16d: R_1 = H; R_2 = 4-F
16e: R_1 = F; R_2 = 4-CN
16f: R_1 = H; R_2 = 4-I
16g: R_1 = F; R_2 = 2-CF$_3$

17

18a: R_1 = H
18b: R_1 = COC$_2$H$_5$

Figure 8-10   Structures of **16a–18b**.

(3000-fold) and NET (1200-fold) [56]. It was also shown in this particular series that changing the benzhydrylic oxygen atom to a nitrogen atom resulted in modest changes in affinity and selectivity for the DAT [57,58]. A more recent study showed that **16c** has higher affinity and more selectivity than **16f** over the SERT using [^{125}I]RTI-55 as a radioligand rather than the formerly used [^{3}H]WIN 35,428 [59]. Furthermore, **16c** was shown to be over 1300-fold selective in inhibiting dopamine over serotonin uptake. Similarly, alkene **17** was originally reported to be a relatively nonselective DAT ligand using [^{3}H]WIN 35,428 and [^{3}H]citalopram as radioligands to label the DAT and SERT, respectively [60]. A more recent study using [^{125}I]RTI-55 to label both the DAT and SERT found **17** to possess subnamolar affinity for the DAT (0.41 nM) and higher selectivity over the SERT than reported previously (100-fold vs. 5-fold) [42]. An additional study has shown that substituents in the N-benzyl group can either increase or decrease bioamine transporter affinity [61]. In the course of examining these analogs, a unique compound was identified [62]. Compound **16g** was found to modulate allosterically certain properties of the human SERT expressed in HEK cells. This result was surprising given it low affinity and selectivity for the SERT [61].

In an effort to create another possible long-acting formulation, several trans-3-hydroxy derivatives of **16d** were synthesized [63]. (R,R)-**18a** was reported to possess high affinity (0.4 nM) for the DAT and good selectivity over both the SERT and the norepinephrine transporter (NET). In addition, (R,R)-**18a** was shown to have 100-fold higher affinity for the DAT than did its enantiomer. Interestingly, conversion of the hydroxyl group in **18a** to a propionyl ester, **18b**, resulted in a loss of more than 1000-fold in affinity for the DAT [63].

Given the clear effects of oxygenation in the 2- and 3-positions of the alkyl chain of GBR piperazines, the effects of oxygenation in the piperidine series was probed [64]. The most SERT-selective ligand in the compounds examined was (S)-**19a** (Fig. 8-11) (330-fold). Compound **19a** was not, however, as selective as the

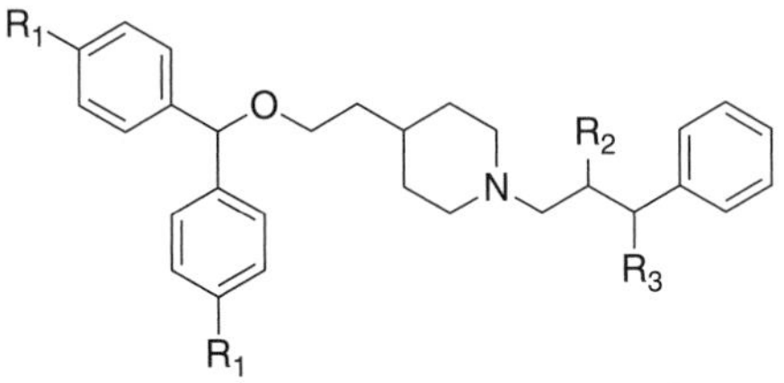

19a: R$_1$ = H; R$_2$ = OH; R$_3$ = H
19b: R$_1$ = F; R$_2$ = OH; R$_3$ = H
19c: R$_1$ = H; R$_2$ = H; R$_3$ = OH

Figure 8-11 Structures of **19a–19c**.

corresponding piperazine (S)-**9d** (900-fold). The highest-affinity DAT ligand identified in this series was (S)-**19b**. 2-Hydroxy analog (S)-**19b** has subnanomolar affinity (K_i = 0.4 nM) and good selectivity over the SERT (200-fold) and NET (900-fold). The most NET-selective ligand identified in the series was (R)-**19c** (>3000-fold).

8.5 CONFORMATIONALLY RESTRICTED ANALOGS

Generally, GBR 12909 analogs are very flexible. In an effort to convert these flexible molecules into more rigid compounds, several new structural templates were developed (Scheme 8-1) [65]. Bioisosteric replacement of the benzhydrylic oxygen

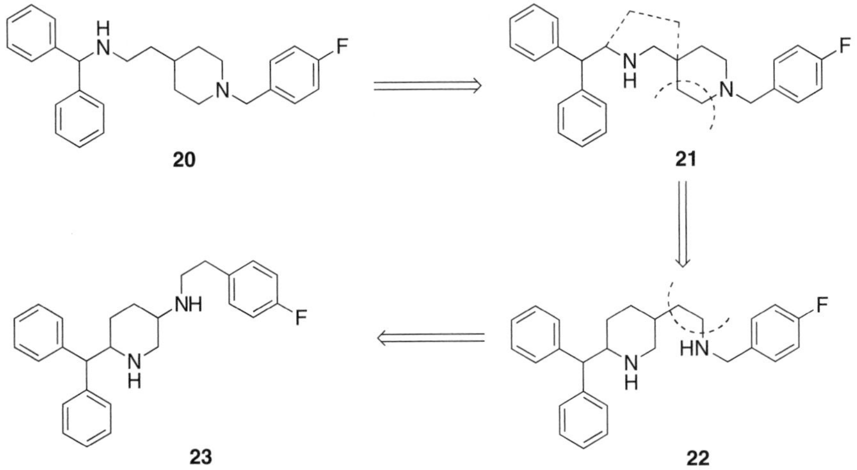

Scheme 8.1 Generation of conformationally restricted analogs from piperidine **20**.

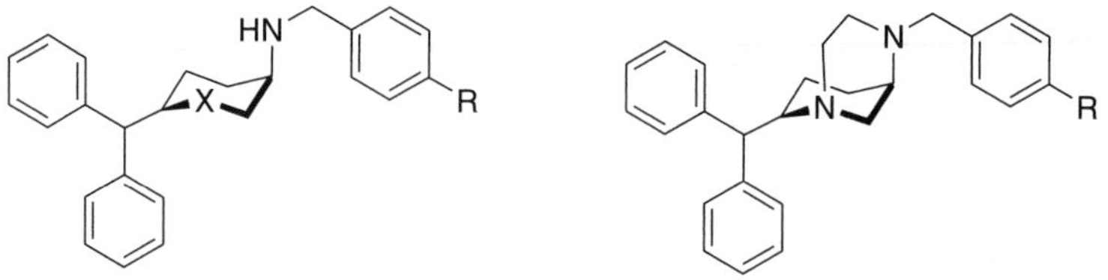

24a: X = NH; R = F
24b: X = NH; R = CN

25a: R = F
25b: R = CN

Figure 8-12   Structures of **24a–25b**.

atom in **16d** with a nitrogen atom creates analog **20**. This change was shown to increase DAT affinity but decrease selectivity over the SERT [57]. Relocating the nitrogen atom to an adjacent position on the ethylene bridge creates analog **21** [43]. This modification, however, resulted in a decrease in DAT affinity and selectivity relative to **20**. Formation of new piperidine ring and ring opening of the existing piperidine ring in **21** leads to template **22** [65]. Further simplification of this structure by removing the ethyl bridge between the piperidine ring and the second amine creates general structure **23**. This modification, however, creates two geometrical isomers, a *cis*-isomer and a *trans*-isomer. Of these two isomers, *cis*-isomer **24a** (Fig. 8-12) was found to exhibit the highest affinity and selectivity for the DAT [65]. This compound was, however, evaluated as a racemate. The more potent (−)-enantiomer of **24a** has the absolute configuration of *S,S* [66]. The most potent analog in this series, (*S,S*)-**24b**, was similar to **1** [66]. *Cis*-isomer **24a** could be further restrained into a rigid bicyclic structure **25a** [67]. This change, however, decreases DAT affinity and selectivity compared to **24a** [67]. A more potent analog is created when the racemate is resolved and the 4-fluoro group is replaced with a methyl ether. However, (*S,S*)-**25b** has 3-fold lower affinity than **1**.

Additional studies have probed replacement of the piperidine nitrogen atom in **24a** with an oxygen atom [68–72]. As seen with **23**, *cis*-isomer **26a** (Fig. 8-13) is more potent than the corresponding *trans*-isomer [68]. It was also shown that DAT affinity could be enhanced by replacement of the 4-fluoro group with a nitro group (**26b**).

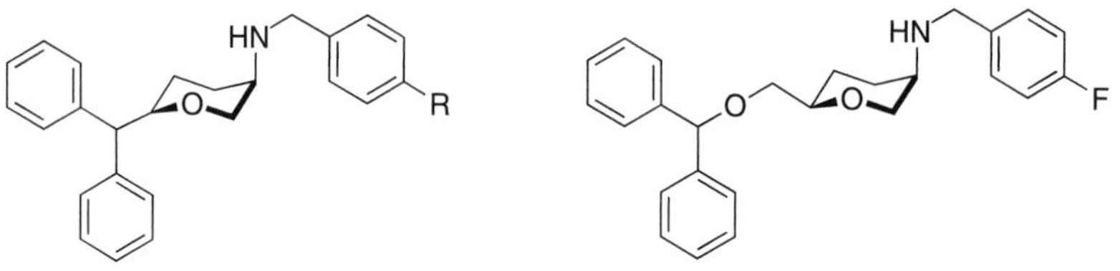

26a: R = F
26b: R = NO$_2$

27

Figure 8-13   Structures of **26a,b** and **27**.

This analog, however, had lower affinity than **1** [69]. Further testing indicated that only the (−)-isomers of **26** exhibited appreciable potency for monoamine transporters [70]. In particular, the most potent analog in the series identified was (*S,S*)-**26b** [72]. Once again, this analog was less potent than **1**. Finally, extension of the 6-position substituent by one carbon (**27**) increased DAT affinity 2-fold compared to **26a** [71].

8.6 QSAR AND MODELING STUDIES OF GBR 12909 ANALOGS

In an effort to better develop GBR 12909 analogs as potential drug abuse medications, several groups have begun to use molecular modeling techniques to gain insight into the structural requirements for their pharmacological activity [67,73–77]. One study used GRID [78]-independent descriptors to obtain fast three-dimensional QSAR models [73]. This work generated significant models for both DAT affinity and DAT selectivity over the SERT.

As mentioned above, GBR 12909-based ligands are very flexible, and prediction of its bioactive conformation is very challenging. In addition, flexible molecules are able to adopt a large number of conformations that hinder the use of three-dimential QSAR techniques such as comparative molecular field analysis (CoMFA) [79]. Recent efforts have identified several techniques that lessen the barriers to using CoMFA with flexible molecules such as **1** [74–76]. These techniques have been used to model the DAT/SERT selectivity of GBR 12909 analogs [77]. The templates for the CoMFA analysis were naphthyl analogs **28a** and **28b** (Fig. 8-14). These templates were chosen as being representative of the regions of three-dimensional space occupied by GBR 12909 analogs [77]. The contour maps predicted several key features to increase DAT/SERT selectivity among GBR 12909 analogs: (1) less bulk near positions 3 and 4 of the 2-naphthyl substituent; (2) more bulk near positions 5, 6, and 7 of the 2-naphthyl substituent, (3) a more positive environment above the naphthyl plane extending axially from position 3, and (4) a more negative environment near position 6 of the 2-napthyl substituent [77]. There are further indications that 3,4-disubstituted phenyl analogs may potentially have improved DAT/SERT selectivity.

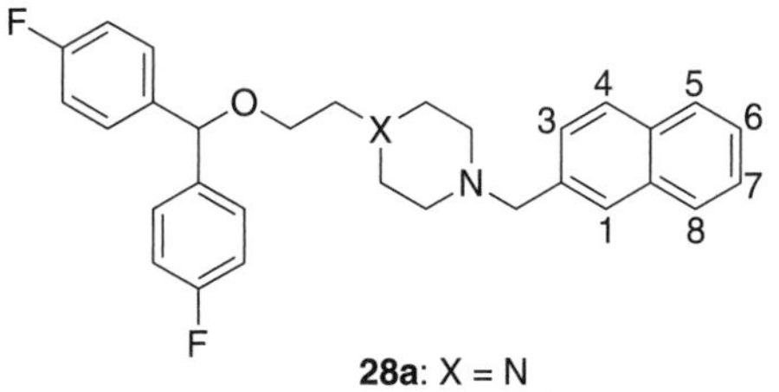

28a: X = N
28b: X = CH

Figure 8-14 Structures of **28a** and **28b**.

8.7   CONCLUSIONS

Since its introduction in 1980, GBR 12909 (**1**) has proven to be a useful pharmacological tool for studying the DAT. Various analogs of **1** have been prepared in an effort to identify the pharmacophore necessary for high affinity and selectivity for the DAT over other monoamine transporters. In addition, **1** and many of its analogs have helped to test the dopamine hypothesis of cocaine and methamphetamine addiction. The structure–activity relationships of GBR piperazines are briefly summarized as follows: (1) removal of the fluorine atoms increases selectivity over the SERT; (2) the distal nitrogen of the piperazine ring is not necessary; (3) the central piperazine ring may be replaced with other nitrogen-containing heterocycles; (4) the optimal chain length for the N-substituent is three carbons; (4) the *p*-substitution is optimal for the phenyl ring; and (5) the phenyl ring may be replaced with heterocycles (Fig. 8-15). The SARs for GBR piperidines are essentially the same, with the exception of the number of carbons for activity in the side chain being 1 to 3 (Fig. 8-16).

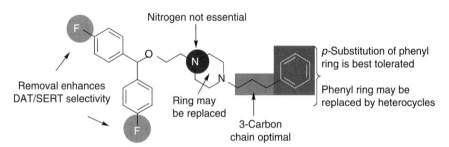

Figure 8-15   Summary of SAR for piperazine analogs. (*See insert for color representation of figure.*)

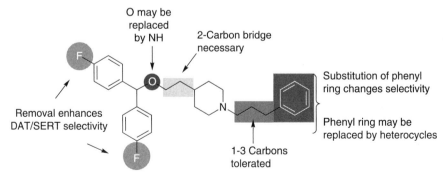

Figure 8-16   Summary of SAR for piperidine analogs. (*See insert for color representation of figure.*)

REFERENCES

1. van der Zee, P., Koger, H. S., Gootjes, J., and Hespe, W. (1980). Aryl 1,4-dialk(en)yl-piperazines as selective and very potent Inhibitors of dopamine uptake. *Eur J Med Chemi*, *15*, 363–370.

2. Preti, A. (2000). Vanoxerine National Institute on Drug Abuse. *Curr Opin Investig Drugs*, *1*, 241–251.

3. Andersen, P. H. (1989). The dopamine inhibitor GBR 12909: selectivity and molecular mechanism of action. *Eur J Pharmacol*, *166*, 493–504.

4. Sharif, N. A., Nunes, J. L., Michel, A. D., and Whiting, R. L. (1989). Comparative properties of the dopamine transport complex in dog and rodent brain: striatal [^{3}H]GBR 12935 binding and [^{3}H]dopamine uptake. *Neurochem Int*, *15*, 325–332.

5. Andersen, P. H. (1987). Biochemical and pharmacological characterization of [^{3}H]GBR 12935 binding in vitro to rat striatal membranes: labeling of the dopamine uptake complex. *J Neurochem*, *48*, 1887–1896.

6. Hirate, K., and Kuribara, H. (1991). Characteristics of the ambulation-increasing effect of GBR-12909, a selective dopamine uptake inhibitor, in mice. *Jpn J Pharmacol*, *55*, 501–511.

7. Rothman, R. B., and Glowa, J. R. (1995). A review of the effects of dopaminergic agents on humans, animals, and drug-seeking behavior, and its implications for medication development: focus on GBR 12909. *Mol Neurobiol*, *11*, 1–19.

8. Kelley, A. E., and Lang, C. G. (1989). Effects of GBR 12909, A selective dopamine uptake inhibitor, on motor activity and operant behavior in the rat. *Eur J Pharmacol*, *167*, 385–395.

9. Rothman, R. B., Mele, A., Reid, A. A., Akunne, H. C., Greig, N., Thurkauf, A., de Costa, B. R., Rice, K. C., and Pert, A. (1991). GBR 12909 antagonizes the ability of cocaine to elevate extracellular levels of dopamine. *Pharmacol Biochem Behav 40*, 387–397.

10. Rothman, R. B., Grieg, N., Kim, A., de Costa, B. R., Rice, K. C., Carroll, F. I., and Pert, A. (1992). Cocaine and GBR 12909 produce equivalent motoric responses at different occupancy of the dopamine transporter. *Pharmacol Biochem Behav*, *43*, 1135–1142.

11. Czoty, P. W., Justice, J. B., Jr., and Howell, L. L. (2000). Cocaine-induced changes in extracellular dopamine determined by microdialysis in awake squirrel monkeys. *Psychopharmacology*, *148*, 299–306.

12. Reith, M. E., Sershen, H., and Lajtha, A. (1981). Binding of [^{3}H]cocaine in mouse brain: kinetics and saturability. *J Receptor Res*, *2*, 233–243.

13. Rothman, R. B., Mele, A., Reid, A. A., Akunne, H., Greig, N., Thurkauf, A., Rice, K. C., and Pert, A. (1989). Tight binding dopamine reuptake inhibitors as cocaine antagonists: a strategy for drug development. *FEBS Lett*, *257*, 341–344.

14. Sogaard, U., Michalow, J., Butler, B., Lund Laursen, A., Ingersen, S. H., Skrumsager, B. K., and Rafaelsen, O. J. (1990). A tolerance study of single and multiple dosing of the selective dopamine uptake inhibitor GBR 12909 in healthy subjects. *Int Clin Psychopharmacol*, *5*, 237–251.

15. Flippen-Anderson, J. L., Deschamps, J. R., George, C., Folk, J. E., Jacobson, A. E., and Rice, K. C. (2002). Twinned 1-{2-[bis(4-fluorophenyl)methoxy]ethyl}-4-(3-phenyl)-propyl)piperazinium chloride (GBR 12909). *Acta Crystallogr E Struct Rep Online*, *58*, o81–o82.

16. Prisinzano, T., Hsin, L. W., Folk, J. E., Flippen-Anderson, J. L., George, C., Jacobson, A. E., and Rice, K. C. (2003). A concise synthesis of (*S*)-(+)-1-(4-{2-[bis(4-fluorophenyl)-methoxy]ethyl}piperazin-1-yl)-2-phenylpropan-2-ol dimaleate. *Tetrahedron Asym, 14,* 3285–3289.

17. Zhang, Y. (1998). The identification of GBR 12909 as a potential therapeutic agent for cocaine abuse. *Med Chem Res, 8,* 66–76.

18. Prisinzano, T., Rice, K. C., Baumann, M. H., and Rothman, R. B. (2004). Development of neurochemical normalization ("agonist substitution") therapeutics for stimulant abuse: focus on the dopamine uptake inhibitor, GBR 12909. *Curr Med Chem CNS Agents, 4,* 47–59.

19. Runyon, S. P., and Carroll, F. I. (2006). Dopamine transporter ligands: recent developments and therapeutic potential. *Curr Top Med Chem, 6,* 1825–1843.

20. Matecka, D., Rothman, R. B., Radesca, L., de Costa, B. R., Dersch, C. M., Partilla, J. S., Pert, A., Glowa, J. R., Wojnicki, F. H., and Rice, K. C. (1996). Development of novel, potent, and selective dopamine reuptake inhibitors through alteration of the piperazine ring of 1-[2-(diphenylmethoxy)ethyl]-and 1-[2-[bis(4-fluorophenyl)methoxy]ethyl]-4-(3-phenylpropyl)piperazines (GBR 12935 and GBR 12909). *J Med Chem, 39,* 4704–4716.

21. Matecka, D., Lewis, D., Rothman, R. B., Dersch, C. M., Wojnicki, F. H., Glowa, J. R., DeVries, A. C., Pert, A., and Rice, K. C. (1997). Heteroaromatic analogs of 1-[2-(diphenylmethoxy)ethyl]- and 1-[2-[bis(4-fluorophenyl)methoxy]ethyl]-4-(3-phenylpropyl)-piperazines (GBR 12935 and GBR 12909) as high-affinity dopamine reuptake inhibitors. *J Med Chem, 40,* 705–716.

22. Lewis, D. B., Matecka, D., Zhang, Y., Hsin, L. W., Dersch, C. M., Stafford, D., Glowa, J. R., Rothman, R. B., and Rice, K. C. (1999). Oxygenated analogues of 1-[2-(diphenyl-methoxy)ethyl]- and 1-[2-[bis(4-fluorophenyl)methoxy]ethyl]-4-(3-phenylpropyl)pipera-zines (GBR 12935 and GBR 12909) as potential extended-action cocaine-abuse therapeutic agents. *J Med Chem, 42,* 5029–5042.

23. Hsin, L. W., Dersch, C. M., Baumann, M. H., Stafford, D., Glowa, J. R., Rothman, R. B., Jacobson, A. E., and Rice, K. C. (2002). Development of long-acting dopamine transporter ligands as potential cocaine-abuse therapeutic agents: chiral hydroxyl-containing derivatives of 1-[2-[bis(4-fluorophenyl)methoxy]ethyl]-4-(3-phenylpropyl)piperazine and 1-[2-(diphenylmethoxy)ethyl]-4-(3-phenylpropyl)piperazine. *J Med Chem, 45,* 1321–1329.

24. Lewis, D., Zhang, Y., Prisinzano, T., Dersch, C. M., Rothman, R. B., Jacobson, A. E., and Rice, K. C. (2003). Further exploration of 1-[2-[bis-(4-fluorophenyl)methoxy]ethyl]piper-azine (GBR 12909): role of N-aromatic, N-heteroaromatic, and 3-oxygenated N-phenyl-propyl substituents on affinity for the dopamine and serotonin transporter. *Bioorg Med Chem Lett, 3,* 1385–1389.

25. Deutsch, H. M., Schweri, M. M., Culbertson, C. T., and Zalkow, L. H. (1992). Synthesis and pharmacology of irreversible affinity labels as potential cocaine antagonists: aryl 1,4-dialkylpiperazines related to GBR-12783. *Eur J Pharmacol, 220,* 173–180.

26. Matecka, D., Rice, K. C., Rothman, R. B., de Costa, B. R., Glowa, J. R., Wojnicki, F. H., Pert, A., George, C., Carroll, F. I., Silverthorn, M. L., Dersch, C. M., Becketts, K. M., and Partilla, J. S. (1995). Synthesis and absolute configuration of chiral piperazines related to GBR 12909 as dopamine reuptake inhibitors. *Med Chem Res, 5,* 43–53.

27. Winfield, L., Izenwasser, S., Wade, D., and Trudell, M. L. (2002). Synthesis and dopamine transporter binding affinity of 2,6-dioxopiperazine analogues of GBR 12909. *Med Chem Res, 11,* 102–115.

28. Rothman, R. B., Lewis, B., Dersch, C., Xu, H., Radesca, L., de Costa, B. R., Rice, K. C., Kilburn, R. B., Akunne, H. C., and Pert, A. (1993). Identification of a GBR 12935 homolog, LR1111, which is over 4,000-fold selective for the dopamine transporter, relative to serotonin and norepinephrine transporters. *Synapse*, *14*, 34–39.

29. Choi, S. W., Elmaleh, D. R., Hanson, R. N., and Fischman, A. J. (1999). Design, synthesis, and biological evaluation of novel non-piperazine analogues of 1-[2-(diphenylmethox-y)ethyl]- and 1-[2-[bis(4-fluorophenyl)methoxy]ethyl]-4-(3-phenylpropyl)piperazines as dopamine transporter inhibitors. *J Med Chem*, *42*, 3647–3656.

30. Choi, S. W., Elmaleh, D. R., Hanson, R. N., and Fischman, A. J. (2000). Novel 3-amino-methyl- and 4-aminopiperidine analogues of 1-[2-(diphenylmethoxy)ethyl]-4-(3-phenyl-propyl)piperazines: synthesis and evaluation as dopamine transporter ligands. *J Med Chem 43*, 205–213.

31. Choi, S. W., Elmaleh, D. R., Hanson, R. N., Shoup, T. M., and Fischman, A. J. (2002). Novel (bisarylmethoxy)butylpiperidine analogues as neurotransmitter transporter inhibi-tors with activity at dopamine receptor sites. *Bioorg Med Chem*, *10*, 4091–4102.

32. Zhang, Y., Rothman, R. B., Dersch, C. M., de Costa, B. R., Jacobson, A. E., and Rice, K. C. (2000). Synthesis and transporter binding properties of bridged piperazine analogues of 1-[2-[bis(4-fluorophenyl)methoxy]ethyl]-4-(3-phenylpropyl)piperazine (GBR 12909). *J Med Chem*, *43*, 4840–4849.

33. Singh, S. (2000). Chemistry, design, and structure–activity relationship of cocaine antagonists. *Chem Rev*, *100*, 925–1024.

34. Zhang, Y., Joseph, D. B., Bowen, W. D., Flippen-Anderson, J. L., Dersch, C. M., Rothman, R. B., Jacobson, A. E., and Rice, K. C. (2001). Synthesis and biological evaluation of tropane-like 1-[2-[bis(4-fluorophenyl)methoxy]ethyl]-4-(3-phenylpropyl)-piperazine (GBR 12909) analogues. *J Med Chem*, *44*, 3937–3945.

35. Bradley, A. L., Izenwasser, S., Wade, D., Klein-Stevens, C., Zhu, N., and Trudell, M. L. (2002). Synthesis and dopamine transporter binding affinities of 3α-benzyl-8-(diarylmethoxyethyl)-8-azabicyclo[3.2.1]octanes. *Bioorg Med Chem Lett*, *12*, 2387–2390.

36. Zhang, S., Izenwasser, S., Wade, D., Xu, L., and Trudell, M. L. (2006). Synthesis of dopa-mine transporter selective 3-diarylmethoxymethyl-8-arylalkyl-8-azabicyclo[3.2.1]octane derivatives. *Bioorg Med Chem*, *14*, 7943–7952.

37. Dutta, A. K., Meltzer, P. C., and Madras, B. K. (1993). Positional importance of the nitrogen atom in novel piperidine analogues of GBR 12909: affinity and selectivity for the dopamine transporter. *Med Chem Res*, *3*, 209–222.

38. Madras, B. K., Reith, M. E., Meltzer, P. C., and Dutta, A. K. (1994). O-526, a piperidine analog of GBR 12909, retains high affinity for the dopamine transporter in monkey caudate-putamen. *Eur J Pharmacol*, *267*, 167–173.

39. Dutta, A. K., Xu, C., and Reith, M. E. (1996). Structure–activity relationship studies of novel 4-[2-[bis(4-fluorophenyl)methoxy]ethyl]-1-(3-phenylpropyl)piperidine analogs: synthesis and biological evaluation at the dopamine and serotonin transporter sites. *J Med Chem*, *39*, 749–756.

40. Glowa, J. R., Fantegrossi, W. E., Lewis, D. B., Matecka, D., Rice, K. C., and Rothman, R. B. (1996). Sustained decrease in cocaine-maintained responding in rhesus monkeys with 1-[2-[bis(4-fluorophenyl)methoxy]ethyl]-4-(3-hydroxy-3-phenylpropyl) piperazinyl decanoate, a long-acting ester derivative of GBR 12909. *J Med Chem*, *39*, 4689–4691.

41. Hsin, L. W., Prisinzano, T., Wilkerson, C. R., Dersch, C. M., Horel, R., Jacobson, A. E., Rothman, R. B., and Rice, K. C. (2003). Synthesis and dopamine transporter affinity of chiral 1-[2-[bis(4-fluorophenyl)methoxy]ethyl]-4-(2-hydroxypropyl)pipera-zines as potential cocaine abuse therapeutic agents. *Bioorg Med Chem Lett, 13,* 553–556.

42. Prisinzano, T., Greiner, E., Johnson, E. M., Dersch, C. M., Marcus, J., Partilla, J. S., Rothman, R. B., Jacobson, A. E., and Rice, K. C. (2002). Piperidine analogues of 1-[2-[bis(4-fluorophenyl)methoxy]ethyl]-4-(3-phenylpropyl)piperazine (GBR 12909): high affinity ligands for the dopamine transporter. *J Med Chem, 45,* 4371–4374.

43. Dutta, A. K., Fei, X. S., Beardsley, P. M., Newman, J. L., and Reith, M. E. (2001). Structure–activity relationship studies of 4-[2-(diphenylmethoxy)ethyl]-1-benzylpiperi-dine derivatives and their N-analogues: evaluation of O- and N-analogues and their binding to monoamine transporters. *J Med Chem, 44,* 937–948.

44. Thurkauf, A., Decosta, B., Berger, P., Paul, S., and Rice, K. C. (1991). Synthesis of tritiated 1-2-(diphenylmethoxy)ethyl-4-3-(3-azidophenyl)propylpiperazine ([³H]-meta azido GBR-12935), a photoaffinity ligand for the dopamine reuptake site. *J Label Compounds Radiopharm, 29,* 125–129.

45. Berger, S. P., Martenson, R. E., Laing, P., Thurkauf, A., Decosta, B., Rice, K. C., and Paul, S. M. (1991). Photoaffinity labeling of the dopamine reuptake carrier protein with 3-azido[³H]GBR-12935. *Mol Pharmacol, 39,* 429–435.

46. Grigoriadis, D. E., Wilson, A. A., Lew, R., Sharkey, J. S., and Kuhar, M. J. (1989). Dopamine transport sites selectively labeled by a novel photoaffinity probe: ¹²⁵I-DEEP. *J Neurosci, 9,* 2664–2670.

47. Dutta, A. K., Reith, M. E., and Madras, B. K. (2001). Synthesis and preliminary characterization of a high-affinity novel radioligand for the dopamine transporter. *Synapse, 39,* 175–181.

48. Vaughan, R. A., Gaffaney, J. D., Lever, J. R., Reith, M. E., and Dutta, A. K. (2001). Dual incorporation of photoaffinity ligands on dopamine transporters implicates proximity of labeled domains. *Mol Pharmacol, 59,* 1157–1164.

49. Czoty, P. W., Ramanathan, C. R., Mutschler, N. H., Makriyannis, A., and Bergman, J. (2004). Drug discrimination in methamphetamine-trained monkeys: effects of monoamine transporter inhibitors. *J Pharmacol Exp Ther, 311,* 720–727.

50. Tomlinson, I. D., Mason, J., Burton, J. N., Blakely, R., and Rosenthal, S. J. (2003). The design and synthesis of novel derivatives of the dopamine uptake inhibitors GBR 12909 and GBR 12935. High-affinity dopaminergic ligands for conjugation with highly fluorescent cadmium selenide/zinc sulfide core/shell nanocrystals. *Tetrahedron, 59,* 8035–8047.

51. Tomlinson, I. D., Mason, J. N., Blakely, R. D., and Rosenthal, S. J. (2006). High affinity inhibitors of the dopamine transporter (DAT): novel biotinylated ligands for conjugation to quantum dots. *Bioorg Med Chem Lett, 16,* 4664–4667.

52. Kimura, M., Masuda, T., Yamada, K., Mitani, M., Kubota, N., Kawakatsu, N., Kishii, K., Inazu, M., Kiuchi, Y., Oguchi, K., and Namiki, T. (2003). Syntheses of novel diphenyl piperazine derivatives and their activities as inhibitors of dopamine uptake in the central nervous system. *Bioorg Med Chem, 11,* 1621–1630.

53. Kimura, M., Masuda, T., Yamada, K., Mitani, M., Kubota, N., Kawakatsu, N., Kishii, K., Inazu, M., Kiuchi, Y., Oguchi, K., and Namiki, T. (2003). Novel diphenylalkyl piperazine

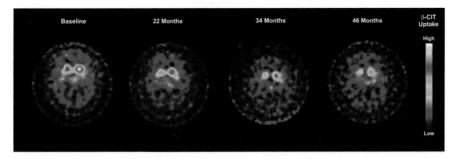

Figure 1-6 In Parkinson's disease DAT binding in the caudate and putamen declines with disease progression, as revealed by [^{123}I]β-CIT SPECT imaging in an early-stage PD patient. (Replicated with permission from the Parkinson Study Group [160].)

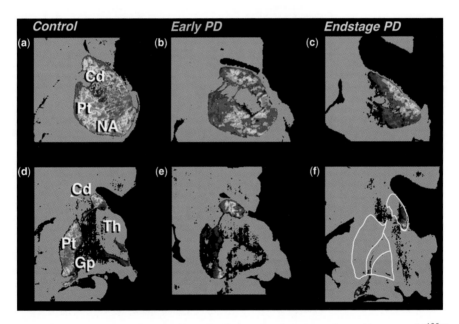

Figure 2-2 Mapping DAT in Parkinson's disease. In vitro autoradiographic maps of [^{125}I] RTI-121 labeling of the DAT in coronal sections of the striatum from a representative age-matched control subject (W, M, 73 years; panels a and d) and an early-stage Parkinson's disease patient (W, M, 69 years; panels b and e) and an advanced end-stage Parkinson's disease patient (W, M, 76 years; panels c and f). The DAT distribution and density are shown in a pseudocolor scale. Red-orange, high densities; yellow-green, intermediate; blue-purple, low). Cd, caudate; GP, globus pallidus; NA, nucleus accumbens; Th, thalamus; Pt, putamen.

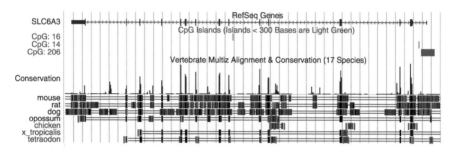

Figure 3-3   Structure of *DAT1*. The gene is displayed with boxes for each of the 15 exons, connected by a thin line representing the introns. The arrowheads within the introns represent the direction of transcription with the first exon on the right. This orientation is based on the convention of displaying the chromosome with the telomere of the p arm to the left. Only a few of the tracks of information from the Genome Browser are shown in this image, including the CpG islands, and conserved regions of sequence with the species noted to the left. (From the UCSC Genome Browser, www.genome.ucsc.edu [125].)

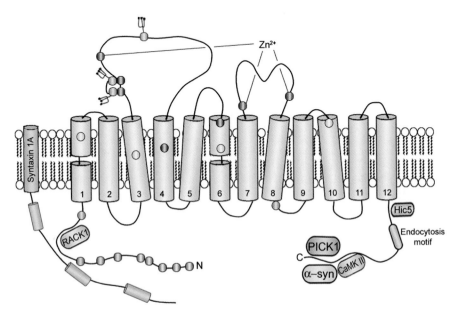

Figure 4-1 DAT posttranslational modifications and protein interaction sites. Schematic diagram of human DAT showing topological organization, transmembrane spanning domains 1 to 12, with TMs 1 and 6 shown as broken helices separated by short regions of random structure, and intracellularly oriented N- and C-termini. Specific sites and modifications represented are: consensus N-linked glycosylation sites at N181, N188, and N205 (orange circles) conjugated to complex carbohydrates (branched structures) terminated by sialic acids (dots); disulfide-bonded cysteines C180 and C189 in EL2 (connected purple circles); zinc-binding site residues H193, H375, and E396 (red circles); dimer and tetramer cross-linking sites at C243 and C306 (purple circles); putative extracellular gate residues R86, Y156, F320, and D471 (yellow circles); putative intracellular gate residues R60 and D436 (green circles); phosphorylation site cluster at serines 2, 4, 7, 12, and 13 (pink circles); and ubiquitylated lysines 19, 27, and 35 (blue circles). N-terminal tail interaction domains are indicated for syntaxin 1A and RACK1. C-terminal tail interaction domains are indicated for PICK1, α-synuclein, CaMKII, Hic-5, and the residue 587–596 endocytosis motif.

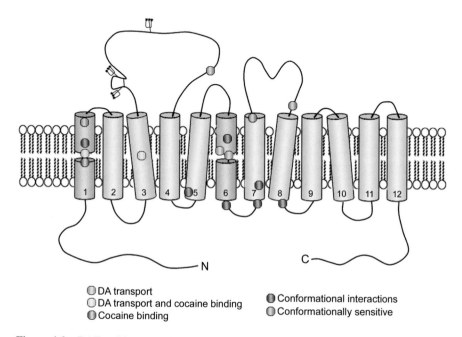

ODA transport
ODA transport and cocaine binding
OCocaine binding

OConformational interactions
OConformationally sensitive

Figure 4-2 DAT residues and domains involved in transport, cocaine binding, and conformational changes. TMs 1 and 6 (pink cylinders) indicate domains irreversibly labeled by cocaine photoaffinity analogs, and TM 3 (orange cylinder) represents a domain essential for high-affinity cocaine binding identified through cross-species chimeras. Residues with support for presence within active sites are shown for dopamine transport (green circles), cocaine binding (red circle), and both transport and cocaine binding (yellow circles). Blue circles indicate residues that are essential for protein structural rearrangements during transport, and orange circles indicate residues that are not essential for activity but undergo conformational movements in the presence of substrates or transport inhibitors. Specific residues indicated are (TM1) F76, D79, W84, C90; (EL2) R219; (TM3) V152; (TM5) K264; (TM6) D313, F320, G323, Y335; (TM7) C342, D345, M371; (EL4) A399; (TM8) D436.

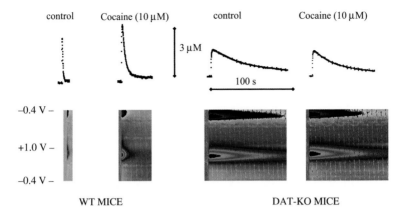

Figure 5-1 Effects of cocaine and genetic deletion of the dopamine transporter on dopamine clearance in core NAc. Single-pulse electrical stimulations were used to evoke dopamine release in slices containing the core NAc from control and DAT-KO mice. Dots are data points, collected every 100 ms. The overflow profile on the left shows a control stimulation, in the absence of drug, in a control mouse. Note that following stimulation, the extracellular level of dopamine returns to baseline levels within 1 second. The second profile shows that 10 µM cocaine markedly slows the return of dopamine to baseline levels, taking approximately 10 seconds to return. The third profile, recorded in a slice from a DAT-KO mouse, shows the dramatic slowing of dopamine clearance from the extracellular space following release. It takes approximately 100 seconds to clear the released dopamine. The fourth profile demonstrates that cocaine has no effect on dopamine clearance in DAT-KO mice. Insets: Below each profile, a pseudocolor plot gives information on voltage on the y-axis, time on the x-axis, and current due to dopamine oxidation and reduction on the z-axis in pseudocolor. Note the prolonged lifetime of dopamine in the extracellular space in the DAT-KO mice (green–blue colors).

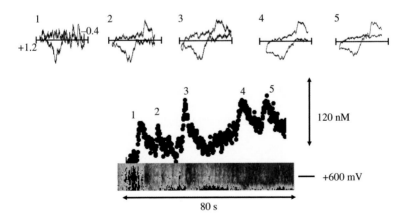

Figure 5-12 Cocaine-induced dopamine transients in the NAc of an anesthetized rat. Representative color plots (below), concentration–time plots (above), and cyclic voltammograms of dopamine (insets) 1 minute following cocaine infusion (1.5 mg/kg, i.v.). There was a significant increase in the frequency and amplitude of dopamine transients after cocaine compared to baseline.

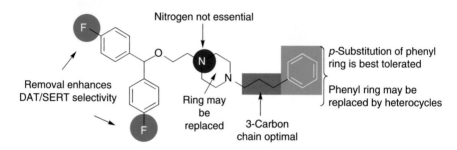

Figure 8-15 Summary of SAR for piperazine analogs.

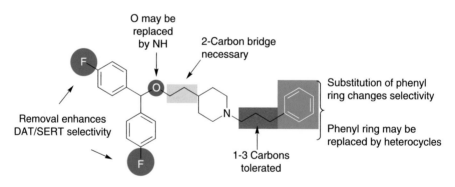

Figure 8-16 Summary of SAR for piperidine analogs.

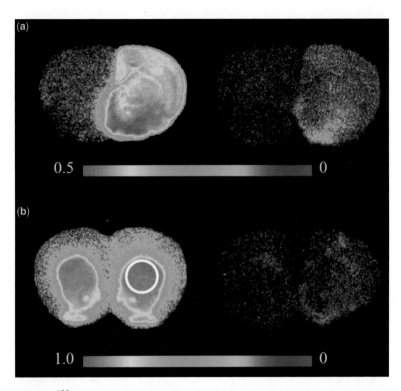

Figure 12-1 [^{125}I]RTI-55 autoradiography: representative autoradiograms of coronal rat brain cryostat sections (level: 10.20 mm anterior to the interaural line) with incubation of 10 pM [^{125}I]RTI-55 under conditions specific for serotonin transporters (a) and DAT (b). The serotonin innervation of the left hemisphere had earlier been destroyed with the selective toxin 5,7-DAT. Total binding is shown to the left and nonspecific binding to the right. The scales show the relative abundance of the two binding components. (Courtesy of Søren Dinesen Østergaard, Aarhus University PET Centre, Aarhus, Denmark.)

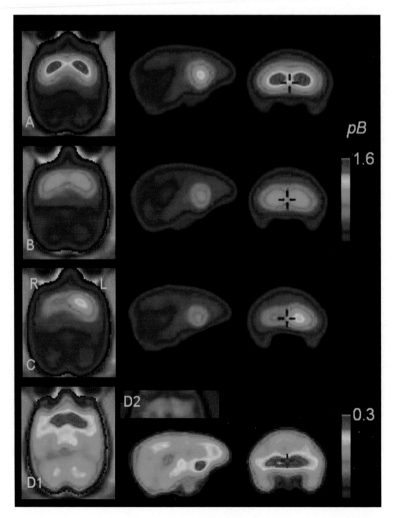

Figure 12-2 Binding potential (*pB*) of [^{11}C]methylphenidate (MP) for catecholamine uptake sites in brain of (A) normal young (*n* = 6) and (B) normal aged (*n* = 4) rhesus monkeys, (C) aged monkeys with unilateral intracarotid infusion of MPTP (*n* = 3), and (D) monkeys of intermediate age with systemic MPTP lesioning (*n* = 6). Mean parametric maps are coregistered to the MR template, shown as gray scale in the horizontal plane images. A sector containing the striatum in the horizontal plane from the MPTP group (D2) is presented with identical intensity scaling as in the intact animals. The cursor in the coronal plane images indicates the position (0,10,10) in the MR coordinate system. (Reproduced with permission from Doudet et al. [50].)

derivatives with high affinities for the dopamine transporter. *Bioorg Med Chem, 11,* 3953–3963.

54. Cao, J., Lever, J. R., Kopajtic, T., Katz, J. L., Pham, A. T., Holmes, M. L., Justice, J. B., and Newman, A. H. (2004). Novel azido and isothiocyanato analogues of [3-(4-phenyl-alkylpiperazin-1-yl)propyl]bis(4-fluorophenyl)amines as potential irreversible ligands for the dopamine transporter. *J Med Chem, 47,* 6128–6136.

55. Dutta, A. K., Coffey, L. L., and Reith, M. E. (1997). Highly selective, novel analogs of 4-[2-(diphenylmethoxy)ethyl]-1-benzylpiperidine for the dopamine transporter: effect of different aromatic substitutions on their affinity and selectivity. *J Med Chem, 40,* 35–43.

56. Dutta, A. K., Davis, M. C., Fei, X. S., Beardsley, P. M., Cook, C. D., and Reith, M. E. (2002). Expansion of structure–activity studies of piperidine analogues of 1-[2-(diphenyl-methoxy)ethyl]-4-(3-phenylpropyl)piperazine (GBR 12935) compounds by altering substitutions in the *N*-benzyl moiety and behavioral pharmacology of selected molecules. *J Med Chem, 45,* 654–662.

57. Dutta, A. K., Xu, C., and Reith, M. E. (1998). Tolerance in the replacement of the benzhydrylic O atom in 4-[2-(diphenylmethoxy)ethyl]-1-benzylpiperidine derivatives by an N atom: development of new-generation potent and selective N-analog molecules for the dopamine transporter. *J Med Chem, 41,* 3293–3297.

58. Dutta, A. K., Davis, M. C., and Reith, M. E. (2001). Rational design and synthesis of novel 2,5-disubstituted *cis*- and *trans*-piperidine derivatives exhibiting differential activity for the dopamine transporter. *Bioorg Med Chem Lett, 11,* 2337–2340.

59. Greiner, E., Prisinzano, T., Johnson, I. E., Dersch, C. M., Marcus, J., Partilla, J. S., Rothman, R. B., Jacobson, A. E., and Rice, K. C. (2003). Structure–activity relationship studies of highly selective inhibitors of the dopamine transporter: *N*-benzylpiperidine analogues of 1-[2-[bis(4-fluorophenyl)methoxy]ethyl]-4-(3-phenylpropyl)piperazine. *J Med Chem, 46,* 1465–1469.

60. Dutta, A. K., Coffey, L. L., and Reith, M. E. (1998). Potent and selective ligands for the dopamine transporter (DAT): structure–activity relationship studies of novel 4-[2-(diphe-nylmethoxy)ethyl]-1-(3-phenylpropyl)piperidine analogues. *J Med Chem, 41,* 699–705.

61. Boos, T. L., Greiner, E., Calhoun, W. J., Prisinzano, T. E., Nightingale, B., Dersch, C. M., Rothman, R. B., Jacobson, A. E., and Rice, K. C. (2006). Structure–activity relationships of substituted *N*-benzyl piperidines in the GBR series: synthesis of 4-(2-(bis(4-fluorophe-nyl)methoxy)ethyl)-1-(2-trifluoromethylbenzyl)piperidine, an allosteric modulator of the serotonin transporter. *Bioorg Med Chem, 14,* 3967–3973.

62. Nightingale, B., Dersch, C. M., Boos, T. L., Greiner, E., Calhoun, W. J., Jacobson, A. E., Rice, K. C., and Rothman, R. B. (2005). Studies of the biogenic amine transporters. XI. identification of a 1-[2-[bis(4-fluorophenyl)methoxy]ethyl]-4-(3-phenylpropyl)piperazine (GBR 12909) analog that allosterically modulates the serotonin transporter. *J Pharmacol Exp Ther, 314,* 906–915.

63. Ghorai, S. K., Cook, C., Davis, M., Venkataraman, S. K., George, C., Beardsley, P. M., Reith, M. E., and Dutta, A. K. (2003). High affinity hydroxypiperidine analogues of 4-(2-benzhydryloxyethyl)-1-(4-fluorobenzyl)piperidine for the dopamine transporter: stereospecific interactions in vitro and in vivo. *J Med Chem, 46,* 1220–1228.

64. Greiner, E., Boos, T. L., Prisinzano, T. E., De Martino, M. G., Zeglis, B., Dersch, C. M., Marcus, J., Partilla, J. S., Rothman, R. B., Jacobson, A. E., and Rice, K. C. (2006). Design and synthesis of promiscuous high-affinity monoamine transporter ligands: unraveling transporter selectivity. *J Med Chem, 49,* 1766–1772.

65. Dutta, A. K., Davis, M. C., and Reith, M. E. A. (2001). Rational design and synthesis of novel 2,5-disubstituted *cis*- and *trans*-piperidine derivatives exhibiting differential activity for the dopamine transporter. *Bioorg Med Chem Lett*, *11*, 2337–2340.

66. Kolhatkar, R. B., Ghorai, S. K., George, C., Reith, M. E., and Dutta, A. K. (2003). Interaction of *cis*-(6-benzhydrylpiperidin-3-yl)benzylamine analogues with monoamine transporters: structure–activity relationship study of structurally constrained 3,6-disubstituted piperidine analogs of (2,2-diphenylethyl)-[1-(4-fluorobenzyl)piperidin-4-ylmethyl]amine. *J Med Chem*, *46*, 2205–2215.

67. Kolhatkar, R., Cook, C. D., Ghorai, S. K., Deschamps, J., Beardsley, P. M., Reith, M. E. A., and Dutta, A. K. (2004). Further structurally constrained analogues of *cis*-(6-benzhydrylpiperidin-3-yl)benzylamine with elucidation of bioactive conformation: discovery of 1,4-diazabicyclo[3.3.1]nonane derivatives and evaluation of their biological properties for the monoamine transporters. *J Med Chem*, *47*, 5101–5113.

68. Zhang, S., Reith, M. E., and Dutta, A. K. (2003). Design, synthesis, and activity of novel *cis*- and *trans*-3,6-disubstituted pyran biomimetics of 3,6-disubstituted piperidine as potential ligands for the dopamine transporter. *Bioorg Med Chem Lett*, *13*, 1591–1595.

69. Zhang, S., Zhen, J., Reith, M. E. A., and Dutta, A. K. (2004). Structural requirements for 2,4- and 3,6-disubstituted pyran biomimetics of *cis*-(6-benzhydryl-piperidin-3-yl)-benzylamine compounds to interact with monoamine transporters. *Bioorg Med Chem*, *12*, 6301–6315.

70. Zhang, S., Zhen, J., Reith, M. E. A., and Dutta, A. K. (2005). Discovery of novel trisubstituted asymmetric derivatives of (2*S*,4*R*,5*R*)-2-benzhydryl-5-benzylaminotetrahydropyran-4-ol, exhibiting high affinity for serotonin and norepinephrine transporters in a stereospecific manner. *J Med Chem*, *48*, 4962–4971.

71. Zhang, S., Zhen, J., Reith, M. E. A., and Dutta, A. K. (2006). Design, synthesis, and preliminary SAR study of 3- and 6-side-chain-extended tetrahydro-pyran analogues of *cis*- and *trans*-(6-benzhydryl-tetrahydropyran-3-yl)-benzylamine. *Bioorg Med Chem*, *14*, 3953–3966.

72. Zhang, S., Fernandez, F., Hazeldine, S., Deschamps, J., Zhen, J., Reith, M. E. A., and Dutta, A. K. (2006). Further structural exploration of trisubstituted asymmetric pyran derivatives (2*S*,4*R*,5*R*)-2-benzhydryl-5-benzylamino-tetrahydropyran-4-ol and their corresponding disubstituted (3*S*,6*S*) pyran derivatives: a proposed pharmacophore model for high-affinity interaction with the dopamine, serotonin, and norepinephrine transporters. *J Med Chem*, *49*, 4239–4247.

73. Benedetti, P., Mannhold, R., Cruciani, G., and Pastor, M. (2002). GBR compounds and mepyramines as cocaine abuse therapeutics: chemometric studies on selectivity using grid independent descriptors (GRIND). *J Med Chem*, *45*, 1577–1584.

74. Gilbert, K., and Venanzi, C. (2006). Hierarchical clustering analysis of flexible GBR 12909 dialkyl piperazine and piperidine analogs. *J Comput Aided Mol Des*, *20*, 209–225.

75. Misra, M., Banerjee, A., Dave, R. N., and Venanzi, C. A. (2005). Novel feature extraction technique for fuzzy relational clustering of a flexible dopamine reuptake inhibitor. *J Chem Inf Modeling*, *45*, 610–623.

76. Fiorentino, A., Pandit, D., Gilbert, K. M., Misra, M., Dios, R., and Venanzi, C. A. (2006). Singular value decomposition of torsional angles of analogs of the dopamine reuptake inhibitor GBR 12909. *J Comput Chem*, *27*, 609–620.

77. Gilbert, K. M., Boos, T. L., Dersch, C. M., Greiner, E., Jacobson, A. E., Lewis, D., Matecka, D., Prisinzano, T. E., Zhang, Y., and Rothman, R. B. (2007). DAT/SERT selectivity of flexible GBR 12909 analogs modeled using 3D-QSAR methods. *Bioorg Med Chem*, *15*, 1146–1159.

78. Goodford, P. J. (1985). A computational procedure for determining energetically favorable binding sites on biologically important macromolecules. *J Med Chem*, *28*, 849–857.

79. Cramer, R. D., Patterson, D. E., and Bunce, J. D. (1988). Comparative molecular-field analysis (CoMFA): 1. Effect of shape on binding of steroids to carrier proteins. *J Am Chem Soc*, *110*, 5959–5967.

9

STRUCTURE–ACTIVITY RELATIONSHIP STUDY OF PIPERIDINE DERIVATIVES FOR DOPAMINE TRANSPORTERS

PRASHANT S. KHARKAR AND ALOKE K. DUTTA

Department of Pharmaceutical Sciences, Wayne State University, Detroit, Michigan

MAARTEN E. A. REITH

Department of Psychiatry, New York University, New York, New York

9.1	Introduction	234
9.2	Piperidine Analogs	237
	9.2.1 Piperidine Molecules Originated from GBR 12909	237
	9.2.2 Conformationally Constrained Piperidine Analogs	238
	9.2.3 Phenylpiperidine Derivatives	243
9.3	Three-Dimensional Pharmacophore-Based Search for Novel DAT Inhibitors	250
9.4	Meperidine Derivatives	251
9.5	Methylphenidate Analogs	252
9.6	CoMFA Study of the Piperidine Analogs of Cocaine	255
References		257

9.1   INTRODUCTION

In the central nervous system (CNS), the biogenic amine neurotransmitters dopamine (DA), serotonin (5-hydroxytryptamine, 5-HT), and norepinephrine (NE) are vital modulators of monoaminergic transmission, involved in a variety of physiological, behavioral, and endocrine functions. Chemical imbalances in these neurotransmitter systems result in the development of brain disorders such as Parkinson's disease, depression, bipolar disorder, attention-deficit/hyperactivity disorder (ADHD), drug abuse, and Tourette's syndrome [1–5]. A complex interplay of the synthesis, storage, release, metabolism, and recapture of these neurotransmitters determine the duration and intensity of monoaminergic signaling [3]. The monoamines released extracellularly undergo enzymatic degradation, dilution by diffusion, and most important, reuptake into the presynaptic neurons via neuronal plasma membrane transporters. The latter process is primarily responsible for the inactivation of released monoamines and is conducted by three presynaptically located transport proteins: the dopamine transporter (DAT), the serotonine transporter (SERT), and the nor-epinephrine transporter (NET) [6–8]. These transporters are actively involved in various pathological processes and implicated in several neurological disorders, such as depression, drugs of abuse, anxiety, mood problems, ADHD, Parkinson's disease, Alzheimer's disease, and schizophrenia [9–14].

   The DAT, SERT, and NET represent established targets for many neuropharma-cological agents, including antidepressants, psychostimulants, and neurotoxins [6,15,16]. They exert their action by inhibiting transporter function, resulting in increased extracellular concentrations of monoamines [6,17]. These transporters belong to a family of Na^+/Cl^--dependent substrate-specific neuronal plasma membrane transporters, which include transporters for γ-aminobutyric acid (GABA), glycine, taurine, betaine, proline, and creatine. The putative structures of these transporters consist of 12 transmembrane domains (TMDs), with both the C- and N-terminal domains located in the cytoplasm. These TMDs are connected by hydrophilic loops. The human DAT, NET, and SERT proteins contain 620, 617, and 630 amino acid residues, respectively. The highest conservation of sequences is found to be within the putative transmembrane domains, whereas N- and C-termini are the least conserved regions [18]. The first crystal structure of a bacterial homolog of monoamine transporters, an amino acid transporter for leucine (LeuT), has recently been reported which demonstrated the existence of leucine and Na^+-binding sites within the transmembrane domains [19]. The mechanism of monoamine reuptake by the transporter is believed to involve sequential binding and cotransport of Na^+ and Cl^- ions [6–8,17]. The driving force for transporter-mediated monoamine uptake is the ionic concentration gradient generated by the plasma membrane Na^+/K^+-ATPase [20,21].

   Cocaine, an alkaloid derived from the *Erythroxylon coca* leaves, has long been recognized for its strong reinforcing activity and abuse potential [22]. Addiction to cocaine is a major problem in our society today, affecting the nation greatly in terms of its economy and associated crime costs. It has also contributed to spreading of HIV infection, due to needle sharing among drug abusers. There is an urgent need for the development of an effective medication for cocaine addiction, due to the current lack of such an effective treatment. Cocaine binds to all three monoamine transporter systems (i.e., DAT, SERT, and NET) in the brain. However, the central mechanism of addiction to cocaine is attributed to its binding to the DAT located pre-synaptically in the dopaminergic neuron [23–25]. Binding of cocaine to the DAT leads to an increased dopamine level in the synapse and is the underlying reason for the stimulatory and perhaps strong reinforcing effects of cocaine [23]. The DAT-binding potencies of several cocainelike compounds correlate well with their potencies, producing self-administration behavior. Furthermore, in a microdialysis study it was demonstrated that cocaine increases dopamine preferentially in the nucleus accumbens in relation to its reinforcing effects [26]. More recently, the dopa-mine hypothesis for cocaine's reinforcing effects was strengthened further by the demonstration that in dopamine transporter knockout (DAT-KO) mice, cocaine and amphetamine increase extracellular dopamine in the nucleus accumbens but not in the caudate putamen [27]. This observation perhaps explains the results of a recent experiment showing self-administration of cocaine by DAT-KO [28], as in these mice cocaine will still block the norepinephrine transporter, which is known to accumulate dopamine efficiently [29] in the nucleus accumbens and thus will increase the extracellular concentration of dopamine in DAT knockouts [30,31]. Finally, in recent PET (positron emission tomography) studies it was very elegantly demon-strated that the subjective effects of cocaine in humans correlate directly with the extent to which cocaine occupies the dopamine transporter [32]. These results further reinforce the dopaminergic hypothesis for cocaine addiction and support a drug development approach targeting the dopaminergic system as a possible avenue by which to develop medications for cocaine addiction [33,34].

Although there is literature reporting a noncompetitive [35] or uncompetitive [36], mechanism for the inhibition of DA uptake by cocaine, the bulk of the literature on cocaine and its analogs points to a competitive mechanism [37,38]. Even so, cocaine may bind to a site different from that for the substrate recognition site on the DAT [39], causing mutually exclusive binding of cocaine and DA according to a competi-tive mechanism [40]. It is often thought that cocaine and DA bind to both overlapping and separate domains on the DAT, and that accordingly, it would be theoretically possible to design agents that bind to the cocaine recognition site and fail to inhibit DA uptake (DA-sparing cocaine antagonists [41]). The concept of partial agonism comes from receptor theories and does not really apply to transporters, but the concept that compounds could interact with the DAT in a way fundamentally different from cocaine and thereby more weakly inhibit DA transport has been advanced [42–44]. The usefulness of cocaine antagonists would probably be limited to the treatment of acute overdose, but there is clear applicability for com-pounds that would be capable of partly mimicking the cocaine response without exhibiting the full reinforcing properties of cocaine (substitution therapy).

A large number of structurally diverse compounds have been developed as DAT inhibitors, with potential applications in the pharmacotherapy of cocaine addiction [45–47]. These compounds can be classified in the following classes: 3-aryltropanes, benztropines, piperidine analogs of tropanes, GBR compounds, methylphenidate,

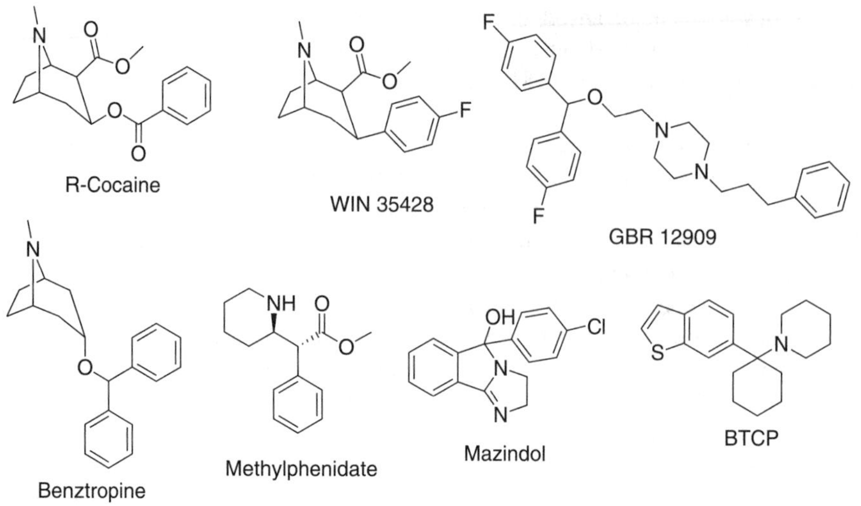

Figure 9-1   Molecular structures of known DAT inhibitors.

mazindol, and phencyclidine analogs. Figure 9-1 shows the representative compounds from each class. The 3-aryltropanes in which the benzoyl group of cocaine is replaced by a substituted aromatic ring (WIN 35,428, Fig. 9-1) represent the most studied group of DAT inhibitors. Extensive structure–activity relationship (SAR) studies involving 3-aryltropanes led to the discovery of highly potent and selective DAT inhibitors [45,47]. These compounds have also demonstrated significant potency in behavioral studies such as locomotor activity, drug discrimination, and self-administration [48]. Benztropine and its analogs, a tropane-based dopamine reuptake inhibitor, have been studied extensively. Benztropine has been used clinically for the treatment of symptoms associated with Parkinson's disease. SAR studies around benztropine molecule generated potent and selective DAT inhibitors [47,49]. The benztropine analogs were generally much less potent than cocaine in locomotor activity studies despite their higher affinity than cocaine for DAT. GBR compounds resulted from the replacement of the tropane ring in benztropine by a piperazine ring. These compounds represent structurally important class of potent and selective dopamine reuptake inhibitors [49]. The conversion of a central piperazine ring to piperidine yielded analogs with a variety of potencies for DAT [50]. Several lines of experimental evidence from behavioral studies strongly suggest that the suitable GBR analogs might be a potential pharmacotherapy for cocaine addiction.

DL-Threo-methylphenidate, a long-known stimulant, is the drug of choice for symptomatic treatment of ADHD. It possesses reinforcing effects equivalent to those of cocaine [51]. Due to its prolonged in vivo half-life compared to that of cocaine (90 minutes vs. 20 minutes for cocaine), its abuse liability in humans appears to be substantially lower [52]. Mazindol, a triple-uptake inhibitor, is currently in therapy for exogenous obesity and as an orphan drug for the treatment of Duchenne muscular dystrophy [53,54]. Although the literature is equivocal about the addictive potential of mazindol [55], it was found to be nonaddictive in clinic. Phencyclidine exhibits complex behavioral profile in animal and human studies with its effects being attributed to its binding to both the DAT and *N*-methyl-D-aspartate (NMDA) ion-channel complex [56]. In vivo, phencyclidine and one of its analogs, BTCP, produced locomotor stimulation like that of cocaine [57].

Despite intense drug development efforts targeted at the DAT to discover new medications for cocaine addiction, none of these compounds has yet made it into the clinic. Progress has been made in finding lead structures for cocaine substitution therapy with compounds possessing slow onset and much longer duration of action. In this chapter we review progress made in the area of piperidine analogs, with particular emphasis placed on several novel leads exhibiting unique behavioral patterns and mechanisms of action.

9.2 PIPERIDINE ANALOGS

Design of piperidine analogs for the DAT originated primarily from structural modifications of GBR and tropane molecules. In this chapter we focus on the description of drug development for the DAT based on the piperidine moiety.

9.2.1 Piperidine Molecules Originated from GBR 12909

In a further SAR study with GBR derivatives, it was demonstrated that only one of the N-atoms in the piperazine ring is required for activity and that the N-atom located distal to the diphenylmethoxy moiety, compound **2** in Figure 9-2, is required for potency and selectivity at the dopamine transporter [50,58]. Consequently, compound **1** with the N-atom proximally located was less potent and selective for the dopamine transporter. In addition, this new piperidine analog was shown to preferentially target the dopamine transporter since it did not show the nonselective

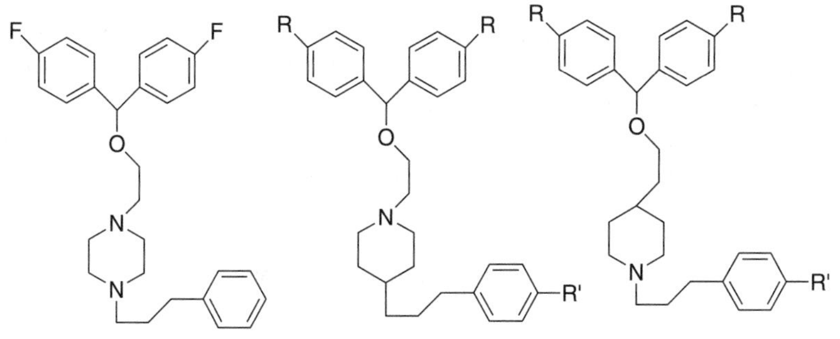

GBR 12909 **1**, R = F, R' = H **2**, R = F, R' = H
3, R = H, R' = F

Figure 9-2 Molecular structures of GBR and related molecules.

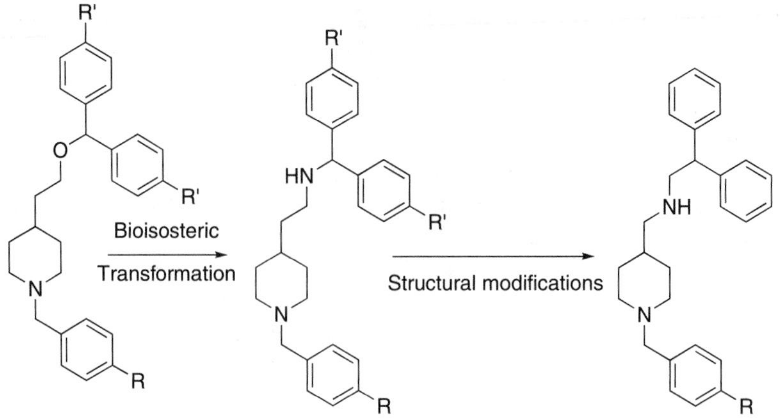

4, R = F; R' = H
5, R = H; R' = H

6, R = F; R' = H
7, R = H; R' = H

8, R = F

Scheme 9-1 Structural modification of piperidine version of GBR molecules into bioisosteric derivatives.

piperazine-binding activity, which contributes as much as 30% to the binding of conventional GBR compounds to brain preparations under commonly used assay conditions [58]. In further SAR studies, it was shown that these piperidine analogs do not quite follow the SARs found in conventional GBR compounds [59,60]. Detailed SAR study on these GBR 12909–based piperidine derivatives is included in Chapter 8. In this chapter our focus is on those piperidine derivatives that evolved from the first-generation flexible piperidine derivatives and represents structurally constrained versions.

Design of structurally constrained piperidine derivatives was initiated from one of the lead derivatives of the flexible piperidine version, where the benzhydryl oxygen atom was replaced by a bioisosteric nitrogen atom as shown in Scheme 9-1. This bioisosteric replacement led to the development of number of potent compounds for DAT as described in Scheme 9-1 and Table 9-1 [61]. Further structural alteration by moving the exocyclic benzhydrilic N-atom to an adjacent position led to derivative **8** [62]. This molecule exhibited potency for the DAT but with reduced selectivity, as it also displayed appreciable affinity for SERT (see Table 9-1). This lead compound was next considered as a precursor template for its conversion into a conformationally constrained molecular template as described next.

9.2.2 Conformationally Constrained Piperidine Analogs

A number of piperidine analogs of GBR 12935 were synthesized and characterized biologically in an effort to develop potent and selective compounds for the DAT [59,60]. Most of these molecules were structurally flexible, due to a higher number of rotatable bonds. With the aim of reducing the structural flexibility and to

TABLE 9-1 Affinity and Selectivity of Drugs at the Dopamine, Serotonin, and Norepinephrine Transporters in Rat Striatum[a]

Compound	DAT, IC_{50} (nM), [^{3}H]WIN 35,428	SERT, IC_{50} (nM), [^{3}H]citalopram	NET, IC_{50} (nM), [^{3}H]nisoxetine	SERT/ DAT
Cocaine	266 ± 37[b]	737 ± 160[b]	3530 ± 550[b]	2.7
GBR 12909	10.6 ± 1.9[b]	132 ± 0[b]	496 ± 22[b]	12
4	15.2 ± 2.8[c]	743 ± 6[c]	—	49
5	17.2 ± 4.7[d]	1920 ± 233[d]	—	113
6	9.37 ± 2.62[e]	585 ± 101[e]	945 ± 20[e]	62
7	4.50 ± 0.64[b]	1560 ± 210[b]	2620 ± 170[b]	347
8	19.7 ± 1.4[b]	137 ± 46[b]	1110 ± 120[b]	7.0

[a]The DAT were labeled with [^{3}H]WIN 35,428, the SERT with [^{3}H]citalopram, and the NET with [^{3}H]nisoxetine. Results are average $\pm$ SEM of three independent experiments assayed in triplicates.
[b]See [62].
[c]See [60].
[d]See [59].
[e]See [61].

investigate the *bioactive conformation* of these molecules, a new design strategy was applied to one of the potent DAT inhibitors [63] (Fig. 9-3). Thus, the formation of a new piperidine ring from the secondary amine and opening of the existing piperidine ring of **8** led to general structure II, which exists in the *cis* and *trans* forms. The newly designed compounds possess two asymmetric centers as opposed to one in the starting structure, thus making the constrained version more complex. Thus, the *cis*- and *trans*-isomers, **9a** and **9b**, represent the conformationally constrained piperidine analogs. The binding affinity results demonstrated higher activity for the ($\pm$)-*cis*-**9a** than for ($\pm$)-*trans*-**9b** (DAT $IC_{50} = 30$ nM vs. 212 nM; see Table 9-2). The *cis*-isomer was 93-fold more selective for the DAT than for SERT and 45-fold more selective than for NET. The *cis*-isomer was about nine times more potent than cocaine at the DAT ($IC_{50} = 30$ nM vs. 266 nM). Moreover, the *cis*-isomer was much more selective for the DAT than was the parent compound **8** (93.7 vs. 6.9), although the *cis*-isomer exhibited a slight drop in potency (30 nM vs. 19.7 nM) [63].

Extensive SAR investigations to develop a receptor-binding profile of this novel 3,6-disubstituted piperidine template resulted in several derivatives with varied binding profiles at the DAT, SERT, and NET. Differently substituted benzyl derivatives, including bioisosteric heterocyclic moieties were introduced on the exocyclic N-atom (see Fig. 9-3) to understand the contributions of steric and electronic factors at this molecular center for interactions with the DAT [64]. Replacement of the 4-fluorophenyl group (**9a**) with the 4-cyanophenyl (**9c**), 3,4-difluorophenyl (**9d**), 4-methoxyphenyl (**9e**), phenyl (**9f**), and 4-hydroxyphenyl (**9h**) groups resulted in potent to moderately potent compounds. Compound ($\pm$)-**9c** with electron-withdrawing substituent was the most potent of these compounds ($IC_{50} = $ 31.5 nM). Separation of the enantiomers from the racemic mixture confirmed the residence of greater DAT binding potency in the ($-$)-isomer ($IC_{50} = 11.3$ nM) compared

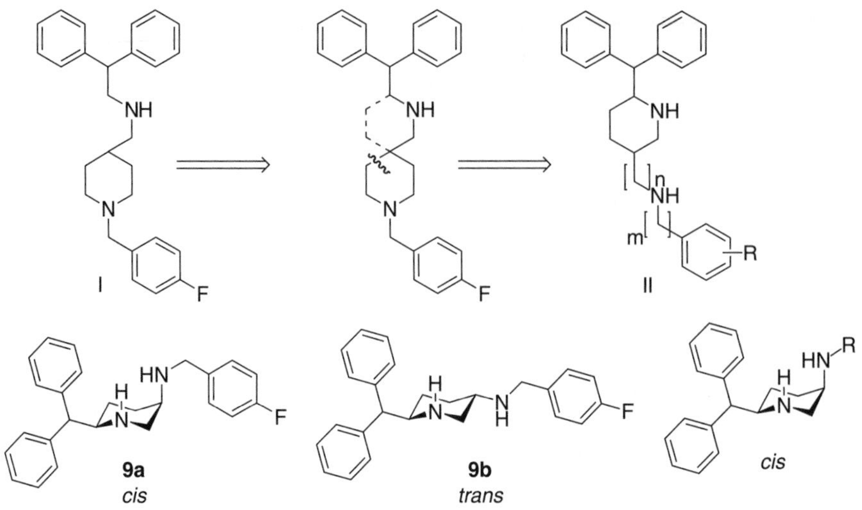

(+)-9a & (−)-9a: R = CH$_2$ 4-FPh
(+)-9c & (−)-9c: R = CH$_2$ 4-CNPh
9d: R = CH$_2$ 3,4-DiFPh
9e: R = CH$_2$ 4-OMePh
9f: R = CH$_2$ Ph
9g: R = CH$_2$ Indol-2-yl
9h: R = CH$_2$ 4-OHPh

Figure 9-3 Design and development of conformationally constrained piperidine derivatives.

to the (+)-isomer (IC$_{50}$ = 109 nM). Replacement of the 4-fluorophenyl group with heterocyclic moieties such as thiophene, benzothiophene, and indole resulted in weakly active to moderately active compounds. Compound **9g**, with 2-indolylmethyl substitution on the N-atom, showed moderate potency at the DAT (IC$_{50}$ = 95.4 nM) and was the most potent among the heterocyclic substitutions. The next effort to extend the present series of 3,6-disubstituted piperidines concentrated around the design of more structurally constrained derivatives of the 3,6-disubstituted piperidines to explore the possible bioactive conformation of the lead structures [65]. Further structural rigidification of this template was done by linking the two nitrogen atoms in the lead structure by an ethylene linker as represented in Figure 9-4, which led to a novel series of 4,8-disubstituted 1,4-diazabicyclo[3.3.1]nonane derivatives. A hypothesis was proposed that a highly conformationally constrained molecule shaped to the complementary binding site of the target will exhibit high affinity due to less entropic penalty encountered (compared to the flexible molecules) resulting from binding to the transporter. A preliminary molecular modeling study was conducted using the bicyclic rigid structure as the template. The local minima of one of the lead structures, **10b**, and the template bicyclic rigid structure were flexibly fitted with low-root mean square values, indicating the best possible fit between these low-energy conformers. This was reasoned to be the probable bioactive

TABLE 9-2 Affinity of Drugs at the Dopamine, Serotonin, and Norepinephrine Transporters in Rat Striatum and in Inhibition of DA Reuptake[a]

Compound	DAT Binding, IC_{50} (nM), [³H] WIN 35,428	SERT Binding, IC_{50} (nM), [³H]citalopram	NET Binding, IC_{50} (nM), [³H]nisoxetine	DAT Uptake, IC_{50} (nM), [³H]DA
Cocaine	266 ± 37^b	737 ± 160^b	$3,530 \pm 550^b$	
GBR 12909	10.6 ± 1.9^b	132 ± 0^b	496 ± 22^b	6.63 ± 0.43^b
8	19.7 ± 1.4^b	137 ± 46^b	$1,110 \pm 120^b$	49.6 ± 7.2^b
(±)-9a	32.5 ± 12.6^c	2220 ± 590^c	1020 ± 70^c	45.7 ± 5.1^c
(±)-9b	212 ± 20^d	1330 ± 102^d	4470 ± 1180^d	106 ± 10^d
(−)-S,S-9a	33.8 ± 5^c	1330 ± 120^c	1420 ± 560^c	53.8 ± 7.4^c
(+)-R,R-9a	229 ± 17^c	3540 ± 640^c	2290 ± 230^c	142 ± 25^c
9c	31.5 ± 5.3^c	1130 ± 250^c	3200 ± 940^c	30.2 ± 5.2^c
9d	63.5 ± 3.2^c	2250 ± 490^c	1590 ± 140^c	77.0 ± 9.7^c
9g	95.4 ± 20.2^c	1800 ± 280^c	1010 ± 490^c	51.0 ± 21.9^c
9f	114 ± 10.6^c	2130 ± 110^c	612 ± 130^c	
9e	47.5 ± 6.2^c	1040 ± 110^c	1110 ± 60^c	40.5 ± 21.8^c
9h	65.9 ± 8.2^c	862 ± 57^c	201 ± 13^c	
(−)S,S-9c	11.3 ± 0.9^c	434 ± 27^c	1670 ± 90^c	9.10 ± 1.86^c
(+)R,R-9c	109.0 ± 16.7^c	1550 ± 420^c	16600 ± 410^c	152 ± 41^c
10a	81.2 ± 11.9^e	8860 ± 730^e	56400 ± 5000^e	56.9 ± 14.7^e
(+)-R,R-10b	329 ± 30^e	6630 ± 1130^e	$>10000^e$	269 ± 4^e
(−)-S,S-10b	22.5 ± 2.1^e	8650 ± 830^e	$>10000^e$	18.4 ± 0.9^e
(±)-11	4.14 ± 0.77^f	2360 ± 500^f	1030 ± 80^f	3.22 ± 1.0^f
(+)-R,R-11	0.46 ± 0.05^f	3600 ± 270^f	1880 ± 230^f	4.05 ± 0.73^f
(−)-S,S-11	56.7 ± 6.5^f	1830 ± 80^f	1550 ± 190^f	38.0 ± 6.0^f
(±)-12	11.2 ± 1.0^f	3310 ± 270^f	2150 ± 660^f	10.1 ± 0.7^f

[a]For binding the DAT was labeled with [³H]WIN 35,428, the SERT with [³H]citalopram, and the NET with [³H]nisoxetine. For uptake by DAT, [³H]DA accumulation was measured. Results are average $\pm$ SEM of three to eight independent experiments assayed in triplicate.
[b]See [62].
[c]See [64].
[d]See [63].
[e]See [65].
[f]See [67].

conformation. In addition, the study was directed to investigating the possible effect of structural rigidity on in vivo activity, as structurally constrained molecules could exhibit favorable pharmacokinetic properties [66]. This SAR study led to the development of potent lead compound (−)-**10a** and (−)-**10b** (Fig. 9-4, Table 9-2), which exhibited high affinity and selectivity for the DAT ($IC_{50} = 22.5$ nM; SERT/ DAT = 384; and NET/DAT > 444; see Table 9-2). It was interesting to note that both (−)-**10b** and the lead compound from the 3,6-disubstituted piperidine series (−)-**9c** exhibited the highest activity in their (S,S)-isomer, indicating a similar regiospecificity requirement for maximum interaction. The potent compounds from the 1,4-diazabicyclo[3.3.1]nonane series exhibited greater selectivity for the

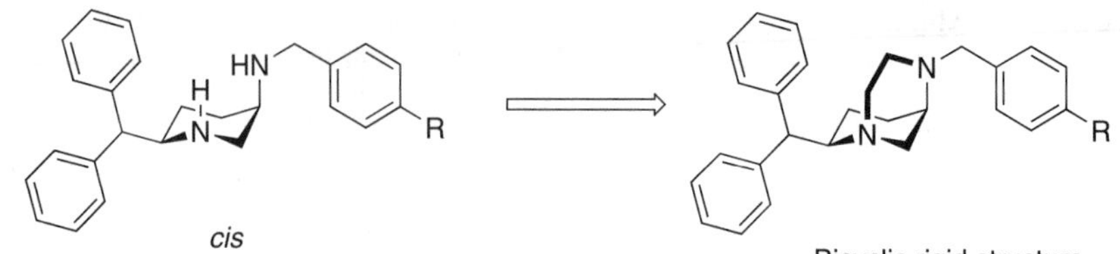

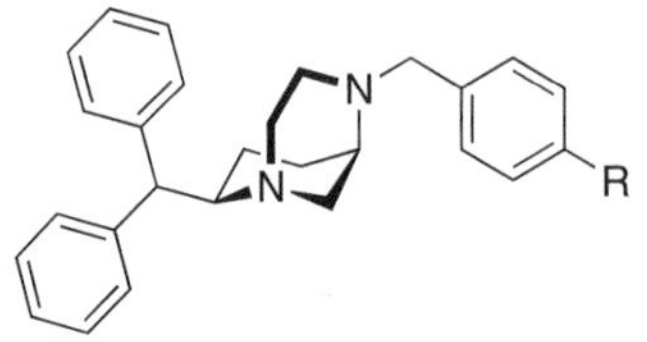

10a; R = F
10b; R = OCH$_3$

Figure 9-4   Design and development of constrained bicyclic piperidine derivatives.

DAT than did their parent compounds (i.e., lesser constrained 3,6-disubstituted piperidines), indicating the effect of structural rigidification on selective interaction with the monoamine transporters. Compounds (−)-**9c** and (−)-**10b** exhibited stimulant activity in locomotor tests in mice in which (−)-**9c** exhibited a slower onset and longer duration of action than did the (−)-**10b**. This indicated the effect of structural rigidity in crossing the blood–brain barrier, as the most constrained (−)-**10b** exhibited an immediate effect in in vivo activity, indicating rapid crossing of the blood–brain barrier. Both compounds occasioned complete cocainelike response in mice trained to discriminate 10 mg/kg cocaine intraperitineally from the vehicle.

In an another effort to introduce an additional substitution in the piperidine ring of the flexible version of GBR analogs, a third functionality added on the central piperidine ring (i.e., a hydroxyl group) led to the development of a series of piperidine-3-ol derivatives represented by the compounds shown in Figure 9-5 [67]. Both the *cis*- and *trans*-3-hydroxy derivatives were synthesized and the racemic trans-compound was

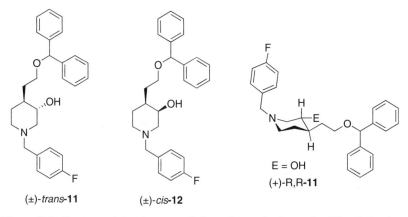

E = OH
(+)-R,R-**11**

(±)-*trans*-**11**   (±)-*cis*-**12**

Figure 9-5   Design and development of *cis*- and *trans*-hydroxy piperidine derivatives.

further resolved into pure enantiomers. The ($\pm$)-*trans*-isomer **11** exhibited a DAT-binding IC_{50} value of 4.14 nM over 11.2 nM for the ($\pm$)-*cis*-**12** isomer (see Table 9-2). The corresponding IC_{50} values for DAT reuptake of [^{3}H]DA were 3.22 and 10.1 nM for the *trans*- and *cis*-isomers, respectively. The two enantiomers of *trans*-**11**, (+)-**11** and (−)-**11**, exhibited marked differential affinities at the DAT, with (+)-**11** being 122-fold more potent than (−)-**11** in inhibiting radiolabeled cocaine analog binding ($IC_{50} = 0.46$ nM vs. 56.7 nM) and nine-fold more active in inhibiting dopamine reuptake ($IC_{50} = 4.05$ nM vs. 38 nM). The compound (+)-**11** was about 22-fold more potent than the standard GBR 12909 at the DAT. In in vivo locomotor activity studies, the ($\pm$)-**11** and (+)-**11** but not (−)-**11** exhibited stimulant activity with a longer duration of effect. All three compounds, the racemate and the enantiomers, within the dose range tested, produced cocainelike responses partially (50%) but incompletely (80%) in mice trained to discriminate 10 mg/kg i.p. cocaine from vehicle. Compound (−)-**11** proved distinct from the (+)-isomer and the racemate, with lack of locomotor activity and the ability to engender (although incompletely) cocainelike responses [67].

9.2.3 Phenylpiperidine Derivatives

Kozikowski et al. [70] revisited molecules belonging to the piperidine-3-carboxylic acid ester class reported originally by Clarke et al. where the authors demonstrated no locomotor stimulatory activity in mice [68,69]. These compounds represent the truncated version of the WIN series of monoamine transporter inhibitors, lacking the ethylene bridge of tropanes [70] (Fig. 9-6). This series of piperidine derivatives preserves the essential pharmacophoric features of the WIN series while enhancing the flexibility of the tropane ring. These piperidine derivatives exhibited substantial activity in both WIN 35,428 binding at the DAT and in the inhibition of [^{3}H]dopamine uptake assays. The *cis*-disubstituted derivative (−)-**13** was about two-fold more potent at the DAT than its trans counterpart (+)-**14** ($IC_{50} =$ 24.8 nM vs. 57.3 nM; see Table 9-3), demonstrating sharp contrast to the data reported in tropane series as the epimerization of the C-2 substituent from β to α has been reported to result in lowering activity 30 to 200-fold [71]. This may be due to the smaller size of the piperidine core relative to tropanes, allowing both the *cis*- and *trans*-isomers to adjust to the binding site on DAT. For *cis*-disubstituted piperidines, a marked enantioselectivity was observed for the (−)-isomer of **13** ($IC_{50} = 85.2$ nM) over the (+)-isomer ($IC_{50} = 5090$ nM), whereas for the trans-disubstituted derivatives, the (+)-isomer was about 5.5-fold more potent in inhibiting dopamine uptake than the (−)-isomer (see Table 9-3). Replacement of the 3-carboxylic acid ester group with 3-alkyl substituent led to an increase in potency for the DAT. The most potent compound (−)-**15** with 3-*n*-propyl substituent exhibited a binding affinity of 3 nM. This derivative was about 33-fold more potent than cocaine in binding to DAT and 29-fold more potent in its inhibition of dopamine uptake. The significant DAT activity and total lack of locomotor activity of these piperidine derivatives may point to their possible usefulness as cocaine antagonists. Further SAR investigations in this series were focused around the steric bulk at the

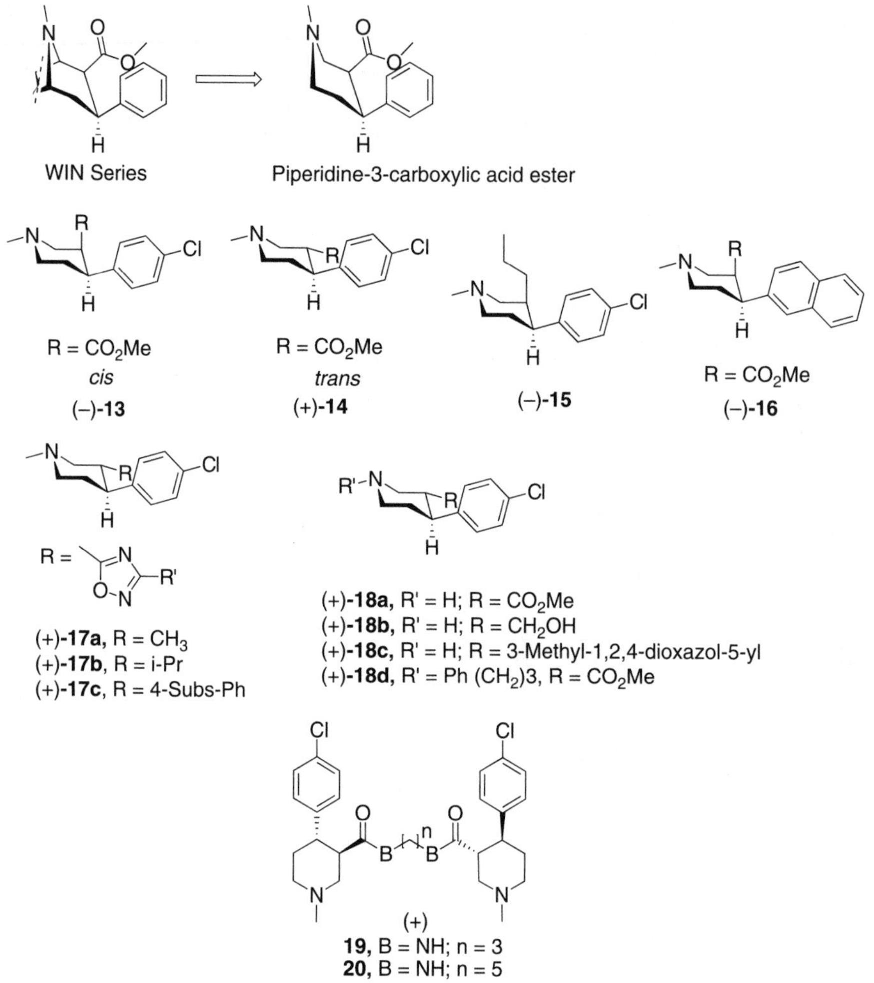

WIN Series

Piperidine-3-carboxylic acid ester

R = CO$_2$Me
cis
(−)-**13**

R = CO$_2$Me
trans
(+)-**14**

(−)-**15**

R = CO$_2$Me
(−)-**16**

R =

(+)-**17a**, R = CH$_3$
(+)-**17b**, R = i-Pr
(+)-**17c**, R = 4-Subs-Ph

(+)-**18a**, R' = H; R = CO$_2$Me
(+)-**18b**, R' = H; R = CH$_2$OH
(+)-**18c**, R' = H; R = 3-Methyl-1,2,4-dioxazol-5-yl
(+)-**18d**, R' = Ph (CH$_2$)3, R = CO$_2$Me

(+)
19, B = NH; n = 3
20, B = NH; n = 5

Figure 9-6   Stereo- and regioisomeric disubstituted phenyl piperidine derivatives derived from tropane structure. Compounds 19 and 20 are representative of bivalent molecules.

4′-position of the 4-phenyl substituent on the piperidine ring [72]. Substitution of the 4-aryl group with 4′-chlorophenyl [compound (−)-**13**] and 2′-naphthyl [compound (−)-**16**, IC$_{50}$ = 23 nM] enhanced potency for the DAT as opposed to the unsubstituted phenyl ring (IC$_{50}$ = 769 nM). Substitution with other groups, such as 1′-naphthyl and 4′-(vinyl, allyl, phenyl, ethyl, cyano)phenyl, led to compounds with lower or no activity at the DAT (see Table 9-3). In an interesting study involving application of the bivalent-ligand concept, novel piperidine-based ligands were synthesized to modulate potency and selectivity at the DAT, SERT, and NET [73]. Compound (+)-**14** was selected as the monomer for the combination of the bivalent ligand design. The representative structure of these bivalent ligands is shown in

Figure 9-6. Overall, only the (+)-isomers exhibited nanomolar affinity for the mazindol binding site on the DAT ($K_i = 8$ to 920 nM) and act as potent DA-reuptake inhibitors ($K_i = 14$ to 341 nM; see Table 9-3). None of the (−)-isomers were highly potent at the DAT. For the (+)-bivalent ligands, increasing the spacer length (number of methylene groups) from 3 (compound **19**, DAT $K_i = 14$ nM) to 8 (DAT $K_i = 341$ nM) resulted in lowering of the DAT and NET potency, whereas a parabolic relationship was observed between the number of methylene groups and the SERT potency with optimum activity when $n = 5$, compound **20**. Compound **20** (DAT $K_i = 39$ nM; SERT $K_i = 7$ nM) exhibits a combined DAT and SERT activity. This compound reduced cocaine's locomotor effects in mice without any effect on locomotion when tested alone. Also, it did not substitute for cocaine in drug discrimination studies in rats at a dose of 1 to 10 mg/kg.

In an attempt to increase the biochemical stability of the piperidine 3-carboxylic acid ester derivatives, bioisosteric replacement of the ester group in (+)-**14** with 3-substituted 1,2,4-oxadiazole moiety resulted in a new series of DAT inhibitors, represented by compounds (+)-**17a–c** in Figure 9-7 [74]. Compound (+)-**17a**, with $R' = CH_3$, which exhibited moderate potency for the DAT (K_i of 201-nM), was the most potent in the series, most closely resembling the lead structure, (+)-**14**. Compound (+)-**17a** exhibited at least two-fold longer duration of action in a locomotor activity test. In case of (+)-isomers, replacement of the 3′-methyl substituent on the oxadiazole ring with groups such as isopropyl, phenyl, 4-substituted phenyl, and so on, resulted in decreased potency at the DAT (see Table 9-3). This is in contrast with the SAR finding for the tropane series: that introduction of the aliphatic and aromatic substituents into the oxadiazole ring leads primarily to changes in selectivity, making compounds more NET or SERT active with mostly constant potency at the DAT [75]. The (−)-isomers of all the synthesized compounds belonging to the 1,2,4-oxadiazole series did not bind to the DAT. Based on these SAR findings and in order to encompass the additional steric restrictions in the pharmacophore models for the DAT-, SERT- and NET-binding sites for piperidine analogs of cocaine with certain 3α substituents, a pharmacophore model featuring a 10-Å conical region extending from the 3-position divided into sterically allowed region from 0 to 5.5 Å and sterically unfavorable or prohibited region from 5.5 to 10 Å has been proposed [74].

Another SAR investigation involving the lead compound (+)-**14** (Fig. 9-6) explored the effect of various alkyl substituents on the piperidine nitrogen [76]. Compared to their N-methyl counterparts, the norpiperidines (compounds (+)-**18a–18c**) showed an increased activity at the SERT and NET with minimal alteration of the DAT potency. Substitution of the piperidine N-atom with phenylalkyl groups [compound **18d** (3-phenylpropyl substitutent), Fig. 9-6] resulted in modest improvements in activity at the SERT, insignificant changes at the NET, and 3.5-fold drop in potency for the DAT ($K_i \geq 810$ nM). Introduction of polar groups in the N-alkyl substituent and removal of piperidine N-atom basicity through formation of amide bonds led to compounds with decreased activity at all the monoamine transporters.

To validate the pharmacophore hypothesis proposed, new series of compounds with different spacer lengths at the 3α-position were prepared [77] (Fig. 9-7).

TABLE 9-3 Affinity, Uptake Inhibitory Activity, and Selectivity of Compounds at the DAT, SERT, and NET

Compound	DAT-Binding IC$_{50}$ (nM)	Uptake^a DAT [^{3}H]DA IC$_{50}$ (nM)	K_i (nM)	SERT [^{3}H]5-HT IC$_{50}$ (nM)	K_i (nM)	NET [^{3}H]NE IC$_{50}$ (nM)	K_i (nM)	SERT/DAT	NET/DAT
Cocaine	102 ± 9[a,b,c]	239 ± 1.0[c]	—						
(−)-13	24.8 ± 1.6[a,b,c]	85.2 ± 2.6[c]							
(+)-14	57.3 ± 8.1[a,b,c]	34.6 ± 3.2[c]							
(−)-15	3.0 ± 0.5[a,b,c]	8.3 ± 0.6[c]							
(−)-16	—	23 ± 1.0[d]	21 ± 0.9[d]	8.2 ± 0.3[d]	7.6 ± 0.2[d]	—	34 ± 0.8[d]	0.36	1.62
(+)-17a	201 ± 2[b,e,f]	—	187.2 ± 3.0[f]	—	5960 ± 80[f]	—	256 ± 16.9[f]	29.5	1.27
(+)-17b	494 ± 103[b,e]	—	391 ± 64.8[f]	—	2920 ± 190[f]	—	778 ± 40.3[f]	7.49	1.99
(+)-17c (R = Ph)	—	—	497.3 ± 95.1[f]	>3000[f]	—	>10000[f]			
(+)-17c (R = 4-Cl/ Br/OMePh)	>3000[b,e,f]	—	>3000[f]	—	>3000[f]	—	>3000[f]		
(+)-18a	—	—	279 ± 98[g]	—	434 ± 50[g]	—	7.9 ± 3[g]	1.56	0.03
(+)-18b	—	—	836 ± 35[g]	—	239 ± 28[g]	—	69 ± 6[g]	0.29	0.08
(+)-18c	—	—	189 ± 24[g]	—	373 ± 4[g]	—	34 ± 6[g]	1.97	0.18
(+)-18d	—	—	810 ± 78[g]	—	3270 ± 160[g]	—	488 ± 30[g]	4.04	0.60
(+)-19	8.4 ± 0.7[e,h,i]	—	14 ± 2.9[i]	—	566 ± 4.1[i]	—	146 ± 8.4[i]	40.43	10.43
(+)-20	103 ± 2.9[e,h,i]	—	39 ± 4.3[i]	—	7 ± 0.6[i]	—	158 ± 15[i]	0.18	4.05
(+)-21	—	—	233 ± 62[j]	—	8490 ± 1430[j]	—	252 ± 43[j]	36	1.1

(+)-22	—	79 ± 8[j]	—	191 ± 13[j]	—	101 ± 18[j]	2.4	1.3
(+)-23	—	68 ± 12[j]	—	255 ± 27[j]	—	31 ± 3[j]	3.8	0.46
(+)-24	—	>3000[f]	—	>3000[f]	—	>3000[f]		
(+)-25	—	44 ± 8[j]	—	32 ± 7[j]	—	52 ± 3[j]	0.73	1.2
(+)-26	—	36 ± 7[j]	—	82 ± 11[j]	—	48 ± 7[j]	2.3	1.3
(+)-27	—	16 ± 1[j]	—	1994 ± 433[j]	—	46 ± 12[j]	125	2.9
(+)-28	—	43 ± 1[j]	—	5188 ± 180[j]	—	356 ± 40[j]	120	8.3
(+)-29	—	16 ± 3[j]	—	2810 ± 256[j]	—	564 ± 4[j]	175	35
30	—	233 ± 62[k]	—	8490 ± 1430[k]	—	252 ± 43[k]	36	1.1
31	—	16 ± 5[k]	—	158 ± 5[k]	—	0.94 ± 0.27[k]	9.88	0.059
32	—	12 ± 1[k]	—	2183 ± 373[k]	—	42 ± 16[k]	182	3.5
33	—	15 ± 5[k]	—	469 ± 36[k]	—	25 ± 6[k]	31	1.67
34	—	9 ± 3[k]	—	87 ± 17[k]	—	29 ± 4[k]	9.67	3.22
35	—	13 ± 3[k]	—	110 ± 45[k]	—	25 ± 2[k]	8.46	1.92
36	—	1 ± 0.2[k]	—	1.1 ± 0.4[k]	—	0.8 ± 0.1[k]	1.1	0.8
37	—	55 ± 8[k]	—	1795 ± 67[k]	—	12 ± 1[k]	32	0.22
38	492 ± 34[e,h,l]	360 ± 25[l]	—	1630 ± 150[l]	—	3860 ± 70[l]	4.53	10.72
39	10.9 ± 1.4[e,h,l]	51 ± 8[l]	—	2380 ± 140[l]	—	177 ± 49[l]	46	3.47
40	—	110 ± 20[m]	—					
41	—	1560 ± 60[n]	—					
42	—	90 ± 10[n]	—					

[a]For uptake by DAT, [3H]DA accumulation was measured; for uptake by SERT, [3H]5-HT accumulation, and for uptake by NET, [3H]NE accumulation, were monitored. [b]For binding, the DAT was labeled with [3H]WIN 35,428. [c]See [70]. [d]See [72]. [e]Represents K_i (nM). [f]See [74]. [g]See [71]. [h]For binding, the DAT was labeled with [3H]mazindol. [i]See [73]. [j]See [77]. [k]See [78]. [l]See [81]. [m]See [82]. [n]See [83].

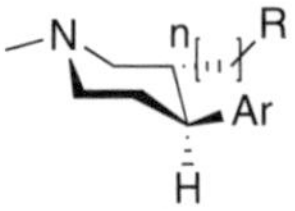

21, Ar = 4-ClPh; n = 0; R = COOMe
22, Ar = 4-ClPh; n = 1; R = COOEt
23, Ar = 4-ClPh; n = 2; R = COOMe
24, Ar = 4-ClPh; n = 0; R = 3-[(1,3-benzodioxol-5-yl)-1,2,4-oxadiazol-5-yl]
25, Ar = 4-ClPh; n = 1; R = 3-[(1,3-benzodioxol-5-yl)-1,2,4-oxadiazol-5-yl]
26, Ar = 4-ClPh; n = 1; R = COO(CH$_2$)$_3$Ph
27, Ar = 4-ClPh; n = 1; R = CONMe$_2$
28, Ar = 4-ClPh; n = 3; R = F
29, Ar = 4-ClPh; n = 3; R = OH

Figure 9-7   Several analogs based on disubstituted phenyl piperidine structure.

Overall, compounds having greater 3α-position flexibility exhibited increased potency at the DAT. Compound **22** with a methylene spacer inserted between the piperidine ring and the ester group showed three-fold gain in DAT potency (K_i = 79 nM) over its no-spacer counterpart [i.e., compound **21** (K_i = 233 nM)]. Further increase in the same spacer length from one to two methylene groups resulted in compound **23** (K_i = 68 nM). Another remarkable increase in DAT potency with the insertion of the spacer was seen with compound **25** (K_i = 44 nM) over compound **24** with no spacer (K_i = >3000 nM). Additional derivatives with increased 3α-flexibility are compounds **26** and **27** with K_i values of 36 and 16 nM, respectively. 3α-Substitution with 3-fluoropropyl (compound **28**) and 3-hydroxypropyl (compound **29**) groups has marked effects on potency and selectivity for DAT. Compound **29** exhibited the highest potency (K_i = 16 nM) and selectivity for the DAT over the SERT (175) and NET (35). Compound **28** with the 3-fluoropropyl substitution resulted in slightly lowered potency (K_i = 43 nM) and selectivity (SERT/DAT = 120, NET/DAT = 8.3). Molecular modeling analysis using molecular superimpositions onto the WIN 35,428 structure as the template confirms the presence of an electrostatic exclusion area (EEA) in addition to the steric exclusion area (SEA) discovered earlier [74]. The elongated or more flexible 3α-substituents adopt a conformation that avoids the EEA and thus results in compounds with substantial potency at the DAT compared to compounds (with no spacer) that would otherwise be inactive. Thus, introducing the flexibility in the 3α-substituent avoids unfavorable interactions with the binding sites of the DAT, SERT, and NET. The presence of EEA may also explain the general observation that the 3β-substituted compounds are more active than the 3α-substituted derivatives. In summary, the flexibility of the 3α-substitutent has a greater impact on the potency than does its distance from the piperidine ring.

In an attempt to design hybrid molecules with varying degrees of potency and selectivity at the monoamine transporters, compound (+)-**30** (nocaine) and modafinil, a non-amphetaminelike wake-promoting agent, were chosen [78]. Nocaine has lower potency and efficacy than cocaine in locomotor tests, although it is fully cocainelike in cocaine-treated animals. Modafinil has a complex mode of action, but it shows lower and selective affinity for the DAT. Figure 9-8 represents the design strategy for hybrid molecules. In particular, the effect of the replacement of the ester group with the methylsulfinylacetamide or methylthiacetamide side chain in the modafinil was explored to reduce the reinforcing properties of nocaine while increasing its half-life. Extensive SAR studies resulted in identification of potent and selective compounds for the DAT, SERT, and NET. Compound **31** with terminal —OH substitution was potent at the NET ($K_i = 0.94$ nM) and DAT ($K_i = 16$ nM). Interestingly, (−)-*cis*-isomer was only twofold less potent at the DAT but 26-fold less potent at the NET [79]. Conversion of sulfide to the sulfoxide (compound **32**) increased potency for the DAT ($K_i = 12$ nM) and substantially reduced potency for the NET ($K_i = 42$ nM). Methylation of the terminal —OH group afforded compound **33** with a K_i value 15 nM, and acetylation provided compound **34** ($K_i = 9$ nM). Among the amide derivatives, compound **35** shows higher potency for the DAT ($K_i = 13$ nM) and compound **36** exhibits 1 nM potency at all three monoamine transporters. The corresponding sulfoxide analogs exhibited reduced potency at DAT, except compound **37** (K_i of 55 nM compared to 85 nM in the sulfide analog). The (−)-and (+)-*cis* analogs of compound **35** exhibited lower potency for the DAT (311 and 215 nM, respectively) and the corresponding (−)- and (+)-*cis* analogs of compound **36** showed DAT K_i values of 13 and 142 nM, respectively (see Table 9-3). Overall, (−)-*cis* and (+)-*trans* analogs exhibited DAT/NET selectivity and (+)-*cis* and (−)-*trans* analogs showed SERT or SERT/NET selectivity [79].

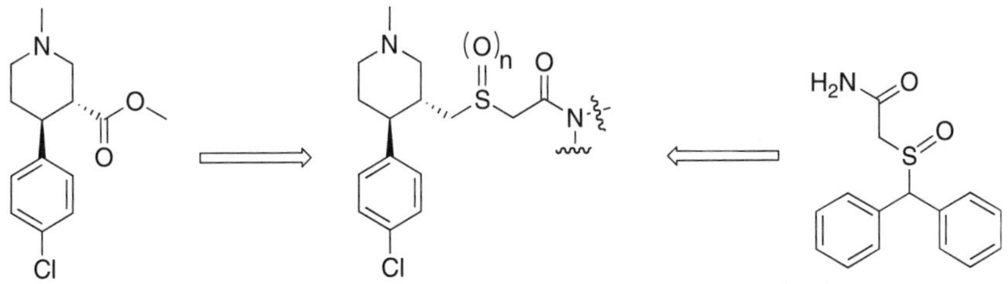

30, Nocaine   Hybrid   Modafinil

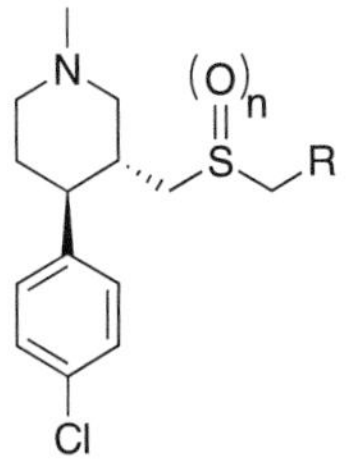

31, R = CH$_2$OH; n = 0
32, R = CH$_2$OH; n = 1
33, R = CH$_2$OMe; n = 1
34, R = CH$_2$OCOCH$_3$; n = 1
35, R = CONHCH$_3$; n = 0
36, R = CONHi-Pr; n = 0
37, R = CONHOH; n = 1

Figure 9-8 Hybrid analogs derived from nocaine and modafinil.

9.3 THREE-DIMENSIONAL PHARMACOPHORE-BASED SEARCH FOR NOVEL DAT INHIBITORS

In an effort to identify novel chemical classes exhibiting DAT potency and selectivity, a three-dimensional pharmacophore was derived based on the three-dimensional structures of cocaine and WIN 35,065-2 [80]. This pharmacophore included features such as the basic nitrogen, carbonyl group, and aromatic ring. The schematic representation of the three-dimensional pharmacophore with the interfeature distances is depicted in Figure 9-9. The pharmacophore-based search of NCI three-dimensional database (206876 structures) identified 4094 initial hits. Further refinement based on additional criteria (i.e., molecular weight, structural diversity) reduced the number of hits to 385. The initial screening using the [^{3}H]mazindol binding assay of 70 (out of 385) compounds identified 13 potential leads exhibiting IC$_{50}$ of <10 μM. These compounds were further evaluated in [^{3}H]DA uptake assay. Compound **38** exhibited reasonable potency at the DAT (binding and uptake K_i values of 492 and 360 nM, respectively), with significant functional antagonism against cocaine and a different in vitro pharmacological profile at the DAT, SERT, and NET. Chemical modifications involving the initial lead structure **38** led to the development of high-affinity analog **39** (binding and uptake K_i values of 11 and 55 nM, respectively; Table 9-3), which mimics partially the effect of cocaine in increasing locomotor activity in mice but lacks the cocainelike discriminative stimulus effect in rats.

Further SAR investigations on the compound **38** led to the synthesis of 16 new analogs with varying functional groups at the 2-, 3-, and 4-positions on each of the two aromatic rings [81]. Out of 16, seven compounds showed reasonable DAT

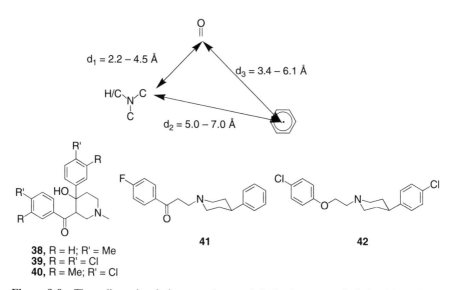

38, R = H; R' = Me
39, R = R' = Cl
40, R = Me; R' = Cl

Figure 9-9   Three-dimensional pharmacophore and the lead compounds derived from three-dimensional pharmacophore search.

potency with K_i values of 0.041 to 0.30 μM and 0.052 to 0.16 μM in [³H]mazindol binding and inhibition of dopamine uptake assays, respectively. Resolution of the racemic mixtures led to pure enantiomers that were evaluated at the DAT, SERT, and NET. In general, (−)-isomers were more potent (in inhibition of DA reuptake) and selective (at the DAT) than their (+)-isomers.

Compound **41** represents another lead structure identified through three-dimensional pharmacophore search which exhibited lower potency at the DAT (K_i = 1.56 μM) but significant functional antagonism. This observation led to SAR studies centered around the N-substituent and the 4-substituents on the piperidine ring of the compound **41** [82]. These studies led to the identification of a reasonably potent analog, compound **42** (K_i = 90 nM), which was 17-fold more potent than the starting lead, compound **41**.

9.4 MEPERIDINE DERIVATIVES

Under some experimental conditions, the binding of cocaine and WIN 35,428 (Fig. 9-1) to the DAT has been reported to exhibit a two-site model consisting of low- and high-affinity components, measured either by binding of radiolabeled ligands to membrane preparations containing the DAT or by inhibition of DA uptake into rat striatal slices [83–85]. There are few DAT inhibitors that discriminate between the low- and high-affinity binding sites on the DAT. Meperidine (**43**, Fig. 9-10), an atypical μ-opioid agonist with some stimulant effect, inhibited [³H]DA uptake into slices to an extent resembling the high-affinity component of the cocaine DA-uptake inhibition curve. Thus, the first plateau of inhibition (20%) observed over a wide concentration range of meperidine was consistent with the high-affinity component of cocaine dopamine-uptake inhibition (∼18%) [86]. In vivo antagonism of the opioid effects of meperidine by the μ-opioid receptor antagonist, naltrexone, resulted in complete substitution of meperidine in cocaine discrimination study in squirrel monkeys trained to discriminate cocaine from saline [86]. These data led to the speculation that meperidine mediates its activity

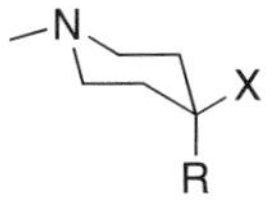

43, Meperidine; X = Ph, R = CO₂Et
44, X = 2-naphthyl, R = CN
45, X = 2-naphthyl, R = CO₂Et
46, X = 3,4-diClPh, R = CO₂Et
47, X = 3,4-diClPh, R = CO₂CH₂Ph
48, X = 3,4-diClPh, R = CH=CH₂
49, X = 3,4-diClPh, R = Et

Figure 9-10 Meperidine analogs developed from SAR studies.

via the high-affinity cocaine-binding site on the DAT as defined in the DA slice uptake experiments.

In follow-up studies, a series of meperidine analogs was synthesized and evaluated [87,88]. The substituents on the phenyl ring had a larger impact on the potency and selectivity of these derivatives for the DAT- and SERT-binding sites as monitored by binding assays. As with meperidine itself, the compounds inhibited [^{3}H]WIN 35,428 monophasically and the following K_i values represent composite affinities at all WIN-binding sites. Overall, meperidine and its analogs were more selective for the SERT compared to the DAT (meperidine, DAT/SERT = 43, Table 9-4) and the esters (compound **45**, DAT K_i = 1.14 μM, SERT K_i = 0.0072 μM) were more potent than their nitrile counterparts (compound **44**, DAT K_i = 2.36 μM, SERT K_i = 0.125 μM). Compound **46** (DAT K_i = 0.125 μM, SERT K_i = 0.019 μM) was the most potent DAT ligand, while compound **45** represented the most potent and selective SERT (DAT/SERT = 158) ligand. Compounds **44** to **46** were more potent at the DAT than meperidine and exhibited dopamine-uptake inhibition curves similar to meperidine [88]. None of the analogs tested produced locomotor effects or substituted for cocaine in drug discrimination studies, possibly due to significant μ-opioid receptor effects. Compounds carrying 3,4-dichlorophenyl and 2-naphthyl moieties were notable in producing two-component inhibition of slice [^{3}H]DA uptake, suggesting that the 3,4-dichlorophenyl and 2-naphthyl moieties in the meperidine system are important for differentiating between the low- and high-affinity components in this uptake system; additionally, these groups play a greater role in enhancing affinity for DAT than for SERT [88].

Further SAR studies centered around compound **46** led to several meperidine analogs varying in potency and selectivity at the DAT and SERT [89]. Slight changes in the ester group at the 4-position of the piperidine ring in compound **46** resulted in significant loss of affinity at the DAT. Moderate to large excess of steric bulk (increase in the size of the ester chain beyond four carbons) was detrimental for the DAT affinity, whereas increase in the size of the ester chain of compound **46** led to increased affinity at the SERT. Compound **47** (SERT K_i = 3.9 nM, DAT/SERT = 760) represents the most SERT-selective meperidine analog reported to date. The N-alkyl-substituted meperidine analogs were less potent at the DAT, corresponding to the N-methyl derivative compound **46**. All other modifications of the ester group resulted in less potent compounds at the DAT and SERT compared to compound **46** except that compounds **48** (SERT K_i = 9 nM) and **49** (SERT K_i = 11 nM) exhibited about two-fold increase in potency and SERT selectivity over the parent compound **46**.

9.5   METHYLPHENIDATE ANALOGS

Methylphenidate was first synthesized in 1944 and was recognized as a stimulant in 1954 [90]. Structure–activity relationship studies around methylphenidate focused primarily on the modifications involving the piperidine and the phenyl rings. Of the two diastereomeric pairs, DL-threo-methylphenidate (compound **50**, Fig. 9-11)

TABLE 9-4 Affinity and Selectivity of the Meperidine Derivatives at the Dopamine, Serotonin, and Norepinephrine Transporters

Compound	DAT, K_i (μM), [3H]WIN 35,428[a]	SERT, K_i (μM), [3H]paroxetine[b]	NET, K_i (μM), [3H]nisoxetine[c]	[3H] DAT Uptake Inhibition, IC_{50} (μM)	DAT/ SERT
43	17.8 ± 2.7[d]	0.413 ± 0.044[d]	—	0.61 ± 2.2 nM (22%)[e]; 12.6 ± 1.2[d]	43
44	2.36 ± 0.66[d]	0.125 ± 0.022[d]	88.1[d,f]	0.37 ± 2.0 nM (11%)[e]; 21.8 ± 1.2[d]	19
45	1.14 ± 0.38[d]	0.0072 ± 0.0001[d]	71.1 ± 9.7[d]	3.19 ± 7.41 nM (16%)[e]; 11.6 ± 1.3[d]	158
46	0.125 ± 0.015[f]	0.0187 ± 0.0026[d]	74.5 ± 5.1[d]	6.01 ± 6.26 nM (22%)[e]; 1.40 ± 1.25[d]	6.7
47	2.970 ± 0.3[g]	0.0039 ± 0.0005[g]	—	—	760
48	0.192 ± 0.031[g]	0.009 ± 0.002[g]	—	—	21
49	0.163 ± 0.038[g]	0.011 ± 0.001[g]	—	—	15

[a]The DAT was labeled with [3H]WIN 35,428 (inhibition was monophasic and K_i values shown represent composite affinities for all WIN binding sites).
[b]The SERT was labeled with [3H]paroxetine.
[c]The NET was labeled with [3H]nisoxetine.
[d]See [89].
[e]Percent of total uptake inhibition.
[f]Percent specific binding at highest dose tested (10 μM).
[g]See [90].

253

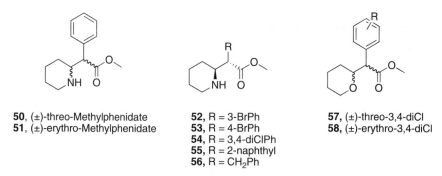

50, (±)-threo-Methylphenidate
51, (±)-erythro-Methylphenidate

52, R = 3-BrPh
53, R = 4-BrPh
54, R = 3,4-diClPh
55, R = 2-naphthyl
56, R = CH₂Ph

57, (±)-threo-3,4-diCl
58, (±)-erythro-3,4-diCl

Figure 9-11   Different methylphenidate analogs developed from SAR studies.

is more potent than the DL-erythro pair (compound **51**) at the DAT. Between the two enantiomers of the DL-threo pair, the D-isomer is about 16 times more potent than the L-isomer at the DAT (see Table 9-5). Among the analogs of methylphenidate bearing substituents at the 2-, 3-, and 4-positions on the phenyl ring, the 2-substituted derivatives were far less potent than those bearing 3- and 4-substituents. This effect appears to be steric in nature, as the increase in the size of the substituent leads to a decrease in the DAT potency [91,92]. The 3- and 4-halo-substituted derivatives were the most active in the series, with bromo-substituted compounds **52** and **53** being the most potent. The 3-substituted derivatives were similar or greater in potency than their 4-substituted counterparts. The 3,4-disubstituted compounds, represented by **54**,

TABLE 9-5   Affinity of the Methylphenidate Derivatives at the Dopamine Transporter[a]

Compound	DAT, IC$_{50}$ (nM), [³H]WIN 35,428 Binding	[³H]DA Uptake Inhibition, IC$_{50}$ (nM)
50	83.0 ± 7.9[b]	224 ± 19[b]
D-Threo-methylphenidate	33	
L-Threo-methylphenidate	540	
52	4.2 ± 0.2[b]	12.8 ± 0.20[b]
53	6.9 ± 0.1[b]	26.3 ± 5.8[b]
54	5.3 ± 0.7[b]	7.0 ± 0.6[b]
55	11 ± 2.5[c]	53 ± 8.0[c]
56	4470 ± 1200[c]	8450 ± 210[c]
57	29.1 ± 5.5[d]	
58	286 ± 10.5[d]	

[a]The DAT was labeled with [³H]WIN 35,428. For uptake by DAT, [³H]DA accumulation was measured. Results are average ± SEM of two to five independent experiments assayed in triplicate.
[b]See [92].
[c]See [95].
[d]See [97].

were similar or less potent than the 3-substituted analogs [91]. N-Methylation of the piperidine ring nitrogen led to 4- to 30-fold less potent compounds at the DAT, suggesting the detrimental effect of the equatorial *N*-methyl substituent toward DAT potency [93].

Replacement of the phenyl ring in methylphenidate by naphthyl groups increased activity at the DAT. Compound **55** with a 2-naphthyl substituent exhibited seven-fold increased binding over the parent compound methylphenidate [94]. The 1-naphthyl analog of compound **55** was much less active. Insertion of a methylene spacer between the phenyl ring and the rest of the methylphenidate molecule, represented by compound **56**, attenuated the binding affinity significantly [95]. Modifications involving the size of the piperidine ring led to a 5- to 10-fold decrease in DAT-binding affinity over methylphenidate [94,95]. Conversion of the piperidine ring into tetrahydropyran moiety resulted in decreased potency, except for compounds **57** and **58**. The threo pair was 10-fold more potent than the erythro pair. The (−)-threo isomer was twice as potent as the (+)-threo isomer, thus exhibiting moderate enantioselectivity [96].

In analogy to the in vitro binding data, D-threomethylphenidate was more active than the L-threo isomer in the locomotor activity testing [97]. When administered intravenously in laboratory animals, methylphenidate appears to have reinforcing effects equivalent to those of cocaine [51]. Its binding affinity for the DAT is almost two-fold greater than that of cocaine. This correlates with its in vivo potency to block dopamine transporter in the human brain. However, it appears to have substantially lower abuse liability in humans [52], possibly due to its longer retention in vivo, owing to its prolonged half-life (90 minutes). Similarly, for several analogs of methylphenidate, a positive correlation was observed between their DAT potency and the extent of generalization in the cocaine drug discrimination assays [98].

9.6 CoMFA STUDY OF THE PIPERIDINE ANALOGS OF COCAINE

A three-dimensional QSAR study using comparative molecular field analysis (CoMFA) of the piperidine analogs of cocaine with flexible 3α-substituents was performed to understand the details of the monoamine transporter–binding sites that accommodate piperidine-based ligands [99]. Previous CoMFA studies focused on the aryltropanes emphasized the favoring of increased negative electrostatic potential in the regions around the 3β-substituent of the tropane ring and the 4'-substituent on the phenyl ring for higher potency in the inhibition of monoamine transporters [100–102]. Recent studies on aryltropanes and piperidinols proposed that the DAT and SERT have a large cavity that can accommodate bulky C-2 substituents of tropanes [103,104], and the size of the substituents at the *para*-position in both phenyl rings of piperidinols is important for inhibition of DA reuptake [81]. The CoMFA model proposed by Lieske et al. indicated that the DAT-binding site of the 3β-substituent of tropanes is barrel-shaped, and hydrophobic interactions make a dominant contribution to the binding [105].

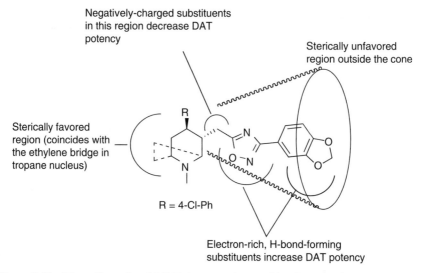

Figure 9-12 Three-dimensional QSAR interpretation resulting from the CoMFA analysis of piperidine analogs of cocaine. The dashed line represents the ethylene bridge present in the tropane nucleus, which coincides with the sterically favored region in the three-dimensional QSAR model. (From Yuan et al. [99].)

For the overlay of the compounds used in the CoMFA study [99], a pharmacophore-based superposition of the fittest conformers was developed and applied successfully. The data set of 42 *trans*-series compounds divided into training (36 compounds) and test (six compounds) sets were used for building two highly predicted CoMFA models. Both models emphasize the importance of steric and electrostatic interactions in the binding of the 3α-substituent of the piperidine-based analogs of cocaine at the DAT. Since two CoMFA models were obtained, there is a possibility of the existence of multiple binding modes for the 3α-substituent. One of these CoMFA models exhibited a large sterically favorable area located close to the piperidine ring (Fig. 9-12). This may partly explain the high activity of tropane-based molecules compared to the piperidine analogs since the ethylene bridge (shown as dashed bonds in Fig. 9-12) in tropane-based compounds would be positioned close to this sterically favorable area. The sterically unfavorable area coincides with the steric exclusion area (SEA) described earlier [74]. Electrostatic contour maps indicate the presence of a positive-charge-favored area near the first atom of the 3α-substituent. The presence of negatively charged substituents near these areas would decrease the DAT potency. The presence of electron density–favoring regions around the 3α-substituent (from the third to the sixth atom) may suggest that DAT has several H-bond donor sites in this area (Fig. 9-12). The binding affinity can be increased by introducing electron-rich groups in this area. In summary, these findings would direct the design and improvement of compounds exhibiting high potency at the DAT.

REFERENCES

1. Carlsson, A. (1987). Perspectives on the discovery of central monoaminergic neurotransmission. *Annu Rev Neurosci*, *10*, 19–40.

2. Greengard, P. (2001). The neurobiology of slow synaptic transmission. *Science*, *294*, 1024–1030.

3. Roth, R. H., and Elsworth, J. D. (1995). In *Psychopharmacology: The Fourth Generation of Progress*, Bloom, F. E., and Kupfer, D. J., Eds. Raven Press, New York, pp. 227–243.

4. Dougherty, D., Bonab, A., Spencer, T., Rauch, S., Madras, B., and Fischman, A. (1999). Dopamine transporter density in patients with attention deficit hyperactivity disorder. *Lancet*, *354*, 2132–2133.

5. Kim, H., Im, J., Yang, S., Moon, D., Ryu, J., Bong, J., Nam, K., Cheon, J., Lee, M., and Lee, H. (1997). Imaging and quantitation of dopamine transporters with iodine-123-IPT in normal and Parkinson's disease subjects. *J Nucl Med*, *38*, 1703–1711.

6. Amara, S., and Kuhar, M. (1993). Neurotransmitter transporters: recent progress. *Annu Rev Neurosci*, *16*, 73–93.

7. Giros, B., and Caron, M. (1993). Molecular characterization of the dopamine transporter. *Trends Pharmacol Sci*, *14*, 43–49.

8. Blakely, R., De Felice, L., and Hartzell, H. (1994). Molecular physiology of norepinephrine and serotonin transporters. *J Exp Biol*, *196*, 263–281.

9. Hahn, M., and Blakely, R. (2002). Monoamine transporter gene structure and polymorphisms in relation to psychiatric and other complex disorders. *Pharmacogenom J*, *2*, 217–235.

10. Heinz, A., Mann, K., Weinberger, D., and Goldman, D. (2001). Serotonergic dysfunction, negative mood states, and response to alcohol. *Alcohol Clin Exp Res*, *25*, 487–495.

11. Klimek, V., Stockmeier, C., Overholser, J., Meltzer, H., Kalka, S., Dilley, G., and Ordway, G. (1997). Reduced levels of norepinephrine transporters in the locus coeruleus in major depression. *J Neurosci*, *17*, 8451–8458.

12. Robertson, D., Flattem, N., Tellioglu, T., Carson, R., Garland, E., Shannon, J., Jordan, J., Jacob, G., Blakely, R., and Biaggioni, I. (2001). Familial orthostatic tachycardia due to norepinephrine transporter deficiency. *Ann N Y Acad Sci*, *940*, 527–543.

13. Miller, G., De La Garza, R., Novak, M., and Madras, B. (2001). Single nucleotide polymorphisms distinguish multiple dopamine transporter alleles in primates: Implications for association with attention deficit hyperactivity disorder and other neuropsychiatric disorders. *Mol Psychiatry*, *6*, 50–58.

14. Ozaki, N., Goldman, D., Kaye, W., Plotnicov, K., Greenberg, B., Lappalainen, J., Rudnick, G., and Murphy, D. (2003). Serotonin transporter missense mutation associated with a complex neuropsychiatric phenotype. *Mol Psychiatry*, *8*, 933–936.

15. Barker, E. L., and Blakely, R. D. (1995). In *Psychopharmacology: The Fourth Generation of Progress*, Bloom, F. E., and Kupfer, D. J., Eds. Raven Press, New York, pp. 321–333.

16. Miller, G., Gainetdinov, R., Levey, A., and Caron, M. (1999). Dopamine transporters and neuronal injury. *Trends Pharmacol Sci*, *20*, 424–429.

17. Reith, M., Xu, C., and Chen, N. (1997). Pharmacology and regulation of the neuronal dopamine transporter. *Eur J Pharmacol*, *324*, 1–10.

18. Torres, G., Gainetdinov, R., and Caron, M. (2003). Plasma membrane monoamine transporters: structure, regulation and function. *Nat Rev Neurosci, 4*, 13–25.

19. Yamashita, A., Singh, S., Kawate, T., Jin, Y., and Gouaux, E. (2005). Crystal structure of a bacterial homologue of Na^+/Cl^--dependent neurotransmitter transporters. *Nature, 437,* 215–223.

20. Rudnick, G., and Clark, J. (1993). From synapse to vesicle: the reuptake and storage of biogenic amine neurotransmitters. *Biochim Biophys Acta, 1144*, 249–263.

21. Gu, H., Wall, S., and Rudnick, G. (1994). Stable expression of biogenic amine transporters reveals differences in inhibitor sensitivity, kinetics, and ion dependence. *J Biol Chem, 269*, 7124–7130.

22. Johanson, C.-E., and Schuster, C. R. (1995). In *Psychopharmacology: The Fourth Generation of Progress*, Bloom, F. E., and Kupfer, D. J., Eds. Raven Press, New York, pp. 1685–1697.

23. Kuhar, M., Ritz, M., and Boja, J. (1991). The dopamine hypothesis of the reinforcing properties of cocaine. *Trends Neurosci, 14*, 299–302.

24. Robinson, T., and Berridge, K. (1993). The neural basis of drug craving: an incentive-sensitization theory of addiction. *Brain Res Rev, 18*, 247–291.

25. Ritz, M., Lamb, R., Goldberg, S., and Kuhar, M. (1987). Cocaine receptors on dopamine transporters are related to self-administration of cocaine. *Science, 237*, 1219–1223.

26. Di Chiara, G., and Imperato, A. (1988). Drugs abused by humans preferentially increase synaptic dopamine concentrations in the mesolimbic system of freely moving rats. *Proc Nat Acad Sci U S A, 85*, 5274–5278.

27. Carboni, E., Spielewoy, C., Vacca, C., Nosten-Bertrand, M., Giros, B., and Di Chiara, G. (2001). Cocaine and amphetamine increase extracellular dopamine in the nucleus accumbens of mice lacking the dopamine transporter gene. *J Neurosci, 21*(RC141), 141–144.

28. Rocha, B., Fumagalli, F., Gainetdinov, R., Jones, S., Ator, R., Giros, B., Miller, G., and Caron, M. (1998). Cocaine self-administration in dopamine-transporter knockout mice. *Nat Neurosci, 1*, 132–137.

29. Yamamoto, B., and Novotney, S. (1998). Regulation of extracellular dopamine by the norepinephrine transporter. *J Neurochemi, 71*, 274–280.

30. Eshleman, A., Carmolli, M., Cumbay, M., Martens, C., Neve, K., and Janowsky, A. (1999). Characteristics of drug interactions with recombinant biogenic amine transporters expressed in the same cell type. *J Pharmacol Exp Ther, 289*, 877–885.

31. Chen, R., Tilley, M., Wei, H., Zhou, F., Zhou, F., Ching, S., Quan, N., Stephens, R. L., Hill, E., Nottoli, T., Han, D., and Gu, H. (2006). Abolished cocaine reward in mice with a cocaine-insensitive dopamine transporter. *Proc Nat Acad Sci U S A, 103*, 9333–9338.

32. Volkow, N., Wang, G., Fischman, M., Foltin, R., Fowler, J., Abumrad, N., Vitkun, S., Logan, J., Gatley, S., Pappas, N., Hitzemann, R., and Shea, C. (1997). Relationship between subjective effects of cocaine and dopamine transporter occupancy. *Nature, 386*, 827–830.

33. Smith, M., Hoepping, A., Johnson, K., Trzcinska, M., and Kozikowski, A. (1999). Dopaminergic agents for the treatment of cocaine abuse. *Drug Discov Today, 4*, 322–332.

34. Giros, B., Jaber, M., Jones, S., Wightman, R., and Caron, M. (1996). Hyperlocomotion and indifference to cocaine and amphetamine in mice lacking the dopamine transporter. *Nature, 379*, 606–612.

35. Missale, C., Castelletti, L., Govoni, S., Spano, P., Trabucchi, M., and Hanbauer, I. (1985). Dopamine uptake is differentially regulated in rat striatum and nucleus accumbens. *J Neurochem, 45*, 51–56.

36. McElvain, J., and Schenk, J. (1992). A multisubstrate mechanism of striatal dopamine uptake and its inhibition by cocaine. *Biochem Pharmacol, 43*, 2189–2199.

37. Wayment, H., Meiergerd, S., and Schenk, J. (1998). Relationships between the catechol substrate binding site and amphetamine, cocaine, and mazindol binding sites in a kinetic model of the striatal transporter of dopamine in vitro. *J Neurochem, 70*, 1941–1949.

38. Wu, Q., Reith, M., Kuhar, M., Carroll, F., and Garris, P. (2001). Preferential increases in nucleus accumbens dopamine after systemic cocaine administration are caused by unique characteristics of dopamine neurotransmission. *J Neurosci, 21*, 6338–6347.

39. Giros, B., Wang, Y., Suter, S., McLeskey, S., Pifl, C., and Caron, M. (1994). Delineation of discrete domains for substrate, cocaine, and tricyclic antidepressant interactions using chimeric dopamine–norepinephrine transporters. *J Biol Chem, 269*, 15985–15988.

40. Reith, M., de Costa, B., Rice, K., and Jacobson, A. (1992). Evidence for mutually exclusive binding of cocaine, BTCP, GBR 12935, and dopamine to the dopamine transporter. *Eur J Pharmacol, 227*, 417–425.

41. Appell, M., Berfield, J., Wang, L., Dunn, W., 3rd, Chen, N., and Reith, M. (2004). Structure–activity relationships for substrate recognition by the human dopamine transporter. *Biochem Pharmacol, 67*, 293–302.

42. Rothman, R. (1990). High affinity dopamine reuptake inhibitors as potential cocaine antagonists: a strategy for drug development. *Life Sci, 46*, PL17–PL21.

43. Katz, J., Kopajtic, T., Agoston, G., and Newman, A. (2004). Effects of N-substituted analogs of benztropine: diminished cocaine-like effects in dopamine transporter ligands. *J Pharmacol Exp Ther, 309*, 650–660.

44. Reith, M., Berfield, J., Wang, L., Ferrer, J., and Javitch, J. (2001). The uptake inhibitors cocaine and benztropine differentially alter the conformation of the human dopamine transporter. *J Biol Chem, 276*, 29012–29018.

45. Singh, S. (2000). Chemistry, design, and structure–activity relationship of cocaine antagonists. *Chem Rev, 100*, 925–1024.

46. Carroll, F. I., Howell, L. L., and Kuhar, M. J. (2002). In *Neurotransmitter Transporters: Structure, Function and Regulation*, 2nd ed., Reith, M. E. A., Ed. Humana Press, Totowa, NJ, pp. 381–482.

47. Dutta, A., Zhang, S., Kolhatkar, R., and Reith, M. (2003). Dopamine transporter as target for drug development of cocaine dependence medications. *Eur J Pharmacol, 479*, 93–106.

48. Carroll, F. (2003). Monoamine transporters and opioid receptors: targets for addiction therapy. *J Med Chem, 46*, 1775–1794.

49. Newman, A., and Kulkarni, S. (2002). Probes for the dopamine transporter: new leads toward a cocaine-abuse therapeutic-A focus on analogues of benztropine and rimcazole. *Med Res Rev, 22*, 429–464.

50. Dutta, A., Meltzer, P., and Madras, B. (1993). Positional importance of the nitrogen atom in novel piperidine analogs of GBR 12909: affinity and selectivity for the dopamine transporter. *Med Chem Res, 3*, 209–222.

51. Bergman, J., Madras, B., Johnson, S., and Spealman, R. D. (1989). Effects of cocaine and related drugs in nonhuman primates. III: self-administration by squirrel monkeys. *J Pharmacol Exp Ther, 251,* 150–155.

52. Volkow, N., Wang, G., Fowler, J., Fischman, M., Foltin, R., Abumrad, N., Gatley, S., Logan, J., Wong, C., Gifford, A., Ding, Y., Hitzemann, R., and Pappas, N. (1999). Methylphenidate and cocaine have a similar in vivo potency to block dopamine transporters in the human brain. *Life Sci, 65,* PL7–PL12.

53. Javitch, J., Blaustein, R., and Snyder, S. (1984). [^{3}H]mazindol binding associated with neuronal dopamine and norepinephrine uptake sites. *Mol Pharmacol, 26,* 35–44.

54. Angel, I., Taranger, M., Claustre, Y., Scatton, B., and Langer, S. (1988). Anorectic activities of serotonin uptake inhibitors: correlation with their potencies at inhibiting serotonin uptake in vivo and [^{3}H]-mazindol binding in vitro. *Life Sci, 43,* 651–658.

55. Chait, L., Uhlenhuth, E., and Johanson, C. (1987). Reinforcing and subjective effects of several anorectics in normal human volunteers. *J Pharmacol Exp Ther, 242,* 777–783.

56. Vignon, J., Cerruti, C., Chaudieu, I., Pinet, V., Chicheportiche, M., and Kamenka, J.-M. (1998). In *Sigma and Phencyclidine-like Compounds as Molecular Probes in Biology,* Domino, E. F., and Kamenka, J.-M., Eds. NPP Books. Ann Arbor, MI, pp. 199–208.

57. Koek, W., Colpaert, F., Woods, J., and Kamenka, J. (1989). The phencyclidine (PCP) analog *N*-[1-(2-benzo[*b*]thiophenyl)cyclohexyl]piperidine shares cocaine-like but not other characteristic behavioral effects with PCP, ketamine and MK-801. *J Pharmacol Exp Ther, 250,* 1019–1027.

58. Madras, B., Reith, M., Meltzer, P., and Dutta, A. (1994). O-526, a piperidine analog of GBR 12909, retains high affinity for the dopamine transporter in monkey caudate-putamen. *Eur J Pharmacol, 267,* 167–173.

59. Dutta, A., Coffey, L., and Reith, M. (1997). Highly selective, novel analogs of 4-[2-(diphenylmethoxy)ethyl]-1-benzylpiperidine for the dopamine transporter: effect of different aromatic substitutions on their affinity and selectivity. *J Med Chem, 40,* 35–43.

60. Dutta, A., Xu, C., and Reith, M. (1996). Structure–activity relationship studies of novel 4-[2-[bis(4-fluorophenyl)methoxy]ethyl]-1-(3-phenylpropyl)piperidine analogs: synthesis and biological evaluation at the dopamine and serotonin transporter sites. *J Med Chem, 39,* 749–756.

61. Dutta, A., Xu, C., and Reith, M. (1998). Tolerance in the replacement of the benzhydrylic O atom in 4-[2-(diphenylmethoxy)ethyl]-1-benzylpiperidine derivatives by an N atom: development of new-generation potent and selective N-analogue molecules for the dopamine transporter. *J Med Chem, 41,* 3293–3297.

62. Dutta, A., Fei, X., Beardsley, P., Newman, J., and Reith, M. (2001). Structure–activity relationship studies of 4-[2-(diphenylmethoxy)ethyl]-1-benzylpiperidine derivatives and their *n*-analogues: evaluation of O-and N-analogues and their binding to monoamine transporters. *J Med Chem, 44,* 937–948.

63. Dutta, A., Davis, M., and Reith, M. (2001). Rational design and synthesis of novel 2,5-disubstituted *cis-* and *trans-*piperidine derivatives exhibiting differential activity for the dopamine transporter. *Bioorg Med Chem Lett, 11,* 2337–2340.

64. Kolhatkar, R., Ghorai, S., George, C., Reith, M., and Dutta, A. (2003). Interaction of *cis-* (6-benzhydrylpiperidin-3-yl)benzylamine analogues with monoamine transporters: structure–activity relationship study of structurally constrained 3,6-disubstituted

piperidine analogues of (2,2-diphenylethyl)-[1-(4-fluorobenzyl)piperidin-4-ylmethyl]amine. *J Med Chem*, *46*, 2205–2215.

65. Kolhatkar, R., Cook, C., Ghorai, S., Deschamps, J., Beardsley, P., Reith, M., and Dutta, A. (2004). Further structurally constrained analogues of *cis*-(6-benzhydrylpiperidin-3-yl)benzylamine with elucidation of bioactive conformation: discovery of 1,4-diazabicyclo[3.3.1]nonane derivatives and evaluation of their biological properties for the monoamine transporters. *J Med Chem*, *47*, 5101–5113.

66. Hruby, V. (1982). Conformational restrictions of biologically active peptides via amino acid side chain groups. *Life Sci*, *31*, 189–199.

67. Ghorai, S., Cook, C., Davis, M., Venkataraman, S., George, C., Beardsley, P., Reith, M., and Dutta, A. (2003). High affinity hydroxypiperidine analogues of 4-(2-benzhydryloxyethyl)-1-(4-fluorobenzyl)piperidine for the dopamine transporter: stereospecific interactions in vitro and in vivo. *J Med Chem*, *46*, 1220–1228.

68. Clarke, R., Gambino, A., Pierson, A., and Daum, S. (1978). (2-exo-3-endo)-2-Aryltropane-3-carboxylic esters, a new class of narcotic antagonists. *J Med Chem*, *21*, 1235–1242.

69. Clarke, R., Daum, S., Gambino, A., Aceto, M., Pearl, J., Levitt, M., Cumiskey, W. R., and Bogado, E. (1973). Compounds affecting the central nervous system. 4: 3β-Phenyltropane-2-carboxylic esters and analogs. *J Med Chem*, *16*, 1260–1267.

70. Kozikowski, A., Araldi, G., Boja, J., Meil, W., Johnson, K., Flippen-Anderson, J., George, C., and Saiah, E. (1998). Chemistry and pharmacology of the piperidine-based analogues of cocaine. Identification of potent DAT inhibitors lacking the tropane skeleton. *J Med Chem*, *41*, 1962–1969.

71. Carroll, F., Lewin, A., Boja, J., and Kuhar, M. (1992). Cocaine receptor: biochemical characterization and structure-activity relationships of cocaine analogues at the dopamine transporter. *J Med Chem*, *35*, 969–981.

72. Tamiz, A., Zhang, J., Flippen-Anderson, J., Zhang, M., Johnson, K., Deschaux, O., Tella, S., and Kozikowski, A. (2000). Further SAR studies of piperidine-based analogues of cocaine. 2: potent dopamine and serotonin reuptake inhibitors. *J Med Chem*, *43*, 1215–1222.

73. Tamiz, A., Bandyopadhyay, B., Zhang, J., Flippen-Anderson, J., Zhang, M., Wang, C., Johnson, K., Tella, S., and Kozikowski, A. (2001). Pharmacological and behavioral analysis of the effects of some bivalent ligand-based monoamine reuptake inhibitors. *J Med Chem*, *44*, 1615–1622.

74. Petukhov, P., Zhang, M., Johnson, K., Tella, S., and Kozikowski, A. (2001). SAR studies of piperidine-based analogues of cocaine. Part 3: oxadiazoles. *Bioorg Med Chem Lett*, *11*, 2079–2083.

75. Carroll, F., Gray, J., Abraham, P., Kuzemko, M., Lewin, A., Boja, J., and Kuhar, M. (1993). 3-Aryl-2-(3′-substituted-1′,2′,4′-oxadiazol-5′-yl)tropane analogues of cocaine: affinities at the cocaine binding site at the dopamine, serotonin, and norepinephrine transporters. *J Med Chem*, *36*, 2886–2890.

76. Petukhov, P., Zhang, J., Kozikowski, A., Wang, C., Ye, Y., Johnson, K., and Tella, S. (2002). SAR studies of piperidine-based analogues of cocaine. 4: effect of N-modification and ester replacement. *J Med Chem*, *45*, 3161–3170.

77. Petukhov, P., Zhang, J., Wang, C., Ye, Y., Johnson, K., and Kozikowski, A. (2004). Synthesis, molecular modeling, and biological studies of novel piperidine-based

analogues of cocaine: evidence of unfavorable interactions proximal to the 3α-position of the piperidine ring. *J Med Chem, 47,* 3009–3018.

78. Zhou, J., He, R., Johnson, K., Ye, Y., and Kozikowski, A. (2004). Piperidine-based nocaine/modafinil hybrid ligands as highly potent monoamine transporter inhibitors: efficient drug discovery by rational lead hybridization. *J Med Chem, 47,* 5821–5824.

79. He, R., Kurome, T., Giberson, K., Johnson, K., and Kozikowski, A. (2005). Further structure–activity relationship studies of piperidine-based monoamine transporter inhibitors: effects of piperidine ring stereochemistry on potency. Identification of norepinephrine transporter selective ligands and broad-spectrum transporter inhibitors. *J Med Chem, 48,* 7970–7979.

80. Wang, S., Sakamuri, S., Enyedy, I., Kozikowski, A., Deschaux, O., Bandyopadhyay, B., Tella, S., Zaman, W., and Johnson, K. (2000). Discovery of a novel dopamine transporter inhibitor, 4-hydroxy-1-methyl-4-(4-methylphenyl)-3-piperidyl 4-methylphenyl ketone, as a potential cocaine antagonist through 3D-database pharmacophore searching: molecular modeling, structure–activity relationships, and behavioral pharmacological studies. *J Med Chem, 43,* 351–360.

81. Wang, S., Sakamuri, S., Enyedy, I., Kozikowski, A., Zaman, W., and Johnson, K. (2001). Molecular modeling, structure–activity relationships and functional antagonism studies of 4-hydroxy-1-methyl-4-(4-methylphenyl)-3-piperidyl 4-methylphenyl ketones as a novel class of dopamine transporter inhibitors. *Bioorg Med Chem, 9,* 1753–1764.

82. Sakamuri, S., Enyedy, I., Kozikowski, A., Zaman, W., Johnson, K., and Wang, S. (2001). Pharmacophore-based discovery, synthesis, and biological evaluation of 4-phenyl-1-arylalkyl piperidines as dopamine transporter inhibitors. *Bioorg Med Chem Lett, 11,* 495–500.

83. Madras, B., Fahey, M., Bergman, J., Canfield, D., and Spealman, R. (1989). Effects of cocaine and related drugs in nonhuman primates. I: [³H]cocaine binding sites in caudate–putamen. *J Pharmacol Exp Ther, 251,* 131–141.

84. Madras, B., Spealman, R., Fahey, M., Neumeyer, J., Saha, J., and Milius, R. (1989). Cocaine receptors labeled by [³H]-2-β-carbomethoxy-3-β-(4-fluorophenyl)tropane. *Mol Pharmacol, 36,* 518–524.

85. Izenwasser, S., Newman, A., and Katz, J. (1993). Cocaine and several sigma receptor ligands inhibit dopamine uptake in rat caudate–putamen. *Eur J Pharmacol, 243,* 201–205.

86. Izenwasser, S., Newman, A., Cox, B., and Katz, J. (1996). The cocaine-like behavioral effects of meperidine are mediated by activity at the dopamine transporter. *Eur J Pharmacol, 297,* 9–17.

87. Lomenzo, S., Izenwasser, S., Gerdes, R., Katz, J., Kopajtic, T., and Trudell, M. (1999). Synthesis, dopamine and serotonin transporter binding affinities of novel analogues of meperidine. *Bioorg Med Chem Lett, 9,* 3273–3276.

88. Lomenzo, S., Rhoden, J., Izenwasser, S., Wade, D., Kopajtic, T., Katz, J., and Trudell, M. (2005). Synthesis and biological evaluation of meperidine analogues at monoamine transporters. *J Med Chem, 48,* 1336–1343.

89. Rhoden, J., Bouvet, M., Izenwasser, S., Wade, D., Lomenzo, S., and Trudell, M. (2005). Structure–activity studies of 3'-4'-dichloromeperidine analogues at dopamine and serotonin transporters. *Bioorg Med Chem, 13,* 5623–5634.

90. Meier, R., Gross, F., and Tripod, J. (1954). Ritalin, a new synthetic compound with specific analeptic components. *Klin Wochenschr*, *32*, 445–450.

91. Deutsch, H., Shi, Q., Gruszecka-Kowalik, E., and Schweri, M. (1996). Synthesis and pharmacology of potential cocaine antagonists. 2: structure–activity relationship studies of aromatic ring-substituted methylphenidate analogs. *J Med Chem*, *39*, 1201–1209.

92. Gatley, S., Pan, D., Chen, R., Chaturvedi, G., and Ding, Y. (1996). Affinities of methylphenidate derivatives for dopamine, norepinephrine and serotonin transporters. *Life Sci*, *58*, 231–239.

93. Froimowitz, M., Deutsch, H., Quing, S., Wu, K.-M., Glaser, R., Adin, I., George, C., and Schweri, M. (1997). Further evidence for a dopamine reuptake pharmacophore: the effect of N-methylation on threo-methylphenidate and its analogs. *Bioorg Med Chem Lett*, *7*, 1213–1218.

94. Deutsch, H., Ye, X., Shi, Q., Liu, Z., and Schweri, M. (2001). Synthesis and pharmacology of site specific cocaine abuse treatment agents: a new synthetic methodology for methylphenidate analogs based on the Blaise reaction. *Eur J Med Chem*, *36*, 303–311.

95. Axten, J., Krim, L., Kung, H., and Winkler, J. (1998). A stereoselective synthesis of *dl*-threo-methylphenidate: preparation and biological evaluation of novel analogs. *J Org Chem*, *63*, 9628–9629.

96. Meltzer, P., Wang, P., Blundell, P., and Madras, B. (2003). Synthesis and evaluation of dopamine and serotonin transporter inhibition by oxacyclic and carbacyclic analogues of methylphenidate. *J Med Chem*, *46*, 1538–1545.

97. Patrick, K., Caldwell, R., Ferris, R., and Breese, G. (1987). Pharmacology of the enantiomers of threo-methylphenidate. *J Pharmacol Exp Ther*, *241*, 152–158.

98. Schweri, M., Deutsch, H., Massey, A., and Holtzman, S. (2002). Biochemical and behavioral characterization of novel methylphenidate analogs. *J Pharmacol Exp Ther*, *301*, 527–535.

99. Yuan, H., Kozikowski, A., and Petukhov, P. (2004). CoMFA study of piperidine analogues of cocaine at the dopamine transporter: exploring the binding mode of the 3α-substituent of the piperidine ring using pharmacophore-based flexible alignment. *J Med Chem*, *47*, 6137–6143.

100. Carroll, F., Gao, Y., Rahman, M., Abraham, P., Parham, K., Lewin, A., Boja, J., and Kuhar, M. (1991). Synthesis, ligand binding, QSAR, and CoMFA study of 3-β-(*p*-substituted phenyl)tropane-2-β-carboxylic acid methyl esters. *J Med Chem*, *34*, 2719–2725.

101. Carroll, F., Mascarella, S., Kuzemko, M., Gao, Y., Abraham, P., Lewin, A., Boja, J., and Kuhar, M. (1994). Synthesis, ligand binding, and QSAR (CoMFA and classical) study of 3β-(3′-substituted phenyl)-, 3β-(4′-substituted phenyl)-, and 3β-(3′,4′-disubstituted phenyl)tropane-2β-carboxylic acid methyl esters. *J Med Chem*, *37*, 2865–2873.

102. Muszynski, I., Scapozza, L., Kovar, K.-A., and Folkers, G. (1999). Quantitative structure–activity relationships of phenyltropanes as inhibitors of three monoamine transporters: classical and CoMFA studies. *Quanti Struct–Activi Relati*, *18*, 342–353.

103. Xu, L., Kulkarni, S., Izenwasser, S., Katz, J., Kopajtic, T., Lomenzo, S., Newman, A., and Trudell, M. (2004). Synthesis and monoamine transporter binding of 2-(diarylmethoxymethyl)-3β-aryltropane derivatives. *J Med Chem*, *47*, 1676–1682.

104. Appell, M., Dunn, W., 3rd, Reith, M., Miller, L., and Flippen-Anderson, J. (2002). An analysis of the binding of cocaine analogues to the monoamine transporters using tensor decomposition 3D QSAR. *Bioorg Med Chem*, *10*, 1197–1206.

105. Lieske, S., Yang, B., Eldefrawi, M., Mackerell, A., Jr., and Wright, J. (1998). (−)-3β-Substituted ecgonine methyl esters as inhibitors for cocaine binding and dopamine uptake. *J Med Chem*, *41*, 864–876.

10

NON-NITROGEN-CONTAINING DOPAMINE TRANSPORTER– UPTAKE INHIBITORS

PETER C. MELTZER

Organix Inc., Woburn, Massachusetts

10.1	Introduction	266
	10.1.1 Leads for the Discovery of Monoamine Transport Ligands	267
	10.1.2 Amines in Endogenous Monoamine Neurotransmitters	268
10.2	Basis for the Design of DAT and SERT Inhibitors	269
	10.2.1 Interaction Between Ligands and Their Presynaptic Binding Sites	269
	10.2.2 Exchange of Nitrogen for Other Heteroatoms in Biological Receptor Ligands	270
	10.2.3 Focus of the Chapter: 8-Heterotropanes	271
10.3	Chemistry and Synthetic Strategies	272
	10.3.1 Synthesis of 3-Aryl-8-oxatropanes	272
	10.3.2 Synthesis of 3-Aryl-8-thiatropanes	275
	10.3.3 Synthesis of 3-Aryl-8-carbatropanes	276
	10.3.4 Synthesis of 3-Diarylmethoxy-8-oxatropanes	278
10.4	Biological Studies	279
	10.4.1 Inhibition of Monoamine Transporters by 8-Heterobicyclo [3.2.1]octanes	283
	10.4.2 Selectivity of Inhibition of the DAT vs. the SERT	286
	10.4.3 Biological Enantioselectivity of Inhibition of the DAT	291
10.5	Pharmacological Studies	291
	10.5.1 Stimulation of Locomotor Activity	293
	10.5.2 Drug Discrimination Studies	296
	10.5.3 Autoradiography	296
	10.5.4 Positron Emission Tomography	296

Dopamine Transporters: Chemistry, Biology, and Pharmacology. Edited by Mark L. Trudell and Sari Izenwasser
Copyright © 2008 John Wiley & Sons, Inc.

10.6 Conclusion   297

Acknowledgments   297

References   298

10.1 INTRODUCTION

Monoamine transport systems such as the dopamine (DAT), serotonin (SERT), and norepinephrine (NET) uptake mechanisms, all located within the membranes of presynaptic neurons, control the concentration of monoamine neurotransmitters in the synapse. These transporters equilibrate excess neurotransmitter that would otherwise impinge on postsynaptic monoamine receptors. Compounds that inhibit the function of these presynaptic transporters have far-reaching pharmacological consequences, both deleterious and therapeutic. The addictive drug cocaine and the therapeutic agent methylphenidate [used for treatment of attention-deficit/hyperactivity disorder (ADHD)] are examples of compounds that inhibit monoamine reuptake. Both possess an amine in their molecular structures. Two questions arise: Is the presence of the amine essential for interaction with presynaptic uptake mechanisms? Furthermore, is this amine a prerequisite for biological and pharmacological activity?

The focus of this chapter is on summarizing attempts to understand the relevance and impact of the presence of an amine on monoamine-uptake inhibitors. The design, biological evaluation, and exploration of nonnitrogen (nonamine)-uptake inhibitors, based on the structural motif of cocaine, are discussed. It will become clear that the presence of an amine is not obligatory for the inhibition of monoamine uptake or the inhibition of exogenous amines such as cocaine. Furthermore, it will become apparent that such nonnitrogen compounds can cross the blood–brain barrier, enter the central nervous system (CNS), and occupy pharmacologically relevant sites in the brain. Most important, these compounds can manifest pharmacological and behavioral effects that are a consequence of interaction with monoamine-binding and monoamine-uptake sites in the striatum.

Monoamine-uptake systems are implicated in diseases that may be a consequence of over- or understimulation of postsynaptic receptors by high or low synaptic concentrations of endogenous dopamine (DA), serotonin (5HT), and norepinephrine (NE). Therefore, compounds that block the DAT have been sought for diseases such as Parkinson's disease, ADHD, Alzheimer's disease, Tourette's syndrome, depression, or cognitive disorders. Compounds that block the SERT or NET have been developed to address diseases such as depression, obsessive-compulsive disorder, and obesity. The transporters also provide molecular targets for the design of medications to address diseases that are a consequence of exogenous neurotoxins that may utilize the transport mechanisms to enter the cell, or addictive substances (cocaine) that owe their CNS activity largely to their ability to block monoamine uptake. The dopamine transporter in the striatum has been implicated primarily, but not exclusively [1–3], in the pharmacology of cocaine addiction [4–6]. Cocaine effectively masks the system that removes monoamine neurotransmitters from the synapse and thus inhibits reuptake of dopamine by the DAT. This then results in an increase in the concentration of synaptic dopamine, with the consequence of increased activation of postsynaptic dopamine receptors. This hyperstimulation of postsynaptic receptors is responsible for the stimulant activity evidenced by cocaine. On the other hand, the reinforcing and addictive properties of cocaine are probably related to its pharmacokinetic profile of extremely rapid onset (1 to 4 minutes) and short duration of action (10 to 15 minutes). Therefore, the search for a safe cocaine-replacement therapeutic agent has been focused largely on the design of compounds that bind with selectivity to the DAT and manifest slow onset and long duration of action in vivo [7].

10.1.1 Leads for the Discovery of Monoamine Transport Ligands

As presented in Chapters 6 and 7, considerable attention has been paid to the class of bicyclo[3.2.1]octanes in the design of prospective medications for cocaine abuse [8–17]. This focus has its historic origins in the structure of cocaine itself as well as the metabolically more stable lead compound, WIN 35,428, a 3β-aryl-2β-carbo-methoxy-8-azabicyclo[3.2.1]octane analog of cocaine (Fig. 10-1) described by Clarke et al. in 1973 [18].

Compounds based on methylphenidate have also been explored for the development of DAT inhibitors that may lead to a cocaine pharmacotherapy [19,20]. The class of aryl-1,4-dialkylpiperazines (GBR analogs; see Chapter 8) has provided potent and pharmacologically active analogs that have been evaluated in models of cocaine addiction. All of these compounds possess an amine, and it has been assumed that binding of these ligands is affected by an ionic bond between the protonated amine and an aspartate residue at the acceptor site (Asp[79]) on the DAT. A fundamental premise that has guided all the work in this area has therefore been that the nitrogen plays a pivotal role in interaction of a ligand with its acceptor site on the monoamine transporters [21,22].

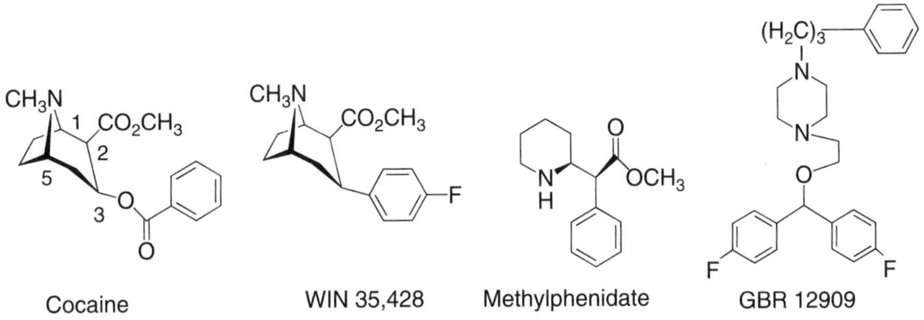

Figure 10-1 Compounds that inhibit monoamine uptake.

Recent studies have led to the conclusion that the actual binding sites on the DAT for these inhibitors may not be identical. Indeed, it is clear [23,24] that dopamine and the 3-aryl-8-aza or 3-aryl-8-oxatropanes* do not occupy identical sites on the DAT. This difference has provided an opportunity to design unique compounds that may function as dopamine-sparing cocaine antagonists [25]. A comparison of a family of 2-carbomethoxy-3-diarylmethoxytropanes with the GBR analogs provided evidence that the SAR of the 2-carbomethoxydiaryl compounds was more like that of the GBR analogs than like the 3-aryltropanes [26]. Comparisons of RTI-55 [27] and the benztropine GA 2-34 indicated that the former labeled transmembrane domain TM 4–7 while the latter labeled TM 1–2 of the DAT [28]. It is therefore possible that the 8-heterotropanes also bind differently at the DAT than do their 8-aza counterparts, notwithstanding great similarities in the SAR of all 8-heterotropanes.

10.1.2   Amines in Endogenous Monoamine Neurotransmitters

The presence of an amine in the endogenous ligands is interesting because it raises fundamental questions about the role of an amine in substrates for transport (dopamine, serotonin, and norepinephrine) (Fig. 10-2) and in compounds that block transport. These questions are particularly relevant when the design of potential monoamine-uptake blocking agents is considered. The monoamine transporters may have evolved to control neurotransmitter concentrations in the synaptic cleft by sequestering them back into presynaptic neurons. These neurotransmitters interact at numerous postsynaptic receptor sites and consequently perform a variety of postsynaptic functions. As an example, the ubiquitous neurotransmitter dopamine is utilized by numerous dopamine receptors. The dopamine affinity for the DAT corresponds to the dopamine concentration found after induced release of dopamine. Does this imply that the DAT is evolutionarily primed to remove excess release of dopamine from the synapse and thus prepare the synapse for the next signal? However, what is the function of the amine on dopamine with regard to its binding to transporters and receptors? The amine may indeed provide a functional handle by which to attach itself to the presynaptic transporters and postsynaptic receptors. It may also have an important role in providing a means of intrasynaptic transport and

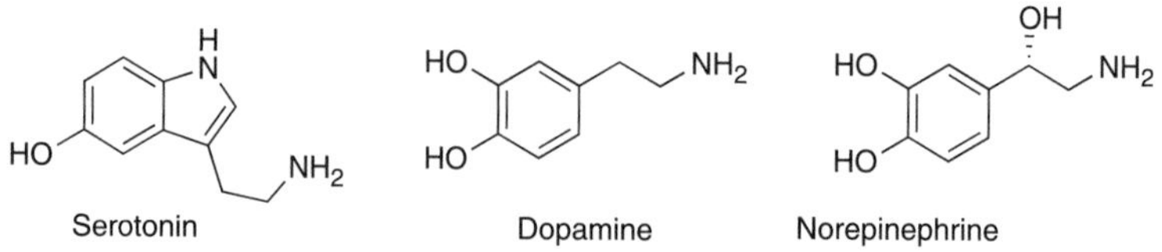

Figure 10-2   Endogenous monoamine neurotransmitters.

*The word *tropane* is used loosely in this chapter. While the term implies an 8-azabicyclo[3.2.1]octane, the terms *8-azatropane*, *8-carbatropane*, *8-thiatropane*, and *8-oxatropane* are used here to indicate that the skeletal structure (bicyclo[3.2.1]octane) is that of the classical tropane but that the 8-position carries the atom N, C, S, or O, respectively.

intracellular neurotransmitter transport in an aqueous medium [29] because pH changes [30] and salt gradients [31,32] probably alter the aqueous and lipid solubility of the neurotransmitter. Among heteroatoms, the nitrogen of an amine is unique in this regard, and neither oxygen nor sulfur is subject to such marked effects upon aqueous solubility as a consequence of pH change.

10.2 BASIS FOR THE DESIGN OF DAT AND SERT INHIBITORS

The amine-containing 3-benzoyltropanes and 3-aryltropanes became of considerable interest [17] with respect to the search for cocaine pharmacotherapies as a consequence of the dopamine hypothesis proposed by Ritz and Kuhar in 1987 [6]. This hypothesis provided a plausible explanation of the manner in which cocaine exercises its effects on the central nervous system and consequently provided a platform from which medicinal chemists could launch a rational foray into the design, synthesis, and evaluation of compounds that could interact with, and thus modulate, the systems responsible for cocaine's pharmacological effects. The long-used and abused naturally occurring tropane, cocaine, served as the early template for design of antagonist medications for cocaine addiction. The class of 3-benzoyltropanes gave rise to numerous studies concerning isomers and analogs of cocaine [33,34]. However, the 3-aryltropanes, reported in 1973 by Clarke et al. [18], in which the labile 3-benzoylester of cocaine had been replaced by a directly linked 3-aryl group, provided the prototypical tropane of later medicinal chemistry and biological efforts: 2β-carbomethoxy-3β-(4-fluorophenyl)-8-azabicyclo[3.2.1]octane (WIN 35,428). This template has served as a springboard from which to explore tropane–DAT interaction.

10.2.1 Interaction Between Ligands and Their Presynaptic Binding Sites

A comprehensive understanding of the ligand–dopamine transporter interaction has been elusive. A focus of our SAR studies has been an exploration of the functional role of the 8-nitrogen in cocaine and its 3-aryltropane analogs. This 8-amine had been proposed to provide an ionic bond [21] between the protonated tertiary amine and an aspartate residue (Asp^{79}) on the DAT [22] (Fig. 10-3). However, quaternary

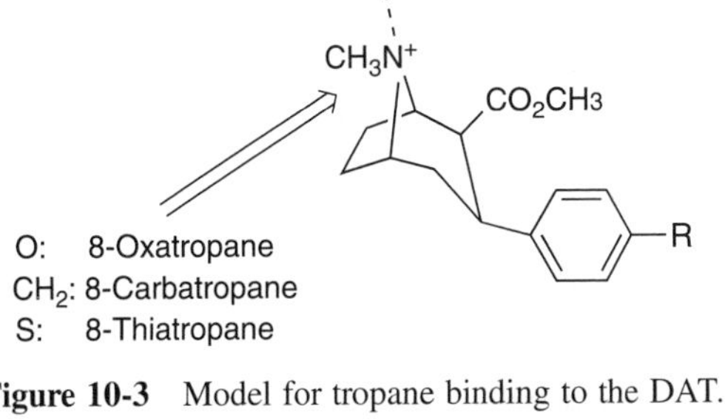

O: 8-Oxatropane
CH_2: 8-Carbatropane
S: 8-Thiatropane

Figure 10-3 Model for tropane binding to the DAT.

cocaine methiodide was found to bind extremely weakly to the DAT [35], implying that such an ionic interaction might result from protonation of the tropane 8-amine by the aspartic acid itself. However, Kozikowski had reported that reduction of nitrogen basicity by introduction of an *N*-sulfonyl group, sufficient to inhibit protonation of the amine under physiological conditions, did not reduce binding potency. Therefore, the importance of a protonated amine was questioned [15]. Consequently, although an ionic bond may play some role in the interaction between the tropanes and their biological receptor, it is clearly not the only possibility. This interaction could also be envisioned to result from hydrogen bonding between an H-donor on the transporter and an H-bond acceptor on the ligand. To evaluate this possibility, we explored 8-oxatropane analogs that could interact by hydrogen bonding but *not* by formation of an ionic bond [29].

Such oxygen for nitrogen replacement could shed light on whether or not the presence of an amine in a bicyclo[3.2.1]octane skeleton is essential for inhibition of monoamine uptake mechanisms. Further, this replacement might have the added advantage of providing compounds of unique biological profiles because the mode of binding might be different from that of their nitrogen counterparts, and their pharmacokinetic profile, biological transport, metabolism, and elimination pathways would probably differ. Upon finding that these 8-oxatropanes were themselves potent inhibitors of monoamine uptake, and therefore that ionic bonding was not a requisite for potent DAT or SERT inhibition, the role of hydrogen bonding in DAT–ligand interaction was explored. The 8-oxygen in these 8-oxatropanes was exchanged with an 8-methylene group [36] (Fig. 10-3). Clearly, such substitution removed all possibility of ionic or hydrogen bonding at the 8-position of the 3-aryl tropanes. The potent inhibitory characteristics observed for these 8-methylene compounds (8-carbatropanes) provided clear evidence that neither hydrogen bonding nor ionic bonding was an essential feature for tight binding and inhibition of monoamine uptake systems. This 8-nitrogen exchange was then extended to the 8-thiatropanes, and these compounds proved similarly potent and selective in their interaction with monoamine-uptake systems. The concept of exchange of nitrogen for another atom was extended to methylphenidate, where it has also been shown that this exchange can result in potent inhibitors of monoamine uptake of varying potencies and selectivities [37].

The surprising potency of these nonnitrogen compounds has demonstrated that with respect to binding, the function of an amine is not necessarily an active binding function (such as hydrogen bonding or electrostatic interaction) but rather, a topological function. It could be that this 8-atom serves to hold the molecule in its rigid three-dimensional conformation and provides, stereospecifically, a molecular volume for recognition or acceptance by a putative "cylindrical" transporter [38]. This fascinating result leads to the speculation that a multitude of molecules will interact with the DAT (and possibly other transporters) provided that they occupy homologous three-dimensional space.

10.2.2 Exchange of Nitrogen for Other Heteroatoms in Biological Receptor Ligands

The substitution of a heteroatom for the nitrogen of a biologically active molecule had been reported previously. Indeed, in 1979, Brownbridge and Chan [39] prepared the 8-oxa analog of cocaine, although no biological data were reported for that compound. Miller et al. [40–44] explored the substitution of sulfur for nitrogen in their studies on D2 ligands. They showed that nitrogen could be replaced by sulfur in these molecules and potency was maintained, albeit much diminished.

Few compounds that enter the CNS and interact with uptake or receptor systems in neurons in the brain, and are devoid of nitrogen, have been discovered. One notable exception is Δ^9-tetrahydocannabinol, which inhibits CB1 receptors in the brain with high affinity [45]. The endogenous ligand of the CB1 receptor is the nitrogen-containing compound anandamide [46]. However, anandamide has the nitrogen in the form of an amide rather than an amine and is therefore not amenable to ionic bonding to a receptor. Therefore, it is not unprecedented that nonnitrogen compounds may bind to and inhibit CNS transporters or receptors, and furthermore, manifest pharmacological activity in vivo.

10.2.3 Focus of the Chapter: 8-Heterotropanes

The classes of nonnitrogen DAT inhibitors discussed in this chapter are the 3-aryl-2-carbomethoxy-8-heterotrop-2-enes, the 3β-aryl-2β-carbomethoxy-8-heterotropanes, and the 3α-aryl-2β-carbomethoxy-8-heterotropanes. The pharmacological focus of our studies has been to design compounds within two mechanistic classes: (1) substitute medication in which the drug is a cocainelike agonist that substitutes for cocaine but provides a substantially different biological pharmacokinetic profile that has much reduced abuse potential (e.g., the clinical use of methadone for treating heroin addiction), and (2) dopamine-sparing cocaine antagonists [22] that block cocaine binding at the DAT while maintaining dopamine trafficking in the synapse [25,47,48].

In a discovery program, the medicinal chemist is confronted by the choice of which targets to synthesize for biological evaluation. This selection is important in the face of limited time and resources and requires the establishment of early priorities: which compounds to synthesize in the beginning stages and which to synthesize at later points in time. The goal is to obtain information expeditiously that will contribute to a fundamental understanding of the SAR of the series. The decision as to which 8-heterotropanes to evaluate was guided by two considerations. First, substantial biological information [16,49,50] had been reported for the 2β-carbomethoxy-3β-(4-fluorophenyl)tropane (WIN 35,428, β-CFT) (Fig. 10-1) and the unsubstituted 2β-carbomethoxy-3β-phenyltropane reported by Clarke et al. [18]. Furthermore, WIN 35,428 had been developed as a positron emission tomography (PET) agent for imaging purposes, had been tritiated to monitor autoradiographic studies [51] and had been radiolabeled and used as a displacement ligand to measure the inhibition constants for ligands at the DAT [52]. WIN 35,428 was the impetus for the design of, among others, the iodinated single-photon emission computed tomography (SPECT) imaging agents RTI-55 [27] and altropane [53,54]. Therefore the 4-fluorophenyl and 4-unsubstituted-aryl analogs within the 8-nonnitrogen series would probably provide important comparative data and were selected for synthesis.

Second, Topliss [55] had described an expeditious means of obtaining information concerning the impact of lipophilicity (π), electronic contribution (σ), and steric effects (Es) on activity with respect to a defined biological assay. The established assays for evaluation of inhibitors of DAT, SERT, and NET were well suited to this widely used approach. Therefore, compounds with certain substituents on the aromatic nucleus were selected for early synthesis (4-H, 4-Cl, 3,4-Cl$_2$, and naphthyl). Notwithstanding these criteria used to guide the early selection of targets, as biology informed chemistry, so targets were changed or new targets were identified to answer emerging biological questions.

10.3   CHEMISTRY AND SYNTHETIC STRATEGIES

A general route to allow rapid access to the three isomers of the 2-carbomethoxy-3-aryl-tropanes (3α-aryl, 3β-aryl, and 2,3-unsaturated-3-aryl) was developed. The classical synthesis [18] of the 8-azatropanes utilized an aryl Grignard addition to the unsaturated anhydroecgonine methyl ester obtained from cocaine (Scheme 10-1). In the case of 8-aza compounds, this provided predominantly the 2β-carbomethoxy-3β-aryl (chair) analogs, along with a minor amount of the 2α-carbomethoxy analogs. In general, the 2α-carbomethoxy analogs were poor inhibitors of the DAT [56]. A considerable disadvantage of this approach was that the 3α-aryl analogs, as well as the 2,3-unsaturated compounds, were unavailable via this route. To address this problem, a more general route was developed for the synthesis of the 8-heterotropanes, and 8-hetero-3-ketoesters [39] provided an excellent starting point for their synthesis. This approach is presented generically in Scheme 10-2.

10.3.1   Synthesis of 3-Aryl-8-oxatropanes

The synthetic routes to the 8-heterobicyclo[3.2.1]octanes were all similar. Therefore, the 8-oxatropanes are discussed in some depth, whereas the 8-thia and 8-carbatropanes are presented in less detail. As mentioned, the 8-oxabicycles could be derived from the ketoester **3** (Scheme 10-3) described by Brownbridge and Chan in 1979 [39] Compound **3** was prepared (45%) from 2,5-dimethoxytetrahydrofuran and

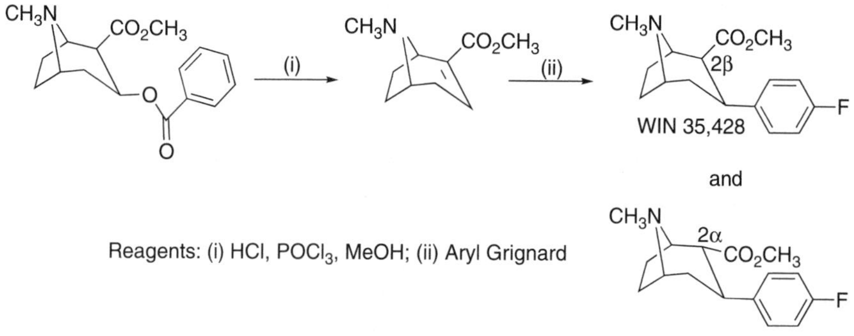

Reagents: (i) HCl, POCl$_3$, MeOH; (ii) Aryl Grignard

Scheme 10-1

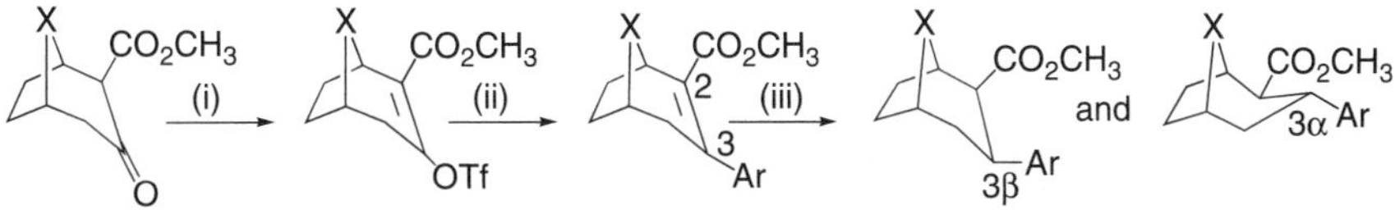

Reagents: (i) NaN(TMS)$_2$, PhNTf$_2$; (ii) ArB(OH)$_2$, Pd$_2$(dba)$_3$ or Pd(PPh$_3$)$_4$; (iii) SmI$_2$

Scheme 10-2

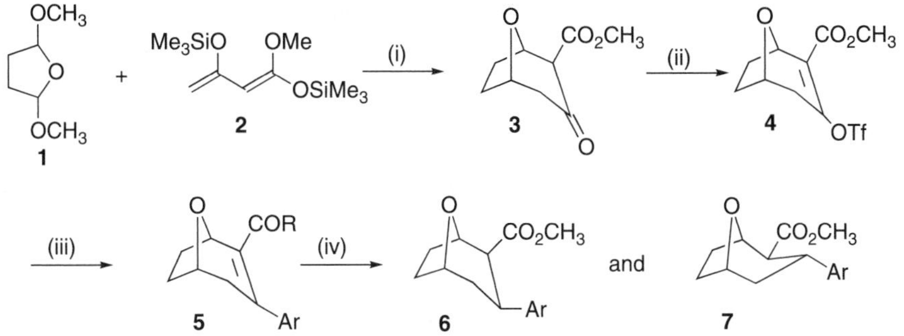

Ar
a. 4-H-Ph
b. 4-F-Ph
c. 4-Cl-Ph
d. 4-Br Ph

Ar
e. 4-I-Ph
f. 4-(3,4-Cl$_2$)-Ph
g. 3-(2-naphthyl)

Reagents: (i) TiCl$_4$; (ii) NaN(TMS)$_2$, PhNTf$_2$; (iii) ArB(OH)$_2$, Pd$_2$(dba)$_3$; (iv) SmI$_2$

Scheme 10-3

1,3-bis(trimethylsiloxy)-1-methoxybuta-1,3-diene [57,58] and titanium tetrachloride. The ketoester **3** then served as the critical intermediate for the preparation of the family of 3-aryl-8-oxatropanes (Scheme 10-3) and enantiomeric resolution (Scheme 10-4) [38]. Thus, ketone **3** was converted to the enol triflate **4** (79%) with *N*-phenyltrifluoromethanesulfonimide and sodium bis(trimethylsilyl)amide in tetrahydrofuran, which was then coupled with the appropriate arylboronic acids [59] by Suzuki coupling [60] in diethoxymethane in the presence of lithium chloride, sodium carbonate, and tris(dibenzylideneacetone)dipalladium(0) to provide aryl octenes **5** (82 to 97%). These 2,3-unsaturated-8-oxatropenes were evaluated biologically and pharmacologically. They were reduced with samarium iodide in tetrahydrofuran/methanol ($-78°$C) [61] to provide the 3β-aryl and 3α-aryl diastereomers **6** and **7**. In general, when trifluoroacetic acid was used to quench the samarium iodide reaction, the major products were the 2β,3α diastereomers **7**, which were isolated in about 50 to 65% yield. The minor products were the 2β,3β compounds **6** isolated in about 14 to 23% yield. With water as the quenching agent, a 1:1 mixture of 3β-aryl and 3α-aryl diastereomers could generally be obtained. The six-membered ring of the

3α-aryl- and 3β-arylbicyclo[3.2.1]octanes adopts different conformations (pseudoboat or chair), depending on the orientation of substitution at C_3. Thus, 3α-aryl compounds exist in a pseudoboat conformation, whereas 3β-aryl compounds adopt a chair conformation. This was evident from ^{1}H NMR spectra. An x-ray crystallographic study of 3α-(3,4-dichlorophenyl)-8-oxatropane confirmed the boat conformation in the solid state for that compound [38].

It is interesting that in the cocaine series [34,61] as well as in the 8-aza and 8-oxa diarylmethoxy family of compounds, the 3α-substituted compounds adopt a chair conformation [62,63]. When the aryl ring is attached directly to C_3, the boat, or pseudoboat, conformation predominates. Presumably, the close proximity of a bulky aryl group to the ring system in the 3-aryl compounds causes conformational selectivity and the aryl-equatorial boat is preferred for 3α-substituted compounds, while the aryl-equatorial chair is preferred for 3β-substituted compounds.

Cocaine binds enantioselectively to the dopamine transporter and the $1R$-isomers of cocaine as well as the classical 8-aza-3-aryl compounds have been found to be the eutomers. However, the fact that the $(1S)$-2β-carbomethoxybenztropine ($1S$-difluoropine) [62]) was a potent and selective ligand for the dopamine transporter, and the $1R$-isomer was not, made it imperative to confirm biological enantioselectivity for this new series of compounds. Therefore, the chiral compounds $(1R)$-**6f** and $(1S)$-**6f**, and $(1R)$-**7f** and $(1S)$-**7f**, were synthesized. A diastereomeric mixture of enol camphanates (**8**) was prepared upon reaction of $(1R/S)$-**3** with (S)-$(-)$-camphanic chloride (57%) (Scheme 10-4), and recrystallization gave the pure diastereomer

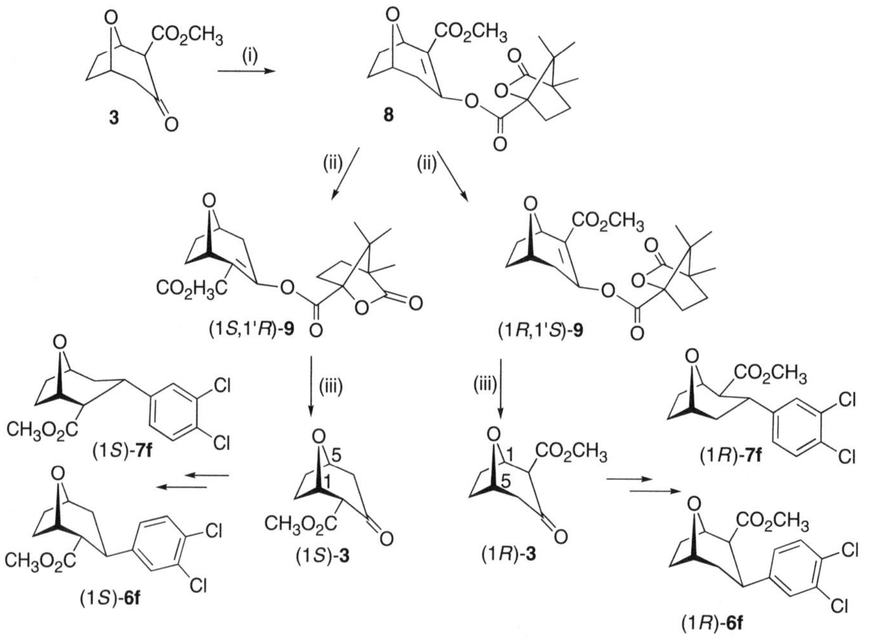

Scheme 10-4

(1R,1'S)-**9** (54%). Analogously, reaction with (R)-(+)-camphanic chloride, and subsequent recrystallization, provided the pure diastereomer (1S,1'R)-**9** (66%).

Hydrolysis of the enantiomerically pure camphanate esters (1R,1'S)-**9** and (1S,1'R)-**9** with LiOH provided (86%) the ketones (1R)-**3** and (1S)-**3**, respectively. Analysis of the ketones (1R)-**3** and (1S)-**3** by chiral high-performance liquid chromotography revealed >96% ee for each enantiomer. These enantiomers were then subjected to the sequence of steps described earlier to obtain the 3α-aryl and 3β-aryl diastereomers of the enantiomerically pure 8-oxatropane analogs. The absolute structure of enantiopure (1R)-**7f** was established by x-ray structural analysis as the (1R)-(−)-boat conformer. Consequently, it was clear that (1S)-**7f** was the 1S-(+)-compound. In turn, this implied that the configurations of the ketone precursors were (1R)-**3** and (1S)-**3**, respectively. The remaining enantiopure compounds were derived from enantiopure ketones (1R)-**3** and (1S)-**3**, and the 3,4-dichloroaryl chair conformers **6f** and **7f** obtained were therefore (1R) and (1S), respectively.

10.3.2 Synthesis of 3-Aryl-8-thiatropanes

The general approach for the synthesis of 8-thiatropanes centered on the ketoester **13** and palladium-catalyzed Suzuki coupling with suitably substituted arylboronic acids (Scheme 10-5) [64,65]. Tropinone (**10**) was quaternized with CH$_3$I to **11** (90%).

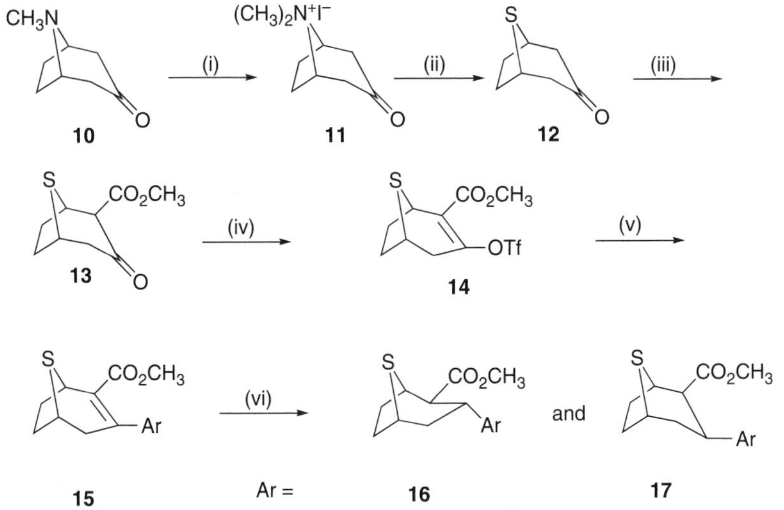

Ar =

a. 4-H-Ph   e. 3-I-Ph
b. 4-F-Ph   f. 3,4-Cl2Ph
c. 4-Cl-Ph   g. 2-Naphthyl
d. 4-Br-Ph

Reagents: (i) CH$_3$I; (ii) Na$_2$S; (iii) LDA, NCCOOCH$_3$; (iv) NaN(TMS)$_2$, PhN(Tf)$_2$; (v) ArB(OH)$_2$, Pd(PPh$_3$)$_4$; (vi) SmI$_2$

Scheme 10-5

Treatment with sodium sulfide [66] then provided the 3-ketobicyclooctane **12** (78%). Introduction of the 2-carbomethoxy group was effected [67] with methylcyanoformate to obtain the ketoester **13** (69%). The 8-hetero ketoesters all existed in an equilibrium mixture of three tautomeric forms: the 2α-carbomethoxy ketoester, the 2β-carbomethoxy ketoester, and the enolester. This tautomerism was of no stereochemical significance because all centers were lost in the ensuing conversion to the enol triflate **14**. However, the existence of the three tautomers complicated their ^{1}H NMR spectra, and therefore these ketoesters were generally characterized as their enol triflates formed upon reaction with N-phenyltriflimide and sodium bis(trimethylsilyl) amide (75%).

The synthesis then followed the route outlined previously for the 8-oxa analogs. Suzuki coupling of the appropriately substituted boronic acids with enol triflate **14** provided **15** (74 to 96%). Samarium iodide reduction [38,61] then provided the saturated compounds as a mixture of **16** and **17** (60 to 90%). In general, a 4 : 1 ratio of the 3α-aryl (**16**) to 3β-aryl (**17**) analogs was obtained, and the compounds were separated by column chromatography to obtain the 3α-aryl (30 to 40%) and 3β-aryl (20 to 35%) 8-thiatropanes. The 3,4-dichlorophenyl analogs were selected for determination of biological enantioselectivity to compare results with those obtained in the 8-aza and 8-oxa series. Intermediate ketoester **13** served as the springboard for resolution via camphanate esters, as described for the 8-oxatropanes (Scheme 10-4).

The absolute stereochemistry of the enantiopure camphanate esters of **13** was established by x-ray crystallography. Diastereomeric excess was established by ^{1}H NMR as the camphanate esters of (1R,1S′)-**13** (de 97%) and (1S, 1R′)-**13** (de 88%), respectively. This optical purity then governed the optical purity of the ketoesters (1R)-**13** and (1S)-**13** generated upon ester hydrolysis, as well as the purity of all subsequent products in the synthetic sequence (Scheme 10-5), because there were no transformations that could effect ring opening and closing, thus altering the chirality at the bridgehead. Subsequent conversion of (1R)-**13** and (1S)-**13** was carried out, as for racemic **13**, to provide the 2,3-enes (1R)-**15f** and (1S)-**15f**. Samarium iodide reduction then gave the 3α-(3,4-dichlorophenyl) compounds (1R)-**16f** and (1S)-**16f** and the 3β-(3,4-dichlorophenyl) compounds (1R)-**17f** and (1S)-**17f**.

10.3.3   Synthesis of 3-Aryl-8-carbatropanes

The synthetic route [36] (Scheme 10-6) provided a racemic mixture of each of the configurational isomers (**24** to **27**) for which biological data were obtained. The critical ketoester **20** was obtained upon treatment of commercially available 3-chlorobicyclo[3.2.1]oct-2-ene (**18**), with sulfuric acid (77%) [68] and subsequent reaction of **19** with methyl cyanoformate in the presence of lithium diisopropylamide (83%). The ketoester **20** was converted to the enol triflate **21** (75%), and Suzuki coupling with appropriate boronic acids then provided the 2,3-unsaturated analogs **22** (54 to 75%). Reduction of **22** with samarium iodide gave **23** (64 to 84%) as a mixture of all four isomers **24** to **27**. The isomers were extremely difficult to separate, and

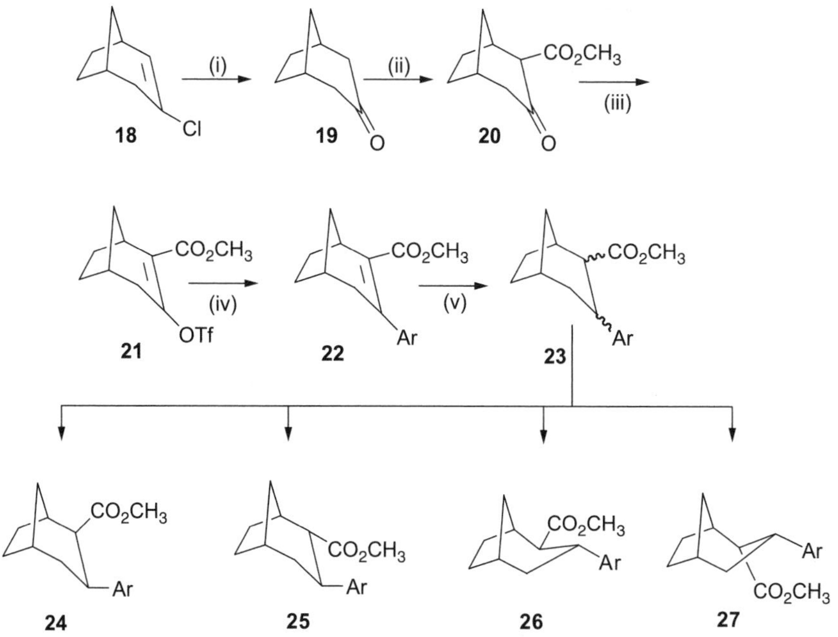

Ar = **a**. C_6H_5 **b**. 4-FC_6H_4 **c**. 3,4-$Cl_2C_6H_3$ **d**. 2-Naphthyl

Reagents: (i) H_2SO_4; (ii) LDA/THF, $CNCOOCH_3$; (iii) $NaN(TMS)_2$, $PhNTf_2$;
(iv) $ArB(OH)_2$, $Pd_2(dba)_3$; (v) SmI_2, CH_3OH

Scheme 10-6

biological evaluation was initially conducted on the isomeric mixtures. The 4-fluoro-phenyl mixture of isomers **23b** and unsubstituted **23a** both presented DAT IC_{50} binding constants of 0.3 to 0.8 μM and were therefore set aside. In sharp contrast, the 3-(3,4-dichlorophenyl) **23c** and 3-(2-naphthyl) **23d** analogs manifested unexpectedly potent binding, and these compounds were separated by gravity column chromatography to provide 2β,3β (**24c**, **24d**), 2α,3β (**25c**, **25d**), 2β, 3α (**26c**, **26d**), and 2α,3α (**27c**, **27d**) 8-carbatropanes in low yields.

The configuration of each isomer was determined by [^1]H NMR studies. An analysis of the multiplicity of the key resonances showed familiar coupling patterns observed previously for the 8-aza and 8-oxa analogs [38,56]. These results provided unequivocal identification of compounds **24c** to **27c** and confirmed that the 3α-aryl-substituted analogs again adopted a pseudo-boat conformation for the six-membered ring, as already demonstrated for the 8-aza and 8-oxa series [38,69]. X-ray structural analyses conducted on four configurational racemates **24c** (2β,3β), **25c** (2α,3β), **25d** (2α,3β), and **27d** (2α,3α) confirmed the configurations determined on the basis of [^1]H NMR studies [36].

10.3.4 Synthesis of 3-Diarylmethoxy-8-oxatropanes

Synthesis of the 8-oxa-3-diarylmethoxytropanes [38] (Scheme 10-7) began with the reduction of keto ester **3**. Thus, treatment of **3** with sodium borohydride provided a mixture of the isomers **28** to **31**. The dominant product of reduction was the enantiomeric pair of 2α-carbomethoxy-3α-hydroxy compounds (**30**). The minor products were **28** and **29**, with very small amounts of **31**. Inversion of **30** was achieved

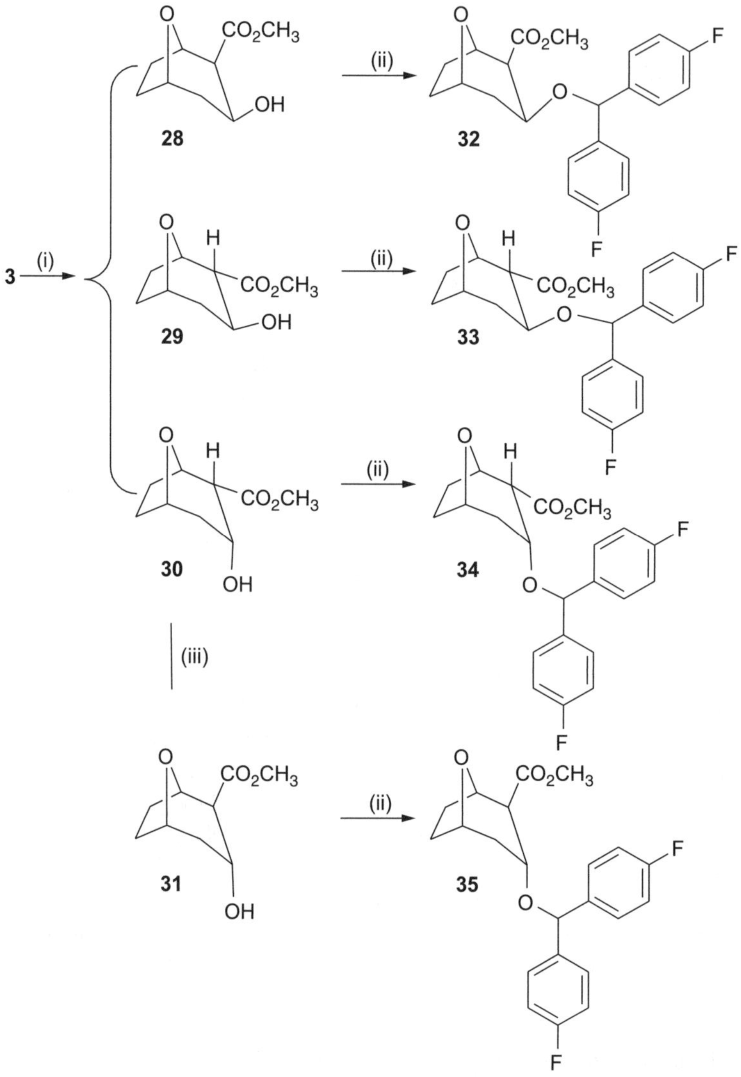

Reagents: (i) $NaBH_4$; (ii) 4,4'-Difluorobenzhydrol, pTSA; (iii) $NaHCO_3$

Scheme 10-7

upon reflux with saturated $NaHCO_3$ to provide 2β-carbomethoxy-3α-hydroxy **31** (21%). Each of the four racemic diastereomers **28** to **31** was then converted to the corresponding 3-diarylmethoxy analogs **32** to **35** by reaction with 4,4′-difluorodibenzhydrol under *p*-TSA catalysis [62]. The 8-oxa diastereomers were characterized by comparison of the ^{1}H NMR spectra of both the free hydroxy compounds (**28** to **31**) and the diarylmethoxy derivatives with the corresponding 8-aza analogs [62]. These results not only identified the four different diastereomers but also confirmed that all were present in the chair conformation.

10.4 BIOLOGICAL STUDIES

The search for a novel pharmacotherapy for the treatment of cocaine addiction has centered primarily around 3-aryltropane analogs of cocaine, 1,4-dialkylpiperazines (GBR compounds), analogs of methylphenidate, and benztropine. An amine is present in all these molecules. Indeed, it had long been assumed that the presence of an amine was essential for potent interaction with biological transporters or receptors. A useful model of the interaction between the class of 8-azabicyclo[3.2.1]-octanes (3-aryltropanes, and cocaine in particular) and the dopamine transporter was presented by Carroll et al. in 1992 [21]. The basis of this model was threefold: First, it was postulated that the tertiary amine of the tropanes provided a means of ionic bonding between the ligand and the DAT. Second, the (1*R*)-stereochemical configuration was assumed to be essential for potent inhibition of the DAT. Third, the 2β-carbomethoxy-3β-aryl (chair) orientation was assumed to be optimum for interaction at an acceptor site on the DAT. This model has since provided a platform for the design of DAT inhibitors by many researchers. However, the development of a useful pharmacological agent for central nervous system disorders such as cocaine addiction requires at least four essential attributes: Specifically, moderate to potent in vitro binding to monoamine uptake sites, ex vivo distribution to appropriate sites in the brain, appropriate blood–brain barrier permeability, and appropriate brain occupancy in vivo. A pharmacological indication of inhibition of the DAT is the observation in vivo of stimulation of locomotor activity. Therefore, the biological and pharmacological evaluation of the 8-heterobicyclo[3.2.1]octanes has included binding, biological, and pharmacological studies.

In the following discussion, the affinities (IC_{50}) of the compounds for the DAT and SERT were determined in competition studies using [^{3}H]WIN 35,428 ([^{3}H] **O-381**) to label the DAT and [^{3}H]citalopram to label the SERT [29,70,71]. Competition studies were conducted with a fixed concentration of radioligand and a range of concentrations of the test drugs. All drugs inhibited [^{3}H]WIN 35,428 and [^{3}H]citalopram binding in a concentration-dependent manner. Selected data are provided in the following tables to illustrate the general themes of SAR within the classes of compounds. The SAR in the four classes of 8-heterobicyclo[3.2.1]octanes (8-N, 8-S, 8-O, 8-C) are compared for the three topologically different tropane skeletons: 2,3-unsubstituted tropanes (Table 10-1), 3α-aryltropanes (Table 10-2), and 3β-aryltropanes (Table 10-3).

TABLE 10-1 Comparison of Aromatic Substituents on the SAR of Binding Affinities of 8-Hetero-3-aryltrop-2-enes[a]

Ar	X = NCH$_3$[b]			X = S			X = O			X = CH$_2$		
	Compound	DAT	SERT	Compound 15	DAT	SERT	Compound 5	DAT	SERT	Compound 22	DAT	SERT
(−)-Cocaine		95	270									
4-H-Ph	O-1449	>2 µM	>10 µM	a	910	>10 µM	a	>10 µM	>10 µM	a	>3 µM	>10 µM
4-F-Ph	O-1104	408	>8 µM	b	220	>10 µM	b	>2 µM	>10 µM	b	390	>10 µM
4-Cl-Ph				c	13	>10 µM	c	238	>10 µM			
4-Br-Ph				d	9.1	>10 µM	d	62	>10 µM			
4-I-Ph				e	6.7	>1 µM	e	68	>4 µM			
3,4-Cl$_2$-Ph	O-1109	1.2	867	f	4.5	>3 µM	f	12.3	>2 µM	c	7.1	>5 µM
3-(2-Naphthyl)	O-1173	2.9	109	g	8.0	>1 µM	g	19.7	704	d	11	>1 µM

[a]IC$_{50}$ (nM) for inhibition of [³H]WIN 35,428 binding (DAT) and [³H]citalopram binding (SERT).
[b](1R)-Enantiomers.

TABLE 10-2 Comparison of Aromatic Substituents on the SAR of Binding Affinities of 8-Hetero-3α-aryltropanes[a]

Structure: X, CO_2CH_3, Ar on tropane skeleton.

Ar	$X = NCH_3$[d]			$X = S$			$X = O$			$X = CH_2$		
	Compound	DAT	SERT	Compound 16	DAT	SERT	Compound 7	DAT	SERT	Compound 26	DAT	SERT
(−)-Cocaine	—	96	270									
WIN 35,428	—	11	160									
4-H-Ph	[b]	101	>5 μM	a	140	>8 μM	a	>1 μM	>10 μM	a	787	>10 μM
4-F-Ph	[b]	21	>5 μM	b	59	>10 μM	b	>1 μM	>10 μM	b	335	na[c]
4-Cl-Ph	[b]	2.4	998	c	11	>1 μM	c	29	816			
4-Br-Ph	—		—	d	6.0	342	d	9.0	276			
4-I-Ph	[b]	2.9	65	e	9.0	70	e	42	72			
3,4-Cl$_2$-Ph	O-1157	0.4	27	f	6.9	99	f	3.1	65	c	13	166
3-(2-Naphthyl)	O-1228	0.6	6.0	g	8.0	36	g	12	18	d	17	64

[a]IC$_{50}$ (nM) for inhibition of [3H]WIN 35,428 binding (DAT) and [3H]citalopram binding (SERT).
[b]Reported by Holmquist et al. [69].
[c]na = not available.
[d](1R)-Enantiomers.

281

TABLE 10-3 Comparison of Aromatic Substituents on the SAR of Binding Affinities of 8-Hetero-3β-aryltropanes[a]

	X = NCH₃[d]			X = S			X = O			X = CH₂		
Ar	Compound	DAT	SERT	Compound 17	DAT	SERT	Compound 6	DAT	SERT	Compound 24	DAT	SERT
(−)-Cocaine	—[b]	95	270									
4-H-Ph	O-381	23	2 μM	a	117	>3 μM	a	>1 μM	>10 μM	a	787	>10 μM
4-F-Ph (WIN 35,428)		11	160	b	38	494	b	546	>2 μM	b	335	na[c]
4-Cl-Ph	O-371	1.4	5.9	c	9.6	33	c	10	107			
4-Br-Ph		1.8	10.6	d	6.0	14	d	22	30			
4-I-Ph	RTI55[b]	1.3	4.2	e	14	10	e	7.0	12			
3,4-Cl₂-Ph	O-401	1.1	2.5	f	5.7	8.0	f	3.4	6.5	c	9.6	33
3-(2-Naphthyl)	O-1229	0.49	2.2	g	16	13	g	7.0	10	d	27	60

[a]IC₅₀ (nM) for inhibition of [³H]WIN 35,428 binding (DAT) and [³H]citalopram binding (SERT).
[b]Reported by Holmquist et al. [69].
[c]na = not available.
[d](1R)-Enantiomers.

General themes are evident among the four heteroatom classes and the three topological classes:

1. The non-nitrogen-bearing bicyclo[3.2.1]octanes have DAT inhibitory potencies similar to those of their 8-azabicyclo[3.2.1]octane progenitors.
2. The most potent inhibitors [3,4-dichlorophenyl and 3-(2-naphthyl) analogs] of the DAT are considerably more potent than cocaine.
3. The least potent compounds at DAT and SERT are the unsubstituted phenyltropanes and their 4-fluorophenyl analogs. The 4-chlorophenyl, 4-bromophenyl, and 4-iodophenyl analogs are generally quite potent inhibitors of the DAT but are less so at SERT.
4. The SAR with respect to substitution on the C_3-aromatic ring is similar in all series.
5. The SAR with respect to topological shape (i.e., 3α-aryl, 3β-aryl, 2,3-ene) is similar among the 8-aza, 8-thia, 8-oxa, and 8-carba classes.

These similarities imply that these compounds probably all bind within very similar domains on the DAT.

10.4.1 Inhibition of Monoamine Transporters by 8-Heterobicyclo [3.2.1]octanes

Compounds in the nonnitrogen series comprise extremely potent inhibitors of the DAT and SERT. The most potent members of these families have binding affinities at the DAT that are very similar to those exhibited by the classical 8-azatropanes. The 3,4-dichlorophenyl- and 3-(2-naphthyl)-substituted bicyclo[3.2.1]octanes present interesting examples. Thus, in the 2,3-ene series (Table 10-1), the 8-aza series **O-1109** (3,4-dichlorophenyl) and **O-1173** [3-(2-naphthyl)] compounds manifested nanomolar affinity for the DAT (IC_{50} values between 1 and 3 nM). The 8-thia (**15f, 15g**) compounds were similarly potent (IC_{50} = 4.5 and 8.0 nM, respectively). The 8-oxa (**5f, 5g**) were about 10-fold weaker than their 8-aza counterparts, however, the 8-carba (**22c, 22d**) were more similar (IC_{50} = 7.1 and 11 nM, respectively) to the 8-thia analogs. Therefore, within the 2,3-unsaturated series, the nonnitrogen compounds were only between 4- and 10-fold weaker than the 8-aza analogs at DAT inhibition. Inhibition of SERT was extremely weak in all 2,3-unsaturated compounds (IC_{50} > 1 μM) with the exception of **O-1173** [3-(2-naphthyl)], which was about two fold more potent than cocaine at SERT.

In the 3α-aryl series (Table 10-2) the 8-aza **O-1157** (3,4-dichlorophenyl) and **O-1228** [3-(2-naphthyl)] manifested subnanomolar DAT IC_{50} values (0.4 and 0.6 nM); their 8-thia (**16f, 16g**), 8-oxa (**7f, 7g**), and 8-carba (**26c, 26d**) analogs manifested DAT IC_{50} values of 3.1 to 17 nM. Therefore these nonnitrogen compounds were about 6- to 30-fold more potent than cocaine and 10-fold weaker than the 8-aza analogs at DAT. These nonnitrogen compounds were moderately potent SERT inhibitors (IC_{50} = 18 to 166 nM). The 3β-aryl series (Table 10-3) has a skeletal structure most similar to that of cocaine and the prototypical tropane WIN 35,428 (**O-381**: $IC_{50} = 11$ nM). In this class, the 3,4-dichlorophenyl **O-401** inhibited the DAT with an $IC_{50} = 1.1$ nM, and the naphthyl analog (**O-1229**) had $IC_{50} = 0.49$ nM. In comparison, the related 8-oxa compounds (**6f, 6g**) also showed single-digit IC_{50} values (3.4 and 7.0 nM, respectively). The 8-thia and 8-carba compounds had similar potencies ($IC_{50} = 5.7$-27 nM) and were about two to threefold weaker than their 8-oxa counterparts. Thus, **17f** (3,4-dichlorophenyl; $IC_{50} = 5.7$ nM) was very similar to **24c** ($IC_{50} = 9.6$ nM), and 8-thia **17g** (DAT $IC_{50} = 16$ nM) was similar to 8-carba **24d** ($IC_{50} = 27$ nM). Although, in general, the binding potencies of the nonnitrogen compounds were weaker than those of their 8-aza progenitors, the IC_{50} values observed for these 8-hetero compounds nevertheless showed an astounding propensity to bind and inhibit the DAT. SERT inhibitory potency among the 3β-aryl analogs ($IC_{50} = 6.5$ to 60 nM) was considerably greater than that of cocaine, and almost equipotent to their DAT affinity.

Our work has concentrated on compounds that bear a 2β-substituent rather than a 2α-substituent. Indeed, in the chemistries that were utilized to prepare the 8-oxa and 8-thia compounds, the 2α-carbomethoxy analogs were generally absent and therefore never evaluated biologically. However, within the 8-carba series, samarium iodide reduction of the 2,3-ene to provide the 3α-aryl and 3β-aryl analogs gave rise to all four products **24** to **27** (Scheme 10-6). Consequently, 2β-carbomethoxy-3β-aryl (**24**), 2α-carbomethoxy-3β-aryl (**25**), 2β-carbomethoxy-3α-aryl (**26**), and 2α-carbomethoxy-3α-aryl (**27**) were all available for biological evaluation.

It had been assumed that C_2 substitution was a requirement for inhibition of the DAT [33,56,72–74]. The dogma of 2β-orientation had guided work in the tropane area, and 2α-compounds have generally not been evaluated [56,75]. The assumption that 2α-substitution leads to considerably less potent DAT inhibitors and therefore less interesting compounds was well founded in earlier work. Clarke et al. [18] had reported that a loss of locomotor stimulant activity occurred in the 3-phenyltropanes when the C_2 substituent was moved from a 2β to a 2α orientation. In a study on cocaine, Reith [33] reported that an equatorial (2α-carbomethoxy) position significantly diminished potency in a behavioral assay compared with the axial 2β-carbomethoxy compound. Ritz [72] confirmed this finding in a ligand-binding assay and reported that the potency ratio between the 2α-carbomethoxy-3-phenyltropane and 2β-carbomethoxy-3-phenyltropane was 1480, thus rendering the 2α-isomer comparatively inactive. In 1989 [52,70], Madras reported that "cocaine receptors" were labeled by 2β-carbomethoxy-3β-(4-fluorophenyl)tropane, WIN 35,428, but not by the 2α-isomer. In the mid-1990s, Carroll [76,77] reported that 3β-phenyl-2α-(3′-methyl-1′,2′,4′-oxadiazol-5′-yl)tropane was a considerably less potent inhibitor of the DAT than the 2β-analog. Within the series of bridge-hydroxylated-8-azatropanes, the 7β-hydroxy-2β-carbomethoxy-3β-(3,4-dichlorophenyl)tropane **O-1164** had a binding potency of $IC_{50} = 1.4$ nM. In contrast, the 7β-hydroxy-2α-carbomethoxy-3β-(3,4-dichlorophenyl)tropane **O-2016** was 35-fold weaker [71].

In contrast, although 2β-carbomethoxytropane, WIN 35,428, had a DAT $IC_{50} = 11$ nM, the 2α-carbomethoxy isomer had an $IC_{50} = 633$ nM. It was therefore considerably weaker than its 2β-counterpart. However, the more potent 8-azatropane,

2β-carbomethoxy-3β-(3,4-dichloro)phenyltropane, **O-401**, and its 2α-carbomethoxy counterpart manifested similar DAT potencies (IC_{50} = 1.1 and 2.0 nM, respectively). Similarly, the 2β-carbomethoxy-3β-(2-naphthyl) and 2α-carbomethoxy-3α-(2-naphthyl) compounds had manifested identical DAT inhibitory potencies (IC_{50} = 0.5 nM) [36]. In 1997, Trudell [12] had reported that the 2α-homologated ester 3β-benzyl-2α-(carbomethoxy)methyl tropane was about equipotent to its 2β analog.

Indeed, among the 8-carbatropanes (Scheme 10-6) there was a limited difference between their DAT inhibitory potencies within the 3,4-dichlorophenyl or 3-(2-naphthyl) series. DAT IC_{50} values ranged from 10 to 42 nM. The range for SERT inhibition was much broader (IC_{50} = 33 to 807 nM) [36]. It is possible that in the search for new pharmacotherapies, 2α-configured compounds have not been explored sufficiently.

Among the 2-carbomethoxybenztropine analogs, the (1S)-8-aza-2-carbomethoxy-3-diarylmethoxytropane **35** was the only potent DAT inhibitor (IC_{50} = 10.9 nM) within the series of eight possible isomers [62] (Table 10-4). However, the SAR of these 2-carbomethoxy-3-diarylmethoxytropanes was different from that of the 3-aryltropanes and resembled more closely the SAR, manifested by the GBR family of compounds [26]. Therefore, it was not entirely surprising that exchange of the 8-aza for an 8-oxa in the diarylmethoxytropanes resulted in a loss of potency at DAT. In fact, all substitutions manifested IC_{50} values greater than 500 nM, at the DAT and SERT. Interestingly, only the 2β,3α compound, analogous to **35**, showed any potency whatsoever (520 nM) [38]. Similarly, in the absence of a C_2-carbomethoxy group, the 8-oxabenztropines showed no affinity for the DAT or SERT [78]. This further supports the notion that these diarylmethoxytropanes bind at a different site on the DAT than do the 3-aryltropanes [26,79,80].

TABLE 10-4   Comparison of 2-Carbomethoxy-8-aza and 8-Oxabenztropines[a]

| Configuration | Compound | X = NCH$_3$ (1S) | | Compound | X = O | |
		DAT	SERT		DAT	SERT
2β,3β	**32**	>3 μM	>2 μM	**O-874**	>10 μM	>2 μM
2α,3β	**33**	>9 μM	>2 μM	**O-872**	>20 μM	>2 μM
2α,3α	**34**	>1 μM	>2 μM	**O-871**	>22 μM	>2 μM
2β,3α	**35**	10.9	>2 μM	**O-873**	520	>2 μM

[a]IC_{50} (nM) for inhibition of [^{3}H]WIN 35,428 binding (DAT) and [^{3}H]citalopram binding (SERT).

10.4.2   Selectivity of Inhibition of the DAT vs. the SERT

Among these classes of compounds are numerous examples of potent SERT inhibitors. In the search for a cocaine pharmacotherapy, we focused initially on DAT inhibition and therefore selected aromatic substituents known to favor DAT inhibitory potency. Furthermore, we learned that among the three topological classes of 2,3-ene, 3α-aryl, and 3β-aryl there was a marked difference in selectivity of DAT inhibition, independent of the identity of the 8-heteroatom. This selectivity was a consequence of their reduced binding potency at SERT compared with that at DAT (Table 10-5). Thus, in a comparison of 3,4-dichlorophenyl-substituted compounds, the 3β-aryl analogs were generally quite potent SERT inhibitors (IC_{50} = 2.5 to 33 nM) and therefore least selective (1.4- to 3.4-fold) for DAT. The 3α-aryl compounds were slightly more DAT selective (8- to 68-fold; SERT IC_{50} = 27 to 166 nM) than were the 3β-aryl compounds. The 2,3-unsaturated compounds were weak SERT inhibitors (IC_{50} = 867 nM to 5 μM) and were therefore most selective (460- to 1000-fold) for DAT [81].

The introduction of 3-(2-naphthyl) favored increased SERT affinity in all classes. Thus, the 3β-(2-naphthyl) and the 3α-(2-naphthyl) compounds had similar DAT inhibitory potencies (3β: IC_{50} = 0.5 to 27 nM; 3α: IC_{50} = 0.6 to 17 nM) and SERT inhibitory potencies (3β: IC_{50} = 2 to 60 nM; 3α: IC_{50} = 6 to 64 nM), and therefore both lacked selectivity. The 2,3-unsaturated-3-(2-naphthyl) analogs had a lower affinity for SERT (IC_{50} = 109 nM to >1 μM) and therefore some selectivity (38- to 163-fold). Therefore, in contrast to the 3,4-dichlorophenyl compounds, these 3-(2-naphthyl) compounds must interact with the acceptor site on the SERT far more effectively than do the 3,4-dichlorophenyl compounds [81].

It is noteworthy that the conformational or topological effects of inhibition of both the DAT and SERT were conserved among these four classes of tropanes and appeared to be independent of which heteroatom was present at the 8-position. This selectivity, which derives from reduced SERT potency, is probably a result of the differences in orientation of the 3-aryl ring. This orientation is controlled by the conformation adopted by the tropane system. While the 3β-aryltropane (the aryl ring is equatorial) undoubtedly favors a chair conformation, as evidenced by NMR and x-ray diffractions studies [36], the conformation within the 3α-aryl class is less certain. Although ^{1}H NMR and x-ray studies indicated that a boat conformation was favored (the aryl ring is again equatorial) [38], there have been examples where the conformation in solution was different from that obtained in the solid state from x-ray diffraction analysis. As example, a (1R)-7β-hydroxytropane (Fig. 10-4) clearly adopted a boat conformation in CDCl$_3$, with an equatorial 3,4-dichlorophenyl group (^{1}H NMR). However, in the solid state, the molecule existed in the chair conformation with an axial 3-aryl substituent. ^{1}H NMR studies conducted in CD$_3$OD/D$_2$O showed a pseudo-boat conformation for the (1S)-3α-aryl analog of this compound [71]. One might conclude that it is questionable whether an extrapolation of structure obtained (^{1}H NMR or x-ray crystallography) outside the influence of the biological medium to a putative three-dimensional structure within the biological system has relevance for the differentiation of biological activity between these chair and boat conformers [71].

TABLE 10-5 Bioenantioselectivity: Binding Affinities of 8-Heterotropanes in the 3-(2-Naphthyl) and 3,4-Dichlorophenyl Series[a]

3,4-Dichlorophenyl series:

X	Compound	DAT	SERT	SERT/DAT	Compound	DAT	SERT	SERT/DAT	Compound	DAT	SERT	SERT/DAT
N (1R)	O-401	1.1	2.5	2.3	O-1157	0.4	27	68	O-1109	1.2	867	723
O (1R)	(1R)-6f	3.3	4.7	1.4	(1R)-7f	2.3	31	13	(1R)-5f	4.6	2.1 µM	460
S (1R)	(1R)-17f	2.0	3.0	1.5	(1R)-16f	4.9	39	8	(1R)-15f	2.5	2.6 µM	1040
CH$_2$ (R/S)	24c	9.6	33	3.4	26c	13	166	13	22c	7.1	5.2 µM	727

3-(2-Naphthyl) series:

X	Compound	DAT	SERT	SERT/DAT	Compound	DAT	SERT	SERT/DAT	Compound	DAT	SERT	SERT/DAT
N (1R)	O-1229	0.49	2.2	4	O-1228	0.6	6.0	10	O-1173	2.9	109	38
O (1R)	(1R)-6g	3.0	5.0	2	(1R)-7g	5.0	10	2.0	(1R)-5g	6.0	223	37
S (1R/S)	17g	16	13	1	16g	8.0	36	4.5	15g	8.0	>1 µM	163
CH$_2$ (R/S)	24d	27	60	2	26d	17	64	3.8	22d	11.0	>1 µM	119

[a]IC$_{50}$ (nM) for inhibition of [^{3}H]WIN 35,428 binding (DAT) and [^{3}H]citalopram binding (SERT).

287

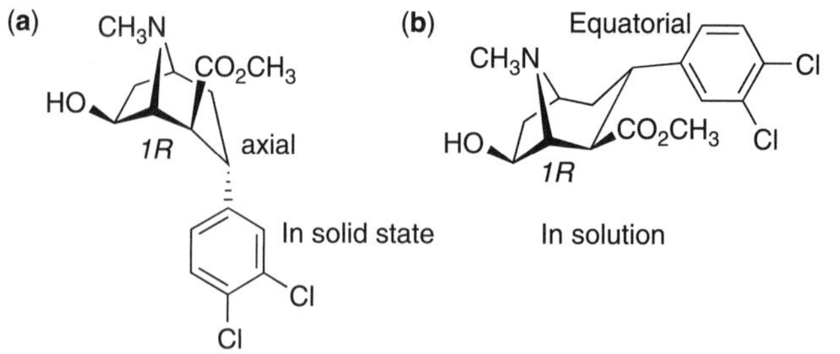

Figure 10-4 Configurations of a 3α-aryltropane in (a) solid form and (b) solution.

This potential conformational flexibility is probably a consequence of a small energy barrier between the chair and the boat, or pseudoboat, conformations for 3α-aryl compounds. No modeling studies that mimic the biological environment of the DAT and SERT have been undertaken. Therefore, one can only speculate on the differences in 3-aryl ring conformation of the fully saturated tropanes within the biological environment. However, this orientation is much more certain within the class of 2,3-unsaturated compounds. Here modeling, x-ray, and ^{1}H NMR studies have confirmed that the skeleton and ring is "flattened" [81].

Consequently, it is evident that the presence of a nitrogen in a DAT or SERT ligand is not a requirement for potent binding. Therefore, ionic binding to a DAT residue (e.g., Asp[79]) is not necessary for DAT inhibition. Furthermore, hydrogen bonding, available to both 8-aza and 8-oxa compounds, is also not essential for DAT inhibition because both the 8-thia and 8-carba analogs inhibit DAT and SERT and neither can engage in hydrogen bonding with an acceptor amino acid on the transport mechanisms. Consequently, binding must occur by a three-dimensional, topological interaction that involves molecular shape and not three-point direct intermolecular interactions. The compounds must orient themselves to optimize interaction with the acceptor site(s) on the DAT, and there must be a "pocket" in which the 3-aryl residue can insert itself. That insertion is probably modulated by the opposite end of the molecule: the 8-hetero atom. In support of this hypothesis, studies had shown that whereas 3,4-dichlorophenyl or 3-(2-naphthyl) substitution was optimum on 8-unsubstituted azatropanes, once bulk was introduced at N_8, the 3,4-dichlorophenyl and 3-(2-naphthyl) compounds lost their potency to inhibit DAT [82]. This may be a consequence of a restricted lipophilic space to accept the larger 3-(3,4-dichlorophenyl) or 3-(2-naphthyl) moieties.

Lieske [83] had explored the dimensions of the aromatic pocket in the 8-aza series and concluded that the binding site was narrow and only slightly greater in diameter than that required for an aromatic ring. An investigation of this aromatic pocket in the 8-oxatropanes confirmed some of its limitations. The orientation of the naphthyl group at C_3 (Table 10-6) proved critical to DAT and SERT binding. Thus the trans-verse 3-(1-naphthyl) (**O-1170**) did not inhibit DAT or SERT; only the linear 3-(2-naphthyl) (**5g**) provided effective inhibition (DAT $IC_{50} = 19.7$ nM). It is interesting that the 1- and 2-naphthyl cocaine analogs (Table 10-7) showed very similar DAT

TABLE 10-6   Comparison of 3-(1-Naphthyl)- and 3-(2-Naphthyl)-8-oxatrop-2-enes[a]

Ar	Compound	DAT	SERT
3-(1-Naphthyl)	O-1170	2 μM	25 μM
3-(2-Naphthyl)	**5g**	19.7	520

[a]IC_{50} (nM) for inhibition of [^{3}H]WIN 35,428 binding (DAT) and [^{3}H]citalopram binding (SERT).

inhibitory potencies ($IC_{50} = 742$ and 327 nM, respectively) [16]. This may be a consequence of rotational flexibility of the C_3-ester linkage and the consequent ability of the aromatic system to rotate to accommodate the aromatic binding pocket of the DAT.

In a further probe of available space (Table 10-8), the linear naphthalenes in the 2,3-ene (**5g**), 3β-aryl (**6g**), and 3α-aryl (**7g**) all inhibited the DAT ($IC_{50} = 7$ to 20 nM) and SERT ($IC_{50} = 10$ to 520 nM), while the "bent" phenanthrenyl compounds (**O-1958**, **O-1970**, **O-1969**) lost considerable DAT potency ($IC_{50} = 50$ to 90 nM), and the "extended" anthracenyl analogs (**O-1537**, **O-1505**) lost all potency at the DAT and SERT [38]. Therefore, there must be a molecular barrier offered by the DAT and SERT at the aromatic acceptor site for these polyaromatic compounds that probe more deeply. However, Trudell [12] had reported that a 3-biaryltropane could inhibit the DAT effectively. This may be analogous to the more rigid "wider" phenanthrene of our studies as well as the cocaine analogs (Table 10-7), and may reflect an additional aromatic acceptor site deeper within the lipophilic pocket. Cini et al. [84] have reported a series of novel 3-aza-6,8-dioxabicyclo[3.2.1]octanes

TABLE 10-7   Comparison of 3-(1-Naphthyl) and 3-(2-Naphthyl) Cocaine Analogs

Ar	DAT
3-(1-Naphthyl)	742
3-(2-Naphthyl)	327

Source: Data from Singh [16].
[a]Displacement (IC_{50} (nM)) of [^{3}H]-cocaine.

TABLE 10-8 Comparison of Binding Affinities of Aryl Probes in the 8-Oxatropanes[a]

Ar	Compound	DAT	SERT	Compound	DAT	SERT	Compound	DAT	SERT
3-(2-Naphthyl)	**5g**	19.7	520	**6g**	7	10	**7g**	12	18
3-(2-Anthracenyl)	O-1537	40 μM	14 μM		na	na	O-1504	50 μM	50 μM
3-(2-Phenanthrenyl)	O-1958	90	200	O-1970	50	39	O-1969	63	67

[a] IC_{50} (nM) for inhibition of [^{3}H]WIN 35,428 binding (DAT) and [^{3}H]citalopram binding (SERT). na, not available.

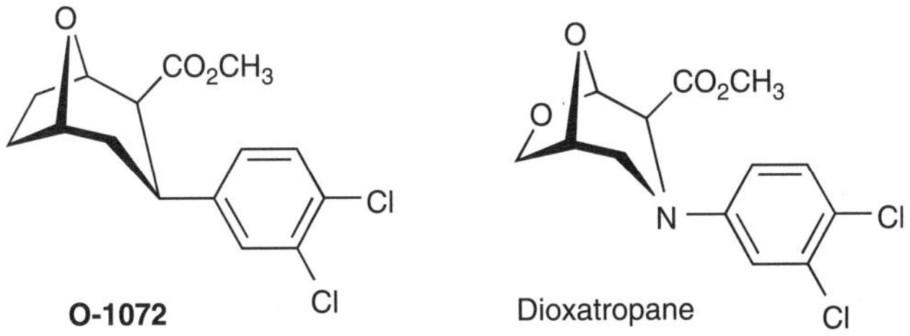

Figure 10-5   Comparison of (1*R*)-**6f** and a dioxa-3-azatropane.

in which an additional oxygen (replacing C_6) and a nitrogen (replacing C_3) were introduced. DAT and SERT affinities were explored and these compounds were found to be less potent inhibitors than the 8-oxatropanes. However, the SAR of DAT inhibition was found to be very similar to that shown by the 8-oxatropanes [38]. The most potent compound proved to be the 3,4-dichlorophenyl analog (Fig. 10-5). A computational comparison of (1*R*)-**6f** and the dioxatropane showed a striking topological similarity, specifically with respect to the bicyclic structure and the orientation of the N_3 aromatic ring. In a modeling study, (1*R*)-**6f** and the dioxatropane could readily be superimposed. However, the electrostatic potential of the bicyclic scaffolds proved quite different. Cini et al. concluded that while the bicyclo [3.2.1]octane topology favored interaction with the DAT and SERT, the charge distribution within the bicyclic system of the dioxatropane may modulate the affinity observed toward DAT and SERT [84].

10.4.3   Biological Enantioselectivity of Inhibition of the DAT

On the assumption that topology is the driving force behind interaction with the DAT and SERT, absolute structure must be expected to play a role in tight binding. Biological and pharmacological enantioselectivity have been reported for (1*R*)-8-azatropanes [33,62,70,85–87]. This enantioselectivity is evident among the 8-hetero-tropanes, albeit to different degrees (Table 10-9). Within the 8-aza series, biological enantioselectivity is particularly marked (408-fold), as evidenced by the 2,3-ene (1*R*)-**O-1109** (DAT $IC_{50} = 1.2$ nM) and (1*S*)-**O-1120** (DAT $IC_{50} = 490$ nM). In contrast, the 8-oxa-2,3-ene **5f** manifested less biological enantioselectivity (12-fold), and the 8-thia-2,3-ene **15f** was least selective (2-fold). Among the 3α-(3,4-dichlorophenyl) and 3β-(3,4-dichlorophenyl) compounds, the 8-thia analogs **16f** and **17f** were less biologically enantioselective (2- and 11-fold) than the 8-oxa compounds **7f** and **6f** (24- and 14-fold). An explanation of this reduction in enantioselectivity is not immediately evident.

10.5   PHARMACOLOGICAL STUDIES

The search for cocaine medication begins with the design and synthesis of molecules that bind and inhibit monoamine-uptake systems, particularly the DAT.

TABLE 10-9 Binding Enantioselectivity at DAT and SERT[a]

Structure (left): CO_2CH_3, 3,4-dichlorophenyl
X = S: **15f**
X = O: **5f**
X = NCH3: **O-1109, 1120**
X = CH2: **22c**

Structure (middle): CO_2CH_3, 3,4-dichlorophenyl
X = S: **16f**
X = O: **7f**
X = NCH3: **O-1157**
X = CH2: **26c**

Structure (right): CO_2CH_3, 3,4-dichlorophenyl
X = S: **17f**
X = O: **6f**
X = NCH3: **O-401**
X = CH2: **24c**

Enantiomer	Compound	IC_{50} (nM) DAT	SERT	SERT/DAT[b]	Compound	IC_{50} (nM) DAT	SERT	SERT/DAT[b]	Compound	IC_{50} (nM) DAT	SERT	SERT/DAT[b]
X = S												
(1R/1S)	**15f**	4.5	>3 µM	800	**16f**	6.9	99	14	**17f**	5.7	8.0	1.4
(1R)	**15f**	2.5	>2 µM	1040	**16f**	4.9	39	8	**17f**	2.0	3.0	1.5
(1S)	**15f**	5.0	>4 µM	800	**16f**	10	139	14	**17f**	22	90	4
Bioenantioselectivity[c]		2.0	1.5			2.0	3.6			11	30	
X = O												
(1R/1S)	**5f**	12	>6 µM	600	**7f**	3.1	64.5	21	**6f**	3.3	6.5	2
(1R)	**5f**	4.6	>2 µM	460	**7f**	2.3	31.0	13	**6f**	3.3	4.7	1.4
(1S)	**5f**	58	>40 µM	805	**7f**	56	>2 µM	51	**6f**	47	58	1.2
Bioenatioselectivity[c]		12.6	22			24	92			14	12	
X = NCH3												
(1R)	**O-1109**	1.2	867	723	**O-1157**	0.4	27	68	**O-401**	1.1	2.5	2.2
(1S)	**O-1120**	490	>2 µM	4.0	na				na			
Bioenantioselectivity[c]		408	2.2									
X = CH2												
(1R/1S)	**22c**	7.1	>5 µM	727	**26c**	13	166	13	**24c**	9.6	33	3.4

Source: X = NCH3 data taken from Meltzer et al. [56], X = O, data from Meltzer et al. [38], X = S, X = CH2 data are for the racemic compounds and are from Meltzer et al. [36].
[a]Inhibition of $[^3H]$WIN 35,428 binding to the DAT and $[^3H]$citalopram binding to the SERT. na, not available; [b]Preference for DAT inhibition over SERT inhibition; [c]Biological enantioselectivity is given as the ratio of IC_{50} distomer/IC_{50} eutomer.

Clearly, 8-azatropanes inhibit the DAT, cross the blood-brain barrier, and label brain regions rich in monoamine transporters in vivo [27,82,88–90] and lead to pharmacological sequellae [17]. It remains to be shown that nonnitrogen tropanes can satisfy all these criteria. It is clear that 8-heterotropanes inhibit DAT and SERT potently and selectively. Moreover, four compounds (8-oxa and 8-carbatropanes: **O-913, O-1072, O-1391, O-1809**) were evaluated in a broad assay of 30 receptors and displayed a complete lack of inhibitory potency at any of these receptors [91]. The question remains: Can these compounds enter the living brain to affect their activities in living animals? A limited number of pharmacological studies have been performed on these 8-heterotropanes. Results are summarized in Table 10-10 and described in the following sections.

10.5.1 Stimulation of Locomotor Activity

The ability to stimulate locomotor activity in rodents provides evidence of drug penetration into the brain. In the case of compounds that selectively inhibit monoamine uptake, such stimulation also provides evidence of inhibition of the DAT in vivo. The 8-oxatropanes had varied stimulatory activities with no clear SAR evident. Thus, there are examples of both stimulants and depressants of locomotor activity. The (1R)-enantiomer of 2-carbomethoxy-3-(3,4-dichlorophenyl)-8-oxabicyclo[3.2.1]oct-2-ene, **O-1059** [(1R)-**5f**], was an interesting example. This compound had substantial DAT inhibitory potency ($IC_{50} = 4.6$ nM) and >460-fold DAT vs. SERT selectivity [38]. In studies of locomotor activity, **O-1059** was a potent stimulant ($ED_{50} = 3.0$ mg/kg), with a duration of action of up to 5.5 hours. In sharp contrast, the racemic mixture **O-1014** [(1R/1S)-**5f**], suppressed locomotor activity ($ID_{50} = 20$ mg/kg) in a time- and dose-dependent manner (duration of action: 2 hours) [92]. The differences between the results for the enantiomerically pure **O-1059** and the racemic mixture **O-1014** are striking and may be of significance to studies presented later.

The enantiomerically pure (**O-1066**) and racemic (**O-913**) 2β-carbomethoxy-3α-(3,4-dichlorophenyl)-8-oxatropanes provided the opposite results. Here the 1R-enantiomer (**O-1066**) failed to stimulate locomotor activity and was a weak depressant ($ID_{50} = 77$ mg/kg), whereas the racemic (**O-913**) was a fairly potent stimulant ($ED_{50} = 6.8$ mg/kg), with an onset time of 1 hour and a duration of action of about 3 hours. The racemic 3β-aryl analog (**O-914**) was also a potent stimulant ($ED_{50} = 8.1$ mg/kg), with an onset of 10 minutes and a duration of action >8 hours. Here the enantiomerically pure isomer (1R)-**O-1072** was also a long-acting stimulant ($ED_{50} = 4.2$ mg/kg; >8 hours).

In the 2β-carbomethoxy-3-(4-dichlorophenyl)-8-oxatropanes, the racemic 3α-aryl analog (**O-915**) suppressed locomotor activity, whereas the racemic 3β-aryl analog (**O-916**) stimulated activity ($ED_{50} = 5.8$ mg/kg). The racemic 2β-carbomethoxy-3-(4-bromophenyl)-8-oxatropane (**O-1597**) and the 3-(4-bromophenyl)-8-oxatrop-2-ene (**O-1584**) both stimulated locomotor activity potently ($ED_{50} = 1.2$ and 7.6 mg/kg, respectively) for a long period (>8 and 2.5 hours, respectively). The 3α-aryl-8-thiatropane **O-2751**, with onset time

TABLE 10-10 Summary of Pharmacological Data

Structure	Compound	Stim. Loc. Act. Mice[a] (ED$_{50}$ mg/kg)	Onset (min)	Duration (min)	DD[b] Rates ED$_{50}$ (mg/kg)	Primate ED$_{50}$ (mg/kg)	DAT IC$_{50}$ (nM)	SERT IC$_{50}$ (nM)
	O-1072 1R	4.2	10	$\geq$480	i.p. 4.1 (f)	10 (p)	3.9	4.7
	O-914 1R/S	8.1	10	$\geq$480	i.p.32.9 (f)	—	3.4	6.5
	O-916 1R/S	5.8	10	$\geq$480	i.p. 7.5 (f) po. 4.6 (f)	2.4–24	10	107
	O-915 1R/S	25 (d)	—	—	Failed	—	28.5	816
	O-1066 1R	77 (d)	10	170	i.p. (p) p.o. (p) (100 mg/kg)	—	2.3	30.3
	O-913 1R/S	6.8	50	180	p.o. (p) (100 mg/kg)	0.29	3.1	64.5

Structure	Compound						
	O-1014 1R/S-**5f**	20 (d)	20	130	—	—	12.3	>2 μM
	O-1059 1R-**5f**	3.0	80	330	i.p. 90.6 (f) p.o. (p) (500 mg/kg)	—	4.6	>2 μM
	O-1597 1R/S	1.2	10	≥480	i.p. 1.2 (f) p.o. 0.76 (f)	—	10	13
	O-1584	7.6	10	150	Failed	—	62	>10 μM
	O-2751	≥45	20	170	—	—	6.9	99

[a]Stimulation of locomotor activity in mice over 8 hours; (d), depression of locomotor activity (ID_{50}).

[b]Drug discrimination : ability to substitute for the discriminative stimulus effects of cocaine (rats: 10 mg/kg; Monkeys (4) 0.4 mg/kg); (f), substituted fully; (p), substituted partially.

of 20 minutes and a duration of action of 170 minutes, was a weak stimulant of loco-motor activity in mice ($ED_{50} > 45$ mg/kg).

10.5.2 Drug Discrimination Studies

The ability of a drug to substitute for the discriminative stimulus effects of cocaine in rats or monkeys provided a measure of how cocainelike the compounds were per-ceived to be. Compound **O-1059** fully substituted for cocaine (10 mg/kg), with an $ED_{50} = 90.6$ mg/kg in an intraperitoneal rat drug discrimination study. Orally, it was less potent and substituted only partially after 360 minutes at a dose of 500 mg/kg [92]. Both compound **O-913**, a stimulant, and **O-1066**, a depressant, in the locomotor activity study, substituted partially for cocaine in the drug discrimi-nation paradigm in rats at a dose of 100 mg/kg. In monkeys, **O-913** was only slightly less potent than WIN 35,428 or cocaine (**O-913**: $ED_{50} = 0.29$ mg/kg; WIN 35,428: $ED_{50} = 0.14$ mg/kg; and cocaine: $ED_{50} = 0.14$ mg/kg) [91].

The 3β-aryl-8-oxatropane analogs **O-914** and **O-1072**, which both stimulated loco-motor activity, also substituted fully for cocaine in rats ($ED_{50} = 4.1$ and 32.9 mg/kg, respectively). In monkeys, **O-1072** substituted partially for cocaine (10 mg/kg). The 4-chloro compound **O-915**, which had suppressed locomotor activity, failed in the drug discrimination paradigm, but its racemic mixture, **O-916**, was active in the rat model (ip.: $ED_{50} = 7.5$ mg/kg and p.o.: $ED_{50} = 4.6$ mg/kg) as well as in the primate model ($ED_{50} = 2.4$ to 24 mg/kg). Among the most potent compounds in the drug discrimination assay was the 4-bromophenyl **O-1597**. It was a potent stimulant of locomotor activity and fully substituted for cocaine in rats ($ED_{50}=$ intraperitoneal (i.p.): 1.2 mg/kg and per kg (p.o.): 0.76 mg/kg).

10.5.3 Autoradiography

In vitro autoradiographic studies were conducted with the $(1R)$-enantiomer **O-1059** [$(1R)$-**5f**]. This compound was tritiated on the methyl ester and the autoradiographic distribution measured in tissue sections of rhesus monkey brain at the level of the striatum [93]. High levels of binding sites were prominent in the caudate and putamen and to a lesser extent in the nucleus accumbens. This distribution of **O-1059** in brain correlated closely to the distribution that had been observed with selective 8-azatropanes such as [^{3}H]WIN 35,428 [94,95], [^{125}I]-altropane [89], and RTI-55 [96].

10.5.4 Positron Emission Tomography

A PET study conducted with the radiolabeled racemic mixture of [^{11}C]$(1R/1S)$-2-carbomethoxy-3-(3,4-dichlorophenyl)-8-oxabicyclo[3.2.1]oct-2-ene **O-1014** [$(1R/1S)$-**5f**] showed that it distributed to the dopamine rich striatum [97]. The cerebellum/striatum ratio was 2.5. This ratio was substantially lower than that seen on the autoradiographic study ($>$50-fold) [93]. However, this could have been a consequence of in vivo metabolism, poor transport, or of the fact that here the racemic compound **O-1014** [(1*R*/1*S*)-**5f**] was used, while the autoradiographic study was conducted with the enantiomerically pure compound (1*R*)-**O-1059**.

Madras et al. recently evaluated certain compounds, including three 8-heterotropanes, as potential pharmacotherapies for Parkinson's disease [98]. Surprisingly, the brain occupancy of the racemic 8-oxa compound **O-1014** [(1*R*/1*S*)-**5f**], the 8-carba compound **O-1231** (**26c**), and the 8-oxa compound **O-1973**, in monkeys, as measured in a PET study in which [^{11}C]WIN 35,428 was displaced by the test compounds rather than by radiolabeling of the test compounds themselves as had been reported for **O-1059** [(1*R*)-**5f**] [91], was less than 15%. Furthermore, there was little impact on amelioration of a variety of Parkinsonian symptoms upon treatment with these compounds. These results may have been affected by the fact that (1*R*/1*S*)-**O-1014** was a racemic mixture and had already been shown to be a depressant in the locomotor activity paradigm. Furthermore, use of a racemic mixture in the displacement of enantiomerically pure [^{3}H]WIN 35,428 may not be an accurate reflection of occupancy by enantiomerically pure (1*R*)-**O-1059** itself.

10.6 CONCLUSION

The search for pharmacotherapies for cocaine addiction and for other diseases in which the inhibition of monoamine neurotransmitters in the CNS is implicated has involved the exploration of diverse molecular classes. The 8-azabicyclo[3.2.1]octanes (8-azatropanes) have played a leading role among those classes that target the dopamine and serotonin transporters. The most clinically advanced example of these 8-azatropanes is RTI-336 [99]. The related class of 8-hetero-(nonnitrogen)-bicyclo[3.2.1]octanes (8-heterotropanes) was developed to explore the functional role of a nitrogen in the tropane skeleton with respect to inhibition of monoamine uptake systems. In addition, once it had been proven that such nonnitrogen tropanes inhibited cocaine binding at the DAT and SERT, the driving question became: Can 8-nonnitrogen tropanes provide biologically and pharmacologically feasible pharmacotherapies? In this chapter we have provided evidence that supports the possibility of development of 8-oxa-, 8-thia-, and even 8-carbatropanes as pharmacotherapies. Indeed, these nonnitrogen tropanes can cross the blood–brain barrier, enter the CNS, and occupy pharmacologically relevant sites in the brain. Most important, these compounds can elicit pharmacological and behavioral effects that are a consequence of interaction with monoamine binding sites in the striatum.

ACKNOWLEDGMENTS

The author acknowledges the contributions of numerous collaborators without whom these studies would not have been possible. Dr. Bertha K. Madras of the New England Regional Primate Research Center, an outstanding collaborator for over a decade, provided all biological data. The chemistry was performed by an enthusiastic and skilled team at Organix: Heather Blanchette, Paul Blundell, Zhengming Chen, Pat Donavan, Hong Huang, Muthusamy Jayaraman, Olga

Kryatova, Shanghao Liu, Sreeletha Panicker, Duy-Phong Pham Huu, Madhusudhan Purushotham, Lokman Torun, Bing Wang, Pinglang Wang, and Yaw Fui Yong. Pharmacological data were provided through the National Institute on Drug Abuse (NIDA) auspices of the University of North Texas Health Science Center (DA7-8076), McLean Hospital (DA7-8073), and the NIDA intramural research program, and by Dr. Madras, New England Regional Primate Research Center (DA06303, DA09462). This work was funded by the NIDA (DA1-8825, DA 011542).

REFERENCES

1. Spealman, R. D. (1993). Modification of behavioral effects of cocaine by selective serotonin and dopamine uptake inhibitors in squirrel monkeys. *Psychopharmacology, 112*, 93–99.

2. Rocha, B. A., Fumagalli, F., Gainetdinov, R. R., Jones, S. R., Ator, R., Giros, B., Miller, G. W., and Caron, M. G. (1998). Cocaine self-administration in dopamine-transporter knockout mice. *Nat. Neurosci, 1* (2), 132–137.

3. Uhl, G. R., Hall, F. S., and Sora, I. (2002). Cocaine, reward, movement and monoamine transporters. *Mol Psychiatry, 7*, 21–26.

4. Kennedy, L. T., and Hanbauer, I. (1983). Sodium sensitive cocaine binding to rat striatal membrane: possible relationship to dopamine uptake sites. *J Neurochem, 34*, 1137–1144.

5. Kuhar, M. J., Ritz, M. C., and Boja, J. W. (1991). The dopamine hypothesis of the reinforcing properties of cocaine. *Trends Neurosci, 14*, 299–302.

6. Ritz, M. C., Lamb, R. J., Goldberg, S. R., and Kuhar, M. J. (1987). Cocaine receptors on dopamine transporters are related to self-administration of cocaine. *Science, 237*, 1219–1223.

7. Rothman, R. B., Mele, A., Reid, A. A., Akunne, H. C., Greig, N., Thurkauf, A., deCosta, B. R., Rice, K. C., and Pert, A. (1991). GBR 12909 antagonizes the ability of cocaine to elevate extracellular levels of dopamine. *Pharmacol Biochem Behav, 40*, 387–397.

8. Carroll, F. I., Lewin, A. H., and Kuhar, M. J. (1998). 3β-Phenyl-2β-substituted tropanes and SAR analysis. *Med Chem Res, 8*, 59–65.

9. Meltzer, P. C., Blundell, P., and Madras, B. K. (1998). Structure—activity relationships of inhibition of the dopamine transporter by 3-arylbicyclo[3.2.1]octanes. *Med Chem Res, 8*, 12–34.

10. Tamagnan, G., Baldwin, R. M., Kula, N. S., Baldessarini, R. J., and Innis, R. B. (2000). Synthesis and monoamine transporter affinity of 2β-carbomethoxy-3β(2″-, 3″- or 4″-substituted) biphenyltropanes. *Bioorg and Med Chem Lett, 10*, 1783–1785.

11. Kelkar, S. V., Izenwasser, S., Katz, J. L., Klein, C. L., Zhu, N., and Trudell, M. L. (1994). Synthesis, cocaine receptor affinity, and dopamine uptake inhibition of several new 2β-substituted 3β-phenyltropanes. *J Med Chem, 37*, 3875–3877.

12. Lomenzo, S. A., Izenwasser, S., Katz, J. L., Terry, P. D., Zhu, N., Klein, C. L., and Trudell, M. L. (1997). Synthesis, structure, dopamine transporter affinity, and dopamine uptake inhibition of 6-alkyl-3-benzyl-2-[(methoxycarbonyl)methyl]tropane derivatives. *J Med Chem, 40*, 4406–4414.

13. Newman, A. H., and Kulkarni, S. S. (2002). Probes for the dopamine transporter: new leads toward a cocaine-abuse therapeutic—a focus on analogues of benztropine and rimcazole. *Med Res Rev*, *22*, 429–464.

14. Newman, A. H. (1998). Novel dopamine transporter ligands: the state of the art. *Med Chem Res*, *8*, 1–11.

15. Kozikowski, A. P., Saiah, M. K. E., Bergmann, J. S., and Johnson, K. M. (1994). Structure–activity relationship studies of *N*-sulfonyl analogs of cocaine: role of ionic interaction in cocaine binding. *J Med Chem*, *37*, 3440–3442.

16. Singh, S. (2000). Chemistry, design, and structure–activity relationship of cocaine antagonists. *Chem Rev*, *100*, 925–1024.

17. Runyon, S. P., and Carroll, F. I. (2006). Dopamine transporter ligands: recent developments and therapeutic potential. *Curr Topics in Med Chem*, *6*, 1825–1843.

18. Clarke, R. L., Daum, S. J., Gambino, A. J., Aceto, M. D., Pearl, J., Levitt, M., Cumiskey, W. R., and Bogado, E. F. (1973). Compounds affecting the central nervous system. 4: 3β-Phenyltropane-2-carboxylic esters and analogs. *J Med Chem*, *16*, 1260–1267.

19. Deutsch, H. M. (1998). Structure–activity relationships for methylphenidate analogs and comparisons to cocaine and tropanes. *Med Chem Res*, *8*, 91–99.

20. Gatley, S. J., Pan, D., Chen, R., Chaturverdi, G., and Ding, Y. S. (1996). Affinities of methylphenidate derivatives for dopamine, norepinephrine and serotonin transporters. *Life Sci*, *58*, 231–239.

21. Carroll, F. I., Lewin, A. H., Boja, J. W., and Kuhar, M. J. (1992). Cocaine receptor: biochemical characterization and structure–activity relationships of cocaine analogues at the dopamine transporter. *J Med Chem*, *35*, 969–981.

22. Kitayama, S., Shimada, S., Xu, H., Markham, L., Donovan, D. H., and Uhl, G. R. (1993). Dopamine transporter site-directed mutations differentially alter substrate transport and cocaine binding. *Proc Natl Acad Sci U S A*, *89*, 7782–7785.

23. Cao, C. J., Young, M. M., Wong, J. B., Mahran, L. G., and Eldefrawi, M. E. (1989). Putative cocaine receptor in striatum is a glycoprotein with active thiol function. *Membr Biochem*, *8*, 207–220.

24. Xu, C., Coffey, L. L., and Reith, M. E. A. (1997). Binding domains for blockers and substrates on the dopamine transporter in rat striatal membranes studied by protection against *N*-ethylmaleimide-induced reduction of [^{3}H]WIN 35,428 binding. *Naunyn-Schmiedebergs Arch Pharmacol*, *355*, 64–73.

25. Meltzer, P. C., Liu, S., Blanchette, H. S., Blundell, P., and Madras, B. K. (2002). Design and synthesis of an irreversible dopamine-sparing cocaine antagonist. *Bioorg and Med Chem*, *10*, 3583–3591.

26. Meltzer, P. C., Liang, A. Y., and Madras, B. K. (1996). 2-Carbomethoxy-3-(diarylmethoxy)-1α*H*,5α*H*-tropane analogs: synthesis and inhibition of binding at the dopamine transporter and comparison with piperazines of the GBR series. *J Med Chem*, *39*, 371–379.

27. Carroll, F., Rahman, M., Abraham, P., Parham, K., Lewin, A., Dannals, R., Shaya, E., Scheffel, U., Wong, D., Boja, J., and Kuhar, M. (1991). [^{123}I]3β-(4-Iodophenyl)tropan-2-carboxylic acid methyl ester (RTI-55), a unique cocaine receptor ligand for imaging the dopamine and serotonin transporters in vivo. *Med Chem Res*, *1*, 289–294.

28. Newman, A. H., Cha, J. H., Cao, J., Kopajtic, T., Katz, J. l., Parnas, M. L., Vaughan, R., and Lever, J. R. (2006). Design and synthesis of a novel photoaffinity ligand for the

dopamine and serotonin transporters based on 2β-carbomethoxy-3β-biaryltropane. *J Med Chem, 49,* 6621–6625.

29. Madras, B. K., Pristupa, Z. B., Niznik, H. B., Liang, A. Y., Blundell, P., Gonzalez, M. D., and Meltzer, P. C. (1996). Nitrogen-based drugs are not essential for blockade of mono-amine transporters. *Synapse, 24,* 340–348.

30. Xu, C., and Reith, M. E. (1996). Modeling of the pH dependence of the binding of WIN 35,428 to the dopamine transporter in rat striatal membranes: Is the bioactive form posi-tively charged or neutral? *J Pharm Exp Ther, 278,* 1340–1348.

31. Chen, N. H., Ding, J. H., Wang, Y. L., and Reith, M. E. (1997). Potential misconceptions in dopamine transporter assays arising from the binding of [^{125}I]RTI-121 to filters: effects of ions and cocaine. *J Neurosci Methods, 75,* 179–186.

32. Wall, S. C., Innis, R. B., and Rudnick, G. (1993). Binding of the cocaine analog 2-β-carbomethoxy-3-β-(4-[^{125}I]iodophenyl)tropane to serotonin and dopamine transporters: different ionic requirements for substrate and 2-β-carbomethoxy-3-β-(4-[^{125}I]iodophenyl)tropane binding. *Mol Pharmacol, 43,* 264–270.

33. Reith, M. E. A., Meisler, B. E., Sershen, H., and Lajtha, A. (1986). Structural requirements for cocaine congeners to interact with dopamine and serotonin uptake sites in mouse brain and to induce stereotyped behavior. *Biochem Pharmacol, 35,* 1123–1129.

34. Carroll, F. I., Coleman, M. L., and Lewin, A. H. (1982). Synthesis and conformational analysis of isomeric cocaines: a proton and carbon-13 nuclear magnetic resonance study. *J Org Chem, 47,* 13–19.

35. Abraham, P., Pitner, J. B., Lewin, A. H., Boja, J. W., Kuhar, M. J., and Carroll, F. I. (1992). N-Modified analogs of cocaine: synthesis and inhibition of binding to the cocaine receptor. *J Med Chem, 35,* 141–144.

36. Meltzer, P. C., Blundell, P., Yong, Y. F., Chen, Z., George, C., Gonzalez, M. D., and Madras, B. K. (2000). 2-Carbomethoxy-3-aryl-8-bicyclo[3.2.1]octanes: potent non-nitrogen inhibitors of monoamine transporters. *J Med Chem, 43,* 2982–2991.

37. Meltzer, P. C., Wang, P., Blundell, P., and Madras, B. K. (2003). Synthesis and evaluation of dopamine and serotonin transporter inhibition by oxacyclic and carbacyclic analogues of methylphenidate. *J Med Chem, 46,* 1538–1545.

38. Meltzer, P. C., Liang, A. Y., Blundell, P., Gonzalez, M. D., Chen, Z., George, C., and Madras, B. K. (1997). 2-Carbomethoxy-3-aryl-8-oxabicyclo[3.2.1]octanes: potent non-nitrogen inhibitors of monoamine transporters. *J Med Chem, 40,* 2661–2673.

39. Brownbridge, P., and Chan, T.-H. (1979). A simple route to the 8-oxabicyclo[3.2.1]octyl and 9-oxabicyclo[3.3.1]nonyl systems: synthesis of the 8-oxa analog of cocaine. *Tetrahedron Lett, 46,* 4437–4440.

40. Hamada, A., Chang, Y. A., Uretsky, N., and Miller, D. D. (1984). Dopamine agonists: comparative actions of amine and sulfonium analogues of dopamine. *J Med Chem, 27,* 675–680.

41. Chang, Y. A., Ares, J., Anderson, K., Sabol, B., Wallace, R. A., Farooqui, T., Uretsky, N., and Miller, D. D. (1987). Dopamine agonists: effects of charged and uncharged analogues of dopamine. *J Med Chem, 30,* 214–218.

42. Harrold, M. W., Chang, Y. A., Wallace, R. A., Farooqui, T., Wallace, L. J., Uretsky, N., and Miller, D. D. (1987). Charged analogues of chlorpromazine and dopamine antagon-ists. *J Med Chem, 30,* 1631–1635.

43. Harrold, M. W., Wallace, R. A., Farooqui, T., Wallace, L. J., Uretsky, N., and Miller, D. D. (1989). Synthesis and D_2 dopaminergic activity of pyrrolidinium, tetrahydrothiophenium, and tetrahydrothiophene analogues of sulpiride. *J Med Chem*, *32*, 874–880.

44. Harrold, M. W., Sriburi, A., Matsumoto, K., Miller, D. D., Farooqui, T., and Uretsky, N. (1993). The interaction of ammonium, sulfonium and sulfide analogues of metoclopramide with the D_2 receptor. *J Med Chem*, *36*, 3166–3170.

45. Sun, H., Mahadevan, A., and Razdan, R. K. (2004). A novel methodology for the synthesis of 1-desoxy-Δ^8-tetrahydrocannabinol (THC analogues). *Tetrahedron Lett*, *45*, 615–617.

46. Razdan, R. K., and Mahadevan, A. (2002). Recent advances in the synthesis of endocannabinoid related ligands. *Chemi Phys Lipids*, *121*, 21–33.

47. Meltzer, P. C., Blundell, P., Liu, S., Blanchette, H. S., and Madras, B. K. (2001). A tropane horse I. Synthesis of a dopamine sparing cocaine antagonist. *Drug Alcohol Depend*, *63*, S94.

48. Madras, B. K., Blundell, P., Liu, S., and Meltzer, P. C. (2001). A tropane horse I. O-1893 interferes with cocaine binding but permits dopamine transport: a cocaine antagonist strategy. *Drug Alcohol Depend*, *63*, S95.

49. Davies, H. M. L., Kuhn, L. A., Thornley, C., Matasi, J. J., Sexton, T., and Childers, S. R. (1996). Synthesis of 3β-aryl-8-azabicyclo[3.2.1]octanes with high binding affinities and selectivities for the serotonin transporter site. *J Med Chem*, *39*, 2554–2558.

50. Newman, A. (1998). Special issue: dopamine transporter. *Med Chem Res*, *8*, 1–113.

51. Madras, B. K., Fahey, M. A., and Kaufman, M. J. (1990). [^{3}H]CFT and [^{3}H]Lu19-005: markers for cocaine receptors/dopamine nerve terminals in Parkinson's disease. *Soc Neurosci Abstr*, *16*, 14.

52. Madras, B. K., Spealman, R. D., Fahey, M. A., Neumeyer, J. L., Saha, J. K., and Milius, R. A. (1989). Cocaine receptors labeled by [^{3}H]2β-carbomethoxy-3β(4-fluorophenyl)tropane. *Mol Pharmacol*, *36*, 518–524.

53. Fischman, A. J., Bonab, A. A., Babich, J. W., Livni, E., Alpert, N. M., Meltzer, P. C., and Madras, B. K. (2001). [(11)C, (127)I] Altropane: a highly selective ligand for PET imaging of dopamine transporter sites. *Synapse*, *39*, 332–342.

54. Fischman, A. J., Bonab, A. A., Babich, J. W., Palmer, E. P., Alpert, N. M., Elmaleh, D. R., Callahan, R. J., Barrow, S. A., Graham, W., Meltzer, P. C., Hanson, R. N., and Madras, B. K. (1998). Rapid detection of Parkinson's disease by SPECT with altropane: a selective ligand for dopamine transporters. *Synapse*, *29*, 128–141.

55. Topliss, J. G. (1977). A manual method for applying the Hansch approach to drug design. *J Med Chem*, *20*, 463–469.

56. Meltzer, P. C., Liang, A. Y., Brownell, A.-L., Elmaleh, D. R., and Madras, B. K. (1993). Substituted 3-phenyltropane analogs of cocaine: synthesis, inhibition of binding at cocaine recognition sites, and positron emission tomography imaging. *J Med Chem*, *36*, 855–862.

57. Danishefsky, S., and Kitahara, T. (1974). A useful diene for the Diels–Alder reaction. *J Am Chem Soc*, *96*, 7807–7808.

58. Chan, T.-H., and Brownbridge, P. (1980). A novel cycloaromatization reaction: regiocontrolled synthesis of substituted methyl salicylates. *J Am Chem Soc*, *102*, 3534–3538.

59. Thompson, W. J., and Gaudino, J. (1984). A general synthesis of 5-arylnicotinates. *J Org Chem*, *49*, 5237–5243.

60. Oh-e, T., Miyaura, N., and Suzuki, A. (1993). Palladium-catalyzed cross-coupling reaction of organoboron compounds with organic triflates. *J Org Chem*, *58*, 2201–2208.

61. Keverline, K. I., Abraham, P., Lewin, A. H., and Carroll, F. I. (1995). Synthesis of the 2β,3α- and 2β,3β-isomers of 3-(*p*-substituted phenyl)tropane-2-carboxylic acid methyl esters. *Tetrahedron Lett*, *36*, 3099–3102.

62. Meltzer, P. C., Liang, A. Y., and Madras, B. K. (1994). The discovery of an unusually selective and novel cocaine analog: difluoropine. Synthesis and inhibition of binding at cocaine recognition sites. *J Med Chem*, *37*, 2001–2010.

63. Newman, A. H., Kline, R. H., Allen, A. C., Izenwasser, S., George, C., and Katz, J. L. (1995). Novel 4'-substituted and 4',4''-disubstituted 3α-(diphenylmethoxy)tropane analogs as potent and selective dopamine uptake inhibitors. *J Med Chem*, *38*, 3933–3940.

64. Meltzer, P. C., Pham-Huu, D.-P., and Madras, B. K. (2004). Synthesis of 8-thiabicyclo[3.2.1]oct-2-enes and their binding affinity for the dopamine and serotonin transporters. *Bioorg and Med Chem Lett*, *14*, 6007–6010.

65. Pham-Huu, D. P., Deschamps, J. R., Liu, S., Madras, B. K., and Meltzer, P. C. (2006). Synthesis of 8-thiabicyclo[3.2.1]octanes and their binding affinity for the dopamine and serotonin transporters. *Bioorg and Med Chem*, *15*, 1067–1082.

66. Parr, A. J., Walton, N. J., Bensalem, S., McCabe, P. H., and Routledge, W. (1991). 8-Thiabicyclo[3.2.1]octan-3-one as a biochemical tool in the study of tropane alkaloid biosynthesis. *Phytochemistry*, *30*, 2607–2609.

67. Majewski, M., DeCaire, M., Nowak, P., and Wang, F. (2000). Enolates of thiabicyclic ketones as stepping stones towards synthesis of sulfur analogs of tropane alkaloids. *Synlett*, *9*, 1321–1323.

68. Jefford, C. W., Gunsher, J., Hill, D. T., Brun, P., Le Gras, J., and Waegell, B. (1988). Bicyclo[3.2.1]octan-3-one. *Organ Synth Coll,*, Vol 6, 142–145.

69. Holmquist, C. R., Keverline-Frantz, K. I., Abraham, P., Boja, J. W., Kuhar, M. J., and Carroll, F. I. (1996). 3α-(4'-Substituted phenyl)tropane- 2β-carboxylic acid methyl esters: novel ligands with high affinity and selectivity at the dopamine transporter. *J Med Chem*, *39*, 4139–4141.

70. Madras, B. K., Fahey, M. A., Bergman, J., Canfield, D. R., and Spealman, R. D. (1989). Effects of cocaine and related drugs in nonhuman primates. I: [^{3}H]Cocaine binding sites in caudate–putamen. *J Pharm Exp Ther*, *251*, 131–141.

71. Meltzer, P. C., Wang, B., Chen, Z., Blundell, P., Jayaraman, M., Gonzalez, M. D., George, C., and Madras, B. K. (2001). Synthesis of 6- and 7- hydroxy-8-azabicyclo[3.2.1]octanes and their binding affinity for the dopamine and serotonin transporters. *J Med Chem*, *44*, 2619–2635.

72. Ritz, M. C., Cone, E. J., and Kuhar, M. J. (1990). Cocaine inhibition of ligand binding at dopamine, norepinephrine and serotonin transporters: a structure–activity study. *Life Sci*, *46*, 635–645.

73. Kozikowski, A. P., Roberti, M., Johnson, K. M., Bergmann, J. S., and Ball, R. G. (1993). SAR of cocaine: further exploration of structural variations at the C-2 center provides compounds of subnanomolar binding potency. *Bioorg Med Chem Lett*, *3*, 1327–1332.

74. Carroll, F. I., Lewin, A. H., Boja, J. W., and Kuhar, M. J. (1993). 3β-(Substituted phenyl) tropane-2 carboxylic acid ester analogs of cocaine: synthesis and biochemical and pharmacological properties. In *Drug Design for Neuroscience*, Kozikowski, A. P., Ed., Raven Press, New York.

75. Carroll, F. I., Gao, Y., Rahman, M. A., Abraham, P., Parham, K., Lewin, A. H., Boja, J. W., and Kuhar, M. J. (1991). Synthesis, ligand binding, QSAR, and CoMFA study of 3β-(p-substituted phenyl)tropane-2β-carboxylic acid methyl esters. *J Med Chem*, *34*, 2719–2725.

76. Carroll, F. I., Gray, J. L., Abraham, P., Kuzemko, M. A., Lewin, A. H., Boja, J. W., and Kuhar, M. J. (1993). 3-Aryl-2-(3′-substituted-1′,2′,4′-oxadiazol-5′-yl)tropane analogues of cocaine: affinities at the cocaine binding site at the dopamine, serotonin, and norepinephrine transporters. *J Med Chem*, *36*, 2886–2890.

77. Kotian, P., Mascarella, S. W., Abraham, P., Lewin, A. H., Boja, J. W., Kuhar, M. J., and Carroll, F. I. (1996). Synthesis, ligand binding, and quantitative structure-activity relationship study of 3β-(4′-substituted phenyl)-2β-heterocyclic tropane: evidence for an electrostatic interaction at the 2β-position. *J Med Chem*, *39*, 2753–2763.

78. Simoni, D., Roberti, M., Rondanin, R., Baruchello, R., Rossi, M., Invidiata, F. P., Merighi, S., Varani, K., Gessi, S., Borea, P. A., Marino, S., Cavallini, S., Bianchi, C., and Siniscalchi, A. (2001). Effects of two-carbon bridge region methoxylation of benztropine: discovery of novel chiral ligands for the dopamine transporter. *Bioorg and Med Chem Lett*, *11*, 823–827.

79. Chen, N., Zhen, J., and Reith, M. E. A. (2004). Mutation of Trp84 and Asp313 of the dopamine transporter reveals similar mode of binding interaction for GBR12909 and benztropine as opposed to cocaine. *J Neurochem*, *89*, 853–864.

80. Vaughan, R. A., Agoston, G. E., Lever, J. R., and Newman, A. H. (1999). Differential binding of tropane-based photoaffinity ligands on the dopamine transporter. *J Neuroscience*, *19*, 630–636.

81. Meltzer, P. C., Blundell, P., Huang, H., Liu, S., Yong, Y. F., and Madras, B. K. (2000). 3-Aryl-2-carbomethoxybicyclo[3.2.1]oct-2-enes inhibit WIN 35,428 binding potently and selectively at the dopamine transporter. *Bioorg and Med Chem*, *8*, 581–590.

82. Meltzer, P. C., Blundell, P., Jones, A. G., Mahmood, A., Garada, B., Zimmerman, R. E., Davison, A., Holman, B. L., and Madras, B. K. (1997). A Technetium-99 m SPECT imaging agent which targets the dopamine transporter in primate brain. *J Med Chem*, *40*, 1835–1844.

83. Lieske, S. F., Yang, B., Eldefrawi, M. E., MacKerell, J. A. D., and Wright, J. (1998). (−)-3β-Substituted ecgonine methyl esters as inhibitors for cocaine binding and dopamine uptake. *J Med Chem*, *41*, 864–876.

84. Cini, N., Danieli, E., Menchi, G., Trabocchi, A., Bottoncetti, A., Raspanti, S., Pupi, A., and Guarna, A. (2006). 3-Aza-6,8-dioxabicyclo[3.2.1]octanes as new enantiopure heteroatom-rich tropane-like ligands of human dopamine transporter. *Bioorg and Med Chem*, *14*, 5110–5120.

85. Sershen, H., Reith, M. E. A., and Lajtha, A. (1980). The pharmacological relevance of the cocaine binding site in mouse brain. *Neuropharmacology*, *19*, 1145–1148.

86. Sershen, H., Reith, M. E. A., and Lajtha, A. (1982). Comparison of the properties of central and peripheral binding sites for cocaine. *Neuropharmacology*, *21*, 469–474.

87. Wang, S., Gai, Y., Laruelle, M., Baldwin, R. M., Scanlet, B. E., Innis, R. B., and Neumeyer, J. L. (1993). Enantioselectivity of cocaine recognition sites: binding of (1S)-and (1R)-2β-carbomethoxy-3β-(4-iodophenyl)tropane (β-CIT) to monoamine transporters. *J Med Chem*, *36*, 1914–1917.

88. Madras, B. K., Fahey, M. A., Kaufman, M. J., Spealman, R. D., Schumacher, J., Isacson, O., Brownell, A.-L., Brownell, G. L., and Elmaleh, D. R. (1991). Cocaine receptor probes in nonhuman and human brain: in vitro characterization and in vivo imaging. *Soc Neurosci Abstr*, *17*, 190.

89. Madras, B. K., Gracz, L. M., Fahey, M. A., Elmaleh, D. R., Meltzer, P. C., Liang, A. Y., Stopa, E. G., Babich, J. W., and Fischman, A. J. (1998). Altropane, a SPECT or PET imaging probe for dopamine neurons. III: Human dopamine transporter in postmortem normal and Parkinson's diseased brain. *Synapse*, *29*, 116–127.

90. Meltzer, P. C., Blundell, P., Zona, T., Yang, L., Huang, H., Fischman, A., and Madras, B. K. (2001). A second generation ^{99m}technetium SPECT agent that provides in vivo images of the dopamine transporter in primate brain. *J Med Chem*, *46*, 3483–3496.

91. Madras, B. K., Fahey, M. A., Miller, G. M., De La Garza, R., Goulet, M., Spealman, R. D., Meltzer, P. C., George, S. R., O'Dowd, B. F., Bonab, A. A., Livni, E., and Fischman, A. J. (2003). Non-amine-based dopamine transporter (reuptake) inhibitors retain properties of amine-based progenitors. *Eur J Pharmacol*, *479*, 41–51.

92. Meltzer, P. C. Unpublished data obtained through NIDA auspices.

93. De La Garza, R., Meltzer, P. C., and Madras, B. K. (1999). Non-amine dopamine transporter probe [3H]tropoxene distributes to dopamine-rich regions of monkey brain. *Synapse*, *34*, 20–27.

94. Canfield, D. R., Spealman, R. D., Kaufman, M. J., and Madras, B. K. (1990). Autoradiographic localization of cocaine binding sites by [^{3}H]CFT ([^{3}H]WIN 35,428) in the monkey brain. *Synapse*, *6*, 189–195.

95. Kaufman, M. J., Spealman, R. D., and Madras, B. K. (1991). Distribution of cocaine recognition sites in monkey brain. I: In vitro autoradiography with [^{3}H]CFT. *Synapse*, *9*, 177–187.

96. Van Dyck, C., Seibyl, J., Malison, R., Laruelle, M., Wallace, E., Zoghbi, S., Zea-Ponce, Y., Baldwin, R., Charney, D., and Hoffer, P. (1995). Age-related decline in striatal dopamine transporter binding with iodine-123-β-CIT SPECT. *J Nucl Med*, *36*, 1175–1181.

97. Madras, B. K., Reith, M. E. A., Meltzer, P. C., and Dutta, A. K. (1994). O-526, a piperidine analog of GBR 12909, retains high affinity for the dopamine transporter in monkey caudate-putamen. *Eur J Pharacol*, *267*, 167–173.

98. Madras, B. K., Fahey, M. A., Goulet, M., Lin, Z., Bendor, J., Goodrich, C., Meltzer, P. C., Elmaleh, D. R., Livni, E., Bonab, A. A., and Fischman, A. J. (2006). Dopamine transporter (DAT) inhibitors alleviate specific Parkinsonian deficits in monkeys: association with DAT occupancy in vivo. *J Pharm and Exp Ther*, *319*, 570–585.

99. Carroll, F. I., Pawlush, N., Kuhar, M. J., Pollard, G. T., and Howard, J. L. (2004). Synthesis, monoamine transporter binding properties, and behavioral pharmacology of a series of 3β-(substituted phenyl)-2β-(3′-substituted isoxazol-5-yl)tropanes. *J Med Chem*, *47*, 296–302.

11

DOPAMINE-RELEASING AGENTS

Bruce E. Blough

Center for Organic and Medicinal Chemistry, Research Triangle Institute, Research Triangle Park, North Carolina

11.1	Introduction	305
11.2	Release Mechanism	307
11.3	Neurotransmitters and Release	308
11.4	Amphetamines	309
11.5	MDMA Analogs and Metabolites	311
11.6	Other Dopamine Releasers	313
11.7	Dopamine Release vs. Norepinephrine Release	313
11.8	Serotonin-Release Modulation	315
11.9	Conclusions	317
References		317

11.1 INTRODUCTION

Dopamine-releasing agents are compounds that induce efflux of dopamine from the dopamine transporter (DAT) containing neurons [1–5]. Through this action, dopamine releasers elevate synaptic levels of neurotransmitter. These increases are similar to changes induced by dopamine-uptake inhibitors, which also elevate synaptic levels of dopamine, but differ mechanistically. Releasers elevate neurotransmitter by direct efflux, whereas uptake inhibitors indirectly elevate neurotransmitter by preventing normal synaptic clearance. Both mechanisms lead to similar stimulant

Dopamine Transporters: Chemistry, Biology, and Pharmacology. Edited by Mark L. Trudell and Sari Izenwasser
Copyright © 2008 John Wiley & Sons, Inc

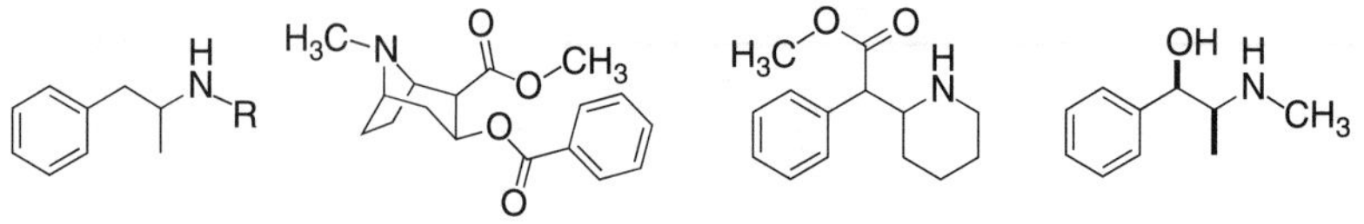

1a, R = H
1b, R = CH₃

Figure 11-1 Dopamine releasers and uptake inhibitors.

behaviors, as evidenced by the parallel behaviors observed between the best known examples (Fig. 11-1): the releasers amphetamine (**1a**) and methamphetamine (**1b**), and the uptake inhibitors cocaine (**2**) and methylphenidate (Ritalin, **3**). All four compounds are highly addictive, presumably due to their action on dopaminergic neurons, and are abused around the globe.

The earliest known dopamine releaser is probably ephedrine (**4**), the active ingredient in naturally occurring plants such as *Ephedra sinica*, also know as the Chinese herb *ma huang* [6,7]. These herbs were used for centuries for a variety of purposes, including stimulation, asthma, decongestion, and appetite suppression. Ephedrine was first isolated and identified in 1885 by Nagai Nagayoshi from *Ephedra vulgaris*. The clinical usefulness of these herbal extracts, and identification of the active ingredient, created interest in the synthesis of ephedrine (and analogs), possibly leading to the first synthesis of amphetamine shortly thereafter, in 1887 (**1a**) [8]. Methamphetamine (**1b**) was first synthesized in 1893 by Nagayoshi, using a route starting with ephedrine. Amphetamines were some of the first medicinally active compounds not derived from natural products. It should be noted that amphetamine, methamphetamine, and several additional analogs have been isolated from *Acacia berlandieri* and *A. rigidula*, a group of shrubs found in Texas [9–11].

Release does not occur just with dopaminergic neurons; the same process is found in serotonergic and noradrenergic neurons [12]. The most commonly discussed serotonin releasers (Fig. 11-2) are the empathogenic/psychedelic 3,4-methylenedioxy-methamphetamine (**5**, MDMA) [13], also known as the street drug "ecstasy," and the anorectic fenfluramine (**6**) [14]. Corresponding serotonin-(re)uptake inhibitors include antidepressants such as paroxetine (Paxil, **7**) and fluoxetine (Prozac, **8**) [15]. Little is known about compounds that induce norepinephrine release, but fenfluramine and metabolites have been found to release norepinephrine [16], and norepinephrine release has been implicated in the stimulant properties of releasers along with their dopaminergic activity [17].

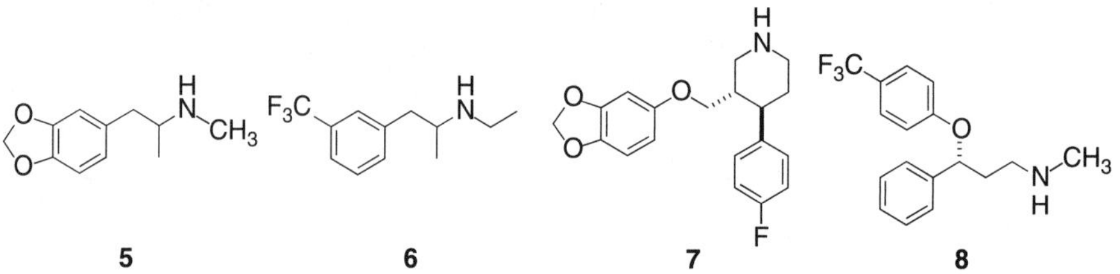

5   6   7   8

Figure 11-2 Serotonin releasers and uptake inhibitors.

11.2 RELEASE MECHANISM

Several groups have focused on the mechanism of action of releasers, due to both the importance of research on abused releasers (amphetamine and MDMA) and to curiosity about transporter mechanisms in general [1,2,18–21]. The most important finding has been that releasers are substrates of the transporters, and that this translocation is required for neurotransmitter efflux [2]. Other mechanistic hypotheses for efflux initiation have included membrane diffusion, internal release of vesicular neurotransmitter, and a chemical chaperoning mechanism. Most evidence now suggests that releasers act as substrates and induce efflux once inside the neuron or in a translocation state past substrate recognition. The delineation between dopamine release and uptake inhibition is important because the kinetics of neurotransmitter change differ dramatically, and these changes result in differences in feedback and receptor/transporter regulation. A releaser dumps neurotransmitter directly into the synapse, regardless of endogenous tone. An uptake inhibitor acts indirectly and relies on normal synaptic vesicular release to increase neurotransmitter, which may be considered a feedback mechanism.

Yu et al. have devised a bioassay screening strategy aimed at binning releasers and uptake inhibitors, and have used this strategy to identify new dopamine releasers [22]. Releasers, such as methamphetamine, are often thought of as having both releaser and uptake inhibitor activity, but the latter is an artifact of the assay. Neurotransmitter uptake inhibition and release assays measure the same endpoint: exogenous neurotransmitter. Each assay independently cannot differentiate the two types of compounds, but when combined, compounds can be binned as uptake inhibitors or releasers. Uptake inhibitors are active in uptake inhibition assays, but not in release (superfusion) assays, since the neurotransmitter is preloaded. The synaptosomes (or cells) in uptake inhibition assays do not undergo the normal neuronal vesicular release process required for an uptake inhibitor to elevate neurotransmitter. On the other hand, releasers are active in both assays, which led people to assume that releasers have both activities. These differences have been exploited to bin compounds as releasers (substrates) or uptake inhibitors (nonsubstrates). An electrochemical method has also been reported by Sonders et al. taking advantage of the electronic differences between compounds that do and do not transport [1].

The transporter-induced efflux mechanism remains controversial [2]. Initially, it was assumed that efflux occurs by a reversal of the process that causes influx. However, recent evidence suggests that this assumption may be incorrect. Evidence suggests that releaser-induced efflux may occur by a transporter oligomerization event followed by transporter-mediated neurotransmitter outflow. Seidel et al. concatenated a serotonin and GABA (r-aminobutyric acid) transporter and loaded cells with GABA [23]. Introduction of a serotonin releaser induced GABA release. Kahlig et al. have recently reported that transporter-mediated efflux may occur by a channel-like mechanism [24].

The term *dopamine releaser* suggests that a compound is able to release a specific neurotransmitter, but this may not be the case. The term is more likely to be meant to refer to compounds that induce neurotransmitter efflux via a dopamine transporter. Similarly, *serotonin releaser* and *norepinephrine releaser* refer more accurately to a compound-inducing efflux via serotonin and norepinephrine transporters, respectively (SERT and NET). The compounds undergoing efflux may not be limited to the primary transporter neurotransmitter, and may include other accumulated neurotransmitters within the neuron or cell. This possibility is intriguing, especially in the case of neurons expressing multiple transporter types. A dopamine releaser acting on a neuron expressing both SERT and DAT may induce both dopamine and serotonin efflux.

11.3 NEUROTRANSMITTERS AND RELEASE

While amphetamine and methamphetamine are widely known as dopamine releasers, neurotransmitters such as dopamine (**9**), norepinephrine (**10**), serotonin (**11**), tyramine (**12**), and β-phenethylamine (**13**) also cause DAT-induced efflux of dopamine [4] (Table 11-1). Dopamine was found to be a DAT releaser, albeit far weaker than amphetamine, with an EC_{50} value of 86.9 nM. β-Phenethylamine ($EC_{50} = 39.5$ nM) was found to be more potent as a DAT releaser, and tyramine ($EC_{50} = 119$ nM) was roughly equipotent. Norepinephrine ($EC_{50} = 869$) and serotonin

TABLE 11-1 Comparison of the DAT-, NET-, and SERT-Releasing Activity of Endogenous Neurotransmitters

Compound	Name	Structure	EC_{50} (nM) DAT	NET	SERT
9	Dopamine		86.9	66.2	Inhibitor
10	Norepinephrine		869	164	Inactive
11	Serotonin		1960	Inhibitor	44.4
12	Tyramine		119	40.6	2775
13	β-Phenethylamine		39.5	10.9	Inactive

Source: DAT, NET, and SERT data from Rothman et al. [4].

(EC$_{50}$ = 1960 nM) were found to be dopamine releasers but were far weaker than dopamine. Self-release also occurs with serotonin (EC$_{50}$ = 44.4 nM) and norepinephrine (EC$_{50}$ = 164 nM).

Neurotransmitter-induced efflux is not limited to the dopamine transporter. Dopamine induces NET release (EC$_{50}$ = 66.2 nM)) more potently than does norepinephrine (EC$_{50}$ = 164 nM), but is a weak serotonin uptake inhibitor (IC$_{50}$ = 6489 nM), suggesting that dopamine is a NET substrate but not a SERT substrate. Tyramine is the only monoamine neurotransmitter that induces neurotransmitter efflux through all three transporters (DAT, NET, and SERT), meaning that it is a substrate of all three transporters. The real significance of these observations is not clear, but may imply that efflux is a common neurotransmitter regulatory mechanism induced by any transporter substrate, possibly as a protective mechanism.

11.4 AMPHETAMINES

The most commonly studied dopamine releasers are amphetamines, including amphetamine (**1a**), methamphetamine (**1b**), and β-phenethylamine (PEA, **13**) [4]. β-Phenethylamine is not a methamphetamine-like stimulant because of rapid metabolism by monoamine oxidase (MAO). However, in the presence of an MAO inhibitor, PEA is a potent dopamine-releasing stimulant, roughly equipotent to amphetamine [25]. The additional α-methyl group found on amphetamines blocks MAO metabolism. Amphetamine-based compounds have been found to be good MAO inhibitors, suggesting that even though both amphetamine and β-phenethylamines bind to MAO, the α-methyl group blocks enzyme function [26].

Table 11-2 shows the dopamine- and norepinephrine-releasing properties of a series of amphetamine analogs varying in substituent and substitution location [27]. The most potent compound in Table 11-2 is S-(+)-amphetamine, which releases dopamine with an EC$_{50}$ of 8.7 nM. The R-(−)-isomer is threefold weaker, at 27.7 nM (EC$_{50}$). Although weaker, a similar trend is seen for the optical isomers of methamphetamine. S-(+)-methamphetamine releases dopamine with an EC$_{50}$ of 24.5 nM, while the R-(−)-isomer was 16-fold less active at 416 nM (EC$_{50}$). Phenethylamine (**13**) is also a potent dopamine releaser, with a potency of 39.5 nM (EC$_{50}$). A one-carbon chain homologation to phenpropylamine (**14**) reduces potency significantly to the micromolar range.

Aromatic substitution changes have a significant effect on amphetamine potencies. Chloro-, fluoro-, and methyl-substituted compounds were found to have dopamine-releasing potencies within an order of magnitude of that of amphetamine. EC$_{50}$ values ranged from 11.8 nM for m-chloroamphetamine (**20**) to 68.2 nM for p-chloroamphetamine (**16**), with the exception of o-methylamphetamine (**23**), which was substantially weaker (EC$_{50}$ = 127 nM). The *meta*-substituted compounds were found to be more potent than their corresponding *ortho* or *para* analogs. The *para*-substituted compounds were the weakest for chloro and fluoro substitution, but the *ortho* analog was the weakest for methyl substitution. Methoxy substitution caused a dramatic decrease in potency. m-Methoxyamphetamine (**22**) was found to have an EC$_{50}$ of

TABLE 11-2 Comparison of the DAT- and NET-Releasing Activity of a Series of Amphetamines

Compound	Structure	EC$_{50}$ (nM)	
		DAT	NET
R-(−)1a		27.7	9.5
S-(+)1a		8.7	10.2
R-(−)1b		416	28.5
S-(+)1b		24.5	12.3
14		1491	222
15		44.1	22.2
16		68.5	23.5
17		51.5	28.0
18		867	166
19		33.3	18.3

(*Continued*)

TABLE 11.2   *Continued*

Compound	Structure	EC$_{50}$ (nM)	
		DAT	NET
20	Cl—C$_6$H$_4$—CH$_2$CH(NH$_2$)CH$_3$	11.8	9.4
21	F—C$_6$H$_4$—CH$_2$CH(NH$_2$)CH$_3$	24.2	16.1
22	MeO—C$_6$H$_4$—CH$_2$CH(NH$_2$)CH$_3$	103	58.0
23	CH$_3$-substituted phenyl, CH$_2$CH(NH$_2$)CH$_3$	127	37.0
24	Cl-substituted phenyl, CH$_2$CH(NH$_2$)CH$_3$	62.4	19.1
25	F-substituted phenyl, CH$_2$CH(NH$_2$)CH$_3$	38.1	24.1
26	OMe-substituted phenyl, CH$_2$CH(NH$_2$)CH$_3$	1478	473

Source: DAT and NET data from Rothman et al. [4]. Some of the data were presented by Blough et al. [27].

103 nM, but the *para* and *ortho* analogs (**18** and **26**) were far worse, generating EC$_{50}$ values of 867 and 1478 nM, respectively. These data show that dopamine-releasing potencies can vary depending on aromatic substituent and point of attachment.

11.5   MDMA ANALOGS AND METABOLITES

MDMA (**5**) is normally labeled a serotonin releaser, but it is also releases dopamine [4] (Table 11-3). This 3,4-methylenedioxy-substituted amphetamine is much weaker than amphetamine as a DAT releaser. MDMA (**5**) releases dopamine with an EC$_{50}$ value of 278 nM, compared to the more potent compound amphetamine (**1a**, EC$_{50}$ = 8.7 nM). Interestingly, the potency of MDMA at the DAT appears to fall between the *meta*- and *para*-methoxyamphetamine analogs, which release with EC$_{50}$ values of 103 and

TABLE 11-3   Comparison of the DAT- and NET-Releasing Activity of MDA, MDMA, and Metabolites

Compound	Name	Structure	EC$_{50}$ (nM) DAT	NET
5	MDMA		278	110
R-(−)-5	(−)-MDMA		3682	564
S-(+)-5	(+)-MDMA		142	136
27	MDA		190	108
R-(−)-27	(−)-MDA		900	287
S-(+)-27	(+)-MDA		98.5	50.0
28	HHMA (N-Me-α-Me-DA)		1729	77
29	HHA (α-Me-DA)		3485	33
30	HMMA		607	625
31	HMA		1450	694

Source: DAT and NET data from Rothman et al. [4]. The metabolite data were presented by Bauman et al. [28].

867 nM, respectively. As was observed with both amphetamine and methamphetamine, the S-(+) isomer (EC_{50} = 142 nM) is more potent than the R-(−) isomer (EC_{50} = 3682). The same trends were observed with the N-demethylated compound MDA (**27**). N-Methylation appears to reduce dopamine-releasing activity, but the effect was not as great as with amphetamine and methamphetamine.

MDMA metabolizes to MDA by N-dealkylation [13]. Both MDMA and MDA metabolize further by conversion of the methylenedioxy group to a catechol, followed by methoxylation of the *meta*-hydroxyl by catechol-*O*-methyltransferase (COMT). The two catechol metabolites, HHMA (**28**) and HHA (**29**), are dopamine analogs and have much weaker activity as dopamine releasers; however, they are both potent releasers of norepinephrine [28]. Methoxylation of the catechol, forming HMMA (**30**) and HMA (**31**), causes a slight increase in DAT activity, but reduces activity dramatically as a norepinephrine releaser and suggests that methoxylation of endogenous catechols is an important regulatory mechanism.

11.6   OTHER DOPAMINE RELEASERS

Several other compounds are known to induce dopamine release, originally identified by their stimulant activity in behavioral assays [4] (Table 11-4). 1-Benzylpiperazine (BZP, **32**) is also an abused stimulant recently scheduled by the DEA [29,30]. BZP has been used in combination with the selective serotonin releaser *m*-trifluoromethylphenylpiperazine (TFMPP, **33**) to create "legal X" [31,32]. Phenmetrazine (**34**), a stimulant known to act by releasing dopamine, was clinically used as an anorectic but removed due to abuse [33]. The N-methylated analog phendimetrazine (Adipost, **35**) is currently used in the clinic as an anorectic and is probably effective due to metabolism to **33**. The appetite suppressant aminorex (**36**) is another compound found to release dopamine and have stimulant activity [4,34,35]. All of these compounds are secondary cyclic amines, and have much weaker activity than that of amphetamine.

Another general class of dopamine-releasing agents are the aminoacetophenones, such as cathinone (**37**), the principal active compound in the stimulant *Catha edulus* (known as Khat) [4,22,36]. Cathinone was found to release both dopamine and norepinephrine. In contrast to the amphetamine analogs, N-methylation to methcathinone (**38**) [4,22,37] increased DAT activity from an EC_{50} value of 83.1 nM to 50 nM [27]. N-Ethylation to **39** caused the compound to become an uptake inhibitor with an IC_{50} of 1067 nM. Interestingly, the latter compound retains norepinephrine-releasing properties, albeit somewhat weaker. As noted earlier, (−)-ephedrine (**4**) is a dopamine releaser and is essentially methcathinone with a reduced ketone.

11.7   DOPAMINE RELEASE VS. NOREPINEPHRINE RELEASE

Dopamine releasers also release norepinephrine, and surprisingly, in all cases more potently [17,38,39]. Most of the trends evident for dopamine release are also true for norepinephrine release, but the differences between compounds tend to be

TABLE 11-4   Comparison of the DAT- and NET-Releasing Activity of Selected Non-Phenethylamine Dopamine Releasers

Compound	Name	Structure	EC_{50} (nM)	
			DAT	NET
32	1-Benzylpiperazine		307	62.2
33	3-Trifluoromethylpiperazine		Inactive	Inactive
34	Phenmetrazine		131	50.4
35	Phendimetrazine		Inhibitor	Inhibitor
36	Aminorex		49.4	26.4
37	Cathinone		83.1	23.6
38	Methcathinone		49.9	22.4
39	N-Ethylaminopropriophenone		Uptake inhibitor	99.3
4	(−)-Ephedrine		236	43.1

Source: DAT and NET data from Rothman et al. [3], Rothman and Baumann [4], Yu et al. [22], and Blough et al. [27].

much smaller. Most historical studies on releasers do not report norepinephrine release, possibly because researchers concentrated on behavioral endpoints, such as stimulant activity, and assumed that such activity was solely a dopaminergic effect. Tseng et al. [40] studied the three methoxyamphetamine analogs (**18**, **22**, **26**) and is the only group known to have included information about the releasing activity of all three transporters. The general trend that dopamine releasers induce NET-induced efflux, and more potently, is an important discovery, due to the potential problems associated with elevations of norepinephrine, such as cardiotoxicity [41]. Concurrent norepinephrine release has also made it difficult to find a truly selective dopamine releaser.

11.8 SEROTONIN-RELEASE MODULATION

Serotonin-releasing activity has been found to modulate the effects of dopamine release [3,4,42]. Compounds releasing dopamine without significant serotonin activity, such as amphetamine, are potent stimulants. Compounds releasing dopamine but with significant serotonin activity are not. This is most evident in the activity of the releaser anorectic Fen-phen, a combination of the dopamine and norepinephrine releaser phentermine (**40**) and serotonin releaser fenfluramine (**41**) [12,43–45] (Table 11-5). Phentermine is a known stimulant, but when combined with fenfluramine does not show significant stimulant activity [46,47]. These observations suggest that the serotonergic activity attenuates stimulant activity and thus the effects of dopamine release. The effects of serotonin release on stimulant activity can also be seen in the development of PAL-287 (**42**) as a potential addiction cessation agent [48]. Initial reports on the activity of PAL-287 suggested that it was inactive

TABLE 11-5   Comparison of the DAT-, NET-, and SERT-releasing Activities of Selected Compounds

			EC_{50} (nM)		
Compound	Name	Structure	DAT	NET	SERT
40	Phentermine		262	39.4	3511
41	Fenfluramine		Inhibitor	739	79.3
42	PAL-287		12.6	11.1	3.4

Source: DAT, NET, and SERT data from Rothman et al. [4] and Rothman et al. [48].

because it failed to induce locomotor stimulation [49]. Rothman et al. [48] have found PAL-287 to be a potent dopamine releaser but to lack potent stimulant activity, presumably due to its activity as a serotonin releaser.

Wee et al. looked at the reinforcing properties of a series of amphetamine analogs [PAL-353 (**21**), PAL-303 (**17**), PAL-314 (**19**), and PAL-313 (**15**)] in rhesus monkeys which varied in DAT- and SERT-releasing potencies (Table 11-6). These compounds vary from 80-fold selective for dopamine over serotonin (PAL-353) to nearly equipotent (PAL-313) [50]. Dopamine-selective compounds were found to be more reinforcing than were less selective compounds. The 18-fold dopamine selective releaser PAL-303 was also found to be a stimulant in rats, similar to amphetamine [51]. PAL-287 (**42**) was also not strongly reinforcing [48]. Similar effects were observed by Negus et al. while studying similar compounds in cocaine- and food-maintained responding experiments in primates [45]. Amphetamine and methamphetamine showed reduction of cocaine but not food self-administration. The addition of serotonergic activity caused reduction in both behaviors.

TABLE 11-6   Comparison of the DAT- and SERT-Releasing Activity of Selected Phenethylamines

			EC_{50} (nM)		DAT
Compound	Name	Structure	DAT	SERT	Selectivity
(+)-1a	(+)Amphetamine		8.0	1756	219
21	PAL-353		24.2	1937	80
17	PAL-303		51.5	939	18
19	PAL-314		33.3	218	6.5
15	PAL-313		44.1	53.4	1.2
42	PAL-287		12.6	3.4	0.27

Source: DAT and SERT data from Rothman et al. [4], Rothman et al. [48], and Wee et al. [50].

11.9 CONCLUSIONS

This review by no means covers all of the recent literature on amphetamines and releasers. Some of the most exciting work is mechanistic in nature and is beyond the scope of this chapter. Absolute releaser potencies may differ between laboratories to a small degree, depending on the specific assay conditions. The intention here was to show how structural changes affect activity based on the same assay conditions. Although dopamine releasers have been known for over a century, modern pharmacology continues to uncover their potential as pharmacotherapies as well as experimental tools for the study of dopaminergic neurons and neural networks.

REFERENCES

1. Sonders, M. S., Zhu, S. J., Zahniser, N. R., Kavanaugh, M. P., and Amara, S. G. (1997). Multiple ionic conductances of the human dopamine transporter: the actions of dopamine and psychostimulants. *J Neurosci*, *17*, 960–974.

2. Sulzer, D., Sonders, M. S., Poulsen, N. W., and Galli, A. (2005). Mechanisms of neurotransmitter release by amphetamines: a review. *Prog Neurobiol*, *75*, 406–433.

3. Rothman, R. B., Blough, B. E., and Baumann, M. H. (2006). Dual dopamine-5-HT releasers: potential treatment agents for cocaine addiction. *Trends Pharmacol Sci*, *27*, 612–618.

4. Rothman, R. B., and Baumann, M. H. (2006). Therapeutic potential of monoamine transporter substrates. *Curr Top Med Chem*, *6*, 1845–1859.

5. Nichols, D. E. (1994). Medicinal chemistry and structure–activity relationships. In *Amphetamine and Its Analogs: Psychopharmacology, Toxicology, and Abuse*, Cho, A. K., and Segal, D. S., Eds. Academic Press, San Diego, CA, pp. 3–41.

6. Chen, K. K., and Schmidt, C. F. (1959). The action and clinical use of ephedrine, an alkaloid isolated from the Chinese drug ma huang; historical document. *Ann Allergy*, *17*, 605–618.

7. Miller, S. C. (2005). Psychiatric effects of ephedra: addiction. *Am J Psychiatry*, *162*, 2198.

8. Edeleanu (1887). *Chem Ber*, *20*, 618.

9. Camp, B. J., and Lyman, C. M. (1956). The isolation of *N*-methyl β-phenylethylamine from *Acacia berlandieri*. *J Am Pharm Assoc* (*Baltimore*), *45*, 719–721.

10. Clements, B. A., Goff, C. M., and Forbes, D. A. (1997). Toxic amines and alkaloids from *Acacia berlandieri*. *Phytochemistry*, *46*, 249–254.

11. Clements, B. A., Goff, C. M., and Forbes, D. A. (1998). Toxic amines and alkaloids from *Acacia rigidula*. *Phytochemistry*, *49*, 1377–1380.

12. Rothman, R. B., and Baumann, M. H. (2002). Serotonin releasing agents: neurochemical, therapeutic and adverse effects. *Pharmacol Biochem Behav*, *71*, 825–836.

13. Baumann, M. H., Wang, X., and Rothman, R. B. (2007). 3,4-Methylene dioxymethamphetamine (MDMA) neurotoxicity in rats: a reappraisal of past and present findings. *Psychopharmacology* (Berl), *189*, 407–424.

14. Baumann, M. H., Ayestas, M. A., Dersch, C. M., Partilla, J. S., and Rothman, R. B. (2000). Serotonin transporters, serotonin release, and the mechanism of fenfluramine neurotoxicity. *Ann N Y Acad Sci*, *914*, 172–186.

15. Sanchez, C., and Hyttel, J. (1999). Comparison of the effects of antidepressants and their metabolites on reuptake of biogenic amines and on receptor binding. *Cell Mol Neurobiol*, *19*, 467–489.

16. Rothman, R. B., Clark, R. D., Partilla, J. S., and Baumann, M. H. (2003). (+)-Fenfluramine and its major metabolite, (+)-norfenfluramine, are potent substrates for norepinephrine transporters. *J Pharmacol Exp Ther*, *305*, 1191–1199.

17. Rothman, R. B., Baumann, M. H., Dersch, C. M., Romero, D. V., Rice, K. C., Carroll, F. I., and Partilla, J. S. (2001). Amphetamine-type central nervous system stimulants release norepinephrine more potently than they release dopamine and serotonin. *Synapse*, *39*, 32–41.

18. Reith, M. A. (2002). *Neurotransmitter Transporters: Structure, Function and Regulation*, 2nd ed. Humana Press, Totowa, NJ.

19. Jones, S. R., Gainetdinov, R. R., Wightman, R. M., and Caron, M. G. (1998). Mechanisms of amphetamine action revealed in mice lacking the dopamine transporter. *J Neurosci*, *18*, 1979–1986.

20. Li, L. B., Cui, X. N., and Reith, M. A. (2002). Is Na(+) required for the binding of dopamine, amphetamine, tyramine, and octopamine to the human dopamine transporter? *Naunyn Schmiedebergs Arch Pharmacol*, *365*, 303–311.

21. Schenk, J. O. (2002). The functioning neuronal transporter for dopamine: kinetic mechanisms and effects of amphetamines, cocaine and methylphenidate. *Prog Drug Res*, *59*, 111–131.

22. Yu, H., Rothman, R. B., Dersch, C. M., Partilla, J. S., and Rice, K. C. (2000). Uptake and release effects of diethylpropion and its metabolites with biogenic amine transporters. *Bioorg Med Chem*, *8*, 2689–2692.

23. Seidel, S., Singer, E. A., Just H., Farhan, H., Scholze, P., Kudlacek, O., Holy, M., Koppatz, K., Krivanek, P., Freissmuth, M., and Sitte, H. H. (2005). Amphetamines take two to tango: an oligomer-based counter-transport model of neurotransmitter transport explores the amphetamine action. *Mol Pharmacol*, *67*, 140–151.

24. Kahlig, K. M., Binda, F., Khoshbouei, H., Blakely, R. D., McMahon, D. G., Javitch, J. A., and Galli, A. (2005). Amphetamine induces dopamine efflux through a dopamine transporter channel. *Proc Natl Acad Sci U S A*, *102*, 3495–3500.

25. Bergman, J., Yasar, S., and Winger, G. (2001). Psychomotor stimulant effects of β-phenylethylamine in monkeys treated with MAO-B inhibitors. *Psychopharmacology* (Berl), *159*, 21–30.

26. Gallardo-Godoy, A., Fierro, A., McLean, T. H., Castillo, M., Cassels, B. K., Reyes-Parada, M., and Nichols, D. E. (2005). Sulfur-substituted α-alkyl phenethylamines as selective and reversible MAO-A inhibitors: biological activities, CoMFA analysis, and active site modeling. *J Med Chem*, *48*, 2407–2419.

27. Blough, B. E., Page, K. M., Partilla, J. S., Budzynski, A. G., and Rothman, R. B. (2005). Structure–activity relationship studies of DAT, SERT, and NET releasers. In *New Perspectives on Neurotransmitter Transporter Pharmacology*. Presentation, Alexandria, VA.

28. Baumann, M. H., Partilla, J. S., Ayestas, M. A., Page, K. M., Blough, B. E., and Rothman, R. B. (2006). Interaction of MDMA and its metabolites at monoamine transporters in rat brain. In *The College on Problems of Drug Dependence*. Presentation, Scottsdale, AZ.

29. Schedules of controlled substances: placement of 2,5-dimethoxy-4-(*n*)-propylthiophenethylamine and *N*-benzylpiperazine into schedule I of the Controlled Substances Act. Final rule. *Fed Reg*, *69*, 12794–12797, 2004.

30. Meririnne, E., Kajos, M., Kankaanpaa, A., and Seppala, T. (2006). Rewarding properties of 1-benzylpiperazine, a new drug of abuse, in rats. *Basic Clin Pharmacol Toxicol*, *98*, 346–350.

31. Baumann, M. H., Clark, R. D., Budzynski, A. G., Partilla, J. S., Blough, B. E., and Rothman, R. B. (2004). Effects of "legal X" piperazine analogs on dopamine and serotonin release in rat brain. *Ann N Y Acad Sci*, *1025*, 189–197.

32. Baumann, M. H., Clark, R. D., Budzynski, A. G., Partilla, J. S., Blough, B. E., and Rothman, R. B. (2005). N-Substituted piperazines abused by humans mimic the molecular mechanism of 3,4-methylenedioxymethamphetamine (MDMA, or "Ecstasy"). *Neuropsychopharmacology*, *30*, 550–560.

33. Rothman, R. B., Katsnelson, M., Vu, N., Partilla, J. S., Dersch, C. M., Blough, B. E., and Baumann, M. H. (2002). Interaction of the anorectic medication, phendimetrazine, and its metabolites with monoamine transporters in rat brain. *Eur J Pharmacol*, *447*, 51–57.

34. Young, R. (1992). Aminorex produces stimulus effects similar to amphetamine and unlike those of fenfluramine. *Pharmacol Biochem Behav*, *42*, 175–178.

35. Gurtner, H. P. (1985). Aminorex and pulmonary hypertension: a review. *Cor Vasa*, *27*, 160–171.

36. Carlini, E. A. (2003). Plants and the central nervous system. *Pharmacol Biochem Behav*, *75*, 501–512.

37. Glennon, R. A., Yousif, M., Naiman, N., and Kalix, P. (1987). Methcathinone: a new and potent amphetamine-like agent. *Pharmacol Biochem Behav*, *26*, 547–551.

38. Zhu, M. Y., Shamburger, S., Li, J., and Ordway, G. A. (2000). Regulation of the human norepinephrine transporter by cocaine and amphetamine. *J Pharmacol Exp Ther*, *295*, 951–959.

39. Schwartz, J. W., Novarino, G., Piston, D. W., and DeFelice, L. J. (2005). Substrate binding stoichiometry and kinetics of the norepinephrine transporter. *J Biol Chem*, *280*, 19177–19184.

40. Tseng, L. F., Menon, M. K., and Loh, H. H. (1976). Comparative actions of monomethoxyamphetamines on the release and uptake of biogenic amines in brain tissue. *J Pharmacol Exp Ther*, *197*, 263–271.

41. Sofuoglu, M., Nelson, D., Babb, D. A., and Hatsukami, D. K. (2001). Intravenous cocaine increases plasma epinephrine and norepinephrine in humans. *Pharmacol Biochem Behav*, *68*, 455–459.

42. Rothman, R. B., and Baumann, M. H. (2006). Balance between dopamine and serotonin release modulates behavioral effects of amphetamine-type drugs. *Ann N Y Acad Sci*, *1074*, 245–260.

43. Rothman, R. B., Blough, B. E., and Baumann, M. H. (2002). Appetite suppressants as agonist substitution therapies for stimulant dependence. *Ann N Y Acad Sci*, *965*, 109–126.

44. Kampman, K. M., Rukstalis, M., Pettinati, H., Muller, E., Acosta, T., Gariti, P., Ehrman, R., and O'Brien, C. P. (2000). The combination of phentermine and fenfluramine reduced cocaine withdrawal symptoms in an open trial. *J Subst Abuse Treat, 19*, 77–79.

45. Negus, S. S., Mello, N. K., Blough, B. E., Baumann, M. H., and Rothman, R. B. (2007). Monoamine releasers with varying selectivity for dopamine/norepinephrine versus serotonin release as candidate "agonist" medications for cocaine dependence: studies in assays of cocaine discrimination and cocaine self-administration in rhesus monkeys. *J Pharmacol Exp Ther, 320*, 627–636.

46. Brauer, L. H., Johanson, C. E., Schuster, C. R., Rothman, R. B., and de Wit, H. (1996). Evaluation of phentermine and fenfluramine, alone and in combination, in normal, healthy volunteers. *Neuropsychopharmacology, 14*, 233–241.

47. Rea, W. P., Rothman, R. B., and Shippenberg, T. S. (1998). Evaluation of the conditioned reinforcing effects of phentermine and fenfluramine in the rat: concordance with clinical studies. *Synapse, 30*, 107–111.

48. Rothman, R. B., Blough, B. E., Woolverton, W. L., Anderson, K. G., Negus, S. S., Mello, N. K., Roth, B. L., and Baumann, M. H. (2005). Development of a rationally-designed, low abuse potential, biogenic amine releaser that suppresses cocaine self-administration. *J Pharmacol Exp Ther, 313*, 1361–1369.

49. Glennon, R. A., Young, R., Hauck, A. E., and McKenney, J. D. (1984). Structure–activity studies on amphetamine analogs using drug discrimination methodology. *Pharmacol Biochem Behav, 21*, 895–901.

50. Wee, S., Anderson, K. G., Baumann, M. H., Rothman, R. B., Blough, B. E., and Woolverton, W. L. (2005). Relationship between the serotonergic activity and reinforcing effects of a series of amphetamine analogs. *J Pharmacol Exp Ther, 313*, 848–854.

51. Marona-Lewicka, D., Rhee, G. S., Sprague, J. E., and Nichols, D. E. (1995). Psychostimulant-like effects of *p*-fluoroamphetamine in the rat. *Eur J Pharmacol, 287*, 105–113.

PART III

PHARMACOLOGY

12

PET/SPECT IMAGING STUDIES OF THE PLASMA MEMBRANE DOPAMINE TRANSPORTER

PAUL CUMMING

Department of Radiology, Ludwig Maximillian University, Munich, Germany

WEIGUO YE

The Russell H. Morgan Department of Radiology and Radiological Science, Johns Hopkins University, Baltimore, Maryland

DEAN F. WONG

The Russell H. Morgan Department of Radiology and Radiological Science, Department of Psychiatry, and Department of Environmental Health Sciences—Bloomberg School of Public Health, Johns Hopkins University, Baltimore, Maryland

12.1	Introduction	324
12.2	Tropanes	328
12.3	Other DAT Ligands	330
12.4	Comparison of Results In Vitro and In Vivo	332
12.5	Kinetic Modeling of DAT Ligands	332
12.6	Clinical DAT Studies	334
	12.6.1 Aging and Genetics	334
	12.6.2 Parkinson's Disease	335
	12.6.3 Other Neuropsychiatric Diseases and Personality Effects	337
	12.6.4 Substance Abuse	338
	12.6.5 Psychostimulant Action and Euphoria	338
12.7	Conclusions	339
	Acknowledgments	339
	References	339

323

12.1 INTRODUCTION

The detection in vivo of plasma membrane dopamine transporters (DATs) by external imaging using positron emission tomography (PET) and single-photon emission computed tomography (SPECT) has been a major landmark in the field of neuropsychopharmacology. Successful imaging of the DAT followed one decade after the advent of PET imaging of dopamine D2 receptors [1], which was the first instance of neurotransmitter imaging in living human brain. During the 1980s, research groups around the world also developed methods for imaging of dopamine D1 and serotonin 5-HT$_2$ receptors. Efforts to detect specifically presynaptic elements of dopamine innervations were initially focused on the dopa decarboxylase substrate [^{18}F]fluorodopa [2], radioactivity from which accumulates within dopamine and serotonin nerve terminals due to the vesicular trapping of its product [^{18}F]fluorodopamine. However, the specific visualization of the DAT in dopamine terminals proved to be a challenging task, since many of the standard radioligands were of relatively low affinity and low specificity, and thus unsuited for imaging studies in vivo.

To address this deficiency, several academic centers entered into collaborations with commercial partners to develop useful DAT ligands. The starting point for DAT tracers was provided by well-established in vitro methods employing β-emitting ([^{125}I] or [^{3}H]) ligands, in conjunction with membrane binding and autoradiographic techniques. By 1993, there had been a great flurry of activity in developing DAT ligands radiolabeled for SPECT ([^{123}I] or [^{99M}Tc]) or PET ([^{11}C] or [^{18}F]) imaging. The SPECT imaging ligands can be prepared from commercially available radioisotopes ([^{123}I] ($t_{1/2} = 13$ hours), or prepared on-site at nuclear medicine and radiological facilities: [^{99m}Tc] ($t_{1/2} = 6$ hours). In contrast, an expensive cyclotron/radiochemistry facility is necessary for the local production of the short-lived PET isotopes [^{11}C] ($t_{1/2} = 20$ minutes) or [^{18}F] ($t_{1/2} = 110$ minutes). Detection of nigrostriatal degeneration in Parkinson's disease and related degenerative disorders has proven to be the major clinical application of DAT imaging studies. As noted above, SPECT ligands have found wide use in the clinical setting, due to the relative ease of their procurement. However, PET imaging offers considerable advantages with respect to the absolute quantitation of DAT, which can be critical in conducting research protocols. These advantages are due to the availability of superior PET instrumentation and to practical/theoretical aspects of attenuation correction and other technical aspects of positron annihilation detection. However, it should be recalled that [^{123}I] compounds can be prepared for SPECT studies at outstandingly high specific activities, sometimes exceeding 10,000 Ci/mmol, such that the injected mass is negligible. In contrast, the relatively lower specific activities typical for [^{11}C]- and [^{18}F]-PET tracers can sometimes present problems related to the administration of a drug at pharmacologically active doses. In this chapter we summarize properties of many of the most successful PET and SPECT radioligands that have been developed for DAT imaging in the past 15 years (Table 12-1). In addition, important clinical findings from DAT imaging studies in the fields of neurology and psychiatry are reviewed in some detail.

TABLE 12-1 Imaging of the Dopamine Transporter

Radioisotope Tracer	Primary Specificity			Animal	Human	References
	DAT	SERT	NET			
[^{99m}Tc]TRODAT-1	×	—	—	Monkeys	—	Huang. W. S., et al. *Nucl Med Commun.* 2003; 24(1):77–83
[^{123}I]β-CIT	×	×	—	—	×	Berding, G., et al. *Nuklearmedizin.* 2003; 42(1):31–38
						Bergström, K. A., et al. *Synapse.* 1995; 19(4):297–300
[^{125}I]β-CIT	×	—	—	Pig	—	Minuzzi, L., et al. *Synapse.* 2006;59:211–219
[^{125}I]β-RIT-121	×	—	—	Monkeys	—	Guilarte, T. R., et al. *Exp Neurol.* 2006;202: 381–390
[^{131}I]β-CIT	×	—	—	Pig and rats	—	Sihver, W., et al. *Nucl Med Biol.* 2007;34(2): 211–219
[^{11}C]β-CIT	×	×	×		×	Laihinen, A. O., et al. *J Nucl Med.* 1995; 36(7):1263–1267
nor-β-CIT	IC$_{50}$ = 0.4 nM	IC$_{50}$ = 4 nM		Monkeys	—	Bergström, K. A., et al. *Eur J Nucl Med.* 1997; 24(6):596–601

(Continued)

TABLE 12.1 *Continued*

Radioisotope Tracer	Primary Specificity				Animal	Human	References
	DAT	SERT	NET				
[^{11}C]FE-CIT	×	—	—		—	×	Antonini, A., et al. *Neurol Sci.* 2002;23 Suppl 2:S51–52
[^{123}I]FP-CIT	×	—	—		Rats	—	Booij, J., et al. *Nucl Med Biol.* 2006;33(3): 409–411
[^{18}F]FP-CIT	×	—	—		—	×	Chaly, T., et al. *Nucl Med Biol.* 1996;23(8): 999–1004
[^{123}I]-IPT	×	—	—		—	×	Pogarell, O., et al. *Eur J Nucl Med Mol Imaging.* 2006;33(4):407–411 Epub 2006 Jan 27
[^{125}I]IPT	×	—	—		Mice	—	*Nucl Med Biol.* 2007; 34(3):239–246
[^{11}C]WIN 35,428	×	—	—		Baboon	×	Wong, D. F., et al. *Synapse.* 1993;15:13, 142
					Mouse	—	Scheffel, U., et al. *Synapse.* 1996;23:61–69
					Rats	—	Kawamura, K., et al. *Ann Nucl Med.* 2003; 17(3):249–253

	DAT	SERT	NET			
[^{11}C]MP	IC$_{50}$ = 82 ± 17 nM X	IC$_{50}$ = 7600 ± 1100 nM —	IC$_{50}$ = 440 ± 69 nM —	—	—	Pan, D., et al. *Eur J Pharmacol.* 1994;264:177–182
[^{11}C]cocaine	X	—	—	—	X	Volkow, N. D., et al. *Biol Psychiatry.* 2005;15;57(6):640–646 Wang, G. J., et al. *Brain.* 2004;127(Pt 11):2452–2458. Epub 2004 Aug 19
[^{11}C]PE2I	X	—	—	Rats	—	Inaji, M., et al. *Cell Transplant.* 2005;14(9):655–663
[^{11}C]brasofensine ([^{11}C]NS2214)	—	—	—	Pig	—	Minuzzi, L., et al. *Synapse.* 2006;59:211–219
[^{125}I]altropane	X	X	—	Monkey	X	Madras, B.K., et al. *Synapse.* 1998;29:93–104

327

12.2 TROPANES

A number of *Erythroxylaceae* species native to the eastern Andes have been culti-vated since ancient times for their psychostimulant properties. The main constituent alkaloid was isolated in 1855 by Friedrich Gaedke, who named it *erythoxyline*. However, the publication of an improved isolation by Albert Neiman in 1859 led to wide acceptance of the name *cocaine*. The bridged ring system of cocaine, and its ester side chain, are amenable to the preparation of diverse radiolabeled *tropane* derivatives, which retain high affinity for the DAT, with varying degrees of selectivity over the plasma membrane transporters of serotonin and noradrenaline. The tropane now constitute a diverse and important class of agents for PET and SPECT imaging studies of monoamine transporters, initially developed in collaboration between the Research Triangle Institute and RBI (Research Biochemicals), together with Johns Hopkins University and other academic institutions.

The fluorophenyl tropane [^{11}C]WIN 35,428 (also known as CFT) was used for the first PET studies in normal human subjects [3] and in patients with Parkinson's disease (PD) [4]. At approximately the same time, a number of SPECT analogs were studied in nonhuman primates: 2β-carbomethoxy-3β-[4-[^{123}I]iodophenyl]-tropane [^{123}I]β-CIT (also known as [^{123}I]RTI-55) [5] and the closely related chlorophenyl compound [^{123}I]IPT [6]. The quantitation of [^{123}I]β-CIT was found to be highly reproducible in the striatum of healthy humans and could sensitively detect the nigrostriatal degeneration of PD patients [7].

The binding properties of the technecium-bearing tropane with [^{99m}Tc]TRODAT to the DAT was characterized in rats [8]. In human SPECT studies with [^{99m}Tc]TRODAT, an optimal binding ratio between striatum and cerebellum occurred a few hours after injection [9]. This tracer has attracted considerable attention around the world because the technecium can be generated on site without the need for a cyclotron. Altropane, an *N*-iodoallyl analog of WIN 35,428, was found to have in vitro properties suitable for evaluation as a SPECT imaging agent. Its affinity in vitro for the DAT (7 nM) was somewhat higher than for WIN 35,428 itself, and it had better selectivity than that of β-CIT for the DAT in preference to serotonin trans-porters [10]. In monkey striatum, the unlabeled E-isomer of altropane was tenfold more potent than the Z-isomer and had a 30-fold preference for DAT over serotonin transporters [11]. In vitro, [^{125}I]altropane bound to single site in monkey striatum with 5 nM affinity and with B_{max} close to that seen previously for the DAT (Table 12-1). The corresponding SPECT ligand [^{123}I]altropane proved useful for the sensitive detection of nigrostriatal degeneration in PD patients [12]. The tropane [^{11}C]NS2214 ([^{11}C]brasofensine) ligand was developed for PET on the basis of its great selectivity for the DAT in vitro. However, this ligand failed to detect the DAT in the striatum of living pigs, which was attributed to a pharmacologi-cal idiosyncrasy of the DAT in porcine brain [13].

Parallel with the development of tropane derivates, PET imaging studies of human and nonhuman primate were carried out using [^{11}C]cocaine itself [14]. Cocaine occurs naturally in two enantiomers; PET studies with the inactive enantiomer (+)-[^{11}C]cocaine reveal it to be hydrolyzed very rapidly in plasma, thus limiting its uptake and binding in living brain [15]. In contrast, the active enantiomer (−)-[^{11}C]cocaine had extensive reversible binding to the DAT in a living striatum, which could readily be quantified relative to the uptake in the cerebellum, a reference region nearly devoid of monoamine transporters. The striatal binding of (−)-[^{11}C]cocaine was largely displaced by pretreatment with more specific DAT ligands [16]. However, displacement studies revealed the additional presence in brain of a high-capacity, low-affinity site that was not displaceable by monoamine-uptake inhibitors. Although very useful for investigations of substance abuse, especially regarding the relationship between behavior and DAT occupancy by cocaine itself [17], [^{11}C]cocaine suffers from lower specificity and affinity than those of other tropane derivatives.

The tropane ligand PE-21 is reported to have considerable selectivity for the DAT in preference to serotonin transporters [18]. In general, however, the tropane class of imaging agents suffers from a general lack of pharmacological specificity, inherited from the parent cocaine, which has little capacity to distinguish the DAT from the plasma membrane transporters of noradrenaline and serotonin. To some extent, this limitation is overcome by the relative abundances and differing anatomical distributions of the transporters in living brain. In particular, tropane binding in the striatum can in the main be attributed to the DAT, which exceeds in abundance the binding components due to transporters of noradrenaline and serotonin [19]. The ambivalence of tropanes for the DA and serotonin transporters can be seen clearly in Fig. 12-1, which shows the binding of [^{125}I]RTI-55 (also known as β-CIT) in the brain of rats with hemi-depletions of the serotonin innervation. In the presence of the selective serotonin transporter ligand citalopram, its binding in vitro is almost entirely selective for the DAT, whereas the addition of a selective DAT ligand to the incubation medium imparts selectivity for serotonin transporters. Furthermore, this image shows that serotonin transporters are very abundant in cerebral cortex, but comprise approximately 25% of the abundance of the DAT in intact rat striatum. Similarly, the preponderance of [^{125}I]β-CIT binding sites in the human amygdala and cerebral cortex could be displaced in vitro with citalopram, whereas 80% of the striatal binding was attributed to the DAT [20].

The regional differences in pharmacological selectivity of tropanes can be a virtue, since [^{125}I]β-CIT has been used to measure serotonin transporters in rodent cerebral cortex [21]. [^{99m}Tc]TRODAT has been used to measure serotonin transporters in the cerebral cortex of nonhuman primates [22], and [^{123}I]β-CIT has been used similarly in human studies [23]. Furthermore, components of the binding of [^{11}C]β-CIT in living thalamus and cortex were vulnerable to competition of inhibitors of either serotonin or norepinephrine transporters [24]. As predicted by results in vitro, cited above, pretreatment of subjects with citalopram unmasks and isolates the DAT component of [^{123}I]β-CIT binding in SPECT studies of the human brain [25], although this approach does not lend itself to routine SPECT examinations. However, the rates of association of [^{123}I]β-CIT to the DA and serotonin transporters are not identical, such that the pharmacological identity of the two main binding components could be distinguished kinetically, at least if entire dynamic time–activity curves and arterial inputs are available.

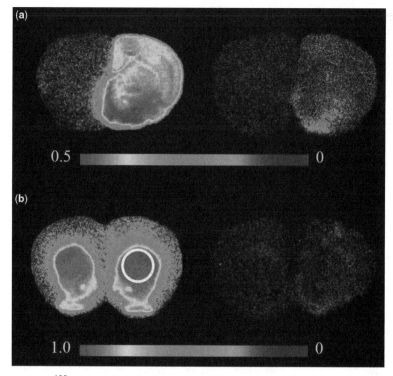

Figure 12-1 [^{125}I]RTI-55 autoradiography: representative autoradiograms of coronal rat brain cryostat sections (level: 10.20 mm anterior to the interaural line) with incubation of 10 pM [^{125}I]RTI-55 under conditions specific for serotonin transporters (a) and DAT (b). The serotonin innervation of the left hemisphere had earlier been destroyed with the selective toxin 5,7-DAT. Total binding is shown to the left and nonspecific binding to the right. The scales show the relative abundance of the two binding components. (Courtesy of Søren Dinesen Østergaard, Aarhus University PET Centre, Aarhus, Denmark.) (*See insert for color representation of figure.*)

12.3 OTHER DAT LIGANDS

The concentration of catecholamine uptake sites, representing the composite of the DA and norepinephrine transporters, can be measured in vitro with [^{3}H]mazindol, a psychostimulant with a tricyclic structure. Displacement studies in vitro show that the great preponderance of its binding in most brain regions is for the DAT [26]; the noradrenergic innervation is relatively sparse in rodent brain. However, the structure of mazindol does not lend itself to preparation as a PET or SPECT ligand. The piperazine compounds [^{3}H]GBR 12783 [27] and [^{3}H]GBR 12935 [28] have been prepared as ligands for DAT in vitro, but have not served as the basis for imaging agents in vivo.

As an alternative PET ligand to the tropanes, the psychostimulant [^{11}C]nomifensine binds reversibly to catecholamine transporters in living human

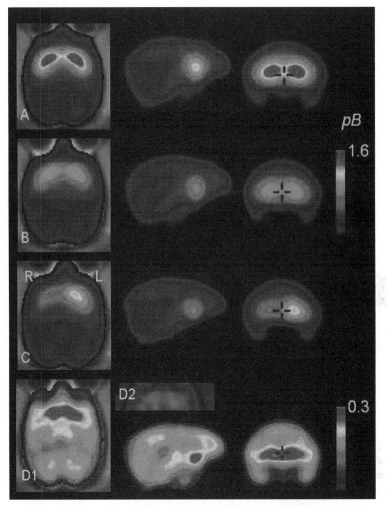

Figure 12-2 Binding potential (*pB*) of [¹¹C]methylphenidate (MP) for catecholamine uptake sites in brain of (A) normal young (*n* = 6) and (B) normal age (*n* = 4) rhesus monkeys, (C) aged monkeys with unilateral intracarotid infusion of MPTP (*n* = 3), and (D) monkeys of intermediate age with systemic MPTP lesioning (*n* = 6). Mean parametric maps are coregistered to the MR template, shown as gray scale in the horizontal plane images. A sector containing the striatum in the horizontal plane from the MPTP group (D2) is presented with identical intensity scaling as in the intact animals. The cursor in the coronal plane images indicates the position (0,10,10) in the MR coordinate system. (Reproduced with permission from Doudet et al. [50].) (*See insert for color representation of figure.*)

brain [29]. Although this binding in striatum is associated almost entirely with DAT, its significant binding in thalamus may reflect the high concentration of norepinephrine transporters in that structure. Similarly, the stimulant drug [¹¹C]methylphenidate, which also recognizes both types of catecholamine transporters, binds mainly to the DAT in striatum of the living mouse [30]. The active enatiomer [^{11}C]d-threo-methylphenidate has particularly high binding in human striatum [31]. This ligand has been employed in PET studies of the effects of MPTP on dopamine innervation (Fig. 12-2) and in clinical studies of Parkinson's disease, discussed below.

12.4   COMPARISON OF RESULTS IN VITRO AND IN VIVO

PET and SPECT studies with DAT ligands purport to measure an index of B_{max}, the absolute abundance of the transporter sites. As such, it is useful to compare the results obtained in living brain with the saturation binding parameters obtained in vitro using DAT ligands (Table 12-2). These results show generally good agreements between the estimates obtained in membrane preparations and by quantitative auto-radiography. As noted above, tropane derivatives can distinguish a high-affinity component constituting 25% of the total specific binding in vitro and a second site of lower affinity. The significance of this heterogeneity remains unclear, but it is likely that only the high-affinity site corresponds to the functional DAT. The concentration of high-affinity sites in membranes from rat striatum is substantially higher when measured at the usual laboratory temperature (20°C) than at body temperature (37°C) [32], suggesting interconversion between active and inactive states of the dopamine transporter.

It is conceivable that endogenous dopamine might compete with radiotracers for binding to DAT sites in living brain. However, the binding of [^{123}I]β-CIT in living monkey striatum was not reduced following treatment with a high dose of levodopa [33]. This result shows that levodopa treatment does not interfere in assay of the DAT in clinical studies of Parkinson's disease. In contrast, amphetamine partially displaced [^{123}I]β-CIT binding in striatum. The interpretation of this finding is uncertain, possible reflecting competition from elevated interstitial dopamine, or direct competition between amphetamine and the SPECT tracer. Conversely, pharmacological inhibition of dopamine release increased the [^{11}C]cocaine binding in monkey striatum, consistent with the presence of competition [30], although the magnitude of this effect was small. In contrast, [^{11}C]methylphenidate binding seemed insensitive to pharmacological-induced changes in dopamine tonus [30]. Thus, the competition model does not seem well established for DAT ligands.

12.5   KINETIC MODELING OF DAT LIGANDS

In any imaging study, the specific binding component, which is proportional to B_{max}, must be isolated from the background presented by nonspecific binding. In principle, nonspecific binding can best be measured after pretreatment with a blocking dose of a competitor, although this approach is difficult to carry out in human studies, due to the risk of toxicity. Alternatively, it has been proposed to use an inactive isomer of the tracer in order to measure nonspecific binding, although this approach fails in the case of [^{11}C]cocaine, due to the rapid metabolism of the inactive enantiomer,

TABLE 12-2 Saturation Binding Parameters for Ligands of the Plasma Membrane DAT in the Striatum of Rat, Monkey, and Human[a]

	$\mathrm{pmol/g}^a$	References
[³H]GBR 12783 (B_{max}), autoradiography, striatum		Leroux-Nicollet, I., and Costentin, J. *J Neural Transm Gen Sect.* 1994;97:93–106
Young rats	420 ± 19	
Aged rats	235 ± 57	
[³H]GBR 12935 (B_{max}, K_d), membranes (#), rat striatum ($n = 3$)	540 ± 48 (2.0 ± 0.3)	Gordon, I. *Brain Res.* 1995; 674:205–210
[³H]mazindol B_{max}, autoradiography, rat		Marshall, J. F., et al. *Neuroscience.* 1990:37:11–21
Dorsal striatum	700	
Ventral striatum	336	
V_{max} [³H]dopamine uptake, synaptosomes, rat		
Dorsal striatum	26 nmol/g/min	
Ventral striatum	16 nmol/g/min	
[³H]mazindol (4 nM), autoradiography, rat ($n = 8$)		Przedborski, S., et al. *Neuroscience.* 1995; 67:631–647
Rostal striatum	374 ± 20	
Nucleus accumbens	217 ± 20	
[³H]mazindole (9.5 nM), autoradiography, rat ($n = 8$)		Cline E. J., et al. *Exp Neurol.* 1995;134:135–149
Lateral striatum	800 ± 20	
Medial striatum	650 ± 20	
[³H]cocaine, B_{max} (K_d), membranes, monkey striatum		Madras, B. K., et al. *J Pharmacol Exp Ther.* 1989;251:131–141
Site 1	28 (19 nM)	
Site 2	283 (1.1 nM)	
[¹²⁵I]RTI, B_{max} (K_d), membranes, human striatum		Staley, J. K., et al. *J Neurochem.* 1994;62:549–556
Site 1	52 (100 pM)	
Site 2	134 (1.8 nM)	
[¹²⁵I]RTI, membranes, B_{max} (K_d), human striatum		Staley, J. K., et al. *Synapse.* 1995; 21:364–372
Site 1	57 (250 pM)	
Site 2	148 (5 nM)	
[¹²⁵I]β-CIT, B_{max} (K_d), membranes, baboon cerebral cortex		Laruelle, M., et al. *J Neurochem.* 1994;62:978–986
	48 (1.5 nM)	
Striatum, site 1	22 (380 pM)	
Striatum, site 2	423 (18 nM)	

[a]Quantities are reported per gram of tissue, assuming where specified (#) a 10% protein content.

cited above. In the case of $[^{123}I]\beta$-CIT, the inactive isomer does enter the brain of living nonhuman primate, but for unknown reasons does not have identical nonspecific binding [34]. In practice, cerebellum is frequently employed as a reference region for constraining the magnitude of the nonspecific binding of tropane ligands in regions of interest. This approach is justified by the very low levels of specific binding present in cerebellum.

Tropanes are inherently lipophilic substances, such that their deposition in living brain approaches the limit established by cerebral perfusion. Consequently, tropanes tend to have high nonspecific binding in living brain; the washout of the SPECT ligand $[^{123}I]\beta$-CIT from nonbinding reference tissues is particularly slow, as is the washout in high-binding regions. Another key issue with respect to modeling is the state of equilibrium obtained for the binding to DAT, which reflects the balance of the rates of association (k_{on}) and dissociation (k_{off}) of the ligand. This state of equilibrium is obtained rapidly with $[^{11}C]$cocaine, but competition studies against $[^{11}C]$cocaine binding in the striatum of living monkey suggested a dissociation rate constant for nonradioactive β-CIT of many hours [31].

Slow kinetics in vivo presents difficulties for quantitation of many tropanes for PET studies if equilibrium is not obtained within 90 minutes, the practical limit of positron emission studies with $[^{11}C]$-labeled compounds. For example, the 4-fluorophenyl analog of $[^{11}C]\beta$-CIT, $[^{11}C]$CFT (also known as $[^{11}C]$WIN 35,428), is rather slow to reach equilibrium [3], which presents difficulties for its quantitation. This practical limitation is overcome through the use of $[^{18}F]$CFT [35], or the fluoropropyl-labeled derivative $[^{18}F]\beta$-CIT-FP, both of which attain equilibrium within a few hours after administration [36], thus within the period of time permitted by the longer physical half-life of $[^{18}F]$. However, the quantitation of these tracers can require prolonged dynamic emission recordings, which may not be tolerated in clinical studies. The very slow kinetics of $[^{123}I]\beta$-CIT do not represent a great impediment to its use in quantitative SPECT studies, since one need only wait until the establishment in striatum of an equilibrium proportional to B_{max} [33]. Indeed, slow kinetics may be a virtue in clinical SPECT studies in that it may be convenient to perform the emission recording on the day following the tracer administration. Then the simple ratio between the radioactivity concentration in the striatum and in the nonbinding reference region serves as an indication of the abundance of DAT. However, this lengthy delay need not be necessary for the quantitation of all SPECT ligands. Ex vivo studies with $[^{125}I]$altropane indicate that a state of equilibrium binding in monkey striatum is obtained within 30 minutes of intravenous injection, with a binding ratio as high as 7 : 1 relative to cerebellum [11].

12.6 CLINICAL DAT STUDIES

12.6.1 Aging and Genetics

Rodent studies indicate age-related reductions in the availability of the DAT (Table 12-1), which can likewise be seen in a $[^{11}C]$methylphenidate-PET study of aged monkey (Fig. 12-2). In populations of neurologically normal humans, the $[^{123}I]\beta$-CIT binding [37] and the $[^{11}C]$methylphenidate binding [38] both declined in striatum by about 8% with each decade of healthy aging. In a semiquantitative SPECT study, $[^{123}I]$FP-CIT binding in human striatum was higher in females than in males, but age-related declines did not show a gender effect [39]. The striatal dopamine concentration in postmortem samples from neurologically normal subjects also declines by about 10% per decade [40], which suggests that there might be a simple relationship between the DAT and dopamine concentration as a function of age. However, the dopamine metabolite ratios in postmortem material are even more increased in the healthy elderly, suggesting the age-related reductions in the DAT may serve to compensate for impaired dopamine storage capacity in the aged brain. These changes are not without consequences for the performance of cognitive and motor tasks. Thus, the reaction time measured in a neuropsychological study of the healthy elderly correlated with the striatal binding of a tropane SPECT ligand, with and without correction for age within the cohort [41].

The expression of the DAT within a healthy population is quite variable, and the factors regulating DAT expression are poorly established. In a study of polymorphism of the human dopamine D_2 receptor TaqI A allele, the A1 allele (which imparts risk for alcoholism) was associated with slightly higher $[^{123}I]\beta$-CIT binding in striatum [42]. The unexpressed sequence of nine or 10 tandem repeats in the human DAT gene has attracted considerable attention because of its association with neuropsychiatric disorders. Heterozygocity of this sequence was associated with a 20% reduction in $[^{123}I]\beta$-CIT binding in a small group of subjects [43], but this finding was not replicated in another study of a larger cohort [44]. In a study of 96 healthy Caucasians scanned with $[^{123}I]\beta$-CIT, the earlier genetic association was replicated, so the finding seems secure. There was only a 10% reduction in DAT binding for subjects with the 10-repeat homozygotes, although this effect survived age correction [45]. Thus, the genetic factors accounting for the majority of variability in DAT expression are not yet defined.

12.6.2 Parkinson's Disease

One of the main uses of PET and SPECT ligands for the DAT is in the differential diagnosis of Parkinson's disease (PD) from related disorders manifesting in nigrostriatal degeneration. Since significant reductions in the DAT occur as a function of normal aging, it is a matter of basic importance to establish the threshold of DAT loss resulting in motor symptoms against the baseline of normal age effects, as has been shown in a $[^{99m}Tc]$TRODAT study [46]. In a microPET study of rats with unilateral dopamine lesions, there was a hyperbolic relationship between the behavioral response to a psychostimulant and the availability of the DAT measured with the tropane ligand $[^{11}C]$PE21. The rotational behavior occurred only in those rats with more than an 80% loss of DAT in the lesioned striatum [47]. Thus, there a rather sharp threshold of dopamine lesions resulting in symptoms of parkinsonism in experimental animals. Similarly, the binding ratio of $[^{123}I]\beta$-CIT in putamen was reduced by approximately 80% in patients with early PD, whereas declines in the putamen were less notable [48]. Furthermore, the binding of [^{11}C]methylphenidate to DAT in human putamen was reduced by as much as 50% without contralateral signs of Parkinson's disease, but mild hemi-parkinsonian symptoms were present when there was a 70% reduction in the DAT [49]. The relatively great reductions in [^{11}C]methylphenidate in early PD led to the speculation that the DAT is down-regulated, as also seems evident in MPTP-lesioned monkeys [50] (Fig. 12-1). As in normal aging, compensatory down-regulation of DAT may enhance the action of dopamine in the partially denervated striatum of patients with early Parkinson's disease.

In clinical SPECT studies, the occurrence of normal DAT levels in striatum almost excludes the diagnosis of degenerative PD. In a study of 150 aged subjects with normal [^{123}I]FP-CIT uptake, only a very few showed a progressive loss of DAT at follow-up, indicating that the remainder suffered from nondegenerative parkinsonism [51]. Since it seems that DAT expression can be modulated, prospective studies of PD progression must consider this as a possible confound. The Parkinson Study Group found putamenal [^{123}I]β-CIT binding to decline by 5% per year in patients with early PD, which is at least five times faster than the decline with normal aging [52]. However, the decline was lower in those patients treated with the direct agonist pramipexole than in those treated with levodopa, suggesting that the direct agonist was neuroprotective. It was noted that this could reflect modulatory action on the treatment on the expression of the DAT rather than a protective action per se [53].

The eventual development of interventative therapies for delaying the progression of PD would be facilitated with DAT ligands of the utmost sensitivity for detecting small reductions in the dopamine innervation. At present it seems uncertain which of the many available SPECT ligands may have optimal properties in regard to sensitivity. For example, [^{123}I]IACFT has been used for detecting early PD [10]. PET studies with [^{18}F]FECNT show that equilibrium striatal binding to DAT is obtained at 90 minutes, with binding ratios of 9 : 1 in healthy subjects and 5 : 1 in patients with early unilateral motor symptoms, and with discernible reductions already evident on the asymptomatic side [54]. Thus, this ligand may serve to identify presymptomatic nigrostriatal degeneration in patients at risk to develop PD.

The main degeneration of PD is evident in the putamen and to a lesser extent in the caudate nucleus. However, extrastriatal monoamine innervations also decline in PD. In a [^{11}C]β-CFT PET study, loss of monoamine transporters could also be seen in the orbitofrontal cortex and amygdala of patients with early PD [55], possibly reflecting loss of serotonin innervations. Similarly, reduced [^{11}C]methyphenidate binding in the anterior cingulate cortex of MPTP-poisoned monkeys may reveal loss of extrastriatal catecholamine fibers (Fig. 12-2).

Restless legs syndrome, which is characterized by episodes of uncontrolled leg movements during sleep, seems in some respects to constitute a PD spectrum disorder, in that it is responsive to clinical treatment with levodopa. However, in a SPECT imaging study with [^{123}I]-IPT, the striatal DAT binding was unaffected in a series of patients with polysomnography-verified restless legs syndrome [56], nor was [^{123}I]β-CIT binding abnormal in another group of subjects [57]. Thus, important declines in the dopamine innervation can be excluded in cases of restless legs syndrome.

12.6.3 Other Neuropsychiatric Diseases and Personality Effects

Striatal $[^{11}C]\beta$-CIT uptake is reduced by 50% in striatum of patients with Huntington's disease [58]; the neuronal degeneration of this disease is restricted to the medium spiny neurons of the striatum. Lesch–Nyhan syndrome, a devastating X-linked metabolic disorder, is associated with even more profound (70%) declines in the striatal binding of $[^{11}C]CFT$ [59]. A SPECT study with $[^{123}I]FP$-CIT shows that striatal DAT is normal in patients with Alzheimer's disease, as distinct from dementia with Lewy bodies (DLBs), in whom striatal DAT is substantially reduced [60]. Thus, SPECT studies of the DAT may not be adequate for the differential diagnosis of PD and DLBs, although reduced DAT specifically in the caudate nucleus was more pronounced in the latter condition [61]. The binding of $[^{11}C]$cocaine was reduced by 20% in striatum of patients with HIV dementia, to an extent correlating with viral load [62].

Despite the historical importance of the dopamine theory of schizophrenia, there have been relatively few investigations of the DAT in that disorder. In a $[^{18}F]CFT$ PET study, DAT binding did not differ between a group of untreated de novo patients with schizophrenia and in healthy control subjects [63], and in a $[^{123}I]\beta$-CIT SPECT study, DAT binding was entirely normal in striatum of a group of 24 patients with schizophrenia [64]. These negative findings were replicated in an even larger cohort of patients in a $[^{99m}Tc]TRODAT$-1 SPECT study. Among those patients suffering predominantly from positive symptoms, the DAT binding declined as a function of symptom severity [65]. In another $[^{99m}Tc]TRODAT$-1 SPECT study, striatal DAT binding was normal in patients with schizophrenia, but there was a lack of the asymmetric binding normally favoring the right striatum [66], a result in line with abnormalities in spontaneous turning bias reported among patients with schizophrenia. In a PET study of chronic patients, striatal $[^{18}F]CFT$ binding was slightly reduced [67], an effect either of medication or of disease progression.

The role of DAT in attention-deficit/hyperactivity disorder (ADHD) has been studied intensively. Initially, elevated striatal DAT was seen in a series of SPECT studies of adolescents with ADHD (see, e.g., [68]). Using $[^{11}C]$altropane, increased DAT was detected in the right caudate nucleus of a large series of well-characterized treatment-naive adults with ADHD [69], which suggests that abnormal regulation of DAT persists through adult life. However, in a recent PET study with $[^{11}C]PE21$ there was no evidence of elevated DAT in unmedicated ADHD patients [70].

The SPECT ligand $[^{123}I]IPT$ showed a twofold increase in striatal DAT in never-medicated children with Tourette's disorder [71]. This group also found a small increase in $[^{123}I]IPT$ binding in adult patients with obsessive-compulsive disorder, although the variability was quite high in the population [72]. There was a substantial increase in striatal $[^{123}I]\beta$-CIT binding in another study of Tourette's syndrome [73]. Overall, available results with ADHD and Tourette's syndrome link elevated DAT with disorders of impulse control.

Chronic treatment with the serotonin transporter inhibitor citalopram increased the striatal binding of $[^{123}I]\beta$-CIT in patients treated for depression [74], suggesting a primary deficiency of dopamine transmission, which was rectified by pharmacological potentiation of serotonin transmission. However, others have reported the presence of increased striatal DAT in a [^{99m}Tc]TRODAT-1 SPECT study of patients with major depression [75]. In contrast, striatal [^{123}I]β-CIT binding was reduced in patients with social phobia [76]. Other studies link the DAT with the expression of specific aspects of personality in normal subjects. Thus, there was a negative correlation between [^{18}F]CFT binding in the right putamen and the score for "detachment" in a personality inventory [77].

12.6.4 Substance Abuse

The striatal binding of [^{123}I]β-CIT was slightly increased in smokers relative to non-smokers. Irrespective of smoking status, there was slightly greater binding of the females subjects in the study [78]. Among alcoholics, there was a reduction in striatal [^{99m}Tc]TRODAT-1 binding [79]. However, [^{11}C]methylphenidate binding was normal in another group of alcoholics [80]. The case for pathologically altered DAT is clearer in studies of methamphetamine toxicity. Thus, striatal [^{11}C]CFT binding was 20% reduced among abstinent methamphetamine users [81]. Similar 20% reductions in striatal [^{11}C]methylphenidate binding were associated with impaired cognitive and psychomotor function in former methamphetamine users [82]. Reduced [^{11}C]CFT binding in the orbitofrontal cortex and amygdala of methamphetamine abusers suggested a basis for these cognitive and behavioral difficulties [83]. Although follow-up studies showed eventual recovery of the DAT following abstinence [84], the possibility must be considered that methamphetamine use may increase the risk for acquiring parkinsonism. Striatal injection of GDNF protected against methamphetamine-evoked declines of [^{11}C]CFT binding in monkeys [85].

The striatal concentration of DAT labeled with [^{123}I]β-CIT was increased among abstinent cocaine abusers [86], consistent with results of a postmortem autoradiographic study with [^{3}H]CFT [19]. Increased DAT expression may serve a homeostatic function in response to chronic cocaine use and may thus contribute to part of the syndrome of withdrawal, if this then evokes a dysphoria during the hypodopaminergic state. Nonetheless, prolonged medication with levodopa or selegiline did not alter striatal [^{123}I]β-CIT binding in patients treated for PD [87]. The latter result may be confounded by impaired homeostatic regulation of DAT expression and function in patients with nigrostriatal degeneration.

12.6.5 Psychostimulant Action and Euphoria

Competition paradigms, cited above, can also be used to assess the occupancy of the DAT associated with abuse of psychostimulants. This competition can be assessed by measuring the dose-dependent displacement of [^{11}C]cocaine by nonradioactive cocaine. In nonaddicted subjects, the extent of euphoria evoked by cocaine correlated positively with the occupancy of striatal [^{11}C]cocaine binding. Greater than 60% occupancy of DAT invariably evoked euphoria [88]. A similar displacement of [^{11}C]cocaine by methylphenidate did not evoke experiences of euphoria in all subjects [89]. This discrepancy returns attention to the topic introduced at the beginning of this chapter: the pharmacological specificity of DAT ligands. Thus, the combined action of cocaine at the DAT and serotonin transporters may be more euphorogenic than the combined action of methylphenidate and the DAT and noradrenaline transporters. However, differing binding kinetics may also contribute to the phenomenology of these psychostimulants.

12.7 CONCLUSIONS

A wide range of ligands have been developed for PET and SPECT studies of the DAT. [^{11}C]Methylphenidate has also been employed for PET studies of the DAT, but suffers from somewhat low specific binding and cross-recognition of the less abundant noradrenaline transporters. Many clinical imaging studies have been carried out with SPECT ligands based on the tropane structure. For the most part, these compounds suffer from a lack of pharmacological selectivity, sharing with [^{11}C] cocaine a mixed affinity for several monoamine transporters. The time required for tropanes to obtain equilibrium binding can be quite long, which is not an obstacle for semiquantitative clinical SPECT studies given the half-life of [^{123}I]. However, rapid equilibration and high specific binding can be obtained with altropane ligands. DAT ligands have been most useful for detecting early stages of nigrostriatal degeneration, and conversely, for revealing an apparent increase in DAT function with impulse-control disorders.

ACKNOWLEDGMENTS

This work was supported by NIH grants DA00412, AA12839, MH78175, and HD24448 (DFW). We thank Farah Ali and Jeff Galecki for their editorial assistance.

REFERENCES

1. Wagner, H. N., Burns, H. D., Dannals, R. F., Wong, D. F., Landstrom, B., Duelfer, T., Ravert, H. T., Links, J. M., Rosenbloom, S. B., Lukas, S. E., Kramer, A. V., and Kuhar, M. J. (1983). Imaging dopamine receptors in the human brain by positron tomography. *Science*, *23*, 1264–1265.

2. Martin, W. R., Palmer, M. R., Patlak, C. S., and Calne, D. B. (1989). Nigrostriatal function in humans studied with positron emission tomography. *Ann Neurol*, *26*, 535–542.

3. Wong, D. F., Yung, B., Dannals, R. F., Shaya, E. K., Ravert, H. T., Chen, C. A., Chan, B., Folio, T. U., et al. (1993). In vivo imaging of baboon and human dopamine transporters by positron emission tomography using [^{11}C]WIN 35,428. *Synapse*, *35*, 428.

4. Frost, J. J., Rosier, A. J., Reich, S. G., Smith, J. S., Ehlers, M. D., Snyder, S. H., Ravert, H. T., and Dannals, R. F. (1993). Positron emission tomographic imaging of the dopamine transporter with ^{11}C-WIN 35,428 reveals marked declines in mild Parkinson's disease. *Ann Neurol*, *34*, 423–431.

5. Shaya, E. K., Scheffel, U., Dannals, R. F., Ricaurte, G. A., Carroll, F. I., Wagner, H. N., Jr., Kuhar, M. J., and Wong, D. F. (1992). In vivo imaging of dopmaine reuptake sites in the primate brain using single photon emission computed tomograph (SPECT) and iodine-123 labeled RTI-55. *Synapse, 10,* 169–172.

6. Malison, R. T., McDougle, C. J., van Dyck, C. H., Scahill, L., Baldwin, R. M., Seibyl, J. P., Price, L. H., Leckman, J. F., and Innis, R. B. (1995). [^{123}I]β-CIT SPECT imaging of striatal dopamine transporter binding in Tourette's disorder. *Am J Psychiatry, 152,* 1359–1361.

7. Seibyl, J., Laurelle, M., van Dyck, C. H., Wallace, E., Baldwin, R., Zoghbi, S., Zea-Ponce, Y., Neumeyer, J. L., Charney, D. S., Hoffer, P. B., and Innis, R. (1996). Reproducibility of iodine-123-β-CIT SPECT brain measurement of dopamine transporters. *J Nucl Med, 37,* 222–228.

8. Kung, M. P., Stevenson, D. A., Ploss, K., Meegalla, S. K., Beckwith, A., Essman, W. D., Mu, M., Lucki, I., and Kunk, H. F. (1997). [^{99m}Tc]TRODAT-1: a novel technetium-99m complex as a dopamine transporter imaging agent. *Eur J Nucl Med, 24,* 372–380.

9. Kung, H. F., Kim, N. H., Kung, M. P., Meegalla, S. K., Plossl, K., and Lee, H. K. (1996). Imaging of dopamine transporters in humans with technetium-99m TRODAT-1. *Eur J Nucl Med, 23,* 1527–1530.

10. Elmaleh, D. R., Fischman, A. J., Shoup, T. M., Byon, C., Hanson, R. N., Liang, A. Y., Meltzer, P. C., and Madras, B. K. (1996). Preparation and biological evaluation of iodine-125-IACFT: a selective SPECT agent for imaging dopamine transporter sites. *J Nucl Med, 37,* 1197–1202.

11. Madras, B. K., Meltzer, P. C., Liang, A. Y., Elmaleh, D. R., Babich, J., and Fischman, A. J. (1998). Altropane, a SPECT or PET imaging probe for dopamine neurons. I: Dopamine transporter binding in primate brain. *Synapse, 29,* 93–104.

12. Fischman, A. J., Bonab, A. A., Babich, J., Palmer, E. P., Alpert, N. M., Elmaleh, D. R., Callahan, R. J., Barrow, S. A., Grahan, W., Meltzer, P. C., Hanson, R. N., and Madras, B. K. (1998). Rapid detection of Parkinson's disease by SPECT with altropane: a selective ligand for dopamine transporters. *Synapse, 29,* 128–141.

13. Minuzzi, L., Olsen, A. K., Bender, D., Arnfred, S., Grant, R., Danielsen, E. H., and Cumming, P. (2006). Quantitative autoradiography of ligands for dopamine receptors and transporters in brain of Gottingen minipig: comparison with results in vivo. *Synapse, 59,* 211–219.

14. Fowler, J. S., Volkow, N. D., Wolf, A. P., Dewey, S. L., Schlyer, D. J., MacGregor, R., Hitzemann, R., Logan, J., Bendriem, B., Gatley, S. J., et al. (1989). Mapping cocaine binding sites in human and baboon brain in vivo. *Synapse, 4,* 371–377.

15. Gatley, S. J., MacGregor, R. R., Fowler, J. S., Wolf, A. P., Dewey, S. L., and Schlyer, D. J. (1990). Rapid stereoselective hydrolysis of (+)-cocaine in baboon plasma prevents its uptake in the brain: implications for behavioral studies. *J Neurochem, 54,* 720–723.

16. Volkow, N. D., Fowler, J. S., Logan, J., Gatley, S. J., Dewey, S. L., MacGregor, R. R., Schlyer, D. J., Pappas, N., King, P., and Wang, G. J. (1995). Carbon-11-cocaine binding compared at subpharmacological and pharmacological doses: a PET study. *J Nucl Med, 36,* 1289–1297.

17. Fowler, J. S., Volkow, N. D., Logan, J., Gatley, S. J., Pappas, N., King, P., and Wang, G. J. (1998). Measuring dopamine transporter occupancy by cocaine in vivo: radiotracer considerations. *Synapse, 28,* 111–116.

18. Emond, P., Garreau, L., Chalon, S., Boazi, M., Besnard, J. C., and Guilloteau, D. (1997). Synthesis and ligand binding of nortropane derivates: N-substituted 2β-carbomethoxy-3β-(4'-iodophenyl) nortropane and N-(3-iodoprop-(2E)-enyl)-2β-carbomethoxy-3β-(3',4'-disubstituted phenyl)nortropane: new high-affinity and selective compunds for the dopamine transporter. *J Med Chem*, *40*, 1366–1372.

19. Little, K. Y., Kirkman, J. A., Carroll, F. I., Clark, T. B., and Duncan, G. E. (1993). Cocaine use increases [3H]WIN 35428 binding sites in human striatum. *Brain Res*, *628*, 17–25.

20. Staley, J. K., Basile, M., Flynn, D. D., and Mash, D. C. (1994). Visualizing dopamine and serotonin transporters in the human brain with the potent cocaine analogue [^{125}I]RTI-55: in vitro binding and autoradiographic characterization. *J Neurochem*, *62*, 549–556.

21. Scheffel, U., Kim, S., Cline, E. J., and Kuhar, M. J. (1994). Occupancy of the serotonin transporter by fluxetine, paroxetine, and sertraline: in vivo studies with [^{125}I]RTI-55. *Synapse*, *16*, 263–268.

22. Dresel, S. H., Kung, M. P., Huang, X. F., Plossl, K., and Kunk, H. F. (1999). Simultaneous SPECT studies of pre- and postsynaptic dopamine binding sites in baboons. *J Nucl Med*, *40*, 660–666.

23. de Win, M. M., de Jeu, R. A., de Bruin, K., Habraken, J. B., Reneman, L., Booij, J., and den Heeten, G. J. (2004). Validity of in vivo [^{123}I]β-CIT SPECT in detecting MDMA-induced neurotoxicity in rats. *Eur Neuropsychopharmacol*, *14*, 185–189.

24. Farde, L., Halldin, C., Muller, L., Suhara, T., Karlsson, P., and Hall, H. (1994). PET study of [^{11}C]β-CIT binding to monoamine transporters in the monkey and human brain. *Synapse*, *16*, 93–103.

25. Ryding, E., Lindstrom, M., Bradvik, B., Grabowski, M., Bosson, P., Traskman-Bendz, L., and Rosen, I. (2004). A new model for separation between brain dopamine and serotonin transporters in ^{123}I-β-CIT SPECT measurements: normal values and sex and age dependence. *Eur J Nucl Med Mol Imaging*, *31*, 1114–1118.

26. Marshall, J. F., O'Dell, S. J., Navarrete, R., and Rosenstein, A. J. (1990). Dopamine high-affinity transport site topography in rat brain: major differences between dorsal and ventral striatum. *Neuroscience*, *37*, 11–21.

27. Leroux-Nicollet, I., and Costentin, J. (1994). Comparison of the subregional distributions of the monoamine vesicular transporter and dopamine uptake complex in the rat striatum and changes during aging. *J Neural Transm Gen Sect*, *97*, 93–106.

28. Richfield, E. K. (1991). Quantitative autoradiography of the dopamine uptake complex in rat brain using [3H]GBR 12935: binding characteristics. *Brain Res*, *540*, 1–13.

29. Salmon, E., Brooks, D. J., Leenders, K. L., Turton, D. R., Hume, S. P., Cremer, J. E., Jones, T., and Frackowiak, R. S. (1990). A two-compartment description and kinetic procedure for measuring regional cerebral [^{11}C]nomifensine uptake using positron emission tomography. *J Cereb Blood Flow Metab*, *10*, 307–316.

30. Gatley, S. J., Volkow, N. D., Fowler, J. S., Dewey, S. L., and Logan, J. (1995). Sensitivity of striatal [^{11}C]cocaine binding to decreases in synaptic dopamine. *Synapse*, *20*, 137–144.

31. Volkow, N. D., Ding, Y. S., Fowler, J. S., Wang, G. J., Logan, J., Gatley, S. J., Schlyer, D., and Pappas, N. (1995). A new PET ligand for the dopamine transporter: studies in the human brain. *J Nucl Med*, *36*, 2162–2168.

32. Laruelle, M., Giddings, S. S., Zea-Ponce, Y., Charney, D. S., Neumeyer, J. L., Baldwin, R. M., and Innis, R. B. (1994). Methyl 3 β-(4-[^{125}I]iodophenyl)tropane-2 β-carboxylate in

vitro binding to dopamine and serotonin transporters under "physiological" conditions. *J Neurochem, 62,* 978–986.

33. Laruelle, M., Baldwin, R. M., Malison, R. T., Zea-Ponce, Y., Zoghbi, S. S., al-Tikriti, M. S., Sybirska, E. H., Zimmermann, R. C., Wisniewski, G., and Neumeyer, J. L (1993). SPECT imaging of dopamine and serotonin transporters with [^{123}I]β-CIT: pharmacological characterization of brain uptake in nonhuman primates. *Synapse, 13,* 295–309.

34. Scanley, B. E., Baldwin, R. M., Laruelle, M., al-Tikriti, M. S., Zea-Ponce, Y., Zoghbi, S., Giddings, S. S., Charney, D. S., Hoffer, P. B., and Wang, S. (1994). Active and inactive enantiomers of 2β-carbomethoxy-3β-(4-iodophenyl)tropane: comparison using homogenate binding and single photon emission computed tomographic imaging. *Mol Pharmacol, 45,* 136–141.

35. Laakso, A., Bergman, J., Haaparanta, M., Vilkman, H., Solin, O., and Hietala, J. (1998). [^{18}F]CFT [(^{18}F)WIN 35,428], a radioligand to study the dopamine transporter with PET: characterization in human subjects. *Synapse, 28,* 244–250.

36. Lundkvist, C., Halldin, C., Ginovart, N., Swahn, C. G., and Farde, L. (1997). [^{18}F]β-CIT-FP is superior to [^{11}C]β-CIT-FP for quantitation of the dopamine transporter. *Nucl Med Biol, 24,* 621–627.

37. van Dyck, C. H., Seibyl, J. P., Malison, R. T., Laruelle, M., Wallace, E., Zoghbi, S. S., Zea-Ponce, Y., Baldwin, R. M., Charney, D. S., and Hoffer, P. B. (1995). Age-related decline in striatal dopamine transporter binding with iodine-123-β-CITSPECT. *J Nucl Med, 36,* 1175–1181.

38. Volkow, N. D., Ding, Y. S., Fowler, J. S., Wang, G. J., Logan, J., Gatley, S. J., Hitzemann, R., Smith, G., Fields, S. D., and Gur, R. (1996). Dopamine transporters decrease with age. *J Nucl Med, 37,* 554–559.

39. Lavalaye, J., Booij, J., Reneman, L., Habraken, J. B., and van Royen, E. A. (2000). Effect of age and gender on dopamine transporter imaging with [^{123}I]FP-CIT SPECT in healthy volunteers. *Eur J Nucl Med, 27,* 867–869.

40. Kish, S. J., Shannak, K., Rajput, A., Deck, J. H., and Hornykiewicz, O. (1992). Aging produces a specific pattern of striatal dopamine loss: implications for the etiology of idiopathic Parkinson's disease. *J Neurochem, 58,* 642–648.

41. van Dyck, C. H., Avery, R. A., Macavoy, M. G., Marek, K. L., Quinlan, D. M., Baldwin, R. M., Seibyl, J. P., Innis, R. B., and Arnsten, A. F. (2007). Striatal dopamine transporters correlate with simple reaction time in elderly subjects. *Neurobiol Aging* (in press).

42. Laine, T. P., Ahonen, A., Rasanen, P., Pohjalainen, T., Tiihonen, J., and Hietala, J. (2001). The A1 allele of the D2 dopamine receptor gene is associated with high dopamine transporter density in detoxified alcoholics. *Alcohol, 36,* 262–265.

43. Heinz, A., Goldman, D., Jones, D. W., Palmour, R., Hommer, D., Gorey, J. G., Lee, K. S., Linnoila, M., and Weinberger, D. R. (2000). Genotype influences in vivo dopamine transporter availability in human striatum. *Neuropsychopharmacology, 22,* 133–139.

44. Martinez, D., Gelernter, J., bi-Dargham, A., van Dyck, C. H., Kegeles, L., Innis, R. B., and Laruelle, M. (2001). The variable number of tandem repeats polymorphism of the dopamine transporter gene is not associated with significant change in dopamine transporter phenotype in humans. *Neuropsychopharmacology, 24,* 553–560.

45. van Dyck, C. H., Malison, R. T., Jacobsen, L. K., Seibyl, J. P., Staley, J. K., Laruelle, M., Baldwin, R. M., Innis, R. B., and Gelernter, J. (2005). Increased dopamine transporter

availability associated with the 9-repeat allele of the SLC6A3 gene. *J Nucl Med*, *46*, 745–751.

46. Mozley, P. D., Scneider, J. S., Acton, P. D., Plossl, K., Stern, M. B., Siderowf, A., Leopold, N. A., Py, L., Alvi, A., and Kung, H. F. (2000). Binding of [^{99m}Tc]TRODAT-1 to dopamine transporters in patients with Parkinson's disease and in healthy volunteers. *J Nucl Med*, *41*, 584–589.

47. Inaji, M., Okauchi, T., Ando, K., Maeda, J., Nagai, Y., Yoshizaki, T., Okano, H., Nariai, T., Ohno, K., Obayashi, S., Higuchi, M., and Suhara, T. (2005). Correlation between quantitative imaging and behavior in unilaterally 6-OHDA-lesioned rats. *Brain Res*, *1064*, 136–145.

48. Innis, R. B., Seibyl, J. P., Scanley, B. E., Laruelle, M., Bi-Dargham, A., Wallace, E., Baldwin, R. M., Zea-Ponce, Y., Zoghbi, S., and Wang, S. (1993). Single photon emission computed tomographic imaging demonstrates loss of striatal dopamine transporters in Parkinson disease. *Proc Natl Acad Sci U S A*, *90*, 11965–11969.

49. Lee, C. S., Samii, A., Sossi, V., Ruth, T. J., Schulzer, M., Holden, J. E., Wudel, J., Pal, P. K., de la Fuente-Fernandez, Calne, D. B., and Stoessl, A. J. (2000). In vivo positron emission tomographic evidence for compensatory changes in presynaptic dopaminergic nerve terminals in Parkinson's disease. *Ann Neurol*, *47*, 493–503.

50. Doudet, D. J., Rosa-Neto, P., Munk, O. L., Ruth, T. J., Jivan, S., and Cumming, P. (2006). Effect of age on markers for monoaminergic neurons of normal and MPTP-lesioned rhesus monkeys: a multi-tracer PET study. *Neuroimage*, *30*, 26–35.

51. Marshall, V. L., Patterson, J., Hadley, D. M., Grosset, K. A., and Grosset, D. G. (2006). Two-year follow-up in 150 consecutive cases with normal dopamine transporter imaging. *Nucl Med Commun*, *27*, 933–937.

52. Parkinson Study Group (2002). Dopamine transporter brain imaging to assess the effects of pramipexole vs. levodopa on Parkinson disease progression. *JAMA*, *287*,1653–1661.

53. Marek, K., Jennings, D., Seibyl, J. (2002). Do dopamine agonists or levodopa modify Parkinson's disease progression? *Eur J Neurol*, *9*(3), 15–22.

54. Davis, M., Votaw, J. R., Bremner, J. D., Byas-Smith, M. G., Faber, T. L., Voll, R. J., Hoffman, J. M., Kilts, C. D., and Goodman, M. M. (2003). Initial human PET imaging studies with the dopamine transporter ligand ^{18}F-FECNT. *J Nucl Med*, *44*, 855–861.

55. Ouchi, Y., Yoshikawa, E., Okada, H., Futatsubashi, M., Sekine, Y., Iyo, M., and Sakamoto, M. (1999). Alterations in binding site density of dopamine transporter in the striatum, orbitofrontal cortex, and amygdala in early Parkinson's disease: compartment analysis for β-CFT binding with positron emission tomography. *Ann Neurol*, *45*, 601–610.

56. Eisensehr, I., Wetter, T. C., Linke, R., Noachtar, S., von Lindeiner, H., Gildehaus, F. J., Trenkwalder, C., and Tatsch, K. (2001). Normal IPT and IBZM SPECT in drug-naive and levodopa-treated idiopathic restless legs syndrome. *Neurology*, *57*, 1307–1309.

57. Michaud, M., Soucy, J.-P., Chabli, A., Lavigne, G., and Montplaisir, J. (2002). SPECT imaging of striatal pre- and postsynaptic dopaminergic status in restless legs syndrome with periodic leg movements in sleep. *J Neurol*, *249*, 1432–1459.

58. Ginovart, N., Lundin, A., Farde, L., Halldin, C., Backman, L., Swahn, C. G., Pauli, S., and Sedvall, G. (1997). PET study of the pre- and post-synaptic dopaminergic markers for the neurodegenerative process in Huntington's disease. *Brain*, *120*, 503–514.

59. Wong, D. F., Harris, J. C., Naidu, S., Yokoi, F., Marenco, S., Dannals, R. F., Ravert, H. T., Yaster, M., Evans, A., Rousset, O., Bryan, R. N., Gjedde, A., Kuhar, M. J., and Breese, G. R. (1996). Dopamine transporters are markedly reduced in Lesch–Nyhan disease in vivo. *Proc Natl Acad Sci U S A, 93*, 5539–5543.

60. O'Brien, J. T., Colloby, S., Fenwick, J., Williams, E. D., Firbank, M., Burn, D., Aarsland, D., and McKeith, I. G. (2004). Dopamine transporter loss visualized with FP-CIT SPECT in the differential diagnosis of dementia with Lewy bodies. *Arch Neurol, 61*, 919–925.

61. Walker, Z., Costa, D. C., Walker, R. W., Lee, L., Livingston, G., Jaros, E., Perry, R., McKeith, I., and Katona, C. L. (2004). Striatal dopamine transporter in dementia with Lewy bodies and Parkinson disease: a comparison. *Neurology, 62*, 1568–1572.

62. Wang, G. J., Chang, L., Volkow, N. D., Telang, F., Logan, J., Ernst, T., and Fowler, J. S. (2004). Decreased brain dopaminergic transporters in HIV-associated dementia patients. *Brain, 127*, 2452–2458.

63. Laakso, A., Vilkman, H., Alakare, B., Haaparanta, M., Bergman, J., Solin, O., Peurasaari, J., Rakkolainen, V., Syvalahti, E., and Hietala, J. (2000). Striatal dopamine transporter binding in neuroleptic-naive patients with schizophrenia studied with positron emission tomography. *Am J Psychiatry, 157*, 269–271.

64. Laruelle, M., bi-Dargham, A., van, D. C., Gil, R., D'Souza, D. C., Krystal, J., Seibyl, J., Baldwin, R., and Innis, R. (2000). Dopamine and serotonin transporters in patients with schizophrenia: an imaging study with [(^{123})I]β-CIT. *Biol Psychiatry, 47*, 371–379.

65. Schmitt, G. J., Frodl, T., Dresel, S., la, F. C., Bottlender, R., Koutsouleris, N., Hahn, K., Moller, H. J., and Meisenzahl, E. M. (2006). Striatal dopamine transporter availability is associated with the productive psychotic state in first episode, drug-naive schizophrenic patients. *Eur Arch Psychiatry Clin Neurosci, 256*, 115–121.

66. Hsiao, M. C., Lin, K. J., Liu, C. Y., Tzen, K. Y., and Yen, T. C. (2003). Dopamine transporter change in drug-naive schizophrenia: an imaging study with ^{99m}Tc-TRODAT-1. *Schizophr Res, 65*, 39–46.

67. Laakso, A., Bergman, J., Haaparanta, M., Vilkman, H., Solin, O., Syvalahti, E., and Hietala, J. (2001). Decreased striatal dopamine transporter binding in vivo in chronic schizophrenia. *Schizophr Res, 52*, 115–120.

68. Cheon, K. A., Ryu, Y. H., Kim, Y. K., Namkoong, K., Kim, C. H., and Lee, J. D. (2003). Dopamine transporter density in the basal ganglia assessed with [^{123}I]IPT SPET in children with attention deficit hyperactivity disorder. *Eur J Nucl Med Mol Imaging, 30*, 306–311.

69. Spencer, T. S., Biederman, J., Madras, B. K., Dougherty, D. D., Bonab, A. A., Livni, E., Meltzer, P. C., Maring, J., Rauch, S., and Fischman, A. J. (2007). Further evidence of dopamine transporter dysregulation in ADHD: a controlled PET imaging study using altropane. *Biol Psychiatry* (in press).

70. Jucaite, A., Fernell, E., Halldin, C., Forssberg, H., and Farde, L. (2005). Reduced midbrain dopamine transporter binding in male adolescents with attention-deficit/hyperactivity disorder: association between striatal dopamine markers and motor hyperactivity. *Biol Psychiatry, 57*, 229–238.

71. Cheon, K. A., Ryu, Y. H., Namkoong, K., Kim, C. H., Kim, J. J., and Lee, J. D. (2004). Dopamine transporter density of the basal ganglia assessed with [^{123}I]IPT SPECT in drug-naive children with Tourette's disorder. *Psychiatry Res, 130*, 85–95.

72. Kim, C. H., Koo, M. S., Cheon, K. A., Ryu, Y. H., Lee, J. D., and Lee, H. S. (2003). Dopamine transporter density of basal ganglia assessed with [^{123}I]IPT SPET in obsessive-compulsive disorder. *Eur J Nucl Med Mol Imaging, 30,* 1637–1643.

73. Malison, R. T., McDougle, C. J., van Dyck, C. H., Scahill, L., Baldwin, R. M., Seibyl, J. P., Price, L. H., Leckman, J. F., and Innis, R. B. (1995). [^{123}I]β-CIT SPECT imaging of striatal dopamine transporter binding in Tourette's disorder. *Am J Psychiatry, 152,* 1359–1361.

74. Kugaya, A., Seneca, N. M., Snyder, P. J., Williams, S. A., Malison, R. T., Baldwin, R. M., Seibyl, J., and Innis, R. B. (2003). Changes in human in vivo serotonin and dopamine transporter availabilities during chronic antidepressant administration. *Neuropsychopharmacology, 28,* 413–420.

75. Brunswick, D. J., Amsterdam, J. D., Mozley, P. D., and Newberg, A. (2003). Greater availability of brain dopamine transporters in major depression shown by [^{99m}Tc]TRODAT-1 SPECT imaging. *Am J Psychiatry, 160,* 1836–1841.

76. Tiihonen, J., Kuikka, J., Bergstrom, K., Lepola, U., Koponen, H., and Leinonen, E. (1997). Dopamine reuptake site densities in patients with social phobia. *Am J Psychiatry, 154,* 239–242.

77. Laakso, A., Vilkman, H., Kajander, J., Bergman, J., Paranta, M., Solin, O., and Hietala, J. (2000). Prediction of detached personality in healthy subjects by low dopamine transporter binding. *Am J Psychiatry, 157,* 290–292.

78. Staley, J. K., Boja, J. W., Carroll, F. I., Seltzman, H. H., Wyrick, C. D., Lewin, A. H., Abraham, P., and Mash, D. C. (1995). Mapping dopamine transporters in the human brain with novel selective cocaine analog [^{125}I]RTI-121. *Synapse, 21,* 364–372.

79. Tiihonen, J., Kuikka, J., Bergstrom, K., Hakola, P., Karhu, J., Ryynanen, O. P., and Fohr, J. (1995). Altered striatal dopamine re-uptake site densities in habitually violent and non-violent alcoholics. *Nat Med, 1,* 654–657.

80. Volkow, N. D., Wang, G. J., Fowler, J. S., Logan, J., Hitzemann, R., Ding, Y. S., Pappas, N., Shea, C., and Piscani, K. (1996). Decreases in dopamine receptors but not in dopamine transporters in alcoholics. *Alcohol Clin Exp Res, 20,* 1594–1598.

81. McCann, U. D., Wong, D. F., Yokoi, F., Villemagne, V., Dannals, R. F., and Ricaurte, G. A. (1998). Reduced striatal dopamine transporter density in abstinent methamphetamine and methcathinone users: evidence from positron emission tomography studies with [^{11}C]WIN-35,428. *J Neurosci, 18,* 8417–8422.

82. Volkow, N. D., Chang, L., Wang, G. J., Fowler, J. S., Leonido-Yee, M., Franceschi, D., Sedler, M. J., Gatley, S. J., Hitzemann, R., Ding, Y. S., Logan, J., Wong, C., and Miller, E. N. (2001). Association of dopamine transporter reduction with psychomotor impairment in methamphetamine abusers. *Am J Psychiatry, 158,* 377–382.

83. Sekine, Y., Minabe, Y., Ouchi, Y., Takei, N., Iyo, M., Nakamura, K., Suzuki, K., Tsukada, H., Okada, H., Yoshikawa, E., Futatsubashi, M., and Mori, N. (2003). Association of dopamine transporter loss in the orbitofrontal and dorsolateral prefrontal cortices with methamphetamine-related psychiatric symptoms. *Am J Psychiatry, 160,* 1699–1701.

84. Volkow, N. D., Chang, L., Wang, G. J., Fowler, J. S., Franceschi, D., Sedler, M., Gatley, S. J., Miller, E., Hitzemann, R., Ding, Y. S., and Logan, J. (2001). Loss of dopamine transporters in methamphetamine abusers recovers with protracted abstinence. *J Neurosci, 21,* 9414–9418.

85. Melega, W. P., Lacan, G., Desalles, A. A., and Phelps, M. E. (2000). Long-term methamphetamine-induced decreases of [^{11}C]WIN 35,428 binding in striatum are reduced by GDNF: PET studies in the vervet monkey. *Synapse, 35,* 243–249.

86. Malison, R. T., Best, S. E., van Dyck, C. H., McCance, E. F., Wallace, E. A., Laruelle, M., Baldwin, R. M., Seibyl, J. P., Price, L. H., Kosten, T. R., and Innis, R. B. (1998). Elevated striatal dopamine transporters during acute cocaine abstinence as measured by [^{123}I]β-CIT SPECT. *Am J Psychiatry, 155(6),* 832–834.

87. Innis, R. B., Marek, K. L., Sheff, K., Zoghbi, S., Castronuovo, J., and Sibyl, J. P. (1999). Effect of treatment with L-dopa/carbidopa or L-selegiline on striatal dopamine transporter SPECT imaging with [^{123}I]β-CIT. *Mov Disord, 14(3),* 436–442.

88. Volkow, N. D., Wang, G. J., Fischman, M. W., Foltin, R. W., Fowler, J. S., Abumrad, N. N., Vitkun, S., Logan, J., Gatley, S. J., Pappas, N., Hitzemann, R., and Shea, C. E. (1997). Relationship between subjective effects of cocaine and dopamine transporter occupancy. *Nature, 386,* 827–830.

89. Volkow, N. D., Wang, G. J., Fowler, J. S., Gatley, S. J., Logan, J., Ding, Y. S., Dewey, S. L., Hitzemann, R., Gifford, A. N., and Pappas, N. R. (1999). Blockade of striatal dopamine transporters by intravenous methylphenidate is not sufficient to induce self-reports of "high." *J Pharmacol Exp Ther, 288,* 14–20.

13

IN VITRO STUDIES OF DOPAMINE TRANSPORTER FUNCTION AND REGULATION

BRIAN R. HOOVER, BRUCE H. MANDT, AND NANCY R. ZAHNISER

Department of Pharmacology and Neuroscience Program, University of Colorado Denver, Aurora, Colorado

13.1	Introduction	348
13.2	Sources of DATs FOR IN VITRO ASSAYS	349
	13.2.1 Brain Tissue Preparations	349
	13.2.2 Heterologous Expression Systems	350
13.3	In Vitro Approaches Used to Study the DAT	350
	13.3.1 Radioligand Binding	350
	13.3.2 mRNA	351
	13.3.3 Protein	355
	13.3.4 Direct Functional Assays	357
	13.3.5 Cell Surface Expression	367
	13.3.6 Trafficking	372
13.4	Conclusions	377
	Acknowledgments	378
	References	378

13.1 INTRODUCTION

Dopamine transporters (DATs) play a key role in limiting dopamine (DA) neurotransmission by transporting DA back into DA neurons, thereby helping to conserve DA for re-release. DATs are expressed exclusively by DA neurons. Since their physiological function is to pump extracellular DA back into DA neurons, they must be present on the neuronal plasma membrane in order to be active. Electron microscopy studies have revealed that in nigrostriatal DA neurons, DATs are localized on the plasma membranes of axon varicosities and terminals in the striatum as well as dendrites in the substantia nigra (SN) [1]. Cell surface DATs are localized both perisynaptically, adjacent to where DA is released from synaptic vesicles, and at a distance from synaptic junctions.

DAT is a member of the 20-member solute carrier gene family 6 (member 3; SLC6A3; www.bioparadigms.org/slc/SLC06.htm), also referred to as the Na^+- and Cl^--dependent neurotransmitter transporter family. The two other plasma membrane monoamine neurotransmitter transporters, the norepinephrine (NE) transporter (NET) and the serotonin (5-HT) transporter (SERT or 5-HTT), also belong to the SLC6 family [2,3]. Uptake mediated by these transporters derives its energy from the cellular electrochemical gradient established by the Na^+-ATPase and is dependent on co-transport of Na^+ and Cl^- ions. DATs have a relatively high affinity ($K_m = \sim 100$ nM in nervous tissue; ~ 1 μM in most heterologous cell expression systems), but low capacity, to transport DA. DATs are selective for DA, as compared to NE and 5-HT, but readily transport tyramine, amphetamines, and the neurotoxins 6-hydroxydopamine and N-methyl-4-phenylpyridinium (MPP^+).

This transporter was first identified by its activity in in vitro uptake assays, which measured the accumulation of [^{3}H]DA in acutely isolated neuronal preparations (e.g., striatal slices and synaptosomes) [4,5]. DAT substrates such as amphetamine and tyramine were found not only to be taken up, but also to induce the release of [^{3}H]DA via reversal of the DAT. Other compounds like cocaine, benztropine, and nomifensine, were found to inhibit the uptake of [^{3}H]DA by the DAT, thereby preventing accumulation of [^{3}H]DA. In vitro [^{3}H]DA uptake assays remain an essential assay by which DAT activity and kinetics are still commonly measured today. Since that time, however, several additional in vitro assays have been developed and used to study the DAT. Many of these assays have utilized electrochemical detection of DA. In addition, the cloning of the DAT in 1991 [6–9] provided a number of new tools, which in turn have allowed new in vitro approaches for studying the DAT to be developed. The use of electrophysiological recording and microscopic techniques in live cells has proven particularly useful. Each of these in vitro assays has its own particular advantages and disadvantages, and each has helped us to more fully appreciate the role and importance of the DAT in terminating DA neurotransmission as well as the complex rapid regulation of the DAT.

In this chapter we review these different in vitro approaches. In addition to giving a descriptive overview of each approach, we point out some of the advantages and disadvantages of each approach and highlight some of the important insights that have come from this work.

13.2 SOURCES OF DATs FOR IN VITRO ASSAYS

13.2.1 Brain Tissue Preparations

Prior to cloning of the DAT, brain was the only source of tissue for in vitro studies of DATs. Brain tissue from a variety of species has been used. Brain synaptosomes, acutely prepared brain slices, and cultured neurons (primary neurons and organotypic slices) have all been used for functional studies. Generally, the most important requirement for measuring DAT function is the viability of the tissue (e.g., it must be alive and healthy).

Brain synaptosomes are resealed nerve endings that are enriched in synaptic domains and presynaptic markers. These are typically prepared at 4°C by gently homogenizing fresh brain tissue in isotonic 0.32 M sucrose buffered at pH 7.4 (10% weight : volume) using 10–20 strokes of a glass–Teflon pestle homogenizer with a clearance of 0.22 μm, followed by differential centrifugation [10]. First, a low-speed spin (1000 g for 10 minutes) is used to sediment nuclei and cell debris (P1 fraction) from the homogenate. The resulting supernatant is then centrifuged at a higher speed (10,000 g for 20 minutes) to isolate the synaptosomes in the second pellet, or P2 fraction. This preparation is distinct from that for synaptoneurosomes, which isolates vesicles in which presynaptic–postsynaptic membrane associations are maintained [11,12]. It should also be noted that as an exception to the fresh tissue rule, synaptosomes have been prepared from cryopreserved frozen human and rat brains and used for [³H]DA uptake assays [13]. Although the K_m values were similar in synaptosomes prepared from fresh and cryopreserved rat striatum, values for maximal uptake velocity (V_{max}) were reduced by ~50% in the cryopreserved tissue, suggesting a loss of active DATs.

Brain slices of 200 to 400 μm that are acutely prepared with a McIlwain tissue chopper or a vibrating microtome are also useful for measuring DAT activity in vitro. With the microtome the brain tissue is submerged in buffer, and the slicing mechanism is gentler than with a tissue chopper. Slice preparations retain some normal neuronal connectivity, although the outside 50 μm of the slice are dead. Based on levels of adenosine and adenine nucleotides and electrophysiological recordings of rat hippocampal slice activity, it appears imperative to allow slices to recover their energy stores by incubating them in oxygenated artificial cerebrospinal fluid (aCSF) buffer at 20 to 34°C for 60 to 90 minutes before using them in in vitro assays [14].

More recently, dissociated rodent mesencephalic neuronal cultures have been used to study DAT function and regulation. Primary cultures are made from either embryonic (E14–E15) or early postnatal (P0–P1) rodent midbrain and can be maintained for several weeks. However, as DAT develops primarily postnatally in rodent brain [15], DAT levels are relatively low in cultured preparations. In our hands, relatively few (often <10%) of our cultured embryonic mesencephalic neurons, co-cultured with striatal neurons, are identified as DA neurons [i.e., positive for tyrosine hydroxylase (TH), the rate-limiting enzyme in DA synthesis]. Strategies using growth factors have been developed with postnatal neurons that increase the number of DA neurons [16,17]. Thus, postnatal neuronal cultures should offer some advantages over embryonic neuronal cultures. In any case, these preparations offer the advantages of ease of access for drug or molecular biological (e.g., RNA interference) treatments and for microscopic visualization. Cultured neurons are normally a useful preparation for viral transfection of mutant constructs; however, DA neurons seem to be more resistant to transfection and expression with viral vectors that have been used successfully with other neuronal preparations (A. Sorkin, personal communication). Thus, this issue remains a challenge for future development.

Organotypic brain slice co-cultures prepared from P0–P1 midbrain and striatum are another type of co-culture preparation that could be useful for studying DAT function since they can be grown in culture for 5–6 weeks [18]. Triple co-cultures of cortex–striatum-midbrain are also used successfully. However, so far, rat organotypic midbrain slice cultures have received only limited use in studying the DAT, having been used to identify a role for the DAT in DA neuronal vulnerability to paraquat toxicity [19].

In addition to functional assays, in vitro radioligand binding assays and Western blots using dead brain tissue preparations such as homogenates or purified synaptosomal membranes have provided much valuable quantitative information about DATs, their pharmacological profile, and their structure–activity relationships. Similarly, radioligand binding coupled with quantitative autoradiography using frozen (10 to 40 μm) brain sections cut with a cryostat, as well as immunohistochemical studies using fixed or frozen brain sections, have provided anatomical information about DAT localization in native brain tissue.

13.2.2 Heterologous Expression Systems

Since the DAT was cloned in 1991, heterologous expression systems have also been used extensively for in vitro functional assays. Cells used have included human and other mammalian cells, insect cells, and frog oocytes. Both cells transiently transfected with DAT and cell lines stably expressing the DAT have proven valuable. Living cells expressing the DAT have allowed many new approaches to be used to study DAT function and have led to an appreciation that cellular trafficking is one of the main mechanisms by which DATs are rapidly regulated. Interestingly, however, an important challenge for the future is exploring how those mechanisms identified in heterologous expression systems relate to DA neurons and brain preparations.

13.3 IN VITRO APPROACHES USED TO STUDY THE DAT

13.3.1 Radioligand Binding

Although radioligand binding assays do not measure DAT function, they are a convenient way to identify and characterize DATs. In addition to binding properties (affinities: K_d for the radioligand or K_i for unlabeled competing ligands), the number of binding sites (B_{max}; assumed to be the total number of DATs) can also be determined accurately. Over the years, several radioligands with high (nM) affinity for the DAT have been developed: notably [3H]mazindol, [3H]WIN 35,428 (2β-carbomethoxy-3β-(4-flurophenyl)tropane or CFT), [125I]RTI-55 (2β-carbomethoxy-3β-(4-iodophenyl)-tropane), [125I]RTI-121 (2β-isopropyl ester-3β-(4-iodophenyl)tropane), and [121I] altropane (*N*-iodoallyl-CFT) [20–22]. Some of these radioligands also have high affinity for the NET and/or the SERT so that nonspecific binding must be defined and measured rigorously. For example, early on, we characterized binding of [3H]nomifensine [23], which, similar to [3H]mazindol [20], has a tenfold higher affinity for NET than for DAT.

Binding assays are most often carried out using tissue preparations that are dead (i.e., homogenates, membranes, and slide-mounted brain sections). Unless purified plasma membranes are used, DATs associated with both intercellular and plasma membranes, assumed to be nonfunctional and functional DATs, respectively, will be measured. Binding to intact frozen/thawed brain sections and analyzing the resulting images with quantitative autoradiographic techniques offers the distinct advantage of defining anatomical specificity along with quantification but still potentially measures functional and nonfunctional DATs.

Binding assays can also be conducted in live cells. All DAT radioligands are lipophilic compounds that readily penetrate membranes so that they also potentially detect internal, as well as cell surface DATs, in intact cells. However, all DAT radioligands are DAT inhibitors that require relatively high millimolar concentrations of Na^+ for binding. We have found that in intact *Xenopus laevis* oocytes expressing the human (h) DAT, changes in [3H]WIN 35,428 binding reflect changes in the number of cell surface DATs [24], most likely because the intracellular Na^+ concentration ($\sim$12 mM) [25] is not high enough to support [3H]WIN 35,428 binding to intracellular DATs. Nonetheless, with any DAT binding assay, it is important to confirm that cell surface binding sites actually do represent functional transporters by conducting functional assays.

13.3.2 mRNA

DAT activity is rapidly regulated by a variety of signaling mechanisms, including DAT substrates, DAT inhibitors, and various signaling cascades [3,26]. Regulation of DAT activity can be accomplished by modulating DAT cell surface expression, affinity for its substrate DA, or other conformational changes. Additionally, total DAT expression levels can be altered, at either the mRNA or protein level. DAT mRNA is expressed only in DA neurons. DAT and TH mRNA expression overlap nearly completely in the principal DA cell bodies of the substantia nigra pars compacta (SNc) and ventral tegmental area (VTA) [27–29]. Accordingly, loss of DA neurons, induced pharmacologically in animal models of Parkinson's disease, is accompanied by a loss of DAT mRNA [30,31].

To measure mRNA expression, several options are available, including Northern blot, in situ hybridization, ribonuclease protection, and polymerase chain reaction assays. Regardless of the method used to measure mRNA, however, the relationship between changes in mRNA expression to changes in protein level and/or function is not known. Therefore, the functional relevance of changes in DAT mRNA is often not clear.

Northern Blotting   Northern blotting allows one to quantify the absolute amount of target mRNA in a given brain region or cells. Total cellular RNA is isolated, linearized, subjected to gel electrophoresis, and transferred to a membrane that is exposed to a radiolabeled DNA probe complementary to the mRNA of interest. Film autoradiography is used to detect the DNA–mRNA complex.

DAT mRNA in the midbrain of rats exposed to seven consecutive daily restraint stress sessions is elevated for >20 hours immediately after stress cessation [32]. This increase in transcript expression is accompanied by increased DAT activity as measured by [^{3}H]DA uptake (see Section 13.3.4). However, others have suggested that Northern blot analysis is not sensitive enough to detect relevant changes in DAT mRNA. For instance, Persico et al. [33] treated animals with chronic amphetamine (7.5 mg/kg twice daily for two weeks) and found only a trend toward increased DAT mRNA 5 to 8 days after withdrawal. An additional limitation is the inability to identity the subtypes of cells expressing the mRNA of interest.

In Situ Hybridization   In situ hybridization is the most commonly utilized method for measuring mRNA expression and corresponding changes in either transcript density/neuron or number of mRNA-expressing neurons in tissue sections. This technique does not allow determination of the absolute amount of mRNA per cell or tissue sample. Instead, in situ hybridization is ideal for comparing relative changes in mRNAs, such as between control samples and samples exposed to various experimental treatments. An antisense DNA or RNA probe, complementary to the target mRNA and often labeled with a radioisotope, is synthesized. As a control, a sense strand is also synthesized; this has the same sequence as the target mRNA (complementary to the antisense strand) and is used to measure nonspecific hybridization of the antisense probe with mRNAs in the tissue section. The hybridized radiolabeled antisense probe/mRNA complex is visualized with either film or liquid emulsion autoradiography. Film autoradiograms allow for the examination and measurement of mRNA expression at the tissue or brain regional level, whereas liquid emulsion autoradiography is used for analysis of mRNA at the single-cell level. The nucleic acid probe can also be labeled nonisotopically and visualized immunochemically or fluorescently, but these variations prohibit quantification of the mRNA expression levels. Limitations of in situ hybridization are its sensitivity of detection, the quality of the oligoprobes/riboprobes used, and the methods used to analyze and quantify the specific autoradiographic signal.

Most studies of DAT mRNA regulation using in situ hybridization have focused on the effects of growth factors or drugs of abuse. Growth factors are a focus because DAT-expressing DA neurons are exquisitely sensitive to perturbation of trophic factor levels, whereas many drugs of abuse interact directly with the DAT. For instance, growth factors may be involved in the reduced striatal DAT mRNA expression per DA neuron observed in mice reared for two months in an enriched vs. a standard environment [34]. Also, chronic intraventricular insulin administration significantly increases DAT mRNA in SNc and VTA [35]. Conversely, hypoinsulinemia, induced by food deprivation (24 to 36 hours), decreases DAT mRNA in SNc and VTA [36]. Accordingly, DAT uptake activity is also decreased. DAT transcription after food restriction is increased, but only when the animals have scheduled access to a sucrose solution (supplementing with food does not produce this effect) [37]. Sucrose access with free-fed rats or nonscheduled sucrose presentation does not alter DAT mRNA levels.

In situ hybridization has been utilized to examine changes in DAT mRNA after acute and chronic exposure to nicotine. For example, rats chronically exposed (4 weeks) to nicotine and cigarette smoke have elevated DAT mRNA in midbrain [38]. Nicotine increases DAT mRNA only in the SNc, whereas smoke exposure increases DAT message in both SNc and VTA. The greatest increases in message are associated with smoke inhalation and are time dependent. A ribonuclease protection assay was also used to examine these changes. In contrast, acute nicotine does not alter DAT mRNA in either SNc or VTA [39].

This technique has also been used to examine changes in DAT mRNA after acute and chronic exposure to the psychomotor stimulants methamphetamine, cocaine, and amphetamine. Rats were allowed to self-administer methamphetamine for 9 hours/ day over 10 days; and after 1 or 30 days of abstinence, DAT mRNA levels were examined [40]. Although there was a trend toward increased DAT mRNA in the SNc, no significant changes were observed in either SNc or VTA. In contrast, with a 10-day cocaine self-administration paradigm, DAT mRNA in the VTA but not the SNc, increased after a withdrawal period of 10 days [41]. Importantly, no changes in DAT mRNA were observed during acute or chronic cocaine self-administration. Similarly, acute cocaine administered by the investigator does not alter DAT mRNA in VTA or SNc [42,43], whereas a "binge" cocaine regimen [3 × 15 mg/kg cocaine intraperitoneally (i.p.) daily] that induces behavioral sensitization decreases DAT mRNA in both brain regions [44]. Non-binge, chronic cocaine treatment also reduces DAT mRNA in midbrain. For instance, two weeks of intravenously delivered cocaine at a dose and infusion schedule similar to cocaine self-administration paradigms decreases DAT mRNA in VTA 10 days posttreatment [45]. Additionally, rats chronically treated with cocaine for 8 days show significant reductions in DAT mRNA in VTA and SNc [46,47]. Any differences in drug effects might be attributed to different routes of administration, dosing schedules, species of rodent used, and/or time periods examined.

DAT mRNA has also been measured with in situ hybridization in order to better understand the role of the DAT in amphetamine-induced behavioral sensitization. In one study rats were given a sensitizing regimen of amphetamine (2.5 mg/kg daily for 5 days) and sacrificed one week later [48]. DAT mRNA was increased in both VTA and SN. Similar increases in DAT mRNA in both brain regions were seen in animals treated with amphetamine for 5 days and sacrificed 3 days later [49]. After two weeks of withdrawal, however, DAT mRNA remained increased only in the VTA.

Nuclease Protection Assays With a nuclease protection assay, mRNA isolated from a specific brain area or cell type is incubated with a complementary RNA or DNA radiolabeled probe. A nuclease is then added to the hybridization solution to digest any single-stranded probe sequences or mRNAs that have not hybridized. In the case of the ribonuclease protection assay, ribonuclease A is the nuclease used and the radiolabeled probe is a complementary RNA sequence. In this way, the complementary regions of the RNA probe/mRNA complex are protected, while the non-complementary sequences are removed. This complex is then subjected to gel electrophoresis followed by film autoradiography to detect the probe, which is a measure of the absolute amount of complementary target mRNA.

Use of nuclease protection assays to measure cocaine-induced changes in DAT mRNA in brain regions suggests, at most, transient down-regulation after repeated exposure. Using a nuclease protection assay, DAT mRNA is reduced in the SN 4 hours after the last injection of a repeated cocaine regimen (twice daily for 6 days) but is not altered after either acute cocaine administration or after 72-hour withdrawal from the repeated treatment [43]. Similarly, a ribonuclease protection assay revealed no change in DAT mRNA in either VTA or SN after a 10-day withdrawal period in rats exposed to binge cocaine for 3 or 14 days [50].

Polymerase Chain Reaction The most sensitive method for detecting and measuring mRNA is the polymerase chain reaction (PCR) technique. However, without the proper control experiments to account for the quantity of transcripts of the PCR product and analysis of changes in unrelated genes (called housekeeping genes), inaccurate conclusions can be drawn. To overcome potential pitfalls, quantitative real-time PCR is the current best option for measuring mRNA levels. With this technique, mRNAs are reverse transcribed, producing double-stranded RNAs (cDNAs) that are then amplified by enzymatic replication. Along with the mRNA of interest (DAT), the expression level of housekeeping genes that are expressed by the same cells at a similar level as the DAT and that should not be altered by the experimental treatment are amplified and used to evaluate and quantitate any changes in the DAT. Fluorescence is used to measure the amount of amplified products in real time.

Only a few studies have utilized PCR to examine changes in DAT mRNA. For example, semiquantitative reverse transcriptase PCR was used to examine relative changes in DAT mRNA in primary embryonic mesencephalic cultures grown in the absence or presence of striatal neurons. This study revealed that levels of DAT mRNA are regulated by extranigral factors (growth factors from target areas) since DAT mRNA increases when mesencephalic neurons are grown in direct contact with striatal neurons [51]. Using quantitative real-time PCR, chronic (12 days), but not acute (48 hours), food restriction increases DAT mRNA in the VTA [52]. This change was attributed to a sensitization of the mesolimbic DA pathway, characterized by an increased clearance of DA in the shell of the nucleus accumbens (NAc) and resulting in reduced basal extracellular DA in vivo. The increased clearance rate could be due to increased DAT transcription.

13.3.3 Protein

Electrophoresis and Western Blotting   Western blotting is a standard technique for examining protein expression. This technique uses polyacrylamide gel electrophoresis (PAGE) to separate proteins. There are two methods of gel electrophoresis: non-denaturing and denaturing. Nondenaturing or native gel electrophoresis separates proteins based on charge and hydrodynamic size. Proteins retain their folded conformation, and consequently, this technique can be used to examine changes in conformation, aggregation, or association with other proteins. However, most studies with DAT have used the second method of gel electrophoresis. This method uses sodium dodecyl sulfate (SDS) to denature proteins and separate them based on molecular mass. In most cases the reducing agent β-mercaptoethanol is also added to disrupt any disulfide bonds. Following SDS-PAGE, proteins are transferred to a blotting membrane, where they can be detected by primary antibodies generated against the protein of interest. Currently, there are about 20 commercially available DAT primary antibodies, which vary depending on species of interest and exact sequence of the amino acids targeted.

For SDS-PAGE, the percent polyacrylamide gel used should be appropriate for the molecular weight of DAT. The mature glycosylated DAT protein is detected between 75 and 100 kDa whereas the immature nonglycosylated DAT protein is detected between 50 and 75 kDa, depending on species. Since DAT trafficking requires a homooligomerized state [53] and DAT can be ubiquitinated as well [54], higher-molecular-weight forms of DAT are also detected with SDS-PAGE. Consequently, a dimeric form of hDAT in porcine aortic endothelial (PAE) cells has been reported around 200 kDa [53,54] and a ubiquitinated hDAT in human cervical carcinoma HeLa cells has been reported between 100 and 150 kDa [54]. Cross-linking hDAT expressed in human embryonic kidney (HEK) 293 cells and running this tissue on nonreducing (absence of β-mercaptoethanol) SDS-PAGE gels also results in an increase in the apparent molecular weight of DAT, from ∼85 to ∼195 kDa, supporting the idea that DAT exists as a dimer or tetramer [55,56]. Furthermore, following multiple methamphetamine injections, higher-molecular-weight DAT complexes (∼170 to 220 kDa) have also been detected in rat striatal tissue run on nonreducing SDS-PAGE [57].

One technical point regarding DAT Western blotting pertains to tissue preparation. Most often, tissue is boiled at 100°C to denature the proteins prior to PAGE, and we have found this to be appropriate for cloned DAT proteins. In contrast, however, we have observed that boiling endogenous brain DAT at 100°C causes DAT protein aggregation. Thus, samples containing native DAT protein from brain should be heated no higher than 75°C prior to running them on SDS-PAGE.

Western blots give valuable information about changes in DAT expression, but by themselves, cannot give fully quantitative information. Immunoreactive bands are commonly visualized by enhanced chemiluminescence on x-ray film and quantitated by densitometry. Thus, inclusion of a standard curve generated with comparable tissue samples allows for semiquantitative analysis because the quantitative response of the chemiluminescence signals and film are then known. When DAT Western blotting has been used to examine overall DAT protein levels, changes in protein levels are assumed to reflect changes in function. Even when changes in overall protein levels are not observed, the possibility exists that DAT redistribution occurs at the membrane, which would not be detected by a simple Western blot. Consequently, Western blotting is frequently used in tandem with other assays of DAT function. In particular, DAT Western blots have been used together with coimmunoprecipitation assays [54] and/or cell surface biotinylation assays [58–61].

Immunohistochemistry   Changes in total DAT protein levels are interpreted as indicative of accompanying changes in function. Although radioligand binding and Western blot analysis are more typically used to measure total numbers of DATs and DAT protein levels, respectively, a few studies have used immunohistochemistry to examine changes in DAT protein expression following chemical lesions of the DA neurons by 1-methyl 4-phenyl 1,2,3,6-tetrahydropyridine (MPTP) or after repeated drug administration. Immunohistochemistry uses an antibody that allows for specific and accurate localization of cellular antigens in tissue sections. Immunocytochemistry is the same process utilized in cell culture systems, such as dissociated neurons or mammalian cell culture. Fixed or frozen tissue is sectioned on a microtome and then incubated sequentially with primary and secondary antibodies. The secondary antibodies can be labeled with either a fluorescent molecule or biotin. For the latter, enzyme-linked reactions allow for the colorimetric visualization of the immunosignal by either light or electron microscopy. Primary concerns with immunhistochemistry are the selectivity of the antibody for the antigen, proper preservation of the antigen, and the sensitivity of the method used to detect the antigen. Also, immunohistochemistry (or immunocytochemistry) is not suited for quantifying the absolute amount of protein expressed in the tissue (or cells). Typically, immunohistochemistry is used to evaluate the presence or absence of the protein of interest, as well as the protein's subcellular localization. This technique can also be used to assess relative changes in a protein's expression based on changes in the optical density of immunostaining between control and experimental groups.

MPTP is a neurotoxin that causes degeneration of nigrostriatal DA neurons and an accompanying loss of DA in the dorsal striatum in primates and mice. In doing so, the toxin replicates many of the clinical symptoms, as well as the primary pathophysiological hallmarks, of Parkinson's disease. Monoamine oxidase B located in glial cells metabolizes MPTP to MPP^+. This ion is selectively taken up by the DAT into DA neurons, where it disrupts mitochondrial function, interrupting oxidative phosphorylation by inhibiting complex I of the mitochondrial electron transport chain. Disruption of the electron transport chain precedes cell death, which is believed to result from the MPP^+-induced impairment in ATP production and free-radical generation [62,63]. Many studies have utilized immunohistochemistry following MPTP-induced DA lesions to assess changes in DAT expression. The general consensus is that DAT expression decreases in a time-dependent fashion days to months after the lesion [64–67]. Several months post-MPTP lesion, however, DAT protein expression levels display at least partial recovery to control levels in parallel to motor recovery [66,67].

DAT immunohistochemistry was also examined in rats after a sensitizing regimen of amphetamine (2 mg/kg daily for 7 days), followed by a 10-day abstinence period. No changes in DAT protein expression were observed in any of the brain regions examined, which included striatum, NAc, and prefrontal cortex [68].

13.3.4 Direct Functional Assays

[³H]DA/DA/[³H]MPP⁺ Uptake   One of the most widely used techniques to study DAT function is uptake of [³H]DA/DA (or the metabolically inert [³H]MPP⁺). Theoretically, uptake can be performed in any tissue containing DATs (i.e., brain synaptosomes, brain slices, cell expression systems, neuronal cultures, etc., see Section 13.2). However, compared to synaptosomes, uptake of [³H]DA into minces of rat striatum is complex, with the normal high-affinity ($K_m = 160$ nM) as well as lower-affinity ($K_m = 4$ μM) uptake component and reduced potencies of DAT inhibitors [69]. Cocaine inhibition of [³H]DA uptake is biphasic in chopped tissue slices of rat striatum and NAc, but monophasic in synaptosomes [70]. The lower-than-expected affinities may be due to diffusion barriers in the slices. Minces/slices also show relatively higher nonspecific [³H]DA uptake, which can result in less accurate uptake/kinetic determinations.

Generally, during preparation, tissue is kept at 4°C to minimize DAT trafficking and then incubated at 34 to 37°C in assay buffer to restore biological activity. The assay buffer should contain a monoamine oxidase inhibitor (e.g., pargyline) to prevent DA degradation during the assay. Following restoration of DAT activity, [³H]DA is added and the tissue is further incubated for a fixed amount of time to assess uptake. Tissue concentration, incubation time, and incubation temperature are important variables; and the assay should be measured under conditions where uptake is linear with both tissue and time. Another important consideration is to define nonspecific [³H]DA uptake since specific DAT-mediated uptake is determined from the difference between total and nonspecific uptake. Nonspecific uptake can be defined in several ways, including [³H]DA uptake at 4°C in Na⁺-free solution or in the presence of a DAT inhibitor. DAT inhibitors such as cocaine, GBR 12909, or benztropine are used at 100 to 1000 times their K_i concentrations to inhibit >99% of the DAT activity. It is important to note that cocaine inhibits DAT, NET, and SERT with essentially the same affinity, so that in brain regions with significant numbers of NETs and SERTs, the more DAT-selective inhibitors GBR 12909 and benztropine are more useful. Assays are terminated by either diluting the assay with ice-cold buffer or returning the assay tubes to a 4°C bath. After stopping uptake, tissue is harvested using rapid filtration over filters with an appropriate pore size by a cell harvester. Uptake is quantified from the radioactive tissue trapped on the filters by liquid scintillation spectrometry. Because of the relative simplicity of this assay, [³H]DA uptake is a good tool to assess the effects of pharmacological and experimental manipulations on DAT function. Uptake assays can also be conducted with unlabeled DA, using high-performance liquid chromatography (HPLC) to measure tissue content, or with real-time in vitro electrochemical detection methods (see the discussion below).

There are two typical ways to utilize uptake assays. The simplest is a "bullet" assay that measures specific uptake with a single [³H]DA concentration. This type of assay is used most appropriately when examining relative changes in DAT function among different conditions. For example, our lab has used this approach to show decreases in hDAT function in PAE cells following phorbol ester or monensin pretreatments [71] and differences in uptake between low- and high-cocaine locomotor responding rats [72]. Bullet uptake gives a snapshot of DAT function but is unable to yield more detailed kinetic information.

To determine the underlying mechanism for differences in DAT function, the second type of uptake assay must be conducted. This assay uses full kinetic analysis of DAT activity to determine the $V_{\max}$ and K_m. A wide range of concentrations (e.g., 0.5 nM to 5 μM) of either [³H]DA by itself or unlabeled DA together with a fixed concentration of [³H]DA are used. A number of labs, including our own, have used this type of assay to study the rapid regulation of DAT. We found that in *Xenopus* oocytes, protein tyrosine kinase (PTK) inhibitors or protein kinase C (PKC) activation with phorbol-12-myristic-13-aceate (PMA) significantly decrease [³H]DA uptake $V_{\max}$ without significantly changing its K_m [73,74]. Changes in maximal velocity without changes in substrate affinity are suggestive of, but do not prove, altered DAT surface expression. Other labs have used kinetic assays to show that in vitro pretreatment of rat striatal synaptosomes with DA, followed by washing, decreases DAT $V_{\max}$ without changing K_m [59]. In contrast, a single cocaine injection (30 mg/kg, i.p.) in rats increases the DAT $V_{\max}$, without changing K_m, in synaptosomes subsequently prepared from NAc [58]. Kinetic analysis of [³H]DA uptake allows for a more detailed analysis of DAT function than single concentration bullet uptake. However, altered $V_{\max}$ values are only suggestive of changes in DAT surface expression and require further confirmation with cell surface biotinylation assays or binding assays using purified plasma membranes to confirm altered surface expression, or fluorescence microscopy to monitor DAT trafficking/localization.

[³H]DA uptake assays have also been used widely to assess the functional effects of DAT mutations. For example, kinetic uptake analyses in DAT–NET chimeras were used to determine discrete structural domains that help to define the pharmacological specificity and kinetic differences between these two catecholamine transporters [75]. With the DAT, Uhl and colleagues [76] have used [³H]DA uptake assays to help characterize more than 100 site-directed single- and multiple-amino acid substitution mutations, providing valuable insights about structure–function relationships of cocaine–DAT interactions. Their work has identified mutants with reduced affinity for cocaine but relatively normal affinity for DA, which may prove useful in development of "cocaine antagonists" that still allow DA uptake.

DAT-Mediated Reverse Transport ([³H]DA/DA/[³H]MPP⁺ Release)   It has been appreciated for more than 30 years that the DAT can transport substrates in an outward as well as an inward direction (see references in [77]). Early evidence for reverse transport came largely from in vitro efflux assays. Thus, reverse transport is readily observed by first preloading brain preparations such as striatal synaptosomes/slices or heterologous cells expressing DAT with [³H]DA (or

[³H]MPP⁺; endogenous DA can also be measured in the overflow buffer by HPLC); next exposing them to amphetamine, tyramine, or higher concentrations of unlabeled DA; and then collecting the bathing medium. This is particularly striking in such cells as hDAT-expressing oocytes because, unlike brain preparations, there is minimal spontaneous release of [³H]DA from cells; thus, little or no [³H]DA release occurs until it is stimulated by addition of the substrate. The early evidence strongly supported the idea that amphetamine-induced [³H]DA release is mediated by DAT because it is Na⁺- but not Ca²⁺-dependent and blocked by DAT inhibitors. More conclusive evidence, however, came from the observation that amphetamine fails to induce DA release in striatal slices of DAT knockout mice [78].

The mechanisms by which amphetamines induce DAT-mediated DA release have been studied extensively [79]. Amphetamines have critical effects on vesicular storage of DA and cause reversal of the vesicular monoamine transporter 2 (VMAT2) in DA neurons, thereby increasing cytosolic DA concentrations [79]. However, our discussion here will be restricted to amphetamine and the plasma membrane DAT. Early studies of amphetamine-stimulated [³H]DA release from striatal minces and homogenates suggested that the DAT releases DA by facilitated exchange diffusion [80,81]. This mechanism depends on asymmetric DAT function, with more efficient inward than reverse transport. Thus, amphetamine binds to the transporter, thereby inducing and stabilizing the substrate-binding conformation of DAT. Next, amphetamine is transported inwardly across the membrane and is released into the cytoplasm, leaving the unoccupied substrate binding site facing the cytoplasm. If excess cytosolic DA is present, it binds to the DAT, is transported in the reverse direction, and is released into the extracellular space. Obviously, the upper limit for this model is the exchange of one molecule of amphetamine for one molecule of DA. Although this mechanism is consistent with many of the observations about amphetamine-induced [³H]DA release, other studies (e.g., showing the strong dependence of DAT-mediated [³H]MPP⁺ release on extracellular Na⁺ that is unrelated to the Na⁺ required for uptake [82]) questioned the simple facilitated exchange diffusion mechanism. Also, more recent evidence using real-time voltammetric and electrophysiological recording methods support involvement of an additional channel-like transport mechanism for amphetamine-induced, DAT-mediated DA release (see [79] and the sections that follow).

Interestingly, PKC has been shown to regulate DAT-mediated DA uptake and release in an asymmetric (i.e., opposite, fashion) [83]. As already mentioned, PKC activation by phorbol esters reduces [³H]DA uptake; this results from increasing DAT endocytosis and decreasing the number of DATs on the plasma membrane (see [3] and Section 13.3.5). Nonetheless, despite a loss of functional DATs, phorbol esters are still able to promote DAT-mediated DA release, most likely by a rapid phosphorylation of the DAT, which would occur more quickly than the altered DAT trafficking [83]. Similarly, inhibitors of PKC block amphetamine-stimulated [³H]DA/DA release but, if anything, increase [³H]DA uptake [84]. These results further emphasize distinctions between DAT-mediated DA uptake and release.

In vivo manipulations can result in changes in DAT activity (uptake and/or reverse transport) that persist in in vitro brain preparations such as synaptosomes and slices. There are a number of examples of this in the literature. One striking example that we found was that administration of a single low dose of cocaine (10 mg/kg, i.p.) to rats resulted in significantly elevated amphetamine-induced release of [³H]DA from their striatal slices which persisted for at least two weeks after the acute drug administration [85]. This long-lasting, but not permanent DAT plasticity depends on DA receptor activation because it was blocked by DA D1/D2 receptor antagonists [86]. This finding was not unique to cocaine, in that we observed similar results following an acute systemic injection of either the DAT inhibitor mazindol or the direct-acting DA receptor agonist apomorphine. Furthermore, in vitro incubation of synaptosomes with drug, followed by washout of the drug, can be a useful model that recapitulates the rapid regulation of DAT function observed after single in vivo drug injections. Notably, in vitro incubation of rat striatal synaptosomes with methamphetamine reduces [³H]DA uptake, but not [³H]WIN 35,428 binding to DAT, in a rapid and reversible fashion, similar to findings after acute administration of methamphetamine to rats and measurement of ex vivo [³H]DA uptake into striatal synaptosomes [87].

Electrochemical Recording of DA Clearance and Efflux   The two catechol ring hydroxyls of DA are easily oxidized at potentials ranging from +0.3 to 0.8 V (vs. a Ag/AgCl reference electrode), resulting in the production of two electrons. DA electrochemistry uses carbon-based microelectrodes as detectors for the transfer of this charge to current [88]. The electrodes are calibrated in solution so that changes in extracellular DA concentrations can be determined, in real time, directly from the currents measured. In terms of DAT-mediated DA clearance/release and its regulation, this approach has been widely used in both in vitro and in vivo studies [88]; we will confine our discussion here to in vitro approaches. Benefits of this approach are miniature-sized electrodes and speed of the measurements. However, the low-nanomolar baseline extracellular concentrations of DA are commonly below the limits of detection. Thus, stimulated DA release and clearance of endogenously released DA or exogenously added DA are measured.

Rotating disk electrode voltammetry (RDEV) is useful for solution measurements of DAT function in dispersed tissue such as brain homogenates and heterologous DAT-expressing cells [89]. RDEV measurement of the clearance of microinjected DA, which is carried out in small volumes (<0.5 mL) with a glassy carbon RDE, has a response time of ~20 ms, making this approach ideal for measuring kinetics and rapid regulation of the DAT. For example, RDEV has shown that in both the dorsal striatum and NAc, DA and two Na^+ ions bind to the DAT randomly before the Cl^- binds and uptake occurs [89]. Simultaneous measurement with RDEV of tyramine uptake and resulting DA efflux in hDAT-HEK 293 cells revealed differential effects of DAT amino acid mutations on inward and outward transport of DA [90]. This electrochemical approach has also been used to evaluate the kinetic mechanisms by which both substrates and competitive inhibitors affect DAT activity. RDEV provided early kinetic evidence for DA D2 receptor-stimulated rapid up-regulation of DAT in striatal synaptosomes [91]. Similarly, the DA D3-preferring agonist (+)-PD 128907 induced a 33% increase in DA uptake V_{max}, whereas the D3 antagonist

GR 103691 reduced uptake, as measured with RDEV in suspensions of rat NAc [92]. It is interesting to note that this type of D2/D3 receptor-induced DAT regulation has not been observed with measurement of [^{3}H]DA uptake into synaptosomes, which may indicate the rapid and transient nature of this effect.

Most often for in vitro electrochemical experiments, acutely prepared brain slices are used. In this case, DA clearance is measured with a carbon fiber microelectrode lowered into the living portion of the slice. Carbon fibers have small diameters (5 to 30 μm) so that there is minimal tissue damage. Compared to in vivo recording, this approach offers the advantages of anatomical precision for electrode placement, precise control of bath concentrations of drugs, and potential use of multiple slices and brain areas from a single animal. A drawback compared to in vivo recording is the disruption of many of the neuronal connections. Often, depending on the orientation of the slice cut, the axons of the relatively long DA neurons are severed. Of course, the health of the tissue slice is also not a trivial concern. For slice experiments, both high-speed chronoamperometry (HSC), a square-wave potential pulse technique, and fast-scan cyclic voltammetry (FSCV), a potential sweep technique, have been used to measure DAT-mediated DA clearance. Both techniques provide relatively high sensitivity with excellent temporal and spatial resolution.

To assess DAT activity, we have used HSC to measure clearance of locally applied exogenous DA in coronal slices of rat midbrain, striatum, and NAc [93,94]. Using a similar in vivo approach, we have demonstrated that clearance of pressure-ejected DA is due predominately to uptake by the DAT in striatum and NAc of urethane-anesthetized rats [95]. For in vitro experiments, after allowing the brain slices (400 μm) to recover their energy stores (see Section 13.2.1), they are placed in a slice chamber and superfused (2 mL/min) with aCSF at ~32°C. For HSC recording and local DA/drug application, we use a Nafion-coated carbon-fiber electrode–micropipette assembly, with the micropipette glued parallel to and at a fixed distance from (150 to 300 μm) the working electrode. Nafion, a cationic polymer, enhances the selectivity of the electrode for DA over ascorbic acid and the major DA metabolite dihydroxyphenylacetic acid (DOPAC). The micropipette contains the DA solution (200 μM DA in 100 mM phosphate buffered saline, pH 7.4, and 100 μM ascorbic acid to prevent oxidation of DA in the barrel). Multibarrel micropipettes are used when drugs are applied locally. After lowering the assembly into the slice, HSC measurements are made continuously at 5 to 10 Hz, with the +0.55-V oxidation potential pulse applied for 100 ms followed by a 100-ms 0.00-V resting potential pulse. During the latter pulse, DA is reduced so that it is not depleted by the sampling. The resulting oxidation and reduction currents are digitally integrated during the last 70 to 90 ms (steady state) of each 100-ms pulse. Control DA clearance signals are elicited, typically at 5-minute intervals, by first zeroing the baseline and then ejecting a finite volume (10 to 100 nL) of DA. Once reproducible oxidation signals are obtained (maximal signal amplitudes equivalent to 0.5 to 2 μM DA), the volume of DA ejected is kept constant for the remainder of the experiment. Various signal parameters (amplitude, decay times, and rate constant) are used to quantitate DA clearance. Unlike for FSCV, HSC oxidation currents do not provide direct identification that the signal is due to DA. However, examining both currents does provide some verification because the ratio of the reduction : oxidation currents for a DA signal should be 0.6 to 0.8. Following a minimum of two reproducible control DA clearance signals ($\leq 15\%$ variation), we introduce drugs into the bath by infusing 100 times the desired final concentration into the aCSF flow. When local drug application is used, drugs are pressure-ejected from a second barrel of the micropipette. We typically eject them 30 to 60 seconds before the DA ejection and at four times the volume and twice the concentration of the DA solution.

HSC coupled with local application of DA in slices of rat SN and striatum has been used to explore potential differences in DA clearance in the somatodendritic vs. the terminal regions of nigrostriatal DA neurons [96]. As expected, based on the greater number of DATs in striatum, clearance of DA occurred more rapidly in striatum than in SN. Nonetheless, DA clearance was inhibited in both regions to the same extent by cocaine and nomifensine and was not affected by either the NET inhibitor desipramine or the SERT inhibitor citalopram. These results suggest that DAT mediates DA clearance, and in a similar fashion, in both somatodendritic and terminal regions of nigrostriatal DA neurons. Using a similar electrochemical approach coupled with electrophysiological recording in SN slices, we also found evidence supporting the idea that DAT is voltage dependent and regulated by DA D2 receptors [93]. Specifically, 15 mM K^+-induced depolarization of the DA neurons slowed the decay time of the DA clearance signal, consistent with inhibition of DAT activity. Superfusion with the D2 receptor antagonist sulpiride (50 μM) produced similar effects. This interesting observation suggested that presynaptic D2 autoreceptors are being activated by DA (endogenous and/or exogenously applied DA), which in turn enhances DAT activity, consistent with the RDEV results mentioned above. However, it also suggested that under our assay conditions DAT activity is tonically up-regulated by D2 autoreceptors. This finding contrasted with our desire for a "pure" assay of DAT function, which we had originally hoped for when developing this HSC approach to measure clearance of locally applied DA.

FSCV measures oxidation and reduction currents in response to a range of applied potentials; typically, potentials are applied at a rate of 300 to 800 V/s to the carbon-fiber electrode and are scanned linearly over a triangular waveform (e.g., -0.4 to $+1.2$ V and back again to -0.4 V) [97]. FSCV is the fastest voltammetric recording technique available. Another advantage of this voltage-sweep method is that the current changes can be resolved over the time and voltage domains, thereby providing an electrochemical "fingerprint" (voltammogram) that can be used to identify and confirm that DA is the primary electroactive species being measured. This is an advantage when measuring clearance of endogenous DA, in particular in brain regions with relatively fewer DA neurons and more NE and/or 5-HT neurons. With this approach endogenous DA release is often evoked with an adjacent stimulating electrode using either single- or multiple-pulse stimulations. Nonlinear curve fitting based on Michaelis–Menton kinetics is used to determine the uptake parameters, K_m and V_{max} [98,99].

In guinea pig midbrain slices, FSCV was used to quantitatively compare DA uptake and diffusion in SN pars compacta (SNc), SN pars reticulata (SNr), and ventral tegmental area (VTA) [100]. As would be expected based on the greater number of DA neuronal cell bodies/dendrites in SNc and VTA than in SNr, DA uptake in SNc and VTA is 10-fold higher than in SNr. Furthermore, DA uptake in SNc and VTA is inhibited to a great extent by the selective DAT inhibitor GBR 12909. Significant desipramine-mediated inhibition of DA uptake was observed in VTA, suggesting that NET also contributes to clearance of DA in VTA. The most striking finding was, however, that diffusion, rather than uptake, is the most important determinant of extracellular DA lifetime in midbrain, which contrasts with striatum, where DAT-mediated uptake predominates.

The question arises as to whether DA release via DAT reversal is important physiologically as well as pharmacologically. This has been controversial. Initial studies using FSCV and HSC in rat SN slices did not find evidence for this type of DA release [93,101]. In contrast, subsequent experiments using HSC and patch-clamp recordings in slices of rat SN showed that nonsynaptic DAT-mediated DA release activates DA D2 receptors, thereby mediating DA neuronal dendrodendritic inhibition [102]. More recent results in mouse midbrain slices, however, suggest that dendritic DA is released from synaptic vesicles in a normal Ca^{2+}-dependent exocytotic manner [103]. Furthermore, DAT played a traditional role by mediating reuptake, which tightly regulated the duration of the resulting D2 receptor-mediated inhibitory postsynaptic current. Thus, to date, the majority of the experimental evidence supports the idea that DAT-mediated DA release, in both somatodendritic and terminal regions of DA neurons, appears to be primarily of pharmacological importance (i.e., when amphetaminelike drugs are present).

Brain slices from a number of different species, including nonhuman primates, have been used for FSCV recording of DA clearance. Cragg and colleagues [104] have studied DAT-mediated DA uptake (GBR 12909–sensitive and desipramine-insensitive) in slices of striatum (caudate–putamen) from the common marmoset, a useful animal model for Parkinson's disease research. They found that both single-pulse DA release and DAT V_{max} were highest in the region of the putamen that is most vulnerable in Parkinson's disease. Also, the regional variation in maximal DA uptake was two- to three-fold greater in marmoset caudate–putamen than in rodent striatum, suggesting greater uptake dynamics in primate brain.

One of the interesting questions that Sarah Jones and her lab have used in vitro FSCV to address is identifying what compensatory changes occur in DA neurons and in the other monoamine transporters in DAT knockout mice. FSCV recordings of the clearance of stimulation-evoked DA in mouse striatal slices show pseudo-first-order rate constants for DA uptake of 16 per second in DAT wild-type mice, 9 per second in heterozygotes, and 0.05 per second in knockout mice, consistent with the expected gene-dosage effect [105]. In contrast to striatal slices from DAT wild-type and heterozygote mice, superfusion of slices from DAT knockout mice with 10 μM cocaine does not alter DA clearance. The extremely slow clearance rate and lack of effect of DAT inhibitors suggest that in striatum of mice lacking DATs, DA is cleared by diffusion. The importance of DAT is underscored because these mice exhibit extensive compensatory changes, including a 95% reduction in DA content and a 75% reduction in DA release, that attempt to normalize DA neurotransmission. More recently, FSCV recording of endogenous DA clearance in slices of NAc from DAT knockout mice has shown that neither NET nor SERT provide compensatory uptake of DA in either the core or shell subregions [106]. This finding is intriguing because cocaine is still reinforcing in these mice, and in intact animals increased DA in the NAc would be expected to be critical for this effect. Since cocaine is no longer reinforcing in DAT/SERT double knockouts [107], it is particularly surprising that SERT does not appear to play a role in clearing DA in NAc of mice lacking the DAT.

In vitro FSCV has been used to evaluate DAT sensitivity to cocaine in NAc slices from rats following 10 days of cocaine binge (high intake, 24-hour access self-administration) and a 1- or 7-day drug deprivation [97]. Cumulative in vitro concentration–response curves (20-minute intervals) were generated for the ability of cocaine (0.03 to 30 μM) to inhibit the apparent K_m for uptake of stimulation-evoked endogenous DA. At both deprivation times, cocaine's maximal effect was markedly reduced to 40% of control. These findings call into question the commonly accepted idea that increased extracellular DA in NAc leads to increased reinforcement. They also highlight the use of a single brain slice to generate a full concentration–response curve.

For recording DAT-mediated DA efflux from neurons or cells, amperometry, where the oxidation potential is held constant throughout the experiment, has been used. Since no information is obtained about the voltammetric properties of the oxidized species detected, this method is used appropriately when the identity of the species being measured is known [88]. This approach was used with the giant DA cell of the snail *Planorbis corneus* to show that amphetamine acts primarily at the synaptic vesicle to redistribute DA into the cytoplasm of the neuron, thereby promoting DAT reverse transport and releasing DA extracellularly [108]. More recently, single-carbon fiber amperometric electrodes have been used in conjunction with patch-clamp recording in the whole-cell configuration (see the next section) to measure amphetamine-induced, cocaine-blocked DAT-mediated DA efflux [109,110]. For these measurements, outside-out membrane-excised patches from HEK 293 cells stably expressing hDAT were used, and the patch was held in close proximity to (~1 μm) the amperometric electrode. A potential of +0.7 V was used to oxidize the DA. This approach allows simultaneous recording of patch and oxidative currents during voltage steps so that current–voltage relationships for the effects of amphetamine on DAT-mediated ionic conductances and DA efflux can be compared. This type of measurement has revealed that amphetamine-induced DA release, via reversal of the DAT, is dependent on both voltage and intracellular Na^+ concentration, with both depolarized potentials and higher Na^+ concentrations interacting to stimulate DA efflux [109]; see the next section for a discussion of other results with this approach.

Electrophysiological Recording of DAT-Associated Currents As already mentioned, several lines of evidence suggest that two Na^+ ions and one Cl^- ion are co-transported, along with one DA^+ molecule, which is positively charged at physiological pH [111–114]. Thus, at a minimum, there should be a net inward flux of two positive charges with each transport cycle. This should result in inward transport-associated currents. Importantly, when we began to explore this possibility, there was already precedence for this type of current being associated with the SLC6 family member SERT [115]. Our calculations, based on overexpression of the cloned hDAT in frog oocytes, suggested that we could use the two-electrode voltage-clamp recording technique to detect these currents [116].

Indeed, a few days after hDAT cRNA injection, we measured small inward currents when hDAT-expressing oocytes, but not water-injected control oocytes, were exposed to DA [116]. Because the currents were relatively small (1 to 100 nA), these were always measured as subtractive currents; that is, the current measured in the presence of drug was subtracted from the immediately preceding current measured in the absence of drug (or vice versa, depending on the convention) [116]. As would be expected for inward currents, these substrate transport-associated currents were voltage-dependent, increasing at more hyperpolarized potentials. These currents also exhibited the concentration–response relationship expected for DA uptake in heterologous cells ($K_m = \sim 2$ μM) and were antagonized by DAT blockers such as cocaine.

However, concurrent measurement of DA-induced currents and [³H]DA uptake in individual hDAT-expressing oocytes revealed that charge movement during substrate transport was greater than would be expected for inward substrate transport with a fixed stoichiometry of two positive charges [116]. Ion channel–like properties, uncoupled from transport, have been described for a number of plasma membrane neurotransmitter transporters [117]. More recently, recordings of authentic single-channel events arising from the *Caenorhabditis elegans* DAT have provided direct evidence that DATs exhibit a true channel mode of conduction [118]. These results help to explain the "excess" depolarizing currents that we and others have observed and suggest that DATs may locally depolarize neurons by this channel mechanism. Indeed, in cultured embryonic midbrain DA neurons, whole-cell patch clamp and perforated-patch recordings have revealed that DAT-mediated conductances increase excitability and firing rates of DA neurons [119]. Interestingly, these DA-stimulated currents have been shown to be carried primarily by anions.

Several other DAT-associated currents have also been identified in hDAT-expressing oocytes. Most surprising to us at the time was our observation that when cocaine was applied to hDAT-expressing oocytes, but not to control oocytes, neither of which had ever been exposed to DA, small outward currents were observed [116]. Subsequent current–voltage (I–V) plots showed that all DAT ligands (DAT blockers at all potentials and DAT substrates at more depolarized potentials) induce outward currents. Results of ion substitution experiments were consistent with the idea that these outward currents were due to blockade of a tonic nonselective cation leak conductance associated with the DAT. In addition to the leak current, we measured transient currents associated with Na^+ binding to DAT and used them in capacitance calculations for PKC trafficking studies [74]. Arachidonic acid stimulates yet another novel DAT-associated cation conductance that is not dependent on Na^+, making it distinct from substrate transport and the other currents described previously [120]. Furthermore, two-electrode voltage clamp recordings in hDAT-expressing oocytes, coupled with flux experiments, have helped to explain the mechanism by which zinc ion (Zn^{2+}) binding to the DAT inhibits DA uptake while simultaneously promoting DAT reversal and substrate efflux [121]. Zn^{2+} does this by enhancing a Cl^- conductance (uncoupled from transport) and thereby causing membrane depolarization. It is certainly intriguing that this conductance is reminiscent of the DA-induced currents in midbrain DA neurons described by Ingram and colleagues [119] that were mentioned above.

Substrate-induced inward currents and substrate blockade of constitutive leak currents, similar to those in hDAT-expressing oocytes, have also been observed with patch-clamp recording in hDAT-expressing mammalian HEK 293 cells [77]. When the magnitudes of DAT-mediated uptake, release, and currents were compared for DA, tyramine and D-amphetamine, release and currents were better correlated, with amphetamine being the most efficacious of the three substrates. DA, on the other hand, was the most efficacious substrate in uptake experiments. The stimulatory effects of Zn^{2+} on amphetamine-induced hDAT release and inward currents support the idea that DAT-mediated release involves ion fluxes and does not occur via the facilitated diffusion mechanism [122].

Amphetamine-induced currents also provided some of the first evidence that in addition to being taken up, causing DAT-mediated efflux, and inducing currents, substrates rapidly reduce DAT activity. Thus, in the 15 minutes following exposure to $2 \mu M$ amphetamine, reduction of the transport-associated currents in FLAG-hDAT expressing HEK cells, along with redistribution of DAT immunofluorescence away from the cell surface, suggested that DAT substrates cause a rapid internalization of plasma membrane DATs [123]. This down-regulation might have evolved as a protective mechanism that would limit the amount of DA taken back up into DA neurons, since DA can contribute to neurotoxicity [124].

Recording of DAT currents provides a real-time measure of DAT activity that can be quantitated for comparison among experimental conditions. Our results showed that $I–V$ plots of DAT ligand-induced currents can be used to unequivocally distinguish between DAT substrates and blockers, which was not always the case with DA uptake assays [116]. Furthermore, our electrophysiological results, as well as results of our studies of DA uptake into voltage-clamped oocytes, confirmed that DAT is an electrogenic transporter (i.e., that its activity is voltage dependent). The finding that DAT activity is lower at more depolarized potentials suggests that DA uptake may be transiently reduced when DA neurons are stimulated, allowing DA to diffuse away from its release site and interact with its receptors. We have further used this approach in hDAT-expressing oocytes to study the rapid down-regulation of DATs by PKC and substrates [74,125,126] and rapid up-regulation by protein tyrosine kinases, DA D2 receptors, and ethanol [24,73,127]. Our electrophysiological recordings of currents and capacitance calculations, coupled with in vitro binding and uptake measurements, have supported the idea that DAT activity can be rapidly (minutes) changed by its trafficking and the resulting changes in its expression at the cell surface.

A recent series of elegant studies using patch-clamp recording coupled with local carbon-fiber amperometric recording at the cell surface to monitor simultaneously DAT-mediated currents and flux further illustrate the power of including the electrophysiological approach in studies of DAT function [109,110,128]. These studies using patch-clamp recording in the outside-out configuration (intracellular side of the membrane patch facing the inside of the patch pipette) have provided novel, mechanistic insights about how amphetamine stimulates DAT-mediated efflux in both DAT-expressing HEK 293 cells and acutely dissociated mouse midbrain neurons. They have shown that amphetamine stimulation of serine phosphorylation in the amino terminus of DAT and intracellular Ca^{2+} are required for both amphetamine-induced DAT currents and DA efflux [128,129]. Furthermore, the results demonstrate that two mechanisms are involved in amphetamine-induced DAT-mediated DA efflux: Besides the slower exchange mechanism that has long been recognized, rapid (ms) bursts of DA efflux occur through a DAT channel-like mode [110]. In addition, these studies also revealed some interesting differences between DAT-mediated uptake and efflux, as well as between the actions of amphetamine and DA.

13.3.5 Cell Surface Expression

Since DATs are functional only at the plasma membrane, the number of DATs on the cell surface of neurons is a primary determinant of DA neurotransmission intensity and duration. DAT cell surface number is regulated by various mechanisms, including pretreatment with DAT substrates, DAT inhibitors, and anesthetics; a variety of receptor signaling cascades, including those initiated by G protein–coupled receptors (GPCRs) and protein kinases; DAT-associated proteins; and post-endoplasmic reticulum (ER) modifications, such as ubiquitination and glycosylation. Additionally, environmental enrichment and aging alter DAT cell surface expression. Two primary methods are used to study DAT cell surface expression: subcellular fractionation and cell surface biotinylation, although each has limitations.

Subcellular Fractionation   Only a few studies have used subcellular fractionation to examine changes in DAT localization on the plasma membrane and/or intracellular compartments. Melikian and Buckley [130] examined the intracellular distribution of the DAT in hDAT-expressing pheochromocytoma (PC12) cells. At steady-state conditions, 63% of the hDAT was intracellular. Furthermore, subcellular fractionation revealed that under both steady-state and PKC-activated conditions, intracellular hDAT was localized to recycling endosomes, compartments distinct from sorting endosomes that express early endosome-associated protein (EEA.1) and the GTPase Rab5. hDAT was presumed to have moved through the sorting endosomes to the recycling endosomes at a rate that did not allow isolation of the transporter to the former [130]. When Wiesinger et al. [131] performed fractionation experiments with hDAT-expressing N2A neuroblastoma (N2A) cells, they found that under steady-state conditions, approximately two-thirds of the intracellular hDAT was co-localized with cellular markers (Rab5A and transferrin receptor) of early/recycling endosomes, with the remainder co-localized with a marker (Rab7) of late endosomes/lysosomes. Iron chelation strongly increased hDAT accumulation in lysosomes of N2A cells after 6 hours of treatment, and by 24 hours, no hDAT was detected in fractions expressing markers of early and recycling endosomes. These results support the idea that iron chelators induce DAT down-regulation by affecting

DAT trafficking rather than DAT synthesis since hDAT mRNA levels were not altered by the chelators [131].

Subcellular fractionation has also been used to examine DAT plasma membrane expression in striatal synaptosomes prepared from rats that had been given increasing doses of nicotine [132]. Fractionation allowed isolation of the DAT in synaptic plasma membrane and vesicular fractions, which did not differ between control and drug-treated groups, despite the fact that nicotine treatment increased the V_{max} of [^{3}H]DA uptake. These results were confirmed with surface biotinylation experiments. Thus, nicotine-induced increases in striatal DAT function occur through a trafficking-independent mechanism.

The primary limitations of subcellular fractionation are that this technique does not allow quantitative separation of plasma membrane and early endosomal components. In addition, fractionation is not sensitive enough to rapidly analyze the internalization and recycling kinetics of DAT [133].

Surface Biotinylation   The predominant method used to measure DAT cell surface expression is surface biotinylation, which uses a membrane-impermeable derivative of biotin (e.g., sulfo-NHS-biotin) that cross-links primary amino residues in the extracellular loops of membrane-localized proteins, including DAT. Following separation with avidin beads, Western blot analysis is used to visualize the total, nonbiotinylated (intracellular) and biotinylated (surface) DATs. DAT surface biotinylation studies have been conducted with nonneuronal cells and synaptosomes prepared from rat striatum. For synaptosomal studies, we have used a modification of the method published by Zhu and colleagues [60].

The DAT undergoes constitutive endocytosis and recycling to the plasma membrane. In hDAT-expressing PC12 cells, reversible biotinylation assays revealed that the hDAT is constitutively internalized at a rate of 3 to 5% per minute [134,135]. Additionally, using biotinylation of rat striatal synaptosomes, we found that 28% of the DAT was localized to the cell surface under basal conditions, suggesting that a large pool of intracellular transporters exists [61]. Our data are in good agreement with the findings in PC12 cells that 37% of the hDAT are localized to the plasma membrane under basal conditions [130]. However, these values are probably an underrepresentation because biotinylation of cell surface proteins does not occur at 100% efficiency. Biotinylation of hDATs in mammalian cells subjected to RNA interference (RNAi)-mediated knockdown of clathrin heavy chain (CHC) and dynamin II demonstrated that constitutive internalization of hDAT occurs mainly through clathrin-coated pits [71].

DAT substrates and inhibitors regulate DAT cell surface expression in an opposing manner. In hDAT-expressing mammalian cells and oocytes, exposure to relatively high concentrations of the DAT substrates DA and amphetamine for minutes reduces DAT cell surface expression, as determined by biotinylation or cell surface–binding experiments [53,59,123,126,136,137], but see [138]. However, a recent study by Johnson et al. [139] suggested that amphetamine's regulation of DAT is biphasic and time-dependent. Thus, in rat striatal synaptosomes, amphetamine exposure very rapidly (within 30 seconds) increased DAT surface expression before causing a slower down-regulation. The initial effect was brief, lasting less than 2.5 minutes, and prevented by cocaine pretreatment. Interestingly, however, this initial effect was not observed with DA pretreatment.

Amphetamine-mediated decreases in DAT cell surface expression involve multiple intracellular signaling pathways. From functional studies, methamphetamine/amphetamine-induced down-regulation of DAT is known to be dependent on PKC [87,126]. Amphetamine's effects on hDAT endocytosis also depend on calcium-calmodulin kinase II (CaMKII) activity [140], whereas activation of phosphatidyl-inositol 3-kinase (PI3K) and Akt signaling by the growth factor insulin block amphetamine's effect [141,142]. Inhibition of Akt activity replicates the effect of amphetamine on hDAT endocytosis [142]. Amphetamine-induced DAT endocytosis is also blocked by pretreating with cocaine [123], Zn^{2+}, or by mutating a tyrosine residue in the third intracellular loop (Y335A) of hDAT [137]. Interestingly, addition of Zn^{2+} to cells expressing this mutated hDAT restores DAT regulation by amphetamine [137]. Besides blocking amphetamine's regulation of hDAT cell surface expression, cocaine by itself also recruits hDAT to the cell surface of heterologous cells [58,143].

The cell surface expression of DAT is also regulated directly by several protein kinases. Among the many participants, PKC has received the most attention. Activation of PKC by the phorbol ester PMA decreases hDAT cell surface expression [71,130,134,138,144]. Additional studies with dominant negative dynamin mutants or RNAi-mediated knockdown of CHC and dynamin II, respectively, demonstrated that PKC-mediated hDAT internalization occurs mainly through clathrin-coated pits [71,138]. Although the identity of the proteins involved in recruiting hDAT into the clathrin-coated pits are unknown, overlapping, nonclassical motifs in the hDAT carboxyl terminus have been implicated in constitutive and PKC-mediated DAT endocytosis [135]. Following PKC-induced internalization and depending on the mammalian cell lines used, hDATs are either recycled back to the membrane or degraded [130,138]. In PC12 cells, the PKC-mediated decrease in hDAT cell surface expression results from both an increase in hDAT endocytosis and a reduction in recycling back to the plasma membrane [134].

Ubiquitination is a recently identified, novel mechanism by which DAT cell surface expression is regulated. Sorkin and colleagues found that activation of PKC triggered ubiquitination of the hDAT, which led to transporter internalization through an endocytotic mechanism mediated by the E3 ubiquitin ligase neural precursor cell expressed, developmentally down-regulated 4-2 (NEDD4-2) [54,145]. Mass spectrometric analysis of purified DAT from hDAT-expressing HeLa cells revealed that the protein ubiquitin was especially abundant following treatment of cells with the PKC activator PMA. Western blot analyses confirmed that PKC activation increased constitutive ubiquitination of hDAT in these cells. One possible role of ubiquitination could be to facilitate DAT internalization and degradation. Indeed, biotinylation assays revealed that PMA-induced ubiquitinated hDAT existed on the cell surface. However, a greater amount of ubiquitinated hDAT was present in the nonbiotinylated samples of hDAT. Thus, hDAT ubiquitination parallels PKC-dependent DAT degradation and may underlie PKC-mediated DAT down-regulation [54].

However, E3 ubiquitin ligase-mediated ubiquitination does not always result in fewer DATs at the cell surface. Co-expression of hDAT and parkin, another E3 ubiquitin ligase, increases DAT cell surface expression in the human dopaminergic neuroblastoma cell line SH-SY5Y [146]. The greater surface expression is thought to be due to increased parkin-mediated ubiquitination and degradation of hDATs that are ER-export deficient due to improper protein folding into their normal, native conformation. Without parkin-mediated ubiquitination, the misfolded DATs still oligomerize with properly folded DATs and interfere with normal transporter trafficking to the plasma membrane [146]. These findings and interpretation agree with previous reports that DAT oligomerization in the ER is necessary for proper post-translational processing and targeting of DAT to the plasma membrane [53,147].

In contrast to PKC, other protein kinases positively regulate DAT cell surface expression and include tyrosine kinases (TKs) and the serine/threonine kinases mitogen-activated protein kinase (MAPK) and PI3K/Akt. TKs are well situated to regulate the DAT since DA neurons express multiple-receptor TKs, including tropomyosin-related kinase B (TrkB) and insulin receptors [148,149]. TrkB receptors are activated through binding of brain-derived neurotrophic factor (BDNF) [150]. TK inhibitors decrease DAT cell surface expression in hDAT-expressing oocytes and rat striatal synaptosomes [61,73]. Constitutive activity of p44/p42, but not p38, MAPKs, and PI3K/Akt, increase DAT cell surface expression [141,142,151,152]. Furthermore, activation of PI3K and Akt by the growth factor insulin up-regulates DAT cell surface expression [141,142].

Efficient export of the DAT from the ER and selective targeting to the plasma membrane, as well as endocytosis of the DAT and processing through endosomes and lysosomes, requires the participation of many proteins as well as post-translational events such as glycosylation. Recent studies have focused on the carboxyl-terminus domain of DAT, not only on the necessity of the carboxyl terminus for efficient and proper targeting of DAT, but also on its role in binding accessory proteins that affect DAT cell surface expression, and thereby function. Cell surface biotinylation assays have been important in these studies, and we summarize some of the important findings from these studies next.

Biotinylation studies demonstrated that hDAT-expressing cells that also overexpress the PDZ domain–containing protein interacting with C kinase-1 (PICK1) contain a greater density of hDATs on the cell surface than do cells expressing hDAT alone [153]. Truncation of the hDAT carboxyl terminus, especially the last three residues, impairs trafficking of hDAT from the ER to the cell surface [147,153]. This finding led the authors to suggest that this was due to a lack of interactions between PDZ-binding sequences in the hDAT carboxyl terminus and PICK1 [147,153]. Indeed, others have found that discrete epitopes in the hDAT carboxyl terminus are essential for the maturation and targeting of hDAT from the ER to the cell surface in HEK 293 and N2A cells [154]. However, surface biotinylation experiments showed that interactions between hDAT carboxyl terminus PDZ-binding sequences and PDZ domain proteins are neither necessary nor sufficient for efficient surface targeting of hDAT. These findings were confirmed with immunofluorescence microscopy [154]. Miranda et al. [155] reported similar findings using cell surface biotinylation assays, having observed that the PDZ-binding motif in the carboxyl terminus of hDAT is not sufficient for ER export. Thus, the hDAT carboxyl terminus has an essential role in ER export, maturation, and surface targeting of hDAT that is independent of PDZ domain interactions [154,155].

Another protein implicated in modulating hDAT cell surface expression via an interaction with the carboxyl terminus of hDAT is the focal adhesion adaptor protein Hic-5 [156]. In heterologous cells, hDAT cell surface expression measured with biotinylation assays was reduced by Hic-5 overexpression. Again, these results were confirmed with immunofluorescence microscopy.

DAT cell surface expression, determined with biotinylation assays, has also been reported to be regulated rapidly by glycosylation, GPCRs, and anesthetics. Single mutation of individual glycosylation sites in hDAT-expressing HEK 293 cells did not affect hDAT trafficking, whereas mutations of all three sites did partially impair hDAT cell surface expression [147]. DA neurons, in which DAT is exclusively localized, express multiple GPCRs, including substance P and DA D2 receptors. Activation of PKC by the phorbol ester PMA triggers DAT internalization (see above). To investigate whether PKC activation under conditions similar to native systems might also regulate hDAT cell surface expression, HEK 293 and N2A cells coexpressing a $G\alpha_q$-coupled human substance P receptor and hDAT were treated with substance P to activate PKC [144]. This work showed that hDAT internalization induced by substance P was mediated predominantly, but not exclusively, by PKC. On the other hand, DA D2 receptors have been reported to up-regulate DAT function. In oocytes coexpressing hDAT and the long form of the D2 receptor, D2 receptor activation increases hDAT function and cell surface binding in a $G_{i/o}$-dependent manner [24]. Similarly, as determined by biotinylation, D2/D3 receptor agonists increase hDAT cell surface expression in EM4 cells coexpressing hDAT and the short form of the D2 receptor [157]. This D2 receptor–mediated regulation is dependent on MEK, but not PI3K, activity. Interestingly, an enzyme-linked immunosorbent assay revealed that hDAT cell surface expression is increased by coexpressing the short form of the D2 receptor in HEK 293 cells [158]. This effect is dependent on direct protein–protein interactions between the DAT and D2 receptor, yet independent of D2 receptor activation. Moreover, biotinylation assays show that this increase in DAT surface expression is due to increased recruitment of the transporter to the membrane, not decreased endocytosis. In contrast, the anesthetic isoflurane induces rapid hDAT internalization in HEK 293 cells [159]. However, only 14% of hDAT was on the cell surface under basal conditions, so that a high degree of biotinylation efficiency was required in order to detect the isoflurane-induced reduction in hDAT cell surface levels.

Cell surface DAT biotinylation studies in synaptosomes have also investigated effects of environmental enrichment and aging on DAT plasma membrane expression in rat brain. An enriched environment was shown to reduce DAT plasma membrane expression selectively in medial prefrontal cortex, with no changes in dorsal striatum or NAc [60]. Striatal synaptosomes prepared from 24-month-old rats ($\sim$72 years human age) possess 30% fewer DATs on their cell surface than do 6-month-old rats ($\sim$18 years human age), suggesting that the number of functional DATs in striatum decreases with aging [160].

Fluorescent Antibodies   So far, to our knowledge, no DAT antibody is available that labels an extracellular domain of the DAT in living cells. There is one antibody that labels the second extracellular loop of DAT, but for this antibody to work, the tissue must be fixed [161–163]. This antibody has been used in conjunction with a fluorescent secondary antibody to examine DAT subcellular localization [164]. As an alternative, DAT constructs have been created with an epitope [e.g., a fluorescent protein, FLAG, hemagglutinin (HA), or histidine (His)] attached to one terminus or the second intracellular loop. In this way, DAT localization and trafficking can be monitored by measuring fluorescence either directly or via fluorescently tagged secondary antibodies to FLAG, HA, or His (see Section 13.3.6).

13.3.6   Trafficking

For the DAT to terminate DA neurotransmission, the transporter must be targeted to the plasma membrane in an efficient and specific manner. Newly synthesized DAT is transported from the ER through the Golgi complex, where the DAT undergoes glycosylation at the extracellular loops, and is then trafficked onto the plasma membrane. DAT export from the ER is the first sorting step of transporter targeting in neurons, and efficient export and trafficking to the membrane requires oligomerization and specific carboxyl terminus sequences [53,147,154,155].

Fluorescence Microscopy   Fluorescence microscopy is used to study DAT trafficking and localization on both the cell surface and within intracellular compartments. Epitope-tagged DATs can be visualized with fluorescently tagged antibodies to the epitope or fluorescent proteins can be attached directly to the DAT. The greatest advantage of using fluorescence microscopy to localize DAT is its applicability for monitoring DAT trafficking in living cells. In addition, fluorescence is a valuable way to complement biochemical analyses (i.e., biotinylation) of DAT localization. A primary limitation of fluorescence microscopy is the inability to quantitate fluorescence signals as a measure of absolute protein levels.

Immunofluorescence visualized with confocal microscopy has been used to investigate the role of glycosylation on hDAT cell surface expression [147,165]. Mutations of single glycosylation sites did not impair hDAT trafficking to the membrane [147,165]. Removal of all three glycosylation sites impaired hDAT trafficking only partially, resulting in a greater retention of intracellular hDATs. Additionally, Li et al. [165] found that non- or partially glycosylated hDATs underwent enhanced endocytosis compared to wild-type hDATs, probably due to the mutants' reduced stability on the cell surface, suggested by the decreased hDAT immunofluorescence on the plasma membrane.

Immunofluorescence has also been utilized to examine the unique contributions that the carboxyl terminus makes to hDAT trafficking. For instance, confocal immunofluorescence microscopy has been used to complement biotinylation experiments, demonstrating that binding of PICK1 to residues in the carboxyl terminus of hDAT increases the level of transporter on the cell surface [153]. Fluorescence microscopy, combined with mutational analyses of the carboxyl and/or amino termini of hDAT, has also clearly demonstrated the importance of the carboxyl terminus in DAT trafficking. Sorkina et al. [71] used this approach to examine the contributions of the DAT termini to constitutive and PKC-mediated hDAT endocytosis. This work showed that whereas the amino terminus was not essential for constitutive or PKC-mediated hDAT internalization, the carboxyl terminus was. Others have also used immunofluorescence and mutational analyses to confirm that residues in the hDAT carboxyl terminus are required for proper targeting to the plasma membrane [154,155].

Studies of the regulation of DAT surface expression under basal conditions, as well as following pretreatments with DAT substrates, DAT inhibitors, PMA, or activation of GPCRs, have utilized fluorescence microscopy. In good agreement with earlier reversible biotinylation studies in hDAT-PC12 cells [134], using immunofluorescence we found that hDAT is constitutively internalized at a rate of $\sim 1.4\%$ per minute in hDAT-expressing PAE cells [71]. Furthermore, localization of hDATs expressed in heterologous cells with either epitope tags or a fluorescent secondary antibody demonstrated that DAT substrates and PKC activation (induced by either PMA or a GPCR) cause an accumulation of intracellular hDAT as it trafficked away from the cell surface [53,71,123,136–138,144,145,164]. Not surprisingly, PKC antagonists increase hDAT recruitment to the plasma membrane [164]. Complementing their biotinylation results, the DAT inhibitor cocaine not only blocks amphetamine's down-regulation of hDAT cell surface expression [123], but also increases hDAT membrane expression in heterologous cells [143]. Furthermore, Zn^{2+}'s abolishment of amphetamine-induced hDAT internalization and its restoration of amphetamine's regulatory effects on mutant hDATs were studied with fluorescence microscopy (see above) [137]. Additionally, fluorescence microscopic analysis of dominant-negative dynamin mutants or examination of cells after RNAi knockdown of CHC and dynamin II established that constitutive hDAT endocytosis, as well as that induced by DAT substrates and PMA occurs through a dynamin- and clathrin-dependent mechanism [71,123,138,145]. Similar analyses showed that neither constitutive nor PMA-induced hDAT endocytosis utilizes lipid rafts [71]. As already mentioned, fluorescence microscopy is a useful way to confirm the results of cell surface biotinylation studies in living cells, as was done using flow cytometry in studies of D2 receptor–mediated increases in DAT surface expression in hDAT-EM4 cells [157]. Similarly, confocal immunofluorescence microscopy was used to show the increased DAT cell surface expression when the D2 receptor and transporter were coexpressed in HEK 293 cells [158].

Immunofluorescence microscopy has been utilized extensively to examine the intracellular pathways mediating constitutive and regulated hDAT endocytosis. For instance, this approach demonstrated that constitutive and PKC-mediated hDAT internalization caused accumulation of hDAT in multiple vesicular compartments that co-localized with markers of early (EEA.1, Rab5, transferrin receptor), sorting (Hrs), and late recycling endosomes (Rab11, Rab7) [53,71,166]. It has also revealed insights into the effects that PKC-induced ubiquitination of DAT has on DAT endocytosis and subcellular localization [54,166]. Based on mass spectrometric analyses, ubiquitin can conjugate to lysine residues in both terminal tails of hDAT. Under basal conditions in HeLa cells coexpressing fluorescently tagged hDAT and ubiquitin, ubiquitin was distributed in the cytosol, endosomes, and nuclei [54]. PKC activation by PMA resulted in ubiquitin accumulation in the endosomes, where it co-localized with hDAT, indicating that endosomes are a primary site of ubiquitinated DAT [54]. Further studies of hDAT ubiquitination, using immunofluorescence and biotinylation in conjunction with site-directed mutagenesis, showed that single or double mutations of the lysine residues identified by mass spectrometry to be ubiquitin conjugation sites did not alter constitutive or PKC-dependent hDAT ubiquitination or hDAT cell surface expression. These results suggest that significant redundancy exists over control of hDAT ubiquitination. Indeed, only simultaneous mutations of three lysines in the hDAT amino terminus produced a significant inhibition of PKC-mediated hDAT ubiquitination and endocytosis. Similar results were observed with the immortalized neuronal cell line $1RB_3AN_{27}$ expressing hDAT. Besides being targeted to early and recycling endosomes following PMA treatment for 5 to 30 minutes, co-localization experiments showed that longer treatment (1 hour) enhanced degradation of hDAT via targeting of internalized hDAT to late endosomes/lysosomes. Taken together, the immunofluorescence microscopy results indicate that hDAT ubiquitination is necessary for PKC-dependent endocytosis of hDAT from the plasma membrane to the early (recycling) and late (degradation) endosomes [166].

FRET To demonstrate and/or confirm direct interactions between DAT and another protein, fluorescence resonance energy transfer (FRET) microscopy has been used. FRET uses the generation of fluorescence signals to detect protein–protein interactions in the range 1 to 10 nm, which typically requires direct interactions between the two proteins. With FRET, energy is transferred in a nonradioactive fashion from a fluorophore in an electrically excited state serving as a donor to an acceptor molecule that is also a fluorophore. This energy transfer depends on the close proximity of the donor and acceptor molecules and occurs only when the acceptor fluorophore's absorption spectrum overlays the emission spectrum of the excited (donor) fluorophore. In this way, FRET is capable of resolving molecular interactions, conformations, associations, and separations between proteins with a spatial resolution beyond that of conventional optical microscopy [167,168]. The primary caveats with FRET are that the absence of FRET does not mean that the proteins of interest do not interact. Also, the ability to observe FRET is dependent on adequate protein expression levels.

FRET microscopy has helped to elucidate mechanisms involved in DAT localization and trafficking, including identifying proteins and regulatory pathways involved. Studies using radiation inactivation [169,170] and cross-linking [55,56] have provided evidence that the hDAT exists as an oligomeric complex in cells. Also, recent co-immunoprecipitation experiments with differently epitope-tagged hDATs and nonfunctional but plasma membrane–localized hDAT mutants strongly suggest that hDATs are olgomerized on the plasma membrane [147]. Sorkina and colleagues [53] used FRET microscopy to examine hDAT oligomerization directly by measuring the energy transfer between hDATs labeled with either cyan or yellow fluorescent proteins (CYP and YFP, respectively) expressed either transiently or stably in PAE cells. These FRET studies showed that hDAT oligomers existed both in intracellular compartments and on the plasma membrane. Furthermore, hDAT oligomerization appeared to occur initially in the ER, and the oligomers were maintained throughout trafficking to the membrane and during sequestration and recycling through endosomes after PMA and amphetamine treatments. Similar findings were observed in the immortalized dopaminergic cell line 1RB$_3$AN$_{27}$. Findings with immunoprecipitation experiments complemented those with FRET. Further FRET analysis of the effects of oligomerization on hDAT trafficking demonstrated that formation of hetero-oligomers between wild-type hDAT and ER export-defective hDAT mutants interfered with the normal delivery of wild-type hDAT to the cell surface [53]. Torres et al. [147] reported similar findings using biotinylation and immunofluorescence assays. More detailed studies using FRET analysis, biotinylation, fluorescent microscopy, and mutagenesis with hDAT-expressing PAE cells, the immortalized neuronal DA cell line 1RB$_3$AN$_{27}$, and primary cultures of rat embryonic midbrain neurons revealed that there are multiple residues in the carboxyl terminus of hDAT that are responsible for export of DAT from the ER [155].

FRET analysis has also been used to demonstrate the direct interaction of ubiquitin and hDAT in endosomes following PKC activation in both hDAT-expressing HeLa and PAE cells [54]. Thus, after PKC activation, endosomes appear to be the major localization site of ubiquitinated hDAT. Although biotinylation studies demonstrated that PMA-induced hDAT ubiquitination was initiated at the cell surface, it was difficult to detect FRET signals on the cell surface given the dispersed localization of hDAT. This was in contrast to endosomes, where accumulated ubiquitinated hDAT was clearly detected with FRET under both basal and PKC-stimulated conditions. hDAT appeared to be degraded rapidly following PKC activation, given the association of hDAT with ubiquitin recognition sorting machinery, including proteins associated with the lysosomal degradation pathway [54].

RNAi RNAi is a naturally occurring posttranscriptional gene-silencing mechanism whereby endogenous mRNAs containing sequences complementary to a double-stranded RNA (dsRNA) trigger are destroyed. Use of short interfering RNAs (siRNAs) mimics the cleavage products of dsRNA and produces sequence-specific gene knockdown in mammalian cells [171,172]. For siRNA studies, proper control experiments include using multiple siRNA sequences that silence only the gene of interest. When possible, rescue experiments should also be used to validate the specificity of the siRNA. In these experiments, mRNA sequences that are not complementary to the siRNA but that still encode the protein of interest are transfected into cells that then undergo siRNA treatment. The siRNA specificity is confirmed if no changes occur in the expression levels of the protein or its mRNA. Additional issues with siRNA include the efficiency of gene knockdown and the half-life of the protein that the target gene encodes.

To date, only a few studies have used siRNA to study DATs, but this is likely to increase in the future [173]. RNAi knockdown of DAT itself has been studied only after in vivo siRNA administration so that subsequent behavioral alterations can be assessed [174,175]. On the other hand, siRNA has been used in vitro to study DAT endocytosis by knocking down proteins involved in endocytosis and recycling and/or degradation of membrane proteins [71]. The effects of siRNA targeting clathrin-coated pit proteins (e.g., CHC or dynamin II) were examined on the constitutive and PKC-mediated internalization of hDAT stably expressed in PAE cells. Although constitutive DAT internalization is PKC-independent, both constitutive and PKC-induced hDAT internalization causes accumulation of transporter in early and recycling endosomes. siRNA knockdown of either CHC or dynamin II abolished hDAT accumulation in these endosomes under basal and PKC-activated conditions, as determined with fluorescence microscopy and surface biotinylation experiments. Thus, siRNA provided additional, strong evidence that clathrin-coated pits serve an essential role in constitutive and PKC-induced hDAT endocytosis [71].

In vitro siRNA has also been used to study the role of ubiquitination in hDAT endocytosis [145]. To monitor DAT trafficking in live cells, a fully functional hDAT mutant that contained an HA epitope tag in the second extracellular loop was expressed in PAE and HeLa cells. In this way, an HA antibody (HA11) was used to monitor hDAT endocytosis and recycling following manipulation of endocytic machinery with siRNA technology. Screening a library of siRNAs targeting endocytic proteins identified several candidates that are necessary for PKC-induced hDAT internalization. Among them was the E3 ubiquitin ligase NEDD4-2. Besides NEDD4-2, siRNA screening also implicated the clathrin coat accessory proteins epsin, Eps15, and Eps15R. The three latter proteins contain ubiquitin-interacting motifs that have been implicated in the recognition of ubiquitinated cargo [176]. siRNA knockdown of NEDD4-2 and the clathrin coat accessory proteins substantially reduced the amount of internalized HA11 signal in endosomes following PKC activation, implicating ubiquitination and the ubiquitin recognition system in PKC-dependent hDAT endocytosis. These siRNA results were confirmed with biotinylation assays. In addition, NEDD4-2 siRNA knockdown also dramatically reduced the amount of ubiquitinated hDAT following PKC activation. Altogether, PKC-induced hDAT ubiquitination and endocytosis was mediated by NEDD4-2 and occurred through the transporter's interactions with adaptor proteins in clathrin-coated pits [145]. These findings complemented and extended other findings from the Sorkin laboratory [54,166].

In vitro RNAi has also been utilized to probe the relationship between hDAT and α-synuclein [177]. RNAi induced a knockdown of α-synuclein by 80% and decreased hDAT cell surface expression in the human neuroblastoma cell line SH-SY5Y. This finding is in agreement with a confocal immunofluorescence study demonstrating that coexpression of α-synuclein and hDAT in mammalian cells increases hDAT cell surface expression via direct interactions between the two proteins [178]. Studies using biotinylation assays, however, reported the opposite relationship and attributed the difference to distinct transfection methods [179]. Importantly, disruption of cell adhesion or the microtubule network, or induction of oxidative stress, reversed α-synuclein's inhibitory effect on DAT function and/or cell surface expression [179,180]. Moreover, formation of a protein–protein complex between α-synuclein and hDAT does not in itself convey inhibition of DAT cell surface expression; and disruption of the microtubule network increases α-synuclein: hDAT complexes while increasing surface biotinylated DAT levels [180].

13.4 CONCLUSIONS

In vitro studies have been critical to our understanding of the mechanisms by which the DAT transports DA into neurons/cells as well as releases DA by reverse transport (Fig. 13-1). These studies have also helped to reinforce the importance of the DAT in limiting DA neurotransmission and mediating effects on DA systems of such psychomotor stimulants as cocaine and amphetamines. [^{3}H]DA/DA/[^{3}H]MPP$^+$ accumulation and drug-induced [^{3}H]DA/DA/[^{3}H]MPP$^+$ efflux remain important functional assays for assessing DAT function and kinetics. However, faster real-time measures of DA clearance using electrochemical recording and DAT-associated currents using electrophysiological recording have provided additional novel insights

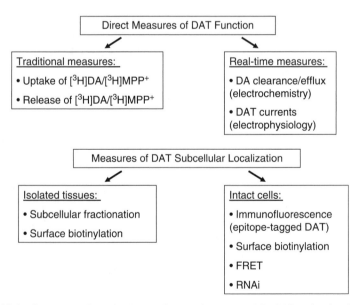

Figure 13-1 Summary of current approaches used to measure DAT function in vitro. Direct measures quantitate uptake and release mediated by the DAT, as well as kinetics of inward transport. Newer, real-time functional measures use electrochemistry and electrophysiology to determine DA clearance/efflux and DAT-mediated currents, respectively. Since the DAT must be localized on the cell surface to be functional and the number of DATs on the cell surface is rapidly regulated by endocytic trafficking, measures of DAT subcellular localization in both isolated tissues and intact cells provide valuable indirect measures of DAT function.

about how the DAT functions, its ion and voltage dependency, its channel-like mode of action, and its rapid regulation. Indeed, recognition that DAT activity is rapidly regulated, often by altering its constitutive trafficking and its presence on the plasma membrane, has lead to more widespread use in in vitro studies of both biochemical approaches and fluorescence microscopy to measure the number of cell surface DATs and to assess changes in its subcellular localization. A recent study from our lab emphasizes the importance of using multiple in vitro approaches to assess potential differences in DAT activity because results from these different assays do not always agree [181]. Although it has already been established that the DAT traffics as a homo-oligomer and that its endocytosis occurs via clathrin-coated pits, most of the specific pathways and players remain to be identified. Also, since many of the mechanisms for the rapid regulation of DAT have been elucidated in model systems, one of the future challenges will be to determine if these same mechanisms are important in DA neurons. Together, the results of both in vitro and in vivo studies of the DAT will lead to a clearer understanding of the roles played by this transporter in both normal and disease states.

ACKNOWLEDGMENTS

We thank members of the N. R. Zahniser and A. Sorkin labs for many helpful discussions, Ms. Melissa Adams for her help preparing this chapter, and NIDA for generous support (DA R37 DA004216, R01 DA014204, and K05 DA015050).

REFERENCES

1. Nirenberg, M. J., Vaughan, R. A., Uhl, G. R., Kuhar, M. J., and Pickel, V. M. (1996). The dopamine transporter is localized to dendritic and axonal membranes of nigrostriatal dopaminergic neurons. *J Neurosci*, *16*, 436–447.

2. Amara, S. G., and Kuhar, M. J. (1993). Neurotransmitter transporters: recent progress. *Annu Rev Neurosci*, *16*, 73–93.

3. Zahniser, N. R., and Doolen, S. (2001). Chronic and acute regulation of Na^+/Cl^--dependent neurotransmitter transporters: drugs, substrates, presynaptic receptors and signaling systems. *Pharmacol Ther*, *92*, 21–55.

4. Snyder, S. H., and Coyle, J. T. (1969). Regional differences in H^3-norepinephrine and H^3-dopamine uptake into rat brain homogenates. *J Pharmacol Exp Ther*, *165*, 78–86.

5. Horn, A. S. (1990). Dopamine uptake: a review of progress in the last decade. *Prog Neurobiol*, *34*, 387–400.

6. Giros, B., El Mestikawy, S., Bertrand, L, and Caron, M. G. (1991). Cloning and functional characterization of a cocaine-sensitive dopamine transporter. *FEBS Lett*, *295*, 149–154.

7. Kilty, J. E., Lorang, D., and Amara, S. G. (1991). Cloning and expression of a cocaine-sensitive rat dopamine transporter. *Science*, *254*, 578–579.

8. Shimada, S., Kitayama, S., Lin, C.-L., Patel, A., Nanthakumar, E., Gregor, P., Kuhar, M., and Uhl, G. (1991). Cloning and expression of a cocaine-sensitive dopamine transporter complementary DNA. *Science*, *254*, 576–578.

9. Usdin, T. B., Mezey, E., Chen, C., Brownstein, M. J., and Hoffman, B. J. (1991). Cloning of the cocaine-sensitive bovine dopamine transporter. *Proc Nat Acade Sci U S A*, *88*, 11168–11171.

10. Salvaterra, P. M., and Matthews, D. A. (1980). Isolation of rat brain subcellular fraction of enriched in putative neurotransmitter receptors and synaptic junctions. *Neurochemi Res*, *5*, 181–195.

11. Johnson, M. W., Chotiner, J. K., and Watson, J. B. (1997). Isolation and characterization of synaptoneurosomes from single rat hippocampal slices. *J Neurosci Methods*, *77*, 151–156.

12. Villasana, L. E., Klann, E., and Tejada-Simon, M. V. (2006). Rapid isolation of synaptoneurosomes and postsynaptic densities from adult mouse hippocampus. *J Neurosci Methods*, *158*, 30–36.

13. Eshleman, A. J., Wolfrum, K., Mash, D. C., Christensen, K., and Janowsky, A. (2001). Drug interactions with the dopamine transporter in cryopreserved human caudate. *J Pharmacol Exp Ther*, *296*, 442–449.

14. Fredholm, B. B., Dunwiddie, T. V., Bergman, B., and Lindstrom, K. (1984). Levels of adenosine and adenine nucleotides in slices of rat hippocampus. *Brain Res*, *295*, 127–136.

15. Le, W. D., Bostwick, J. R., and Appel, S. H. (1992). Use of [^{3}H]-GBR 12935 to measure dopaminergic nerve terminal growth into the developing rat striatum. *Dev Brain Res*, *67*, 375–377.

16. Burke, R. E., Antonelli, M., and Sulzer, D. (1998). Glial cell line–derived neurotrophic growth factor inhibits apoptotic death of postnatal substantia nigra dopamine neurons in primary culture. *J Neurochem*, *71*, 517–525.

17. Pothos, E., Larsen, K. E., Krantz, D. E., Liu, Y.-J., Haycock, J. W., Setlik, W., Gershon, M. D., Edwards, R. H., and Sulzer, D. (2000). Synaptic vesicle transporter expression regulates vesicle phenotype and quantal size. *J Neurosci*, *20*, 7297–7306.

18. Kerr, J. N., and Plenz, D. (2002). Dendritic calcium encodes striatal neuron output during up-states. *J Neurosci*, *22*, 1499–1512.

19. Shimizu, K., Matsubara, K., Ohtaki, K., and Shiono, H. (2003). Paraquat leads to dopaminergic neural vulnerability in organotypic midbrain culture. *Neurosci Res*, *46*, 523–532.

20. Javitch, J. A., Strittmatter, S. M., and Snyder, S. H. (1985). Different visualization of dopamine and norepinephrine uptake sites in rat brain using [^{3}H]mazindol autoradiography. *J Neurosci*, *5*, 1513–1521.

21. Madras, B. K., Meltzer, P. C., Liang, A. Y., Elmaleh, D. R., Babich, J., and Fischman, A. J. (1998). Altropane, a SPECT or PET imaging probe for dopamine neurons. I: dopamine transporter binding in primate brain. *Synapse*, *29*, 93–104.

22. Chen, N.-H., Wang, Y.-L., and Reith, M. E. A. (1997). Binding of the [^{125}I]3β-(iodophenyl)tropan-2β-carboxylic acid isopropyl ester to the dopamine transporter at a physiologically relevant temperature: mutually exclusive binding and different ionic requirements for various uptake blockers and substrates. *Synapse*, *25*, 155–162.

23. Dubocovich, M. L., and Zahniser, N. R. (1985). Binding of the dopamine uptake inhibitor [³H]nomifensine to striatal membranes. *Biochem Pharmacol*, *34*, 1137–1144.

24. Mayfield, R. D., and Zahniser, N. R. (2001). Dopamine D_2 receptor regulation of the dopamine transporter expressed in *Xenopus laevis* oocytes is voltage-independent. *Mol Pharmacol*, *59*, 113–121.

25. Barish, M. E. (1983). A transient calcium-dependent chloride current in the immature *Xenopus* oocyte. *J Physiol (Lond)*, *342*, 309–325.

26. Gulley, J. M., and Zahniser, N. R. (2003). Rapid regulation of dopamine transporter function by blockers, substrates and presynaptic receptor ligands. *Eur J Pharmacol*, *479*, 139–152.

27. Augood, S. J., Westmore, K., McKenna, P. J., and Emson, P. C. (1993). Co-expression of dopamine transporter mRNA and tyrosine hydroxylase mRNA in ventral mesencephalic neurones. *Mol Brain Res*, *20*, 328–334.

28. Lorang, D., Amara, S. G., and Simerly, R. B. (1994). Cell-type-specific expression of catecholamine transporters in the rat brain. *J Neurosci*, *14*, 4903–4914.

29. Blanchard, V., Raisman-Vozari, R., Vyas, S., Michel, P. P., Javoy-Agid, F., Uhl, G., and Agid, Y. (1994). Differential expression of tyrosine hydroxylase and membrane dopamine transporter genes in subpopulations of dopaminergic neurons of the rat mesencephalon. *Mol Brain Res*, *22*, 29–38.

30. Drandarevski, N., Marburger, A., Walther, D., Reum, T., Uhl, G., and Morgenstern, R. (2001). Dopaminergic mRNA expression in the intact substantia nigra of unilaterally 6-OHDA-lesioned and grafted rats: an in situ hybridization study. *J Neural Trans*, *108*, 141–151.

31. Xu, Z., Cawthon, D., McCastlain, K. A., Slikker, Jr., W., and Ali, S. F. (2005). Selective alterations of gene expression in mice induced by MPTP. *Synapse*, *55*, 45–51.

32. Copeland, B. J., Neff, N. H., and Hadjiconstantinou, M. (2005). Enhanced dopamine uptake in the striatum following repeated restraint stress. *Synapse*, *57*, 167–174.

33. Persico, A. M., Schindler, C. W., Brannock, M. T., Gonzalez, A. M., Surratt, C. K., and Uhl, G. R. (1993). Dopaminergic gene expression during amphetamine withdrawal. *Neuroreport*, *4*, 41–44.

34. Bezard, E., Dovero, S., Belin, D., Duconger, S., Jackson-Lewis, V., Przedborski, S., Piazza, P. V., Gross, C. E., and Jaber, M. (2003). Enriched environment confers resistance to 1-methyl-4-phenyl-1,2,3,6-tetrahydropyridine and cocaine: involvement of dopamine transporter and trophic factors. *J Neurosci*, *23*, 10999–11007.

35. Figlewicz, D. P., Szot, P., Chavez, M., Woods, S. C., and Veith, R. C. (1994). Intraventricular insulin increases dopamine transporter mRNA in rat VTA/substantia nigra. *Brain Res*, *644*, 331–334.

36. Patterson, T. A., Brot, M. D., Zavosh, A., Schenk, J. O., Szot, P., and Figlewicz, D. P. (1998). Food deprivation decreases mRNA and activity of the rat dopamine transporter. *Neuroendocrinology*, *68*, 11–20.

37. Bello, N. T., Sweigart, K. L., Lakoski, J. M., Norgren, R., and Hajnal, A. (2003). Restricted feeding with scheduled sucrose access results in an upregulation of the rat dopamine transporter. *Am J Physiol Regul Integrat Comp Physiol*, *284*, R1260–R1268.

38. Li, S., Kim, K. Y., Kim, J. H., Kim, J. H., Park, M. S., Bahk, J. Y., and Kim, M. O. (2004). Chronic nicotine and smoking treatment increases dopamine transporter mRNA expression in the rat midbrain. *Neurosci Lett*, *363*, 29–32.

39. Ferrari, R., Novere, N. L., Picciotto, M. R., Changeux, J. P., and Zoli, M. (2001). Acute and long-term changes in the mesolimbic dopamine pathway after systemic or local single nicotine injections. *Eur J Neurosci*, *15*, 1810–1818.

40. Shepard, J. D., Chuang, D. T., Shaham, Y., and Morales, M. (2006). Effect of methamphetamine self-administration on tyrosine hydroxylase and dopamine transporter levels in mesolimbic and nigrostriatal dopamine pathways of the rat. *Psychopharmacology*, *185*, 505–513.

41. Arroyo, M., Baker, W. A., and Everitt, B. J. (2000). Cocaine self-administration in rats differentially alters mRNA levels of the monoamine transporters and striatal neuropeptides. *Mol Brain Res*, *83*, 107–120.

42. Spangler, R., Zhou, Y., Maggos, C. E., Zlobin, A., Ho, A., and Kreek, M. J. (1996). Dopamine antagonist and "binge" cocaine effects on rat opioid and dopamine transporter mRNAs. *Neuroreport*, *7*, 2196–2200.

43. Xia, Y., Goebel, D. J., Kapatos, G., and Bannon, M. J. (1992). Quantitation of rat dopamine transporter mRNA: effects of cocaine treatment and withdrawal. *J Neurochem*, *59*, 1179–1182.

44. Burchett, S. A., and Bannon, M. J. (1997). Serotonin, dopamine, and norepinephrine transporter mRNAs: heterogeneity of distribution and response to "binge" cocaine administration. *Mol Brain Res*, *49*, 95–102.

45. Cerruti, C., Pilotte, N. S., Uhl, G., and Kuhar, M. J. (1994). Reduction in dopamine transporter mRNA after cessation of repeated cocaine administration. *Mol Brain Res*, *22*, 132–138.

46. Letchworth, S. R., Daunais, J. B., Hedgecock, A. A., and Porrino, L. J. (1997). Effects of chronic cocaine administration on dopamine transporter mRNA and protein in the rat. *Brain Res*, *750*, 214–222.

47. Letchworth, S. R., Sexton, T., Childers, S. R., Vrana, K. E., Vaughan, R. A., Davies, H. M. L., and Porrino, L. J. (1999). Regulation of rat dopamine transporter mRNA and protein by chronic cocaine administration. *J Neurochem*, *73*, 1982–1989.

48. Shilling, P. D., Kelsoe, J. R., and Segal, D. S. (1997). Dopamine transporter mRNA is up-regulated in the substantia nigra and the ventral tegmental area of amphetamine-sensitized rats. *Neurosci Lett*, *236*, 131–134.

49. Lu, W., and Wolf, M. E. (1997). Expression of dopamine transporter and vesicular monoamine transporter 2 mRNAs in rat midbrain after repeated amphetamine administration. *Mol Brain Res*, *49*, 137–148.

50. Maggos, C. E., Spangler, R., Zhou, Y., Schlussman, S. D., Ho, A., and Kreek, M. J. (1997). Quantitation of dopamine transporter mRNA in the rat brain: mapping, effects of "binge" cocaine administration and withdrawal. *Synapse*, *26*, 55–61.

51. Perrone-Capano, C., Tino, A., Amadoro, G., Pernas-Alonso, R., and di Porzio, U. (1996). Dopamine transporter gene expression in rat mesencephalic dopaminergic neurons is increased by direct interaction with target striatal cells in vitro. *Mol Brain Res*, *39*, 160–166.

52. Lindblom, J., Johansson, A., Holmgren, A., Grandin, E., Nedergard, C., Fredriksson, R., and Schioth, H. B. (2006). Increased mRNA levels of tyrosine hydroxylase and dopamine transporter in the VTA of male rats after chronic food restriction. *Euro J Neurosci*, *23*, 180–186.

53. Sorkina, T., Doolen, S., Galperin, E., Zahniser, N. R., and Sorkin, A. (2003). Oligomerization of dopamine transporter visualized in living cells by fluorescence resonance energy transfer microscopy. *J Biol Chem*, *278*, 28274–28283.

54. Miranda, M., Wu, C. C., Sorkina, T., Korstjens, D. R., and Sorkin, A. (2005). Enhanced ubiquitylation and accelerated degradation of the dopamine transporter mediated by protein kinase C. *J Biol Chem, 280*, 35617–35624.

55. Hastrup, H., Karlin, A., and Javitch, J. A. (2001). Symmetrical dimer of the human dopamine transporter revealed by cross-linking Cys-306 at the extracellular end of the sixth transmembrane segment. *Proc Nat Acad Sci U S A, 98*, 10055–10060.

56. Hastrup, H., Sen, N., and Javitch, J. A. (2003). The human dopamine transporter forms a tetramer in the plasma membrane: cross-linking of a cysteine in the fourth transmembrane segment is sensitive to cocaine analogs. *J Biol Chem, 278*, 45045–45048.

57. Baucum, A. J., Rau, K. S., Riddle, E. L., Hanson, G. R., and Fleckenstein, A. E. (2004). Methamphetamine increases dopamine transporter higher molecular weight complex formation via a dopamine- and hyperthermia-associated mechanism. *J Neurosci, 24*, 3436–3443.

58. Daws, L. C., Callaghan, P. D., Moron, J. A., Kahlig, K. M., Shippenberg, T. S., Javitch, J. A., and Galli, A. (2002). Cocaine increases dopamine uptake and cell surface expression of dopamine transporters. *Biochem Biophys Res Commun, 290*, 1545–1550.

59. Chi, L., and Reith, M. E. A. (2003). Substrate-induced trafficking of the dopamine transporter in heterologously expressing cells and in rat striatal synaptosomal preparations. *J Pharmacol Exp Ther, 307*, 729–736.

60. Zhu, J., Apparsundaram, S., Bardo, M. T., and Dwoskin, L. P. (2005). Environmental enrichment decreases cell surface expression of the dopamine transporter in rat medial prefrontal cortex. *J Neurochem, 93*, 1434–1443.

61. Hoover, B. R., Everett, C. V., Sorkin, A., and Zahniser, N. R. (2007). Rapid regulation of dopamine transporters by tyrosine kinases in rat neuronal preparations. *J Neurochem, 101*, 1258–1271.

62. Dauer, W., and Przedborski, S. (2003). Parkinson's disease: mechanisms and models. *Neuron, 39*, 889–909.

63. Smeyne, R. J., and Jackson-Lewis, V. (2005). The MPTP model of Parkinson's disease. *Mol Brain Res, 134*, 57–66.

64. Kurosaki, R., Muramatsu, Y., Watanabe, H., Michimata, M., Matsubara, M., Imai, Y., and Araki, T. (2003). Role of dopamine transporter against MPTP (1-methyl-4-phenyl-1,2,3,6-tetrahydropyridine) neurotoxicity in mice. *Metab Brain Dis, 18*, 139–146.

65. Kurosaki, R., Muramatsu, Y., Kato, H., and Araki, T. (2004). Biochemical, behavioral and immunohistochemical alterations in MPTP-treated mouse model of Parkinson's disease. *Pharmacol Biochem Behav, 78*, 143–153.

66. Jakowec, M. W., Nixon, K., Hogg, E., McNeill, T., and Petzinger, G. M. (2004). Tyrosine hydroxylase and dopamine transporter expression following 1-methyl-4-phenyl-1,2,3,6-tetrahydropyridine-induced neurodegeneration of the mouse nigrostriatal pathway. *J Neurosci Res, 76*, 539–550.

67. Petzinger, G. M., Fisher, B., Hogg, E., Abernathy, A., Arevalo, P., Nixon, K., and Jakowec, M. W. (2006). Behavioral motor recovery in the 1-methyl-4-phenyl-1,2,3,6-tetrahydropyridine-lesioned squirrel monkey (*Saimiri sciureus*): changes in striatal dopamine and expression of tyrosine hydroxylase and dopamine transporter proteins. *J Neurosci Res, 83*, 332–347.

68. Armstrong, V., Reichel, C. M., Doti, J. F., Crawford, C. A., and McDougall, S. A. (2004). Repeated amphetamine treatment causes a persistent elevation of glial fibrillary acidic protein in the caudate–putamen. *Eur J Pharmacol*, *488*, 111–115.

69. Near, J. A., Bigelow, J. C., and Wightman, R. M. (1988). Comparison of uptake of dopamine in rat chopped tissue and synaptosomes. *J Pharmacol Exp Ther*, *245*, 921–927.

70. Izenwasser, S., Rosenberger, J. G., and Cox, B. M. (1992). Inhibition of [^{3}H]dopamine and [^{3}H]serotonin uptake by cocaine: comparison between chopped tissue slices and synaptosomes. *Life Sci*, *50*, 541–547.

71. Sorkina, T., Hoover, B. R., Zahniser, N. R., and Sorkin, A. (2005). Constitutive and protein kinase C–induced internalization of the dopamine transporter is mediated by a clathrin-dependent mechanism. *Traffic*, *6*, 157–170.

72. Briegleb, S. K., Gulley, J. M., Hoover, B. R., and Zahniser, N. R. (2004). Individual differences in cocaine- and amphetamine-induced activation of male Sprague–Dawley rats: contribution of the dopamine transporter. *Neuropsychopharmacology*, *29*, 2168–2179.

73. Doolen, S., and Zahniser, N. R. (2001). Protein tyrosine kinase inhibitors alter human dopamine transporter activity in *Xenopus* oocytes. *J Pharmacol Exp Ther*, *296*, 931–938.

74. Zhu, S.-J., Kavanaugh, M. P., Sonders, M. S., Amara, S. G., and Zahniser, N. R. (1997). Activation of protein kinase C inhibits uptake, currents and binding associated with the human dopamine transporter expressed in *Xenopus* oocytes. *J Pharmacol Exp Ther*, *282*, 1358–1365.

75. Buck, K. J., and Amara, S. G. (1994). Chimeric dopamine-norepinephrine transporters delineate structural domains influencing selectivity for catecholamines and 1-methyl-4-phenylpyridinium. *Proc Nat Acad Sci U S A*, *91*, 12584–12588.

76. Uhl, G. R., and Lin, Z. (2003). The top 20 dopamine transporter mutants: structure-function relationships and cocaine actions. *Eur J Pharmacol*, *479*, 71–82.

77. Sitte, H. H., Huck, S., Reither, H., Boehm, S., Singer, E. A., and Pifl, C. (1998). Carrier-mediated release, transport rates, and charge transfer induced by amphetamine, tyramine, and dopamine in mammalian cells transfected with the human dopamine transporter. *J Neurochem*, *71*, 1289–1297.

78. Giros, B., Jaber, M., Jones, S. R., Wightman, R. M., and Caron, M. G. (1996). Hyperlocomotion and indifference to amphetamine in mice lacking the dopamine transporter. *Nature*, *379*, 606–612.

79. Sulzer, D., Sonders, M. S., Poulsen, N. W., and Galli, A. (2005). Mechanisms of neurotransmitter release by amphetamines: a review. *Prog Neurobiol*, *75*, 406–433.

80. Arnold, E. B., Molinoff, P. B., and Rutledge, C. O. (1977). The release of endogenous norepinephrine and dopamine from cerebral cortex by amphetamine. *J Pharmacol Exp Ther*, *202*, 544–557.

81. Fischer, J. B., and Cho, A. K. (1979). Chemical release of dopamine from striatal homogenates: evidence for an exchange diffusion model. *J Pharmacol Exp Ther*, *208*, 203–209.

82. Pifl, C., and Singer, E. A. (1999). Ion dependence of carrier-mediated release in dopamine or norepinephrine transporter-transfected cells questions the hypothesis of facilitated exchange diffusion. *Mol Pharmacol*, *56*, 1047–1054.

83. Gnegy, M. E. (2003). The effect of phosphorylation on amphetamine-mediated outward transport. *Eur J Pharmacol*, *479*, 83–91.

84. Kantor, L., and Gnegy, M. E. (1998). Protein kinase C inhibitors block amphetamine-mediated dopamine release in rat striatal slices. *J Pharmacol Exp Ther*, *284*, 594–598.

85. Peris, J., and Zahniser, N. R. (1987). One injection of cocaine produces a long-lasting increase in [^{3}H]-dopamine release. *Pharmacol Biochem Behav*, *27*, 533–535.

86. Peris, J., and Zahniser, N. R. (1989). Persistent augmented dopamine release after acute cocaine requires dopamine receptor activation. *Pharmacol Biochem Behav*, *32*, 71–76.

87. Sandoval, V., Riddle, E. L., Ugarte, Y. V., Hanson, G. R., and Fleckenstein, A. E. (2001). Methamphetamine-induced rapid and reversible changes in dopamine transporter function: an in vitro model. *J Neurosci*, *21*, 1413–1419.

88. Borland, L. M., and Michael, A. C. (2006). An introduction to electrochemical methods in neuron-science. In *Electrochemical Methods for Neuroscience*, Borland, L. M., and Michael, A. C., Eds. CRC Press, Boca Raton, FL, pp. 1–15.

89. Schenk, J. O., Wright, C., and Bjoklund, N. (2005). Unraveling neuronal dopamine transporter mechanisms using rotating disk electrode voltammetry. *J Neurosci Methods*, *143*, 41–47.

90. Chen, N., and Justice, J. B., Jr. (2000). Differential effect of structural modification of human dopamine transporter on the inward and outward transport of dopamine. *Mol Brain Res*, *75*, 208–215.

91. Meiergerd, S. M., Patterson, T. A., and Schenk, J. O. (1993). D_2 receptors may modulate the function of the striatal transporter for dopamine: kinetic evidence from studies in vitro and in vivo. *J Neurochem*, *61*, 764–767.

92. Zapata, A., and Shippenberg, T. S. (2002). D_3 receptor ligands modulate extracellular dopamine clearance in the nucleus accumbens. *J Neurochem*, *81*, 1035–1042.

93. Hoffman, A. F., Zahniser, N. R., Lupica, C. R., and Gerhardt, G. A. (1999). Voltage-dependency of the dopamine transporter in the rat substantia nigra. *Neurosci Lett*, *260*, 105–108.

94. Gulley, J. M., Larson, G. A., and Zahniser, N. R. (2006). Using high-speed chronoamperometry with local dopamine application to assess dopamine transporter function. In *Electrochemical Methods for Neuroscience*, Borland, L. M., and Michael, A. C., Eds. CRC Press, Boca Raton, FL, pp. 83–102.

95. Cass, W. A., Zahniser, N. R., Flach, K. A., and Gerhardt, G. A. (1993). Clearance of exogenous dopamine in rat dorsal striatum and nucleus accumbens: role of metabolism and effects of locally applied uptake inhibitors. *J Neurochem*, *61*, 2269–2278.

96. Hoffman, A. F., Lupica, C. R., and Gerhardt, G. A. (1998). Dopamine transporter activity in the substantia nigra and striatum assessed by high-speed chronoamperometric recordings in brain slices. *J Pharmacol Exp Ther*, *287*, 487–496.

97. Mateo, Y., Lack, C. M., Morgan, D., Roberts, D. C. S., and Jones, S. R. (2005). Reduced dopamine terminal function and insensitivity to cocaine following cocaine binge self-administration and deprivation. *Neuropsychopharmacology*, *30*, 1455–1463.

98. Wightman, R. M., Amatore, C., Engstrom, R. C., Hale, P. D., Kristensen, E. W., Kuhr, W. G., and May, L. J. (1988). Real-time characterization of dopamine overflow and uptake in the rat striatum. *Neuroscience*, *25*, 513–523.

99. Jones, S. R., Garris, P. A., Kilts, C. D., and Wightman, R. M. (1995). Comparison of dopamine uptake in the basolateral amygdaloid nucleus, caudate–putamen, and nucleus accumbens of the rat. *J Neurochem, 64*, 2581–2589.

100. Cragg, S. J., Nicholson, C., Kume-Kick, J., Tao, L., and Rice, M. E. (2001). Dopamine-mediated volume transmission in midbrain is regulated by distinct extracellular geometry and uptake. *J Neurophysiol, 85*, 1761–1771.

101. Cragg, S. J., Rice, M. E., and Greenfield, S. A. (1997). Heterogeneity of electrically-evoked dopamine release and uptake between substantia nigra, ventral tegmental area, and striatum. *J Neurophysiol, 77*, 863–873.

102. Falkenburger, B. H., Barstow, K. L., and Mintz, I. M. (2001). Dendrodendritic inhibition through reversal of dopamine transport. *Science, 293*, 2465–2470.

103. Beckstead, M. J., Grandy, D. K., Wickman, K., and Williams, J. T. (2004). Vesicular dopamine release elicits an inhibitory postsynaptic current in midbrain dopamine neurons. *Neuron, 42*, 939–946.

104. Cragg, S. J., Hille, C. J., and Greenfield, S. A. (2000). Dopamine release and uptake dynamics within nonhuman primate striatum in vitro. *J Neurosci, 20*, 8209–8217.

105. Jones, S. R., Gainetdinov, R. R., Jaber, M., Giros, B., Wightman, R. M., and Caron, M. G. (1998). Profound neuronal plasticity in response to inactivation of the dopamine transporter. *Proc Nat Acad Sci U S A, 95*, 4029–4034.

106. Mateo, Y., Budygin, E. A., John, C. E., Banks, M. L., and Jones, S. R. (2004). Voltammetric assessment of dopamine clearance in the absence of the dopamine transporter: no contribution of other transporters in core or shell of nucleus accumbens. *J Neurosci Methods, 140*, 183–187.

107. Sora, I., Hall, F. S., Andrews, A. M., Itokawa, M., Li, X.-F., Wei, H.-B., Wichems, C., Lesch, K.-P., Murphy, D. L., and Uhl, G. R. (2001). Molecular mechanism of cocaine reward: combined dopamine and serotonin transporter knockouts eliminate cocaine place preference. *Proc Nat Acad Sci U S A, 98*, 5300–5305.

108. Sulzer, D., Chen, T. K., Lau, Y. Y., Kristensen, H., Rayport, S., and Ewing, A. (1995). Amphetamine redistributes dopamine from synaptic vesicles to the cytosol and promotes reverse transport. *J Neurosci, 15*, 4102–4108.

109. Khoshbouei, H., Wang, H., Lechleiter, J. D., Javitch, J. A., and Galli, A. (2003). Amphetamine-induced dopamine efflux: a voltage-sensitive and intracellular Na^+-dependent mechanism. *J Biol Chem, 278*, 12070–12077.

110. Kahlig, K. M., Binda, F., Khoshbouei, H., Blakely, R. D., McMahon, D. G., Javitch, J. A., and Galli, A. (2005). Amphetamine induces dopamine efflux through a dopamine transporter channel. *Proc Nat Acad Sci U S A, 102*, 3495–3500.

111. Krueger, B. K. (1990). Kinetics and block of dopamine uptake in synaptosomes from rat caudate nucleus. *J Neurochem, 55*, 260–267.

112. McElvain, J. S., and Schenk, J. O. (1992). A multisubstrate mechanism of dopamine uptake and its inhibition by cocaine. *Biochem Pharmacol, 43*, 2189–2199.

113. Gu, H., Wall, S. C., and Rudnick, G. (1994). Stable expression of biogenic amine transporters reveals differences in inhibitor sensitivity, kinetics and ion dependence. *J Biol Chem, 269*, 7124–7130.

114. Berfield, J. L., Wang, L. C., and Reith, M. E. A. (1999). Which form of dopamine is the substrate for the human dopamine transporter: the cationic or the uncharged species? *J Biol Chem, 274*, 4876–4882.

115. Mager, S., Min, C., Henry, D. J., Chavkin, C., Hoffman, B. J., Davidson, N., and Lester, H. A. (1994). Conducting states of a mammalian serotonin transporter. *Neuron, 12,* 845–859.

116. Sonders, M. S., Zhu, S.-J., Zahniser, N. R., Kavanaugh, M. P., and Amara, S. G. (1997). Multiple ionic conductances of the human dopamine transporter: the actions of dopamine and psychostimulants. *J Neurosci, 17,* 960–974.

117. Sonders, M. S., and Amara, S. G. (1996). Channels in transporters. *Curr Opin Neurobiol, 6,* 294–302.

118. Carvelli, L., McDonald, P. W., Blakely, R. D., and DeFelice, L. J. (2004). Dopamine transporters depolarize neurons by a channel mechanism. *Proc Nat Acad Sci U S A, 101,* 16046–16051.

119. Ingram, S. L., Prasad, B. M., and Amara, S. G. (2002). Dopamine transporter–mediated conductances increase excitability of midbrain dopamine neurons. *Nat Neurosci, 5,* 971–978.

120. Ingram, S. L., and Amara, S. G. (2000). Arachindonic acid stimulates a novel cocaine-sensitive cation conductance associated with the human dopamine transporter. *J Neurosci, 20,* 550–557.

121. Meinild, A.-K., Sitte, H. H., and Gether, U. (2004). Zinc potentiates an uncoupled anion conductance associated with the dopamine transporter. *J Biol Chem, 279,* 49671–49679.

122. Pifil, C., Rebernik, P., Kattinger, A., and Reither, H. (2004). Zn^{2+} modulates currents generated by the dopamine transporter: parallel effects on amphetamine-induced charge transfer and release. *Neuropharmacology, 46,* 223–231.

123. Saunders, C., Ferrer, J. V., Shi, L., Chen, J., Merrill, G., Lamb, M. E., Leeb-Lundberg, L. M. F., Carvelli, L., Javitch, J. A., and Galli, A. (2000). Amphetamine-induced loss of human dopamine transporter activity: an internalization-dependent and cocaine-sensitive mechanism. *Proc Nat Acad Sci U S A, 97,* 6850–6855.

124. Rabinovic, A. D., Lewis, D. A., and Hastings, T. G. (2000). Role of oxidative changes in the degeneration of dopamine terminals after injection of neurotoxic levels of dopamine. *Neuroscience, 101,* 67–76.

125. Doolen, S., and Zahniser, N. R. (2002). Conventional protein kinase C isoforms regulate human dopamine transporter activity in *Xenopus* oocytes. *FEBS Lett, 516,* 187–190.

126. Gulley, J. M., Doolen, S., and Zahniser, N. R. (2002). Brief, repeated exposure to substrates down-regulates dopamine transporter function in *Xenopus* oocytes in vitro and rat dorsal striatum in vivo. *J Neurochemi, 83,* 400–411.

127. Mayfield, R. D., Maiya, R., and Zahniser, N. R. (2001). Ethanol potentiates the function of the human dopamine transporter expressed in *Xenopus* oocytes. *J Neurochem, 79,* 1070–1079.

128. Gnegy, M. E., Khoshbouei, H., Berg, K. A., Javitch, J. A., Clarke, W. P., Zhang, M., and Galli, A. (2004). Intracellular Ca^{2+} regulates amphetamine-induced dopamine efflux and currents mediated by the human dopamine transporter. *Mol Pharmacol, 66,* 137–143.

129. Khoshbouei, H., Sen, N., Guptaroy, B., Johnson, L., Lund, D., Gnegy, M. E., Galli, A., and Javitch, J. A. (2004). N-Terminal phosphorylation of the dopamine transporter is required for amphetamine-induced efflux. *PLOS Biol, 2,* 387–393.

130. Melikian, H. E., and Buckley, K. M. (1999). Membrane trafficking regulates the activity of the human dopamine transporter. *J Neurosci, 19,* 7699–7710.

131. Wiesinger, J. A., Buwen, J. P., Cifelli, C. J., Unger, E. L., Jones, B. C., and Beard, J. L. (2007). Down-regulation of dopamine transporter by iron chelation in vitro is mediated by altered trafficking, not synthesis. *J Neurochem*, *100*, 167–179.

132. Middleton, L. S., Apparsundaram, S., King-Pospisil, K. A., and Dwoskin, L. P. (2007). Nicotine increases dopamine transporter function in rat striatum through a trafficking-independent mechanism. *Eur J Pharmacol*, *554*, 128–136.

133. Zahniser, N. R., and Sorkin, A. (2004). Rapid regulation of the dopamine transporter: role in stimulant addiction? *Neuropharmacology*, *47*, 80–91.

134. Loder, M. K., and Melikian, H. E. (2003). The dopamine transporter constitutively internalizes and recycles in a protein kinase C–regulated manner in stably transfected PC12 cell lines. *J Biol Chem*, *278*, 22168–22174.

135. Holton, K. L., Loder, M. K., and Melikian, H. E. (2005). Nonclassical, distinct endocytic signals dictate constitutive and PKC-regulated neurotransmitter transporter internalization. *Nat Neurosci*, *8*, 881–888.

136. Kahlig, K. M., Javitch, J. A., and Galli, A. (2004). Amphetamine regulation of dopamine transport: combined measurements of transporter currents and transporter imaging support the endocytosis of an active carrier. *J Biol Chem*, *279*, 8966–8975.

137. Kahlig, K. M., Lute, B. J., Wei, Y., Loland, C. J., Gether, U., Javitch, J. A., and Galli, A. (2006). Regulation of dopamine transporter trafficking by intracellular amphetamine. *Mol Pharmacol*, *70*, 542–548.

138. Daniels, G. M., and Amara, S. G. (1999). Regulated trafficking of the human dopamine transporter: clathrin-mediated internalization and lysosomal degradation in response to phorbol esters. *J Biol Chem*, *274*, 35794–35801.

139. Johnson, L. A., Furman, C. A., Zhang, M., Guptaroy, B., and Gnegy, M. E. (2005). Rapid delivery of the dopamine transporter to the plasmalemmal membrane upon amphetamine stimulation. *Neuropharmacology*, *49*, 750–758.

140. Wei, Y., Williams, J. M., Dipace, C., Sung, U., Javitch, J. A., Galli, A., and Saunders, C. (2007). Dopamine transporter activity mediates amphetamine-induced inhibition of Akt through a Ca^{2+}/calmodulin-dependent kinase II–dependent mechanism. *Mol Pharmacol*, *71*, 835–842.

141. Carvelli, L., Moron, J. A., Kahlig, K. M., Ferrer, J. V., Sen, N., Lechleiter, J. D., Leeb-Lundberg, L. M., Merrill, G., Lafer, E. M., Ballou, L. M., Shippenberg, T. S., Javitch, J. A., Lin, R. Z., and Galli, A. (2002). PI 3-kinase regulation of dopamine uptake. *J Neurochem*, *81*, 859–869.

142. Garcia, B. G., Wei, Y., Moron, J. A., Lin, R. Z., Javitch, J. A., and Galli, A. (2005). Akt is essential for insulin modulation of amphetamine-induced human dopamine transporter cell-surface redistribution. *Mole Pharmacol*, *68*, 102–109.

143. Little, K. Y., Elmer, L. W., Zhong, H., Scheys, J. O., and Zhang, L. (2002). Cocaine induction of dopamine transporter trafficking to the plasma membrane. *Mol Pharmacol*, *61*, 436–445.

144. Granas, C., Ferrer, J., Loland, C. J., Javitch, J. A., and Gether, U. (2003). N-terminal truncation of the dopamine transporter abolishes phorbol ester– and substance P receptor–stimulated phosphorylation without impairing transporter internalization. *J Biol Chem*, *278*, 4990–5000.

145. Sorkina, T., Miranda, M., Dionne, K. R., Hoover, B. R., Zahniser, N. R., and Sorkin, A. (2006). RNA interference screen reveals an essential role of NEDD4–2 in dopamine transporter ubiquitination and endocytosis. *J Neurosci*, *26*, 8195–8205.

146. Jiang, H., Jiang, Q., and Feng, J. (2004). Parkin increases dopamine uptake by enhancing the cell surface expression of dopamine transporter. *J Biol Chem*, *279*, 54380–54386.

147. Torres, G. E., Carneiro, A., Seamans, K., Fiorentini, C., Sweeney, A., Yao, W.-D., and Caron, M. G. (2003). Oligomerization and trafficking of the human dopamine transporter: mutational analysis identifies critical domains important for the functional expression of the transporter. *J Biol Chem*, *278*, 2731–2739.

148. Numan, S., and Seroogy, K. B. (1999). Expression of trkB and trkC mRNAs by adult midbrain dopamine neurons: a double-label in situ hybridization study. *J Comp Neurol*, *403*, 295–308.

149. Figlewicz, D. P., Evans, S. B., Murphy, J., Hoen, M., and Baskin, D. G. (2003). Expression of receptors for insulin and leptin in the ventral tegmental area/substantia nigra (VTA/SN) of the rat. *Brain Res*, *964*, 107–115.

150. Huang, E. J., and Reichardt, L. F. (2003). Trk receptors: roles in neuronal signal transduction. *Annu Rev Biochem*, *72*, 609–642.

151. Lin, Z., Zhang, P.-W., Zhu, X., Melgari, J.-M., Huff, R., Spieldoch, R. L., and Uhl, G. R. (2003). Phosphatidylinositol 3-kinase, protein kinase C, and MEK1/2 kinase regulation of dopamine transporters (DAT) require N-terminal DAT phosphoacceptor sites. *J Biol Chem*, *278*, 20162–20170.

152. Moron, J. A., Zakharova, I., Ferrer, J. V., Merrill, G. A., Hope, B., Lafer, E. M., Lin, Z. C., Wang, J. B., Javitch, J. A., Galli, A., and Shippenberg, T. S. (2003). Mitogen-activated protein kinase regulates dopamine transporter surface expression and dopamine transport capacity. *J Neurosci*, *23*, 8480–8488.

153. Torres, G. E., Yao, W.-D., Mohn, A. R., Quan, H., Kim, K.-M., Levey, A. I., Staudinger, J., and Caron, M. G. (2001). Functional interaction between monoamine plasma membrane transporters and the synaptic PDZ domain-containing protein PICK1. *Neuron*, *30*, 121–134.

154. Bjerggaard, C., Fog, J. U., Hastrup, H., Madsen, K., Loland, C. J., Javitch, J. A., and Gether, U. (2004). Surface targeting of the dopamine transporter involves discrete epitopes in the distal C terminus but does not require canonical PDZ domain interactions. *J Neurosci*, *24*, 7024–7036.

155. Miranda, M., Sorkina, T., Grammatopoulos, T. N., Zawada, W. M., and Sorkin, A. (2004). Multiple molecular determinants in the carboxyl terminus regulate dopamine transporter export from endoplasmic reticulum. *J Biol Chem*, *279*, 30760–30770.

156. Carneiro, A. M., Ingram, S. L., Beaulieu, J.-M., Sweeney, A., Amara, S. G., Thomas, S. M., Caron, M. G., and Torres, G. E. (2002). The multiple LIM domain-containing adaptor protein Hic-5 synaptically colocalizes and interacts with the dopamine transporter. *J Neurosci*, *22*, 7045–7054.

157. Bolan, E. A., Kivell, B., Jaligam, V., Oz, M., Jayanthi, L. D., Han, Y., Sen, N., Urizar, E., Gomes, I., Devi, L. A., Ramamoorthy, S., Javitch, J. A., Zapata, A., and Shippenberg, T. S. (2007). D_2 receptors regulate dopamine transporter function via an ERK 1/2-dependent and PI3 kinase–independent mechanism. *Mol Pharmacol*, *71*, 1222–1232.

158. Lee, F. J. S., Pei, L., Moszczynska, A., Vukusic, B., Fletcher, P. J., and Liu, F. (2007). Dopamine transporter cell localization facilitated by a direct interaction with the dopamine D2 receptor. *EMBO J*, *26*, 2127–2136.

159. Byas-Smith, M. G., Li, J., Szlam, F., Eaton, D. C., Votaw, J. R., and Denson, D. D. (2004). Isoflurance induces dopamine transporter trafficking into the cell cytoplasm. *Synapse*, *53*, 68–73.

160. Salvatore, M. F., Apparsundaram, S., and Gerhardt, G. A. (2003). Decreased plasma membrane expression of striatal dopamine transporter in aging. *Neurobiol Aging*, *24*, 1147–1154.

161. Ciliax, B. J., Heilman, C., Demchyshyn, L. L., Pristupa, Z. B., Ince, E., Hersch, S. M., Niznik, H. B., and Levey, A. I. (1995). The dopamine transporter: immunochemical characterization and localization in brain. *J Neurosci*, *15*, 1714–1723.

162. Ciliax, B. J., Drash, G. W., Staley, J. K., Haber, S., Mobley, C. J., Miller, G. W., Mufson, E. J., Mash, D. C., and Levey, A. I. (1999). Immunocytochemical localization of the dopamine transporter in human brain. *J Comp Neurol*, *409*, 38–56.

163. Hersch, S. M., Yi, H., Heilman, C. J., Edwards, R. H., and Levey, A. I. (1997). Subcellular localization and molecular topology of the dopamine transporter in the striatum and substantia nigra. *J Comp Neurol*, *388*, 211–227.

164. Pristupa, Z. B., McConkey, F., Liu, F., Man, H. Y., Lee, F. J., Wang, Y. T., and Niznik, H. B. (1998). Protein kinase–mediated bidirectional trafficking and functional regulation of the human dopamine transporter. *Synapse*, *30*, 79–87.

165. Li, L.-B., Chen, N., Ramamoorthy, S., Chi, L., Cui, X.-N., Wang, L. C., and Reith, M. E. A. (2004). The role of N-glycosylation in function and surface trafficking of the human dopamine transporter. *J Biol Chem*, *279*, 21012–21020.

166. Miranda, M., Dionne, K. R., Sorkina, T., and Sorkin, A. (2007). Three ubiquitin conjugation sites in the amino terminus of the dopamine transporter mediate protein kinase C–dependent endocytosis of the transporter. *Mol Biol Cell*, *18*, 313–323.

167. Berney, C., and Danuser, G. (2003). FRET or no FRET: a quantitative comparison. *Biophysi J*, *84*, 3992–4010.

168. Sekar, R. B., and Periasamy, A. (2003). Fluorescence resonance energy transfer (FRET) microscopy imaging of live cell protein localizations. *J Cell Biol*, *160*, 629–633.

169. Berger, S. P., Farrell, K., Conant, D., Kempner, E. S., and Paul, S. M. (1994). Radiation inactivation studies of the dopamine reuptake transporter protein. *Mol Pharmacol*, *46*, 726–731.

170. Milner, H. E., Beliveau, R., and Jarvis, S. M. (1994). The in situ size of the dopamine transporter is a tetramer as estimated by radiation inactivation. *Biochim Biophysi Acta*, *1190*, 185–187.

171. Hannon, G. J. (2002). RNA interference. *Nature*, *418*, 244–251.

172. Dykxhoorn, D. M., Novina, C. D., and Sharp, P. A. (2003). Killing the messenger: short RNAs that silence gene expression. *Nat Rev Mol Cell Biol*, *4*, 457–467.

173. Davidson, B. L., and Boudreau, R. L. (2007). RNA interference: a tool for querying nervous system function and an emerging therapy. *Neuron*, *53*, 781–788.

174. Thakker, D. R., Natt, F., Husken, D., Maier, R., Muller, M., van der Putten, H., Hoyer, D., and Cryan, J. F. (2004). Neurochemical and behavioral consequences of widespread gene knockdown in the adult mouse brain by using nonviral RNA interference. *Proc Nat Acad Sci U S A*, *101*, 17270–17275.

175. Salahpour, A., Medvedev, I. O., Beaulieu, J. M., Gainetdinov, R. R., and Caron, M. G. (2007). Local knockdown of genes in the brain using small interfering RNA: a phenotypic comparison with knockout animals. *Biol Psychiatry*, *61*, 65–69.

176. Hicke, L., Schubert, H. L., and Hill, C. P. (2005). Ubiquitin-binding domains. *Nat Rev Mole Cell Biol*, *6*, 610–621.

177. Fountaine, T. M., and Wade-Martins, R. (2007). RNA interference-mediated knockdown of α-synuclein protects human dopaminergic neuroblastoma cells from MPP$^+$ toxicity and reduces dopamine transport. *J Neurosci Res*, *85*, 351–363.

178. Lee, F. J. S., Liu, F., Pristupa, Z. B., and Niznik, H. B. (2001). Direct binding and functional coupling of α-synuclein to the dopamine transporters accelerate dopamine-induced apoptosis. *Fed Ame Soci Exp Biol J*, *15*, 916–926.

179. Wersinger, C., Prou, D., Vernier, P., and Sidhu, A. (2003). Modulation of dopamine transporter function by α-synuclein is altered by impairment of cell adhesion and by induction of oxidative stress. *Fed Ame Soci Exp Biol J*, *17*, 2151–2163.

180. Wersinger, C., and Sidhu, A. (2005). Disruption of the interaction of α-synuclein with microtubules enhances cell surface recruitment of the dopamine transporter. *Biochemistry*, *44*, 13612–13624.

181. Gulley, J. M., Everett, C. V., and Zahniser, N. R. (2007). Inbred Lewis and Fischer 344 rat strains differ not only in novelty- and amphetamine-induced behaviors, but also in dopamine transporter activity. *Brain Res*, *1151*, 32–45.

14

IN VIVO STUDIES OF DOPAMINE TRANSPORTER FUNCTION

Jane B. Acri

Addiction Treatment Discovery Program, Division of Pharmacotherapies and Medical Consequences of Drug Abuse, National Institute on Drug Abuse, Bethesda, Maryland

14.1	Introduction	391
14.2	DAT Knockout Studies	392
14.3	Studies Using Multiple Transporter Ligands	395
14.4	Studies of Specific Transporter Ligands	401
	14.4.1 GBR 12909	402
	14.4.2 Cocaine Analogs	404
	14.4.3 Benztropine and Analogs	414
	14.4.4 Mazindol	418
	14.4.5 Bupropion	420
	14.4.6 Methylphenidate	421
14.5	Conclusions	425
	References	426

14.1 INTRODUCTION

Since 1987, when it was first proposed that the dopamine transporter (DAT) was the site of cocaine binding in the brain [1], the dopamine transporter became one of the major targets of efforts to develop medications to treat cocaine dependence. It has also become a major focus of research to understand the actions of cocaine, and to

Dopamine Transporters: Chemistry, Biology, and Pharmacology. Edited by Mark L. Trudell and Sari Izenwasser

that end, much of our understanding of the role of the DAT is in reference to the actions of cocaine. Although cocaine also binds to the serotonin transporter (SERT) and the norepinephrine transporter (NET), it is the potency of cocaine at the DAT and its actions there that have been most definitively linked to its behavioral stimulant and euphoric effects. As a result, this chapter will focus on the dopamine transporter with regard to the behavioral effects of cocaine.

This review will take several approaches to presenting in vivo studies of dopamine transporter function. The first is to examine the effects of compounds in animals where the dopamine transporter has been removed through genetic manipulation. Although results of studies examining DAT knockout animals are confounded by several variables, primarily the development of neuroadaptations by other systems, these results have been intriguing and controversial.

The second part of the chapter will examine studies in which cocaine and other transporter ligands have been studied to determine the relationship of behavioral effects to binding, uptake, or selectivity for the DAT as a way of determining some of the transporter requirements for the transduction of behavioral effects. Although these studies have shed some light on the question of the role of the DAT in the mediation of cocainelike effects, there have emerged many exceptions in each analysis that are unexplained by the relationships that are described. For this reason, a third section will describe some of the behavioral effects of various structural series of DAT ligands, with particular focus on those that are not cocainelike.

14.2 DAT KNOCKOUT STUDIES

Technologies permitting targeted disruption of the mouse DAT gene and the resulting generation of homozygotic and heterozygotic knockout (KO) mice have provided a novel and groundbreaking method of studying DAT function. Rather than examining the effects of compounds interacting with the DAT, knockout studies attempt to elucidate transporter function by determination of what behaviors are altered by its absence. This indirect or "subtractive" method of study has been plagued by the remarkable ability of damaged neuronal systems to adapt in order to maintain homeostasis, a basic tenet of biological systems related to evolution and species survival. Our deepening understanding of pervasive developmental neuroadaptations to maintain homeostasis has increasingly informed our early simplistic interpretations of the results of DAT knockout studies.

Early knockout studies elucidated the role of the transporter in regulating dopaminergic neurotransmission. Homozygotes were viable but had severely altered dopaminergic systems and were behaviorally five times more active than wild-type mice in tests of locomotor activity, despite lower synaptic dopamine and down-regulated postsynaptic receptors. However, given the role of the DAT in clearing dopamine from the synapse, it was not surprising that dopamine persisted over 100-fold longer in the extracellular space. It was not surprising that both cocaine and amphetamine were unable to increase locomotor activity in knockout mice, further confirming the role of the transporter in the locomotor stimulant actions of psychostimulants [2,3].

It was, however, quite surprising when several groups reported that DAT knockout mice continued to show a place preference to cocaine and could be trained to self-administer cocaine. Rocha and colleagues [4] were the first to report that mice lacking the DAT could still be trained to self-administer cocaine using standard operant lever-pressing techniques. Although it was noted that homozygotic knockout mice required significantly more training sessions to acquire self-administration up to criterion values, once acquired, their self-administration of cocaine was dose-dependent and did not differ between wild-type and knockout genotypes. Both groups also showed a typical "extinction burst" of responses when cocaine was no longer available and extinguished within 3 days, and both groups reinstated self-administration when cocaine became available by pressing the opposite lever [4]. Similarly, another laboratory reported no disruption in conditioned place preference to cocaine in another strain of knockout mice, despite the absence of gene expression in striatum [5]. These studies used a "biased" design with a two-compartment chamber in which naive wild-type animals have preferences for a particular side. This preference can be reversed by pairing with cocaine, causing the animals to "prefer" the initially non-preferred side. Using this paradigm, wild-type mice showed a preference for the cocaine-paired side at conditioning doses of 5 and 10 mg/kg. Heterozygous mice also showed preference for the cocaine-paired side, but only with a training dose of 10 mg/kg. It was somewhat of a surprise, then, that homozygous mice also showed a robust preference for the cocaine-paired side at 10 mg/kg cocaine, resulting in the testing of two cohorts, testing on separate occasions, and confirmation of genotype. A place preference to methylphenidate also was conditioned in the DAT-knockout (KO) animals using a dose of 5 mg/kg methylphenidate [5].

Both Rocha and colleagues [4] and Sora and colleagues [5] suspected that because cocaine "reward" seemed to be intact in DAT knockout mice, that despite early research studies implicating the DAT in the reinforcing effects of cocaine [1], perhaps the rewarding effects of cocaine were actually mediated through cocaine's interactions with the serotonin system. However, when cocaine was tested for its ability to condition a place preference in homozygous serotonin transporter knockout mice, cocaine reward not only was intact but was enhanced significantly compared to wild-type mice, providing some of the first hints that developmental neuroadaptations were occurring in transporter knockout animals [5].

Microdialysis and voltammetry studies have been used to study the effects of deletions of the DAT gene ex vivo and in vivo. These studies have confirmed that in striatal slices, there is a 300-fold increase in the amount of time that dopamine is present in the extracellular space in DAT KO mice, confirming a decreased clearance rate. In addition, the number of dopamine molecules released in response to an electrical pulse was only 75% of that seen in slices from wild-type mice [6]. Similarly, no-net flux microdialysis confirmed a fivefold elevation in extracellular dopamine in DAT-KO mice compared to wild-type mice [7]. These results confirm the role of the DAT in terms of dopamine clearance from the synapse, but also highlight the compensatory mechanisms that may have occurred as a result of the transporter absence.

The next series of studies aimed at elucidating the role of the dopamine transporter in cocaine reward used norepinephrine transporter (NET) knockout mice [8] and combined dopamine and serotonin knockouts [9]. In norepinephrine knockout mice, not only could a preference for cocaine be conditioned using place-preference techniques, but the preference for cocaine in NET-KO mice was enhanced significantly relative to that which could be conditioned in the wild-type mice. In addition, the percent change in dopamine levels in response to cocaine was indistinguishable in the two genotypes, despite lower basal levels of dopamine in the NET knockouts [8]. This study confirmed that cocaine reward could not be attributed to actions and any one monoamine transporter, and strongly suggested an inhibitory role of the NET in cocaine reward.

The double-knockout study was designed to measure the relative contributions of the serotonin and dopamine transporters by inclusion of all permutations of heterozygotes in the analysis. Results showed that with both transporters deleted in double-knockout animals, double-homozygous mice showed no preference for cocaine, suggesting that both transporters were involved in cocaine reward as measured by place preference. These studies also examined double knockouts where one gene copy of either the DAT or SERT was present. In the double knockouts that were heterozygous with regard to the DAT, place preference was retained, whereas when one gene copy of the SERT was present, no place preference was seen. These studies confirmed that although no single monoamine transporter is absolutely necessary for cocaine reward in mice that develop in the absence of the transporter, some transporters are more important than others. Despite the apparent redundancy in the system, the DAT seems to have a greater role than the other transporters in cocaine reward [9].

Additional experiments have confirmed the abnormal reward systems in transporter knockout animals, confirming the redundancy, overlap, and balance in monoamine systems. Hall and colleagues [10] have reported that fluoxetine, a selective serotonin-reuptake inhibitor, which is not rewarding in wild-type mice, is strongly preferred in homozygous DAT knockout mice. Similarly, nisoxetine, a selective NET inhibitor that is not rewarding in wild-type mice, is significantly preferred in homozygous DAT knockouts. In contrast, in homozygous NET knockout mice, fluoxetine, but not nisoxetine is rewarding, whereas in homozygous SERT knockout mice, both fluoxetine and nisoxetine fail to condition a place preference. Finally, in SERT/NET double knockouts, cocaine produces an enhanced reward relative to wild-type mice [10]. These studies highlight the complexity of monoaminergic systems and the danger in over-interpretation of studies using knockout animals. The results of lifelong gene deletions on development clearly result in the recruitment of compensatory mechanisms and neuroadaptations that may obfuscate the interpretation of the studies in which these animals are used.

Another strategy to examine the role of the dopamine transporter in mediation of the effects of cocaine was the development of "knock-in" mice with another type of mutant transporter [11]. To that end, mutations at various loci in the DAT gene were generated with the intention of retaining normal function with regard to the uptake of dopamine, while reducing the ability of cocaine to bind and inhibit dopamine reuptake. In these mutant mice, referred to as DAT-CI (dopamine transporter cocaine-insensitive), there were normal levels of DAT mRNA as compared to wild-type mice, but the DAT-CI mice were 89-fold less sensitive to inhibition of dopamine uptake by cocaine than wild-type mice, confirming the lack of sensitivity to cocaine. The were other differences, however, including reductions in whole brain content of dopamine and a reduced rate of dopamine uptake relative to wild-type mice. Using in vivo microdialysis, DAT-CI mice failed to show increases in extracellular dopamine produced by cocaine as seen in wild-type animals, but showed higher basal levels of dopamine. In addition, mutant mice were more active, but showed no locomotor response to cocaine. Finally, DAT-CI mice failed to condition a place preference to cocaine when trained at doses of 5 and 20 mg/kg, unlike wild-type mice, which showed robust preference at both doses. Interestingly, amphetamine was tested as a positive control and was able to produce a place preference, suggesting that reward mechanisms were intact and normal [11]. These results were interpreted to suggest that cocaine-induced elevations in extracellular dopamine, increases in locomotor activity, and reward are all actions mediated through the dopamine transporter in intact animals. Because these mice have functional dopamine transporters that are insensitive to cocaine, compensatory mechanisms and neuroadaptations have not developed to the same extent as in knockouts, and therefore, elucidation of the function of the DAT with regard to the mediation of the behavioral effects of cocaine is not confounded.

14.3 STUDIES USING MULTIPLE TRANSPORTER LIGANDS

Since the "cocaine receptor" was identified as the DAT [1], a number of investigators have examined transporter ligands in behavioral studies and attempted to understand the behavioral differences in terms of interactions with the dopamine transporter. A selection of these studies will be reviewed below.

Early attempts to link transporter binding to behavioral effects of cocaine were an attempt to understand the significance of the cocaine recognition site and the behavioral effects of cocaine, with the expectation that potencies for binding and the generation of behavioral stimulation would be related. To this end, Spealman and colleagues [12] studied the effects of monoamine-uptake inhibitors with varying selectivity for dopamine, serotonin, and norepinephrine transporters on schedule-controlled behavior. In these studies, squirrel monkeys were trained on a fixed-interval schedule of stimulus shock avoidance that is sensitive to stimulant effects of drugs, which were administered intravenously (i.v.), resulting in similar times of onset of effects. Mazindol, methylphenidate, nomifensine, GBR 12909, and bupropion had behavioral effects that were similar to those of cocaine, although all compounds were less potent than cocaine in these studies. In contrast, desipramine and citalopram had effects that differed qualitatively from those of cocaine, presumably because of greater selectivity for norepinephrine and serotonin transporters, respectively, as compared to cocaine. The order of potencies of the compounds that produced cocainelike behavioral effects were compared with potencies to displace [³H]cocaine in previous studies using rat and monkey brain tissue and were found to be highly correlated. Given that cocaine binds to all three monoamine transporters, results cannot be attributed solely to binding at the DAT; however, results were suggestive of that conclusion because of the lack of cocainelike effects produced by desipramine and citalopram. However, the authors cautiously conclude that effects at other cocaine binding sites such as histamine and muscarinic sites cannot be ruled out [12].

A similar study examining the relationship between transporter selectivity and behavioral stimulant effects was conducted using cocaine and several novel analogs: RTI-177, FECNT, and RTI-112 [13]. In addition to examining the effects of compounds under a fixed-interval schedule of stimulus shock termination in squirrel monkeys, in vivo microdialysis was used to examine changes in extracellular dopamine in the caudate nucleus that were produced by administration of the compounds under study. Two compounds with low affinity for the norepinephrine transporter, RTI-177 and FECNT, produced dose-dependent stimulation of behavior, but with a slower onset and longer duration of action than cocaine. In vivo microdialysis revealed elevations in extracellular dopamine that corresponded in time to behavioral effects. Interestingly, RTI-112 had little selectivity for the DAT and also bound to the norepinephrine and serotonin transporters with high affinity. This compound failed to produce significant stimulant effects on behavior, and its effects on extracellular dopamine were variable across subjects, despite the fact that its binding to the DAT was similar to that of both RTI-177 and FECNT, suggesting the possibility that transporter selectivity is an important determinant of behavioral and neurochemical effects of transporter ligands [13].

Cline and colleagues [14] studied cocaine and a group of closely related analogs, including WIN 35,428, RTI-31, RTI-32, RTI-51, and RTI-55, for their effects on locomotor activity in mice and their effects on binding to the DAT as measured by in vivo accumulation of [³H]WIN 35,428 in mouse striatum. The results of these studies indicated a correspondence between in vivo binding and stimulation of locomotor activity, with the analogs being approximately 20-fold more potent than cocaine in both binding and locomotor activity. The results were also thought to reflect degree of transporter occupancy, since the maximum displacement of [³H]WIN 35,428 occurred at the same doses that produced the greatest increases in locomotor activity. Overall, results were interpreted as being consistent with identification of the DAT as the site of action of locomotor stimulant effects of cocaine and its analogs [14].

Vaugeois and colleagues [15] also attempted to equate the level of DAT occupancy in the striatum with the ability of compounds to stimulate locomotor activity using [³H]GBR 12909 as a label for the DAT. A group of structurally distinct transporter ligands were studied, and the 20-minute period in which maximal effects on locomotor activity occurred was used for comparison. Compounds were administered in doses corresponding to the ED_{50} doses for the in vivo occupancy of striatal binding sites, and the locomotor stimulant effects were measured. Mazindol, nomifensine, and cocaine showed little stimulation at ED_{50} doses for transporter occupancy, whereas GBR 12783, BTCP, and amineptine were robust stimulants at these doses. This study concluded that a given level of transporter occupancy does not result in a constant or consistent increase in locomotor activity. However, the authors point out that the effects of uptake inhibitors on levels of extracellular dopamine are dependent on neuron firing rate, and for GBR 12783 there was a significant correlation between baseline spontaneous locomotor activity and response to drug, whereas effects of releasing agents such as *d*-amphetamine do not depend on neuron firing rate. Other confounding variables that may interfere with the relationship between transporter occupancy and behavioral effects were proposed, but not elucidated by this paper [15].

A related study that examined dopamine uptake inhibitors of different structural categories was done by Izenwasser and colleagues [16]. Compounds were examined for displacement of [³H]WIN 35,428 and effects on locomotor activity, and some compounds were observed to displace [³H]WIN 35,428 by binding to two distinct sites. The competition curves that were best fit by a two-site model included diclofensine, WIN 35,428, nomifensine, WIN 35,981, WIN 35,065-2, cocaine, and bupropion. The compounds best fit by a one-site model included RTI-31, BTCP, GBR 12909, mazindol, and benztropine, among others. For all compounds analyzed together, there was a significant correlation between affinity for the DAT and the ED_{50} for stimulation of locomotor activity. When only cocaine and its close structural analogs were included in the analysis, the correlation was considerably stronger, and when they were removed from the analysis, there was no correlation among the remaining compounds. And interestingly, for compounds where binding was best characterized by a two-site model, there was no correlation between binding to the low-affinity site and stimulation of locomotor activity, but there was a significant correlation between binding affinity at the high-affinity site and the ED_{50} values for stimulation of locomotor activity. These results were interpreted to suggest that there is not a simple relationship between binding to the dopamine transporter and stimulation of activity except as has been reported previously in the case of close structural analogs of cocaine [14]. For uptake inhibitors that are structurally unrelated to cocaine, the lack of correlation could be due to differences in pharmacokinetic properties or other pharmacological effects that at are, at this time, unknown. In addition, the significance of the two-site model, which suggests that some compounds may bind to the DAT in a different manner, is unknown [16].

Additional studies examining the stimulation of locomotor activity in rodents by dopamine-uptake inhibitors have examined the rates of transporter occupancy produced by different compounds as important mediators of their behavioral effects. Cocaine, GBR 12909, and several structural analogs of benztropine were studied in assays of locomotor activity time course in mice for a period of 8 hours [17]. Additionally, the rates of transporter occupancy by [¹²⁵I]RTI-121 were measured at different time points to capture the relationship between stimulation of locomotor activity and transporter occupancy. All compounds increased locomotor activity, although maximal stimulation by cocaine was greater than that of the benztropine analogs but not significantly different from that of GBR 12909. In contrast, the displacement of [¹²⁵I]RTI-121 by the highest dose of cocaine was significantly lower than that produced by the highest doses of each of the other compounds. Across the time points at which in vivo binding was measured, there was a significant positive correlation between displacement of $[^{125}I]$RTI-121 and the locomotor stimulant effects of all compounds, but for cocaine, the increases in locomotor activity were greater than would have been predicted by displacement. Overall, the results indicated slower rates of displacement (or slower association rates to the dopamine transporter) produced by GBR 12909 and the benztropine analogs, which were related to the differences in the behavioral effects of the compounds as compared to cocaine [17].

A number of studies have examined DAT ligands with regard to the transduction of cocainelike discriminative stimulus effects, and similarly attempted to equate the transduction of behavioral effects to binding affinity, potency to inhibit uptake, transporter occupancy, and transporter selectivity.

As early as 1988, it was observed that when cocaine was trained as a discriminative stimulus, compounds that fully substituted for it may share a common binding site. Further, it was noted that for drugs that fully substituted (diethylpropion, methylphenidate, phenmetrazine, and phentermine), cross-tolerance could be measured following chronic administration of cocaine. In contrast, a compound that did not substitute for acute cocaine, fenfluramine, similarly did not substitute following chronic cocaine, suggesting that the lack of substitution was not simply an issue of potency but was related to a specific binding site [18], although that site had not yet been identified.

A comprehensive study conducted in 1990 in which rhesus monkeys were trained to recognize the discriminative stimulus effects of cocaine similarly concluded that only "indirect agonists" of dopamine such as GBR 12909, mazindol, nomifensine, and bupropion produced full substitution for cocaine, whereas compounds that inhibit the reuptake of norepinephrine (tomoxetine and nisoxetine) or serotonin (fluoxetine) do not substitute for cocaine. Similarly, compounds acting at postsynaptic dopamine receptors such as SKF 38393 and quinpirole did not produce full substitution, leading to the conclusion that blockade of dopamine reuptake rather than stimulation of postsynaptic receptors is sufficient to produce the discriminative stimulus effects of cocaine [19].

Broadbent and colleagues [20] extended this work by equating the potency of compounds to substitute for cocaine with their potencies at the dopamine transporter. In this study, compounds that fully substituted for cocaine in rats trained to discriminate cocaine from saline included nomifensine, GBR 12909, and bupropion, but not the norepinephrine blockers nisoxetine, desipramine, or imipramine. Although only a limited number of compounds were studied, the authors were among the first to conclude that the discriminative stimulus effects of cocaine were transduced mainly through blockade of the dopamine transporter and that potency to substitute for cocaine corresponded to affinity at the transporter. Importantly, the authors did acknowledge that the compounds that substituted for cocaine were, like cocaine, not necessarily selective for the DAT, but nonetheless, it was the blockade of dopamine transport that appeared to be most relevant [20].

Cunningham and Callahan [21] also trained cocaine as a discriminative stimulus and found that monoamine-uptake inhibitors that were not primarily dopamine-uptake inhibitors (desipramine and fluoxetine) did not substitute for the discriminative stimulus effects of cocaine, whereas a more dopamine-selective compound, GBR 12909, fully substituted for the discriminative stimulus effects of cocaine. Despite the lack of substitution, it was reported that low doses of desipramine and fluoxetine, like low doses of GBR 12909, could enhance the discriminative stimulus effects of cocaine and reduce its ED_{50} to produce full substitution. These results confirm the importance of the dopamine transporter in the full transduction of cocainelike effects, but suggest that actions at other monoamine transporters can have a modulatory role [21].

Similar modulation of the discriminative stimulus effect of cocaine by other monoamine-uptake inhibitors has been described by others [22–26]. These studies are generally consistent with the conclusion that blockade of norepinephrine or serotonin reuptake cannot reproduce the discriminative stimulus effects of cocaine but can have a positive modulatory effect on compounds that do fully substitute.

Other investigators have attempted to understand the relationship between discriminative stimulus effects of cocaine and blockade of the dopamine transporter by examination of compounds where binding to the transporter can be resolved into high- and low-affinity sites. Katz and colleagues [27] examined a group of compounds for displacement of [^{3}H]WIN 35,428 binding in rat caudate–putamen as a measure of DAT affinity, and compared that with compound ED_{50} values for substitution for a 10 mg/kg dose of cocaine in rats. The compounds where binding was best fit by a two-site model included WIN 35,428, nomifensine, WIN 35,981, WIN 35,065-2, methylphenidate, cocaine, and bupropion, whereas the compounds where binding was best fit by a one-site model included RTI-31, RTI-55, RTI-121, RTI-32, GBR 12909, and mazindol. Of the compounds that fully substituted for cocaine, all had affinity for the dopamine transporter as measured by displacement of [^{3}H]WIN 35,428 binding. Whereas K_i (expressed at $K_{0.5}$ to include both one-site binding and the combination of high- and low-affinity sites) values were highly correlated with ED_{50} values for compounds binding to one site, K_{hi} and K_{lo} values were also correlated with ED_{50} values to produce full substitution for cocaine. When K_{hi} values were considered in an analysis that included $K_{0.5}$ values, there was also a significant relationship, in contrast to K_{lo} values in combination with $K_{0.5}$ values, which when analyzed together were not related to ED_{50} values for cocaine substitution. The authors conclude that because the relationship of discriminative stimulus effects was more strongly related to binding to the high-affinity site than the $K_{0.5}$ for one-site binding or binding to the low-affinity site, it would appear that discriminative stimulus effects of cocaine may be mediated by the high-affinity binding site to a greater extent than by the low-affinity site in cases where compounds bind to two sites [27].

As in studies of stimulant and discriminative stimulus effects, a number of studies have examined the self-administration of cocaine and other dopamine transporter ligands in an attempt to understand the function of the dopamine transporter with regard to the ability of dopaminergic compounds to act as reinforcers. The first and most important of these studies was a watershed event in the drug abuse research community because it definitively established the site of cocaine's action at the dopamine transporter [1].

In 1987, Ritz and colleagues [1] evaluated 18 compounds that bind to monoamine transporters and analyzed the K_i values for binding at dopamine, serotonin, and norepinephrine transporter sites in terms of their relative potency as compared to cocaine, where possible. Then, either from their own studies or from literature values, the relative behavioral potency of 11 of the 18 compounds that had been tested and were self-administered were compared to the relative potencies at the three transporter sites. Using multiple regression analysis to determine the contribution of binding to each of the monoamine transporters alone and in combination with the others, it was concluded that binding to the dopamine transporter was both necessary and sufficient for producing drug reinforcement or for maintaining drug self-administration [1]. Thus, binding to the DAT appeared to be the determinant of cocaine reinforcement.

Another set of studies using squirrel monkeys examined reinforcing effects of monoamine inhibitors as well as their behavioral stimulant effects in light of potency to inhibit binding of [^{3}H]cocaine in monkey caudate–putamen. Results examining WIN 35,065-2, nomifensine, cocaine, WIN 35,981, methylphenidate, GBR 12909, and bupropion were generally in agreement with those of Ritz et al. [1] in that the relative potencies of compounds to maintain self-administration behavior and to produce behavioral stimulant effects were generally related to the relative potency to displace [^{3}H]cocaine binding with only minimal exceptions [28].

Several other studies have examined self-administration of cocaine or other monoamine-uptake inhibitors in light of transporter occupancy. One study compared RTI-113, a long-acting analog of cocaine, on self-administration and DAT occupancy [29]. In this study, RTI-113 was substituted for cocaine in rhesus monkeys trained to self-administer cocaine, and at least one dose produced self-administration above saline levels. Like cocaine, the dose–effect function for RTI-113 could generally be described as an inverted U. In two of three monkeys, RTI-113 maintained rates of responding that were similar to those produced by cocaine, suggesting that they were equipotent in terms of efficacy for the production of reinforcing effects. DAT occupancy was measured in the same monkeys using position emission tomography (PET) imaging and [^{18}F]FECNT as a DAT label. Relative to cocaine, transporter occupancy was higher in the three monkeys tested, reaching 99% transporter occupancy at the maximum self-administered dose for two monkeys and 94% occupancy at the maximum self-administered dose for the third. In contrast, cocaine produced occupancy of 65%, 70%, and 76% in the three monkeys. When RTI-113 was administered as a pretreatment, it decreased responding for cocaine at transporter occupancies of 72 to 78%. The reason for the high levels of transporter occupancy to both maintain self-administration and to reduce cocaine self-administration are unclear, but are hypothesized to be related to the long duration of action of RTI-113 [29]. The limited number of compounds studied makes it impossible to draw conclusions as to the relationship between transporter occupancy and reinforcing effects.

Wee and colleagues [30] examined the effects of several cocaine analogs with a slow onset, including WIN 35,428, RTI-31, and RTI-51 for their effects on self-administration using a progressive-ratio schedule in rhesus monkeys. In this paradigm, in consideration of the slow rates of onset, a 1-hour timeout period between injections was used, and all compounds served as reinforcers with the following maximum number of injections: cocaine > WIN 35,428 > RTI-31 > RTI-51. In vivo binding studies were conducted using tissue from rat striatum, and in light of the self-administration data, equipotent doses of the compounds were administered and were found to displace approximately 25% of [^{3}H]WIN 35,428– labeled transporters. The time required to achieve approximately 25% displacement varied across compounds, with cocaine at 5.8 minutes, WIN 35,428 at 22.4 minutes, RTI-31 at 30.8 minutes, and RTI-51 at 44.1 minutes after i.v. injection. Importantly, using this type of analysis there was a positive relationship between relative reinforcing efficacy and faster displacement of [^{3}H]WIN 35,428 binding, suggesting that compounds with a slower onset will be less reinforcing [30].

Roberts and colleagues [31] examined a group of cocaine analogs including PTT to assess their ability to act as reinforcers in light of their selectivity for inhibiting dopamine vs. serotonin reuptake. In this study, rats were trained to self-administer cocaine on a progressive ratio schedule of reinforcement, and then a range of doses of seven different compounds were substituted for cocaine in different groups of rats to assess their reinforcing efficacy. Four of the compounds, WF11 (PTT), WF23, WF24, and WF55, were self-administered, whereas three compounds, WF31, WF54, and WF60, were not. Two of the compounds that were not self-administered had IC$_{50}$ values for displacing binding at the DAT of less than 20 nM and were fairly potent uptake inhibitors. The authors also constructed ratios of selectivity for binding and inhibiting uptake at the DAT relative to the SERT, and found that the strongest relationship was not between breakpoint for self-administration and affinity for the DAT, but with breakpoint and the selectivity ratio for DAT/SERT binding. The authors qualify their finding with restatement that only a small number of compounds were tested and that contributions of duration of action, bioavailability, lipophilicity, and absorption were not controlled in these studies. Nonetheless, the analysis highlights the importance of selectivity, and not just binding potency, as an important factor [31].

14.4 STUDIES OF SPECIFIC TRANSPORTER LIGANDS

Hundreds of DAT ligands have been synthesized for evaluation as potential "blockers" or "antagonists" of cocaine; and hundreds of others have been synthesized as potential "substitute" or "agonist" therapies that partially reproduce some of the effects of cocaine. The search for the "dopamine-sparing cocaine antagonist" was prompted by a great deal of intriguing work that suggested that cocaine and dopamine bind to distinct and dissociable sites on the transporter. As a result, most of the in vivo effects of DAT ligands are described with reference to cocaine, either by comparison to cocaine or by their ability to block the behavioral effects of cocaine.

There are few non-overlapping classifications of DAT inhibitors. In the following sections, compounds will be described either with reference to chemical class or as structural analogs, although to some extent these classifications may overlap. This chapter will focus by default on compounds for which there is the greatest amount of published data, which in most cases is the "parent" compound of a set of analogs.

14.4.1 GBR 12909

GBR 12909 has received a great deal of study both for being a prototype of an extensive series of analogs developed by a number of different medicinal chemists, and importantly because GBR 12909 itself long been under consideration as a potential cocaine therapeutic. The reasons for its serious consideration as a "substitute" medication for cocaine dependence are based on the results of extensive behavioral studies, which have highlighted its differences from the effects of cocaine in animal models.

GBR 12909 produces a dose-dependent and long-lasting stimulation of locomotor activity [17,32,33], unlike the effects of cocaine, which last only 30 minutes. These effects mirror the onset and duration of effects of GBR 12909 and cocaine on extracellular dopamine as measured using in vivo microdialysis in rats [34] and monkeys, where cocaine's effects peak in the first 10 minutes and last 40 to 60 minutes, whereas the effects of GBR 12909 on dopamine peak at 50 minutes postinjection and do not return to baseline by 120 minutes postinjection [35]. In addition to its acute locomotor stimulation, following subchronic treatment of one injection of 20 mg/kg every 2 days for 14 days in rats, there was evidence of sensitization to the stimulant effects of GBR 12909, which lasted up to 7 weeks following the treatment [32].

Chronic treatment with a continuous infusion of GBR 12909 or another selective dopamine uptake inhibitor RTI-117 produced different patterns of locomotor activity than cocaine, with peak effects occurring on the second day of treatment, followed by a significant decrease in effect on subsequent days. Cocaine, however, produced significant increases immediately upon initial exposure and levels of activity remained high for several days with repeated injections [36]. Unlike cocaine, which had no effect on dopamine transporter density, chronic treatment with GBR 12909 produced significant decreases in [3H]WIN 35,428 binding in the caudate–putamen and nucleus accumbens, and these effects persisted for up to 20 days [37]. This suggests that GBR 12909 and cocaine may regulate the DAT in different ways.

In rats and monkeys, GBR 12909 produces full dose-dependent substitution for the discriminative stimulus effects of cocaine [19,21,38–41]. In most of these studies, the potency of GBR 12909 was only slightly lower than that of cocaine.

Like cocaine, GBR 12909 maintains i.v. self-administration in rats and monkeys trained to self-administer cocaine. In rats, GBR 12909 was self-administered similarly to cocaine using both fixed-ratio (FR) and progressive-ratio (PR) schedules [42]. Breakpoints on the PR schedule were similar to those of cocaine, but on the FR schedule, the inter-infusion interval was directly related to dose and was approximately three times longer than that established previously for equivalent doses of cocaine [42], consistent with GBR 12909's longer duration of action. Tella and colleagues [43] have reported that whereas GBR 12909 maintains self-administration behavior for a few weeks as reported by other investigators, when examined over a longer period of time, it fails to maintain self-administration as consistently as cocaine [43]. In squirrel monkeys trained to self-administer GBR 12909 and cocaine under a second-order fixed-interval schedule, responding was maintained, but cocaine was approximately threefold more potent that GBR 12909 [44]. Similarly, results from a different laboratory using similar techniques in squirrel monkeys reported that GBR 12909 was robustly self-administered but was approximately four times less potent than cocaine [45]. In rhesus monkeys, increasing doses of cocaine or GBR 12909 per injection led to increased rates of lever pressing, but GBR 12909 produced lower maximum rates than cocaine, and this occurred following higher doses of GBR 12909 than of cocaine [46]. Stafford and colleagues [47] examined the self-administration of GBR 12909 in rhesus monkeys under a progressive ratio schedule. They reported more variability across subjects than typically seen for cocaine self-administration, but reported robust and reliable self-administration similar to cocaine in two out of four monkeys that were tested [47]. Pretreatment with GBR 12909 resulted in dose-dependent decreases in cocaine maintained responding for high doses of cocaine, but these decreases were associated with rate decreases rather than increased length of the postreinforcement pause, suggesting that the reinforcing efficacy of cocaine was not diminished [46].

A great deal of attention has been focused on GBR 12909's ability to reduce cocaine self-administration in rats and monkeys. In rats trained to self-administer cocaine, GBR 12909 dose-dependently produced a downward shift in the descending limb of the cocaine dose–effect curve. When examined as a function of time, GBR 12909 at doses of 3.0 and of 5.6 mg/kg produce significant and dose-dependent decreases in the number of infusions of 1 mg/kg cocaine. This was in contrast to more serotonin- and norepinephrine-selective uptake inhibitors, which had little effect on cocaine self-administration in this paradigm [48].

In rhesus monkeys, a pivotal series of studies suggested that pretreatment with GBR 12909 could decrease responding for various doses of cocaine without affecting responding for food reinforcers [49]. In these studies, a number of experimental manipulations were incorporated to control for effects on rate and effects of cocaine. In the first experiment, pretreatment with GBR 12909 at doses of 1.7 to 3.0 mg/kg decreased rates of cocaine self-administration while having minimal effects on rates of responding for food. In a second experiment, a dose of 1.7 mg/kg GBR 12909 was evaluated for effects on three i.v. doses of cocaine and was found to decrease rates of responding at doses ranging from 10 to 100 µg/kg per infusion. These doses were on the descending limb of the cocaine dose–effect curve, and there was little effect on response rates maintained by food reinforcers in animals self-administering these doses of cocaine. An additional study was designed to control for rate-dependent effects of drugs by manipulating the cocaine dose to produce comparable response rates for food and cocaine. Similarly, in these studies GBR 12909 showed a selective effect to decrease rates of responding for doses of 5.6 to 10 µg/kg per injection of cocaine, with only minimal effects on response rates for food at the highest dose (3.0 mg/kg) of GBR 12909. When animals self-administered a high unit dose of cocaine (56 µg/kg per injection), GBR 12909 still reduced cocaine self-administration, but under these experimental conditions, response rates for food reinforcers also were decreased. In a final experiment, GBR 12909 was shown to decrease rates of responding across the dose–effect curve of cocaine, with little effect on responding for food [49].

Based on experiments where 1.0 to 3.0 mg/kg GBR 12909 selectively reduced responding for cocaine vs. a food reinforcer, PET studies were conducted to determine the degree of DAT occupancy required to produce this effect. In these experiments, the DAT was labeled with [^{11}C]WIN 35,428 in two baboons that underwent a series of PET scans following saline, 1.0, 3.0, and 10.0 mg/kg GBR 12909. The results indicated that the doses of GBR 12909 that produced significant reductions in cocaine self-administration also produced a high degree of DAT occupancy. The authors concluded that at least 70% occupancy was needed to reduce cocaine self-administration [50], and suggested that these results be used to guide human dose requirements of GBR 12909 [51].

The development of GBR 12909 as a potential cocaine treatment medication was halted following phase 1 human safety studies conducted by the National Institute on Drug Abuse (NIDA). Two studies were conducted in which cardiovascular effects were deemed problematic by the Food and Drug Administration (FDA) and in need of further assessment.

14.4.2 Cocaine Analogs

This categorization of dopamine uptake inhibitors is broad and inclusive of compounds from different structural series and compounds synthesized by different chemists. It will focus on the more familiar compounds for which there are the greatest number of source documents, which include PTT, WIN 35,428, and RTI-336.

PTT   2β-Propanoyl-3β-(4-tolyl)tropane, otherwise known as PTT or WF-11, is one of the more familiar exemplars of this category of uptake inhibitors and about which a great deal of literature has been generated. PTT was hypothesized to have greater metabolic stability than cocaine by removal of both ester moieties, which would give it a longer duration of action in vivo. The locomotor stimulant effects of PTT were first compared to those of cocaine following i.p. injection in habituated Fischer rats. The highest dose of PTT (3.0 mg/kg) produced over twice as much horizontal activity as the highest dose of cocaine (30 mg/kg), leading to the conclusion that PTT was 10 to 20 times more potent than cocaine [50]. PTT also produced more than three times the stereotypic behavior produced by cocaine. In time-course studies, the locomotor response to cocaine reached a maximum at 30 minutes and returned to baseline levels within 3 hours. In contrast, the locomotor response to PTT reached a maximum at 90 minutes and was still significantly greater than saline control at 5 hours, presumably because of increased bioavailability due to its greater metabolic stability compared to cocaine. However, it was also noted that PTT produced intense and distinctive patterns of stereotypy not observed with cocaine [50].

Subsequent studies in the same laboratory confirmed and extended these results by demonstrating a similar locomotor effect of PTT following i.v. administration, with the addition of an analysis of cerebral metabolism using 2-[^{14}C]deoxyglucose methods to measure local cerebral glucose metabolism. Results showed widespread elevations in glucose utilization in portions of the mesocorticolimbic and nigrostriatal regions, with alterations in the hippocampus, locus coeruleus, and dorsal raphe noted. One unique finding was that a low dose of PTT (0.1 mg/kg) decreased metabolic activity in the nucleus accumbens and olfactory tubercle, a finding that distinguishes PTT from cocaine and other stimulants, such as methylphenidate and amphetamine [52].

Another study evaluated the locomotor stimulant effects of both cocaine and PTT and correlated the effects with measurements of extracellular dopamine in the nucleus accumbens of adult male Fischer rats. These studies replicated the 20- to 30-fold potency difference for the stimulation of locomotor activity and found that the dynamics of extracellular dopamine also differed. PTT was approximately 30-fold more potent than cocaine in terms of the maximal increase and differed in the time course. PTT reached a maximal effect at 20 to 30 minutes post i.p. administration, whereas cocaine effects peaked at 10 minutes; further, the effects of PTT persisted for 6 hours, as compared to a 2-hour effect for cocaine [53].

In drug discrimination tests in rhesus monkeys, PTT fully substituted for the discriminative stimulus effects of cocaine and was approximately 10-fold more potent. In addition, the effects of PTT persisted for approximately 8 hours [54].

The first studies examining the reinforcing effects of PTT and its effects on cocaine self-administration were done in rats trained to self-administer a dose of 0.33 mg/kg per infusion of cocaine in 3-hour sessions using an FR-10 schedule. In these studies a dose of 0.03 mg/kg per infusion of PTT maintained self-administration behavior measured as number of infusions in rats that did not differ from cocaine control, suggesting that under these experimental conditions, PTT has a reinforcing efficacy similar to that of cocaine [55]. However, it was observed that the temporal pattern of responding maintained by PTT involved low-rate responding followed by rapid responding 2 hours into the session, compared to the extremely regular rates of control responding engendered by 0.33 mg/kg per infusion of cocaine. It was also noted that higher doses of PTT were less effective in maintaining self-administration than the dose of 0.03 mg/kg per infusion [55].

Effects of pretreatment doses of 1.0 and 3.0 mg/kg i.p. PTT on cocaine self-administration were also evaluated. There was a dose-dependent effect to decrease cocaine self-administration during the first 2 hours of the session, which used an FR-10, 20-second timeout procedure. In a separate study, the effects of PTT on responding for food were evaluated, and both cocaine and the higher dose of PTT significantly reduced responding for food under these experimental conditions [55].

Another study in rats was conducted using a progressive ratio schedule, and in this study, PTT was found to be robustly self-administered at breakpoints as high as those reached by rats self-administering cocaine, and the most efficacious dose of PTT to maintain high breakpoints was 0.3 mg/kg per infusion. However, it was noted that the time course of effects differed, with the maximal ratio for self-administration reached approximately 15 hours into the experimental session. In contrast, maximal breakpoints for cocaine are reached within 2 hours [31].

A number of additional studies have examined PTT self-administration as well as the effects of PTT on cocaine self-administration in rhesus monkeys. The effects of PTT on cocaine self-administration were examined using two different doses of cocaine and three different pretreatment doses of PTT [54]. In these studies, which used an FI-5 schedule of reinforcement over a 4-hour session, PTT at doses of 0.1 and 0.3 mg/kg were found to significantly decrease self-administration rates and the total intake of two infusion doses of cocaine, 0.03 and 0.1 mg/kg per infusion. In addition, PTT itself was shown to be ineffective as a reinforcer at doses ranging from 0.001 to 0.1 mg/kg, and maintained response rates that did not differ from those of saline [54].

Another study from the same laboratory was conducted to determine the effects of a fixed-ratio (FR) reinstatement schedule and various timeout periods (TOs) on the reinforcing effects of PTT in rhesus monkeys [56]. These studies, which again administered doses of 0.003 to 0.03 mg/kg per infusion of PTT, also concluded that while an infusion dose of 0.1 mg/kg of PTT maintained some degree of self-administration behavior that was significantly greater than that of saline, response rates were also significantly lower than rates maintained by cocaine. PTT was, however, more potent than cocaine in decreasing response rates for food [56].

Another series of studies from the same laboratory examined the reinforcing effects of PTT in naive and cocaine-experienced rhesus monkeys using an FR-30 schedule of food reinforcement and a 3-hour session. In these studies, first saline was substituted, then PTT, cocaine, and then PTT again, to determine effects on rates of responding. Neither saline nor PTT at doses of 0.001 to 0.03 mg/kg per infusion were self-administered before or after substitution with cocaine, which, as expected, produced robust self-administration at doses from 0.0003 to 0.3 mg/kg per infusion [57]. When sessions were extended to 22 hours in duration, PTT doses of 0.001 to 0.03 mg/kg per infusion maintained self-administration in three of four monkeys that had a brief history of cocaine self-administration. The maximum number of reinforcers earned at the peak of the PTT dose–effect function was less than the number of reinforcers earned for 0.03 mg/kg per infusion of cocaine [57].

Additional studies in rhesus monkeys examined the relative reinforcing effects of PTT compared to cocaine under progressive-ratio self-administration and under choice procedures. While PTT was self-administered under PR schedules, cocaine availability produced a greater number of injections and resulted in higher breakpoints than did PTT availability. In addition, the dose of PTT that maintained the highest breakpoints, 0.01 mg/kg per injection, was not preferred over cocaine [58].

In one of the few instances in which dopamine-uptake inhibitors have been tested in reinstatement procedures, it was reported that when using a within-subject experimental design in rhesus monkeys trained to self-administer cocaine, (−)PTT at doses ranging from 0.001 to 0.1 mg/kg robustly reinstated cocaine self-administration and was approximately 10-fold more potent than cocaine itself in triggering reinstatement [59]. Doses producing reinstatement were lower than those previously reported to reduce cocaine self-administration.

CFT (WIN 35,428) and CPT (WIN 35,065)   Analogs of cocaine were synthesized by Sterling Winthrop Research Institute in the early 1970s in the hope of finding a useful stimulant or antidepressant [60]. Some of the most potent behavioral stimulants that were discovered at this time were later known as WIN 35,428 and WIN 35,065. These compounds were 64 and 16 times as active as cocaine, respectively, in terms of their potency to increase locomotor activity in mice, and WIN 35,428 was described as lasting for more than 5 hours. Like cocaine, both compounds were reported to reverse reserpine-induced eyelid ptosis in mice, in addition to producing running, jumping, biting, and squeaking. In addition, WIN 35,428 was reported to have only about 15% of the local anesthetic activity of cocaine in an intradermal test in guinea pigs [60].

These compounds were characterized and compared to cocaine in several tests using schedule-controlled behavior in squirrel monkeys. In these studies, one group of monkeys was trained in a multiple fixed-interval fixed-ratio (FR = 30) schedule of stimulus-shock termination. Another group of monkeys was trained on a multiple FI FR schedule of food presentation, and the last group was trained on a fixed-ratio (FR = 30) schedule of food presentation. After training was complete and responding had stabilized, drug experiments were conducted to compare the effects of cocaine, WIN 35,428, and WIN 35,065-2 (the active enantiomer of WIN 35,065).

The results of these studies indicated that although the three compounds affected schedule-controlled behavior in a qualitatively similar manner, they could be distinguished by differences in potency, in the onset of action, and in the duration of effects. In all measures evaluated, the dose–effect functions of the WIN analogs compared to cocaine suggested that they were considerably more potent than cocaine by 3- to 10-fold. In addition, under the multiple schedule of stimulus shock termination, the effects of cocaine were evident within 10 to 14 minutes, whereas in contrast, the maximal effects of the two WIN analogs were not apparent until about 40 to 45 minutes after injection [61]. Differences in duration of effect were measured by altering pretreatment times in the FR schedule of food presentation. Cocaine is typically administered only 5 minutes before the session and is ineffective in decreasing rates of responding when administered more than 120 minutes before the session. In contrast, decreases in responding following administration of the WIN analogs was evident following a 360-minute pretreatment interval, suggesting a much longer behavioral effect. Overall, this early paper documented a cocainelike behavioral profile, with differences in potency and time course [61].

Another study examined the same two WIN analogs, WIN 35,428 and WIN 35,065, on locomotor activity in mice and rotational behavior in rats with 6-hydroxy-dopamine lesions of the substantia nigra. WIN 35,428 and WIN 35,065 were confirmed to be potent stimulants of locomotor activity and were more similar in potency to d-amphetamine as compared to cocaine. In addition, the stimulant effects could be dose-dependently blocked by the postsynaptic receptor antagonists haloperidol and pimozide. In terms of rotational behavior in lesioned rats, both WIN analogs were similarly potent compared to d-amphetamine and were more potent than cocaine, nomifensine, or mazindol. In addition, rotational behavior could be blocked by administration of haloperidol [62].

In other tests of locomotor activity, WIN 35,428 was found to stimulate activity in rats with an ED_{50} of 0.416 mg/kg [16], in CD-1 mice with an ED_{50} of 1.55 μmol/kg [14] and was reported to stimulate activity in BALB/cBy at a dose of 5.0 μmol/kg [63]. In this study, WIN 35,065-2 also produced locomotor stimulation at a dose of 17 μmol/kg and was more potent than cocaine [63].

In a subsequent study examining the ability of WIN 35,065-2 to produce locomotor sensitization in BALB/cBy mice, i.p. injections of WIN 35,065-2 and cocaine were given for periods of 3 or 4 days [64]. Although both compounds produced sensitization on day 3, the doses that had seemed comparable for degree of stimulation, 6.23 mg/kg WIN 35,065-2 and 25 mg/kg cocaine, were not comparable in their effects on sensitization on day 4. Subsequently, dose–effect curves were established for sensitization on day 3, and whereas both compounds produced an inverted V-shaped function of dose, the peak dose of WIN 35,065-2 was lower than expected. While a dose of 25 mg/kg cocaine was optimal for producing sensitization as measured on day 3, a dose of 3.1 mg/kg WIN 35,065-2 produced maximal locomotor sensitization. It is not known if the effects of higher doses of both compounds that did not result in sensitization were the result of tolerance to locomotor effects, sensitization of stereotyped nonambulatory behavior, or other unknown effects, but results emphasized the importance of dose in the determination of sensitization in cocaine analogs [64].

It was also reported that WIN 35,065-2, like cocaine, was unable to produce behavioral sensitization to its locomotor stimulant effects when injected directly into the ventral tegmental area. In these studies, rats receiving four daily injections of cocaine (1.5, 5.0, or 15 nmol/side into the VTA did not show sensitization to a challenge dose of 15 mg/kg i.p. cocaine one week later, but repeated injection of another dopamine uptake inhibitor, GBR 12935 (15 nmol/side), did produce sensitization to cocaine. In contrast, WIN 35,065-2 (15 nmol/side) was more similar to cocaine itself in that it did not produce sensitization to a challenge dose of cocaine [65].

WIN 35,428 also was evaluated and compared to cocaine for its ability to produce a conditioned taste aversion (CTA) in rats, in a test of the hypothesis that duration of action is related to the production of conditioned taste aversion in rodents [66]. It was also tested and compared to cocaine for its effects on fixed-interval responding in rats. These studies used male hooded rats, and in the CTA experiments, flavored drinking solutions were paired with drug injections or vehicle for a total of four pairings each. Two days later, animals were tested with both flavored solutions present. Results indicated that WIN 35,428 and cocaine produced the same degree of conditioned taste aversion as measured by the mean fluid intake of the paired solutions, but WIN 35,428 was more potent than cocaine, with a dose of 3.2 μmol/kg WIN 35,428 producing an effect similar to cocaine at a dose of 100 μmol/kg. In addition, the effect did not appear to be related to duration of drug action, since effects of WIN 35,428 to reduce rates of responding were prolonged (longer than 150 minutes) compared to cocaine, effects of which returned to control levels after 90 minutes [66].

In tests of drug discrimination, WIN 35,428 produced full substitution for the discriminative stimulus effects of cocaine in rats [40,67,68], and in squirrel monkeys, where it was found to be 3.5 times more potent than cocaine [69]. In this study,

WIN 35,428 (CFT) was compared to both cocaine and CCT (also known as RTI-31), which was similarly found to be 5.4 times more potent than cocaine [69]. In pigeons trained to discriminate cocaine, both WIN 35,428 and WIN 35,065-2 produced full substitution for the discriminative stimulus effects of cocaine and were both more potent than cocaine itself [70].

Finally, WIN 35,428 (CFT) was tested for its ability to maintain self-administration in squirrel monkeys. It was found to be robustly self-administered at rates that were similar to those of cocaine, and at doses that were lower, consistent with its higher in vitro potency and greater potency in other behavioral measures [71]. WIN 35,428 (CFT) was concluded to be approximately six-fold more potent than cocaine [72]. Other WIN analogs tested for their ability to maintain self-administration in squirrel monkeys were tested under a 5-minute fixed interval (FI) schedule of i.v. drug injection. Both WIN 35,065-2 (tested at doses from 0.6 to 64.3 µg/kg per injection) and WIN 35,981 (tested at doses from 0.8 to 245.5 µg/kg per injection) were found to be as effective as cocaine in maintaining responding in squirrel monkeys under the FI schedule. However WIN 35,065-2 was found to be 3 to 10 times more potent than cocaine, and both cocaine and 35,065-2 were 3 to 10 times more potent than WIN 35,981. The enantiomer of WIN 35,065-2, which is WIN 35,065-3, failed to maintain responding at any dose tested (0.6-6427.7 µg/kg per injection) [71].

WIN 35,428 was also tested for its ability to maintain cocaine self-administration in rhesus monkeys. It was compared to cocaine and "slow-onset" cocaine analogs, RTI-31 and RTI-51, and correlated self-administration with rates of in vivo binding to striatal tissues in rats. In monkeys trained under a progressive ratio schedule with a 1-hour timeout between injections, all compounds functioned as positive reinforcers, with the maximum number of injections produced by cocaine > WIN 35,428 > RTI-31 > RTI-51. In these studies, the relative reinforcing efficacy of the compounds tested was negatively correlated with rate of DAT binding, such that slower displacement of [^{3}H]WIN 35,428 from striatum was correlated with decreased reinforcing efficacy [30].

There is one study examining the effects of pretreatment of WIN 35,428 on cocaine self-administration in rats. In this study, male Sprague–Dawley rats were trained to self-administer cocaine on an FR-1 schedule, and response requirements were gradually increased to FR-5 [73]. Testing was conducted when responding had reached stable levels during daily 2-hour sessions using a training dose of 0.5 mg/kg per infusion of cocaine. Rats were pretreated with doses of 0.1 to 1.0 mg/kg WIN 35,428 with a pretreatment time of 30 minutes and effects on the entire inverted-U-shaped dose–effect curve of cocaine were examined. WIN 35,428 decreased responding maintained by intermediate and high doses of cocaine, and slightly increased responding maintained by lower doses. Using a reinstatement paradigm, WIN 35,428 had a cocainelike priming effect to reinstate previously extinguished responding. In addition, WIN 35,428 increased the ability of low doses of cocaine to reinstate responding, suggesting additive effects through the same mechanism [73].

Several of the WIN analogs, including WIN 35,065-2, WIN 35,428 (CFT), and RTI-55, have been evaluated using in vivo binding and imaging procedures. Time course and regional binding of [^{3}H]WIN 35,065-3 and [^{3}H]WIN 35,428 were evaluated following i.v. injection in mice. Regional accumulation of both ligands peaked at approximately 60 minutes, was displaceable by known inhibitors of DAT, and ED$_{50}$ values for displacement of [^{3}H]WIN 35,428 by DAT uptake inhibitors correlated well with in vitro IC$_{50}$ values [67]. In imaging studies in baboons using [^{123}I]RTI-55 and single-photon emission computed tomography (SPECT), WIN 35,428 and cocaine caused a rapid dose-dependent displacement of both striatal and hypothalamic binding [67].

Reith et al. [74] studied the effects of cocaine, WIN 35,428, WIN 35,065-2, WIN 35,140, and WIN 35,065-3 on sniffing and biting in BALB/cBy mice using intracerebroventicular drug administration. Results were analyzed to produce a sniffing threshold that was positively correlated with inhibition of cocaine binding in mouse striatum, but results for each compound were not described individually [74].

RTI 55 (β-CIT) and RTI 31 (CCT)   Another series of compounds that are closely related to cocaine and the WIN analogs are RTI-55 (β-CIT) and RTI-31 (CCT), as well as a number of Research Triangle Institute (RTI) compounds first described by Boja and colleagues [75,76]. The first behavioral analysis of these compounds attempted to correlate potency to stimulate locomotor activity with accumulation of [^{3}H]WIN 35,428 in striatum. In these studies, CD-1 mice were first habituated to a locomotor activity arena and then injected i.p. with doses of saline, WIN 35,428, RTI-31 (CCT), RTI-32, RTI-51, or RTI-55 (β-CIT), and then tested for 40 minutes using cumulative dosing techniques. For in vivo binding, mice were injected i.v. in the tail vein with test compounds, and then injected with [^{3}H]WIN 35,428 [14]. Mice were later sacrificed and brain tissues analyzed for radioactivity content in striatum vs. cerebellum. Binding was characterized as a ratio of the two tissues and presented as percent of control of [^{3}H]WIN 35,428 binding. All compounds were locomotor stimulants that were approximately 20-fold more potent than cocaine, and did not differ from each other, with ED$_{50}$ values ranging from 0.68 μmol/kg for RTI-51 to 2.18 μmol/kg for RTI-32, and all were more potent than cocaine in inhibiting [^{3}H]WIN 35,428 binding, with approximately the same rank order of potency [14]. It was also noted that all compounds other than cocaine and RTI-55 produced decreased locomotor activity and increased stereotypy at the highest doses, although it was hypothesized that the same effects would have occurred for cocaine and RTI-55 if the doses had been high enough. Results suggested that peak effects on locomotor activity occurred at nearly complete occupancy of transporter sites, based on doses producing maximal effects for each measure [14].

In behavioral studies, RTI-55 was found to stimulate locomotor activity with greater potency than cocaine, WIN 35,428, and RTI-32. It was slightly less potent than RTI-51 and RTI-31, but other than cocaine, the compounds were not statistically different from each other [14]. As mentioned above, these compounds showed the same rank order of potency in stimulating locomotor activity and inhibition of [^{3}H]WIN 35,428 accumulation [14]. In another study, RTI-55, RTI-121, and RTI-130 were compared to cocaine for their ability to stimulate locomotor activity in mice and to bind to the DAT. RTI-55, RTI-121, and RTI-130 were all more potent that cocaine in binding to the dopamine transporter and inhibiting its function. In addition, they were more potent and longer-lasting than cocaine in stimulating locomotor activity in mice. Relative potencies were 55-, 13-, and 31-fold greater than cocaine for RTI-55, RTI-121, and RTI-130, respectively, and effects lasted up to 10 hours, whereas cocaine effects typically last only 2 hours. Stereotypy occurred at the highest doses of RTI-55, RTI-121, and RTI-130 [77].

RTI-55, RTI-31, RTI-32, and WIN-35,428 were also tested for their ability to substitute for the discriminative stimulus effects of cocaine in rats trained to discriminate cocaine injections from saline. All of the compounds were found to produce full, dose-dependent substitution for the discriminative stimulus effects of cocaine in rats [68]. In a separate study, RTI-15 was also reported to fully substitute for the discriminative stimulus effects of cocaine in rats, although unlike a number of other cocaine analogs, it did not stimulate locomotor activity [78].

RTI-55 was tested for its ability to substitute for the discriminative stimulus effects of cocaine in two rhesus monkeys trained to discriminate doses of intravenous cocaine from saline. This study also used a time-course paradigm to determine the duration of discriminative stimulus effects. It was reported that at a pretreatment time of 10 minutes, doses of 0.012 and 0.025 mg/kg produced full substitution for the training dose of cocaine, and that these effects lasted for 4 hours and continued to produce responses on the cocaine-paired lever. When the pretreatment time was increased to 8 hours, a dose of 0.05 mg/kg was required to produce 100% cocaine lever responses. The authors concluded that RTI-55 was approximately eightfold more potent than cocaine in producing a cocainelike discriminative stimulus effect [79].

RTI-55 also was tested for its ability to maintain self-administration in rhesus monkeys trained to self-administer 0.03 mg/kg per injection of cocaine on an FR-10 schedule. In this study, RTI-55 maintained responding above saline levels at doses of 0.0007 and 0.0015 mg/kg per injection of RTI-55 and was over 10-fold more potent than cocaine. The authors concluded that RTI-55 is cocainelike in the production of discriminative stimulus effects and that it functioned as a positive reinforcer in monkeys [79].

RT1 55 (β-CIT) was also studied using in vivo binding techniques to determine its suitability as a radiolabeled ligand for the dopamine transporter. One of these studies examined distribution of binding sites in various regions of the mouse brain. Results indicated that [^{125}I]RTI-55 binds to dopamine transporters with high uptake into the brain. Specific binding in striatum increased up to 8 hours, and decreased by 24 hours. Further, binding in striatum and olfactory tubercles was saturable with increasing doses of unlabeled RTI-55, and was inhibited by increasing doses of WIN 35,428 and RTI-31. [^{125}I]RTI-55 also labeled serotonin transporters in other regions of the brain such as the hypothalamus, and it was concluded that RTI-55 is a useful tool for labeling both sites [80].

RTI-55 was prepared as [^{123}I]β-CIT to evaluate it as a SPECT radiotracer in baboons. These studies showed the highest activities in striatal regions that peaked at 179 minutes and lasted up to another 190 minutes ($t_{1/2} = 27$ hours). The second-highest activity was seen in the hypothalamus, which peaked at 45 minutes and washed out more rapidly ($t_{1/2} = 294$ minutes). Cocaine and CFT both produced rapid displacement of $[^{123}I]\beta$-CIT from both sites, in contrast to citalopram, which binds only to serotonin transporters and displaced 57% of hypothalamic binding but only 5% of striatal activity. These results are consistent with the conclusion that $[^{123}I]\beta$-CIT is labeling both dopamine and serotonin reuptake sites [81]. A similar study examined distribution of $[^{123}I]\beta$-CIT following intravenous injection of four baboons, again confirming the labeling of both dopamine and serotonin transporters in nonhuman primate brain [82].

Another primate study conducted using the same ligand, $[^{123}I]\beta$-CIT, was done to characterize the pharmacokinetics, distribution, dosimetry, and regional brain distribution as a foundation for quantitative SPECT imaging of the dopamine and serotonin transporters in vivo. These studies concluded that $[^{123}I]\beta$-CIT was labeling regions with high densities of dopamine and serotonin transporters. In addition, two metabolites appeared with 30 minutes in both baboons and vervet monkeys, but they were not further characterized. Overall, the favorable radiation dosimetry, high brain uptake, and specific uptake in appropriate brain regions suggested that $[^{123}I]\beta$-CIT had promise as a SPECT imaging agent, with potential use in the noninvasive measurement of dopaminergic terminal activity in neurological conditions such as Parkinson's disease [83].

A number of other analogs of the phenyltropane series have been prepared and characterized in locomotor activity, including RTI-31, RTI-98, and RTI-113 [84], RTI-51, RTI-55, RTI-108, RTI-112, RTI-113, RTI-116, RTI-120, RTI-121, RTI-126, RTI-139, RTI-141, RTI-150, RTI-171, RTI-177, RTI-199, RTI-204, and RTI-219 [85]. All of the compounds named above were more potent than cocaine, and several were longer acting [85]. In addition, several other papers have described related series [86–89], none of which will be described further, with the exception of one compound, RTI-336 [90].

The time course of stimulant effects of RTI-336 was determined in squirrel monkeys trained under a fixed-interval (FI), 300-second order schedule of stimulus termination. A dose of RTI-336 (0.3, 1.0, and 1.7 mg/kg), cocaine (0.3 mg/kg), or saline was administered i.v. 5 seconds before the experimental session. This time-course study indicated that RTI-336 at doses of 0.3 and 1.0 mg/kg had a relatively long duration of action relative to a dose of 0.3 mg/kg cocaine. Response rates were significantly different from saline at 1.0 mg/kg, and a dose–effect determination produced a U-shaped function for stimulation of behavior that is typical of psychostimulants [91].

RTI-336 was evaluated in a drug discrimination procedure where male CD albino rats (Sprague–Dawley derived) were trained to discriminate 10 mg/kg cocaine from saline in a two-lever food-reinforced procedure. RTI-336 was administered at doses ranging from 1 to 10 mg/kg i.p. in rats trained under a fixed-ratio 10 (FR-10) procedure, and results were calculated as the percentage of rats choosing the cocaine-associated lever for the first fixed ratio. Full substitution was achieved at a dose of 10 mg/kg RTI-336, with 88% of rats choosing the cocaine-associated lever and completing the first FR, resulting in an ED_{50} of 5.6 mg/kg. RTI-336 was also tested in a time-course paradigm following oral administration of doses of 2.5 to 10 mg/kg. Full substitution was measured following 10 mg/kg at 45, 90, and

180 minutes after administration. The ED_{50}, calculated using the first period in which full substitution was measured, was 3.54 to 5.8 mg/kg, depending on the vehicle used [86,92]. These results are consistent with oral bioavailability and a long duration of action, from 45 to 180 minutes, in producing a full generalization effect.

To determine the effects of RTI-336 on cocaine self-administration, male Sprague–Dawley rats were trained to self-administer cocaine in intravenous infusions of 0.5 mg/kg per infusion of cocaine, using a final ratio requirement of FR-5. The effects of RTI-336 were evaluated in six rats self-administering the training dose of 0.5 mg/kg per infusion, and in a second group of six that were self-administering a lower dose of 0.25 mg/kg per infusion. Rats were pretreated with RTI-336 by oral gavage 15 minutes before the start of the self-administration sessions, to determine the effects of RTI-336 on cocaine self-administration. Doses of RTI-336 ranged from 5.4 to 53.8 mg/kg and dose-dependently reduced cocaine self-administration for both the 0.25- and 0.5-mg/kg cocaine infusions. Statistically significant reductions of cocaine self-administration were observed at the highest doses of RTI-336, suggesting that RTI-336 can reduce cocaine self-administration in rats [92].

Effects of RTI-336 on cocaine self-administration in squirrel monkeys was determined in monkeys initially trained under a second-order schedule of stimulus termination. Subsequently, the method of reinforcement was changed such that the termination resulted in the injection of cocaine. After establishing a full dose–effect curve for cocaine self-administration, various doses of RTI-336 were evaluated. The effects of RTI-336 on responding maintained on a second-order schedule of cocaine self-administration indicate that RTI-336 reduced response rates significantly compared to cocaine alone. Response rates for RTI-336 did not differ from rates of responding for saline. These results were consistent with reduced reinforcing efficacy for RTI-336 compared to cocaine [13].

The effect of pretreatment with RTI-336 on cocaine self-administration was also characterized in four rhesus monkeys trained to self-administer cocaine under a multiple schedule of i.v. drug or food delivery. Doses of 0.1 and 0.3 mg/kg per infusion of cocaine were selected for drug pretreatment experiments because they were positioned on the peak and descending limb of the dose–effect curve, respectively. RTI-336 produced dose-dependent reduction in cocaine self-administration in all subjects and was equally effective in reducing self-administration of the 0.1- and 0.3-mg/kg doses of cocaine. The mean ED_{50} dose of RTI-336 for pooled data from both groups was 1.07 ± 0.17 mg/kg [93]. Food-maintained behavior was suppressed at the same dose of RTI-336 that suppressed cocaine-maintained behavior [93].

RTI-336 was evaluated for its ability to maintain self-administration in monkeys trained to self-administer cocaine. In these studies, four rhesus monkeys were trained to self-administer cocaine, and each subject was allowed to self-administer several doses of RTI-336 in random order to establish a dose–effect curve. Both cocaine and RTI-336 reliably maintained drug self-administration at levels greater than saline in all subjects. The dose–effect curves could be described as the inverted-U-shaped function typical of self-administration curves for psychomotor stimulants, but rates of responding for RTI-336 were lower than those produced by cocaine across a range of doses for most subjects. The unit dose of RTI-336 that maintained peak rates of responding, which could be construed as the maximally reinforcing dose for the group, was 0.10 ± 0.06 mg/kg per injection [93].

In vivo microdialysis studies were conducted to examine the effects of RTI-336 compared to cocaine on dopamine dynamics in the caudate of squirrel monkeys. Results indicated that cocaine produced an immediate increase of dopamine, which was cleared completely in approximately 70 minutes. Dopamine levels peaked about 40 minutes after administration of RTI-336 and were present throughout the experiments. Thus, a dose of 1 mg/kg RTI-336 produced changes in dopamine dynamics that differed from a dose of 1 mg/kg cocaine, with a slower onset of increase in extracellular dopamine and a longer duration of action than cocaine [91].

Dopamine transporter occupancy by RTI 336 was determined using PET imaging in monkeys used in self-administration studies to determine the level of DAT occupancy associated with behaviorally relevant doses of RTI-336. Results indicated that percent DAT occupancy at the ED_{50} dose for RTI-336 that reduced cocaine self-administration was 90 ± 5.3. When the dose of RTI-336 that maintained peak responding was administered as an i.v. bolus of the total dose each animal received, the percent DAT occupancy was 62 ± 12 [93]. Peak plasma concentrations of RTI 336 were measured and were observed at 0.5 to 1.0 hour postinjection. At 2 hours postinjection, plasma concentrations were approximately 70% of peak values for the group [93].

To characterize the in vivo rate of uptake and clearance from the brain, both cocaine and RTI-336 were labeled with the PET isotope, [^{11}C]. PET neuroimaging studies were conducted in three conscious female rhesus monkeys, and effects were measured for a period of 90 minutes after single i.v. bolus infusions. The primary region of interest was the putamen, with the cerebellum analyzed as a reference. Cocaine reached peak brain concentrations within the first 5 to 10 minutes following i.v. administration, with an average peak time of 12.5 ± 0.0 minutes. Cocaine concentrations dropped markedly within 40 to 50 minutes. In contrast, uptake of RTI-336 continued over the 90-minute period of image acquisition with an average time to peak concentration of 66.7 ± 6.7 minutes and with sustained levels for the entire duration of the 90-minute session. The pharmacokinetic profile of RTI-336 differed from that of cocaine and was hypothesized to play a role in limiting the reinforcing efficacy of RTI-336 [93].

14.4.3 Benztropine and Analogs

Benztropine is a drug that has been in clinical use for many years under the brand name Cogentin and has been used as an adjunct therapy to reduce motor side effects associated with neuroleptic treatment with antipsychotic medications. Benztropine has a mixed pharmacology with potent action as an antimuscarinic, antihistaminergic, and as a dopamine-uptake inhibitor. It has a behavioral profile that is distinctly different from that of cocaine and other dopamine-uptake inhibitors, and initially that was assumed to be the result of its actions at other receptors. However, more recent work with analogs of benztropine that are more selective for the dopamine transporter have called some of those assumptions into question and have further highlighted the potential of this class as cocaine treatment agents.

Benztropine itself has been compared with cocaine and GBR 12935 in side-by-side experiments to determine its differences from more "classical" dopamine-uptake inhibitors using measures of fixed-interval responding, locomotor activity, drug discrimination, and convulsions [94]. All compounds increased rates of responding and decreased quarter-life values in fixed-interval responding. However, benztropine differed from cocaine and GBR 12935 in that it did not increase timeout responding. All compounds increased locomotor activity, but the increases produced by benztropine were smaller than those produced by cocaine or GBR 12935. Both benztropine and GBR 12935 at higher doses decreased the locomotor effects of cocaine. In rats trained to discriminate 10 mg/kg cocaine from saline, GBR 12935 produced full dose-dependent substitution for the discriminative stimulus effects of cocaine, but benztropine did not, resulting in a maximum of approximately 40% cocaine lever responding. In contrast to GBR 12935, benztropine did not potentiate the effects of a low dose of cocaine. Finally, in contrast to GBR 12935, benztropine did not potentiate the convulsant effects of a dose of 55 mg/kg cocaine in mice, which normally produces convulsions in approximately 27% of mice under control conditions. These results suggest that benztropine has effects that differ from both cocaine and GBR 12935 [94].

Following the initial characterization of benztropine as different from cocaine, in addition to clinical studies that have long described it as "noneuphorigenic," a number of benztropine analogs have been synthesized and evaluated for their behavioral effects alone and in combination with cocaine. One early paper examined 4'Cl-benztropine and several analogs for binding and uptake, but only 4'-Cl-benztropine was tested in vivo. This compound was chosen for study because it bears structural similarities to both cocaine and GBR 12909 and has high affinity to the dopamine transporter. In these studies, 4'-Cl-benztropine did not stimulate locomotor activity and did not substitute for the discriminative stimulus effects of cocaine, despite its ability to inhibit dopamine uptake via the DAT. Its analogs were also described as unlike cocaine in their behavioral effects [95].

Both 4'-Cl benztropine and 3'-Cl benztropine were considerably less potent than cocaine in stimulating locomotor activity despite higher affinity at the DAT. In addition, 3'-Cl benztropine produced full substitution for the discriminative stimulus effects of cocaine, whereas 4'-Cl benztropine did not. This was surprising because 3'-Cl benztropine had higher affinity than 4'-Cl benztropine at the muscarinic M1 receptor, and muscarinic activity has often been assumed to account for the differences in the behavioral profile of benztropine compared to cocaine [96].

Additional structure–activity relationships have been conducted, and a number of other analogs have been studied. In one study of 3α-diphenylmethoxytropane analogs, 11 compounds stimulated locomotor activity, and their ED_{50} values were inversely correlated with binding at the DAT. For compounds that depressed locomotor activity, there was no relationship between potency to decrease locomotor activity and binding to the transporter. In addition, none of the compounds fully substituted for 10 mg/kg cocaine at greater than 70% unless the pretreatment time was extended. At a pretreatment interval of 30 to 90 minutes, 4′4″-diF-benztropine produced a full, dose-dependent substitution for the discriminative stimulus effects of cocaine. Several analyses were conducted to determine if potency at muscarinic receptors could mask or block cocainelike effects, but there was no evidence to support masked or blocked effects, or disruption in the ability to respond. There also was no relationship between affinity for the dopamine transporter and the production of cocainelike behavioral effects, as reported previously [15,16]. The authors hypothesize that the 3α-diphenylmethoxytropane analogs of benztropine might be accessing different binding domains on the DAT compared to cocaine and analogs of WIN 35,428, and that this might possibly account for the behavioral differences [97].

Further studies of the in vivo effects of 4′-Cl-benztropine were conducted to determine the time course of behavioral effects and neurochemical activation as compared to cocaine. In addition, the effects of both drugs and their interactions were examined using both local and systemic administration. Results indicated that cocaine was a more efficacious stimulant of locomotor activity than 4′-Cl-benztropine, and benztropine had a slower onset and longer duration of action than cocaine in rats [98]. There were no differences in the production of stereotypy. These effects occurred in parallel to the effects of each compound on extracellular dopamine as measured by in vivo microdialysis in the nucleus accumbens, where cocaine had an immediate effect that returned to baseline in 120 minutes, whereas 4′-Cl-benztropine did not reach a peak effect until 60 minutes after i.p. injection, and lasted the entire 120 minutes. When locally perfused into the nucleus accumbens, 4′-Cl-benztropine was significantly more potent and longer lasting in its effects on dopamine overflow, but neither cocaine nor 4′-Cl-benztropine produced behavioral stimulation when locally infused to one side. When infused bilaterally, both 4′-Cl-benztropine and cocaine stimulated locomotor activity, but not to a degree greater than the stimulant effects measured following i.p. administration. 4′-Cl-benztropine potentiated the nucleus accumbens dopamine and the locomotor responses to a low i.p. dose of cocaine (10 mg/kg) administered 2 hours later. In addition, pretreatment with 4′-Cl-benztropine potentiated the effects of cocaine to produce stereotypy and convulsions and potentiated the discriminative stimulus effects of cocaine. The authors conclude that 4′-Cl-benztropine has a different behavioral profile than that of cocaine, and that the difference in time course may be related to slower binding and dissociation kinetics at the dopamine transporter rather than a slower distribution to the brain. Finally, they highlight differences between 4′-Cl-benztropine and GBR 12909, which is capable of blunting the effects of cocaine on dopamine overflow, highlighting the differences in the effects produced by compounds from two different chemical classes of dopamine uptake inhibitors [98].

Both 4′-Cl-benztropine and cocaine produced increases in dopamine levels, with cocaine reaching higher peak levels but 4′-Cl-benztropine producing more sustained effects in both regions of the accumbens. In the caudate, peak levels produced by each compound were similar, but cocaine's effects were shorter-lived than those of 4′-Cl-benztropine. In the medial prefrontal cortex, 4′-Cl-benztropine was not only longer lasting than cocaine, but was more efficacious at all time points. The authors conclude that while the significance of this difference is unclear, it may be a factor related to the behavioral differences that have been described [99].

Several studies have examined the reinforcing effects benztropine analogs in rhesus monkeys. The first of these studies assessed the effects of 3′-Cl-benztropine and 4′-Cl-benztropine in monkeys trained to self-administer cocaine using a fixed-ratio (FR10) schedule or reinforcement. Results suggested that despite high-affinity binding to the DAT, these compounds were weak positive reinforcers, in that they maintained self-administration behavior whereas the parent compound, benztropine, did not [100]. A subsequent study examined these analogs as well as several compounds with known reinforcing effects using methods designed to analyze and compare the degree of reinforcing efficacy. Results indicated that under an FR-25 schedule, both 3′-Cl-benztropine and 4′-Cl-benztropine had reinforcing effects, as did GBR12909 and other known reinforcing compounds. Another analog, 3′,4″-diCl-benztropine was not self-administered. Under a progressive-ratio schedule, the rank order of efficacy was cocaine > GBR 12909 > 3′-Cl-benztropine > 4′-Cl-benztropine ≫ 3″, 4′-diCl-benztropine, suggesting that in these studies, high affinity for the DAT did not predict level of reinforcing efficacy [101].

Several other evaluations of benztropine analogs have been described [102,103] and analyzed with regard to muscarinic [103,104] and histaminergic binding [105]. The only individual compounds that have been analyzed in depth are JHW 007 [106], AHN 1-055 and AHN 2-005 [17,107]. AHN 1-055 and AHN 2-005 were less potent than cocaine and GBR 12909, and had a slower onset of action to stimulate locomotor activity. In addition, although all compounds displaced [^{125}I]RTI-121, cocaine was the least potent, and maximal displacement of [^{125}I]RTI-121 was obtained with cocaine at 30 minutes after injection, with GBR 12909 at 125 minutes after injection, and for AHN 1-005 and AHN 2-005 150 and 240 minutes after injection, respectively. At all time points, there was a significant correlation between displacement of [^{125}I]RTI-121 binding and locomotor stimulant effects of cocaine, GBR 12909, AHN 1-055, and ANH 2-005. It was concluded that the slower association rate of the benztropine analogs was responsible for their slower onset of locomotor stimulation for the two compounds tested [17]. The same compounds were compared to cocaine using in vivo microdialysis to determine the effects on dopamine levels in the nucleus accumbens of rats. AHN 1-055 and AHN 2-005 produced sustained increases in dopamine compared to cocaine, and maximum effects were measured at 2 hours and 1 hour, respectively, compared to cocaine, which produced a maximum effect on dopamine elevation at 10 minutes. Results support the hypothesis that the differential effects of these compounds compared to cocaine are attributable to different kinetic profiles involving a slow onset followed by a long duration of action [107].

The kinetics of AHN 1-055 and AHN 2-005 and the different profile of behavioral effects led to the hypothesis that these compounds were less rewarding than cocaine. The hypothesis was tested using a conditioned place preference paradigm in rats, in which AHN 1-055, AHN 2-005, and JHW 007 were compared to cocaine and atropine for their ability to produce a place preference to the environment in which they were administered. As expected, cocaine produced a robust place preference, but atropine did not. AHN 1-055 at doses up to 3.0 mg/kg did not produce a place preference at 15, 45, or 90 minutes after injection. In contrast, AHN 2-005 did not produce a place preference at 15 or 90 minutes, but did at 45 minutes after injection. JHW 007 did not produce a place preference at 15 or 90 minutes, but at 45 minutes the time spent in the drug-paired environment approached significance. The lack of consistent effects of the benztropine analogs on place conditioning is consistent with a reduced profile of rewarding effects in this pharmacological class of dopamine-uptake inhibitors [108].

JHW 007 is the first in the series of analogs of benztropine that are not only devoid of cocainelike effects but that has been reported to block the effects of cocaine. JHW 007 at doses of 1 to 10 mg/kg had no effect on locomotor activity in mice through an 8-hour measurement period, but when administered as a pretreatment at 10 mg/kg 270 minutes prior to cocaine, was able to block the locomotor stimulant effect of 10 to 60 mg/kg of cocaine in mice. In contrast, a related analog, AHN 2-005, did not block the locomotor stimulant effects of cocaine. In addition to blocking effects of cocaine on locomotor activity, a pretreatment dose of 10 mg/kg JHW 007 administered 4.5 hours prior to cocaine was able to shift the cocaine dose–effect curve for discriminative stimulus effects approximately threefold to the right. The differences in the effects of JHW 007 compared to cocaine were hypothesized to be the result of its slower association rate in the brain; however, its actions to block the locomotor and discriminative stimulus effects of cocaine may be the result of other pharmacological factors, such as affinity for sigma and histamine receptors and higher potency for the inhibition of DAT uptake as opposed to DAT binding [106].

14.4.4 Mazindol

Mazindol is a sympathomimetic amine that is marketed as an anorexigenic by several companies and is the major representative of a class of dopamine uptake inhibitors. For a time, it was the tritiated ligand used to measure binding at the DAT and was one of the compounds tested in the seminal study by Ritz and colleagues [1] that established the DAT as the site of cocaine's actions. In that study, a group of compounds structurally related to cocaine and amphetamine were compared to cocaine in terms of relative potency to inhibit [3H] mazindol binding at the DAT and relative potency to function as reinforcers in drug self-administration experiments. This paper was quite influential, and led to the expectation that all DAT inhibitors would have cocainelike effects and behavioral potency predicted by affinity at the DAT. As more classes of DAT inhibitors have been evaluated, it seems to be the case now that this assumption now holds true only for closely related analogs of cocaine [15,16].

One of the earliest papers describing behavioral effects of mazindol was published in 1976, shortly after mazindol (Sanorex) was introduced in the United States as an anorexigenic. This study examined the ability of three doses of mazindol to maintain self-administration in monkeys trained to self-administer cocaine. It was reported that doses of 50 and 100 μg/kg per injection of mazindol resulted in a mean number of injections that was statistically different from saline. Further, the inverse relationship between dose and mean number of injections was consistent with findings with other psychostimulants under conditions of limited daily access [109].

The behavioral effects of mazindol and other DAT inhibitors, including cocaine, were evaluated in monkeys trained to respond under a schedule of stimulus-shock termination [12]. In these studies, mazindol (and a number of other DAT inhibitors) had behavioral effects that were qualitatively similar to those of cocaine, producing dose-related increases in response rates up to a dose, beyond which responding was decreased. Mazindol was only slightly less potent than cocaine in producing this effect. Using the results of these studies and data from other published studies, the authors found a significant positive correlation between the relative potencies of 15 different drugs for producing cocainelike increases in response rates and for displacing [3H] cocaine in brain tissue. Therefore, mazindol was again one of the compounds that was used to conclude that the behavioral effects of cocaine were largely attributable to its effects at the DAT [12].

A study by the same authors evaluating some of the same compounds in tests of self-administration was conducted. Mazindol (0.03 to 3.0 mg/kg per injection), unlike cocaine, bupropion, GBR 12909, methylphenidate, and nomifensine, maintained self-administration in only two of the four monkeys studied. In the other two monkeys, only erratic responding occurred over the range of doses tested, and in one monkey, injections of 0.3 mg/kg produced profuse salivation, self-mutilation, and vocalizations interpreted as distress calls. In the monkeys who reliably self-administered mazindol, the rates of responding were only 60 to 70% of those of cocaine, although the ED_{50} values suggested that mazindol was approximately three to five times more potent. These results were interpreted as consistent with previous reports that the reinforcing effects of mazindol can vary among individual subjects [110].

A number of studies have examined the ability of mazindol and various other compounds to substitute for the discriminative stimulus effects of cocaine, and mazindol has been uniformly described as producing full substitution for the discriminative stimulus effects of cocaine in rats and monkeys [40,111,112]. It also has been reported to reduce rates of responding for both cocaine and food reinforcement when administered as a pretreatment [112,113].

Mazindol also was evaluated, with a group of other dopamine uptake inhibitors, for its interactions with cocaine in a sensitization paradigm. In this study, compounds were evaluated for their effects on locomotor activity in rats, and results were compared to the effects of the same compounds on locomotor activity following previous exposure to a sensitizing dose of cocaine, 40 mg/kg i.p. Cocaine, mazindol, RTI-55, and GBR 12909 all stimulated locomotor activity, and for all compounds with the exception of GBR 12909, a sensitized response was measured after preexposure to cocaine. In terms of this evaluation, mazindol was similar to cocaine and RTI-55 in its effects [114].

A number of analogs of mazindol have been prepared and evaluated in vitro [115–118]. One of the few analogs of mazindol that has been described in detail and compared to mazindol is mazindane, a prodrug form of mazindol [119]. Both mazindol and mazindane produced long-lasting stimulation of locomotor activity in mice, with ED_{50} values of 3.9 and 2.5 mg/kg i.p., respectively. At 25 mg/kg, mazindane was toxic, whereas mazindol was devoid of toxic side effects at doses up to 50 mg/kg. Both compounds produced full substitution for the discriminative stimulus effects of cocaine in rats, with mazindane approximately fourfold more potent than mazindol. In rhesus monkeys trained to discriminate cocaine from saline, mazindol and mazindane were equipotent [119]. Although the authors suggested that mazindane could be useful alternative to mazindol, it does not appear that further work with this compound was carried out.

14.4.5 Bupropion

Another dopamine-uptake inhibitor that is typically studied as a "standard" DAT inhibitor is bupropion, currently marketed as the antidepressant Wellbutrin and as the smoking cessation aid, Zyban. Like cocaine, it is a locomotor stimulant with an ED_{50} value of 12.5 mg/kg in Swiss Webster mice. When compared to a group of dopamine-uptake inhibitors structurally related to cocaine, the binding potency of bupropion was not related to its potency to stimulate locomotor activity [16]. In squirrel monkeys trained on a fixed-interval schedule of shock avoidance, bupropion, like other dopamine-uptake inhibitors (mazindol, methylphenidate, and nomifensine), increased rates of responding in a manner that was similar to the effects of cocaine [12].

Bupropion also produced full substitution for the discriminative stimulus effects of cocaine in rats [20,27,120–122], pigeons [123], and rhesus monkeys [19]. Bupropion substituted fully for the discriminative stimulus effects of d-amphetamine in rhesus monkeys [124], and fully substituted for the effects of GBR 12909 in squirrel monkeys [38]. When bupropion itself was trained as a discriminative stimulus in rats, generalization was produced by stimulants and monoamine-uptake inhibitors, including amphetamine, cocaine, caffeine, methylphenidate, nomifensine, and viloxazine [125,126].

In squirrel monkeys trained under a second-order fixed-interval schedule of i.v. cocaine self-administration, bupropion and other uptake inhibitors, including methylphenidate, GBR 12909, and nomifensine, maintained comparable rates of responding and were self-administered [127]. In baboons trained to self-administer cocaine under a fixed ratio 80 or 160 with a 3-hour timeout, bupropion, along with uptake inhibitors diclofensine and nomifensine, maintained self-administration behavior above the level of vehicle, and some doses of each compound maintained levels of behavior similar to that maintained by cocaine [121]. In another study, bupropion was compared with nomifensine for its ability to be self-administered by rats. In drug-naive rats, self-administration behavior was acquired and maintained using bupropion or nomifensine as a reinforcer at doses of 3 and 1 mg/kg per infusion. At lower doses, self-administration behavior was acquired but not maintained. Self-administration of bupropion dose-dependently up-regulated dopamine transporters in the caudate–putamen and nucleus accumbens, whereas nomifensine self-administration had no effect on DAT levels, pointing out the heterogeneity in the effects of transporter inhibitors [43].

Despite the fully cocainelike profile of bupropion in animal models, it typically does not produce a cocainelike discriminative stimulus effect in humans [128], and it is not generally abused.

14.4.6 Methylphenidate

Like mazindol and bupropion, methylphenidate is a dopamine-uptake inhibitor that has been in clinical use for many years, and despite similarities to cocaine, has not been widely abused. A recent review article found 60 studies related just to the abuse potential of methylphenidate in animals and humans [129]. It has also been used commonly as a "standard" dopamine-uptake inhibitor in studies determining the contribution of the uptake inhibitor to the behavioral effects of cocaine and other ligands, so there are a number of studies in which its effects are assessed, although it is not the focus of the studies. Finally, a number of analogs of methylphenidate have been prepared and evaluated in vitro. However, the only methylphenidate-related compound that appears to have been evaluated in vivo is the (−)enantiomer. For these reasons, a comprehensive review of the in vivo effects of methylphenidate analogs is not presented here.

Methylphenidate is a locomotor stimulant that produces sensitization to its own effects and to those of cocaine [130]. Methylphenidate-induced locomotor activity can be reduced by treatment with nisoxetine [131], and behavioral sensitization is accompanied by decreases in the amplitude of sensory-evoked field responses in the VTA, which is interpreted as electrophysiological sensitization [132]. Because of the prevalent use of methylphenidate for the treatment of attention-deficit/hyperactivity disorder in children, many groups have studied the effects of methylphenidate in rodents at different ages. Niculescu and colleagues [133] reported that postweaning mice showed less total activity than that of both periadolescent and adult mice at a dose of 10 mg/kg methylphenidate [133]. A 7-day cocaine treatment regimen produced locomotor sensitization in adult and periadolescent mice, whereas methylphenidate produced sensitization only in adult mice, and postweaning, periadolescent, and adult mice did not show sensitization to the locomotor-stimulating effects of amphetamine in this model [133]. Torres-Reveron and Dow-Edwards [134] also reported complex interactions of age and chronic methylphenidate treatment, but concluded that methylphenidate produced cross-sensitization to cocaine regardless of age at which methylphenidate was administered. Brandon and colleagues [135] reported that treatment of adolescent rats with moderate doses of methylphenidate (5 and 10 mg/kg for 5 or 7 days) enhanced the psychomotor response to cocaine, as measured by different challenge doses of cocaine (0 to 30 mg/kg) in adulthood.

In addition to effects on locomotor activity, one study examined the neuroadaptations produced by methylphenidate compared to cocaine and other dopamine-uptake inhibitors. Using a treatment paradigm in adolescent rats that was known to produce sensitization to cocaine in adulthood, Brandon and Steiner [136] reported that acute methylphenidate produced dose-dependent increases in early gene expression (c-fos and zif 268), and repeated administration increased the expression of dynorphin.

Following pretreatment with methylphenidate, cocaine-induced expression of c-fos and zif 268, as well as of Substance P, was attenuated throughout the striatum. This was interpreted to suggest that methylphenidate produces neuroadaptations similar to those produced by cocaine [136]. Izenwasser and colleagues [137] compared the effects of chronic methylphenidate with the effects of chronic cocaine on locomotor activity and transporter number using two different methods of drug administration. The authors reported that unlike cocaine, methylphenidate produced neither sensitization to repeated injections, nor tolerance to a continuous infusion; however, like cocaine, continuous infusion had no effect on the number of dopamine transporters in the nucleus accumbens or the rostral caudate-putamen. In contrast to cocaine, repeated daily injections of methylphenidate significantly decreased the number of transporters in the rostral caudate–putamen, suggesting that methylphenidate produces different neuroadaptations than those produced by cocaine [137].

Stimulant effects of methylphenidate were also assessed in squirrel monkeys trained on a fixed-interval schedule of stimulus-shock termination, and the effects of methylphenidate, as well as the dopamine-uptake inhibitors mazindol, nomifensine, and bupropion, were considered qualitatively similar to those of cocaine. This study noted a close correspondence between the potencies of compounds to produce cocainelike behavioral effects with the in vitro potency of compounds to displace tritiated cocaine in the caudate–putamen [12]. In another analysis of the stimulant effects of methylphenidate, rats were trained to respond for food reinforcement on a progressive-ratio schedule, and drug effects on these response rates were assessed. In these animals, methylphenidate, in addition to amfonelic acid, caffeine, oxolinic acid, nomifensine, and d-amphetamine, all produced increased rates of responding, whereas nonstimulant compounds from other classes did not increase response rates. The authors conclude that the paradigm is sensitive to psychomotor stimulant effects of drugs such as methylphenidate [138].

Discriminative stimulus effects of methylphenidate in cocaine-trained animals have uniformly been reported to be cocainelike. In rats trained to discriminate a small dose of cocaine, methylphenidate fully substituted for the discriminative stimulus effects of cocaine [139]. In rats trained to discriminate 10 mg/kg cocaine, methylphenidate, like the uptake inhibitors or releasers diethylpropion, phenmetrazine, and phentermine, produced full, dose-dependent substitution for cocaine. If rats were treated chronically with cocaine, cross-tolerance was observed such that doses of methylphenidate required to produce full substitution were approximately twofold higher, as they were for phenmetrazine and phentermine, suggesting a common mechanism [18]. In rats trained to discriminate cocaine or BTCP from saline, methylphenidate, like mazindol, indatraline, and GBR 12909, fully substituted for both compounds [111]. In a study that compared substitution for the discriminative stimulus effects of cocaine with potencies to bind at the dopamine transporter, Katz and colleagues [27] reported that methylphenidate fully substituted for cocaine in rats, and its binding, like that of cocaine, nomifensine, and several WIN compounds, was better fit by a two-site model. Of compounds that were best fit by a two-site model, binding at the high affinity site appeared to be most closely related to potency to substitute for cocaine [27]. Another study that examined both substitution for cocaine and the nature of additive interactions with cocaine reported that methylphenidate, like GBR 12909, indatraline, nomifensine, mazindol, methamphetamine, *d*-amphetamine, methcathinone, cathinone, fencamfamine, and phentermine, fully substituted for the discriminative stimulus effects of cocaine and shifted its dose–effect curve to the left. While isobolographic analysis indicated that interactions were generally additive, it was noted that methamphetamine and *d*-amphetamine were more potent than the other compounds studies in shifting the cocaine curve to the left [140]. In squirrel monkeys trained to recognize methamphetamine injections of 0.32 mg/kg, methylphenidate produced dose-related responding on the methamphetamine-associated lever and fully substituted for the discriminative stimulus effects of methamphetamine, as did cocaine. In addition to discrimination, extracellular dopamine was measured in the caudate–putamen, and doses of cocaine and methylphenidate that produced full substitution produced comparable increases in extracellular dopamine of about 250% of control. However, methamphetamine doses that produces only 42% methamphetamine lever responding produced similar increases in caudate dopamine, suggesting that methamphetamine lever responding was not the direct result of specific dopamine levels and that other neurochemical drug actions were also involved [141]. In squirrel monkeys trained to discriminate GBR 12909 from saline, methylphenidate, like cocaine, mazindol, and bupropion, produced full substitution, whereas inhibitors of norepinephrine and serotonin or compounds that bind to postsynaptic dopamine receptors did not [38].

Methylphenidate has been studied in a number of self-administration paradigms in animals and in humans. In squirrel monkeys, methylphenidate at doses ranging from 0.01 to 0.3 mg/kg per injection maintained comparable rates and patterns of responding compared to cocaine, bupropion, and GBR 12909. The relative potency of compounds to maintain self-administration was generally consistent with the relative potencies of the same compounds to displace tritiated cocaine binding in the caudate–putamen [127]. In baboons, methylphenidate was compared to cocaine and secobarbital for its ability to maintain self-administration on a progressive-ratio schedule. While both methylphenidate and cocaine were self-administered, cocaine produced higher breaking points at a comparable dose of 0.4 mg/kg, whereas manipulation of doses of either compound had little effect on breaking points. In contrast, increasing doses of secobarbital produced higher breaking points within the same subjects [142,143]. In a study in rhesus monkeys trained to lever-press for food or cocaine reinforcement on a multiperiod session, methylphenidate, like morphine, oxymorphone, codeine, pentazocine, and *d*-amphetamine, maintained responding at rates significantly greater than those maintained by saline. In contrast, cyclazocine, naloxone, levallorphan, scopolamine, chlorpromazine, and fenfluramine were among the drugs that did not maintain responding [144]. Methylphenidate also was self-administered by rhesus monkeys in a choice procedure in which methylphenidate was paired with saline, cocaine, or in which different doses were paired. In general, higher doses of methylphenidate and cocaine were preferred over low doses, and when equal doses of cocaine and methylphenidate were paired, no preference was seen. However, response rates maintained by cocaine were generally higher than those maintained by methylphenidate. Thus, whereas methylphenidate and cocaine were both reinforcing, the absolute reinforcing efficacy is considered in light of both response rates and concurrently available drugs [145]. A study in rhesus monkeys that compared the reinforcing efficacy of methylphenidate, cocaine, and a number of novel psychostimulants under a progressive-ratio schedule concluded that methylphenidate was equal to cocaine in this paradigm, and in general, the potency to maintain a peak breakpoint correlated with DAT affinity [146].

Several studies have examined the effects of prior treatment with methylphenidate on acquisition of cocaine self-administration. In rats, prior treatment with methylphenidate (5 or 20 mg/kg per day) for 9 days significantly decreased the latency for acquisition of cocaine self-administration [147]. Brandon and colleagues [135] similarly reported that exposure to low-dose methylphenidate in adolescence increased cocaine self-administration in adulthood.

Methylphenidate has been examined in rat reinstatement procedures, where rats are trained to self-administer cocaine, and following some period of maintenance, drug-reinforced responding is extinguished. In such a paradigm, noncontingent priming injections of methylphenidate, as well as cocaine, amphetamine, and caffeine, produced a reinstatement of responding for cocaine. Amphetamine was the most efficacious compound in producing reinstatement, followed by cocaine and methylphenidate, which were equipotent [148]. In another study evaluating the effect of methylphenidate on reinstatement, a conditioned place-preference paradigm in mice was utilized in which a place preference to cocaine was conditioned and then subsequently, extinguished. Priming injections of cocaine, methamphetamine, or methylphenidate reinstated a previously extinguished preference to the cocaine-paired compartment, whereas a priming injection of phencyclidine was ineffective. It was concluded that methamphetamine and methylphenidate substituted for the reinforcing cue of cocaine [149], which is consistent with its pharmacological similarity to cocaine.

Microdialysis studies have attempted to differentiate effects of dopamine-uptake inhibitors in terms of measurement of extracellular concentrations of dopamine and its metabolites in rats. When infused directly into the striata of awake rats, methylphenidate, like d-amphetamine, GBR 12909, cocaine, nomifensine, bupropion, and benztropine, all increased dopamine levels in a dose-dependent manner, but only d-amphetamine also increased dopamine metabolite concentrations [150]. Like GBR 12909 and benztropine, methylphenidate, had biphasic effects on dialysate concentrations of dopamine when applied at 1000 μm concentrations, whereas cocaine, nomifensine, and bupropion produced monophasic increases in extracellular dopamine, suggesting concentration-dependent differences in effects produced by different transporter ligands.

Like other dopamine-uptake inhibitors, analogs of methylphenidate have been prepared and evaluated for binding [151–154], but only a few compounds have been evaluated in vivo. Davids and colleagues [155] evaluated the effects of d-, l-, and dl-threo-methylphenidate on locomotor activity in intact and 6-OHDA-lesioned rats. Both d- and dl-methylphenidate stimulated activity in intact rats and inhibited activity in 6-OHDA-lesioned rats, whereas l-methylphenidate had no effect in either group. d-Methylphenidate was 3.3 times more potent than the racemate in inhibiting activity in the lesioned rats, but these effects could be blocked by pretreatment with *l*-methylphenidate, suggesting that the potency of *d*-methylphenidate may be more than double that of the racemate [155]. Two other derivatives of methylpheni- date, *p*-bromomethylphenidate or *p*-methoxymethylphenidate, were tested with methylphenidate to determine their effects on conditioned place preference in rats. All compounds produced a significant place preference, and extracellular dopamine concentrations in the striatum were increased similarly by all compounds. Pretreatment with *p*-methoxymethylphenidate failed to block the increases in extra- cellular dopamine and the locomotor stimulant effects of a 20 mg/kg dose of cocaine. The authors conclude that the methylphenidate derivatives were generally similar to methylphenidate and other psychostimulant drugs [156]. Schweri and col- leagues [157] reported on a series of 11 methylphenidate analogs, all but one of which was more potent than the parent compound in displacing tritiated WIN 35,428 binding to the dopamine transporter. All but two of the analogs produced full substitution for cocaine, and among those compounds that substituted, a positive correlation was observed between the relative potencies for binding to the dopamine transporter and substitution for cocaine in rat drug discrimination. When several of the analogs were given in combination with cocaine, substitution was enhanced. The authors conclude that because of the similarly between the methylphenidate analogs and cocaine itself, the analogs may have potential as substitution therapies [157] but were not subsequently evaluated.

14.5 CONCLUSIONS

This chapter has attempted to elucidate the role of the dopamine transporter by describing (1) results of studies of DAT mutant animals where a subtractive analysis is employed to examine the behavioral effects of ligands in the absence of the trans- porter, (2) in vivo studies using primarily correlational analyses of potencies of mul- tiple ligands in binding to the transporter and transduction of behavioral effects, and (3) behavioral studies of some of the most commonly known transporter ligands and their chemical series. The relevance of the DAT to the behavioral effects of cocaine is indisputable; whether ligands that bind to the transporter and do not fully recapitulate the behavioral effects of cocaine can be found and developed as cocaine treatment medications remains to be determined.

Although the preponderance of evidence remains consistent with the hypothesis that binding and inhibition of dopamine uptake at the DAT is primarily responsible for mediating the stimulant, discriminative stimulus, and reinforcing effects of cocaine, the existence of compounds that do not share the behavioral effects of cocaine despite high affinity binding to the DAT are intriguing anomalies. The reasons for the differences are not clear.

The issue of selectivity for the DAT vs. NET and SERT has been much discussed and may or may not be important in the mediation or blockade of cocainelike effects. Knockout studies as well as the human abuse liability data in marketed drugs suggest that compounds with significant binding at the NET and SERT may not be reinfor- cing; however, cocaine itself has actions at the NET and SERT and is highly reinforcing. In addition, marketed compounds with actions at the DAT and other transporters have not been effective as cocaine medications in clinical trials. Compounds that bind primarily to the DAT with additional interactions with the NET and SERT have been evaluated in animals and for the most part are fully cocainelike, although there are exceptions. Whether it is possible to fine-tune the balance of effects at the three transporters to create a compound that is capable of blocking effects of cocaine is unknown. Truly DAT-selective ligands have yet to be found, although there are some that are 100-fold selective for the DAT over the NET and SERT by binding, but not by uptake. One might expect a truly DAT-selective compound to fully reproduce the effects of cocaine and possibly be even more reinforcing, although early evidence does not support this prediction. However, a complete analysis of the in vitro selectivity of any compound would first require that compounds be evaluated at all monoamine transporters in the same assay system, whether human clones or using selective ligands in brain tissue. Second, to attribute the behavioral effects of compounds to their binding at a single site, a full analysis of the binding profile of a compound at a multitude of sites would be necessary to ensure that compounds did not exert behavioral effects through other or additional receptors or biological systems. Finally, it would be necessary to know the full metabolic profile of each compound in terms of the production of active metabolites, and the complete pharmacological profile of each active metabolite that might contribute to in vivo pharmacology would have to be elucidated. Although this type of analysis is often conducted for compounds in development, it is prohibitive to even consider such analyses in pharmacology studies examining ligands across chemical structures or structural classes. For these reasons, analyses of selectivity are typically quite limited.

In addition, in examination of in vivo effects of DAT ligands, differences in behavioral effects may occur as a secondary effect of time-course considerations, which may result from poor absorption, poor brain penetrability, or slow rates of binding to transporters. In addition, compounds may have differences in solubility, lipophilicity, and distribution that affect their behavioral profile. Recognition of the effects of multiple factors prevents simple comparisons of the effects of transporter ligands and attributions to a single pharmacological action at the transporter.

REFERENCES

1. Ritz, M. C., Lamb, R. J., Goldberg, S. R., and Kuhar, M. J. (1987). Cocaine receptors on dopamine transporters are related to self-administration of cocaine. *Science, 237,* 1219–1223.

2. Giros, B., Jaber, M., Jones, S. R., Wightman, R. M., and Caron, M. G. (1996). Hyperlocomotion and indifference to cocaine and amphetamine in mice lacking the dopamine transporter. *Nature, 379,* 606–612.

3. Jaber, M., Jones, S., Giros, B., and Caron, M. G. (1997). The dopamine transporter: a crucial component regulating dopamine transmission. *Mov Disord, 12,* 629–633.

4. Rocha, B. A., Fumagalli, F., Gainetdinov, R. R., Jones, S. R., Ator, R., Giros, B., Miller, G. W., and Caron, M. G. (1998). Cocaine self-administration in dopamine-transporter knockout mice. *Nat Neurosci*, *1*, 132–137.

5. Sora, I., Wichems, C., Takahashi, N., Li, X. F., Zeng, Z., Revay, R., Lesch, K. P., Murphy, D. L., and Uhl, G. R. (1998). Cocaine reward models: conditioned place preference can be established in dopamine- and in serotonin-transporter knockout mice. *Proc Nat Acad Sci U S A*, *95*, 7699–7704.

6. Gainetdinov, R. R., Jones, S. R., and Caron, M. G. (1999). Functional hyperdopaminergia in dopamine transporter knock-out mice. *Biol Psychiatry*, *46*, 303–311.

7. Jones, S. R., Gainetdinov, R. R., Jaber, M., Giros, B., Wightman, R. M., and Caron, M. G. (1998). Profound neuronal plasticity in response to inactivation of the dopamine transporter. *Proc Natl Acad Sci U S A*, *95*, 4029–4034.

8. Xu, F., Gainetdinov, R. R., Wetsel, W. C., Jones, S. R., Bohn, L. M., Miller, G. W., Wang, Y. M., and Caron, M. G. (2000). Mice lacking the norepinephrine transporter are supersensitive to psychostimulants. *Nat Neurosci*, *3*, 465–471.

9. Sora, I., Hall, F. S., Andrews, A. M., Itokawa, M., Li, X. F., Wei, H. B., Wichems, C., Lesch, K. P., Murphy, D. L., and Uhl, G. R. (2001). Molecular mechanisms of cocaine reward: combined dopamine and serotonin transporter knockouts eliminate cocaine place preference. *Proc Nat Acad Sci U S A*, *98*, 5300–5305.

10. Hall, F. S., Li, X. F., Sora, I., Xu, F., Caron, M., Lesch, K. P., Murphy, D. L., and Uhl, G. R. (2002). Cocaine mechanisms: enhanced cocaine, fluoxetine and nisoxetine place preferences following monoamine transporter deletions. *Neuroscience*, *115*, 153–161.

11. Chen, R., Tilley, M. R., Wei, H., Zhou, F., Zhou, F. M., Ching, S., Quan, N., Stephens, R. L., Hill, E. R., Nottoli, T., Han, D. D., and Gu, H. H. (2006). Abolished cocaine reward in mice with a cocaine-insensitive dopamine transporter. *Proc Nat Acad Sci U S A*, *103*, 9333–9338.

12. Spealman, R. D., Madras, B. K., and Bergman, J. (1989). Effects of cocaine and related drugs in nonhuman primates. II: Stimulant effects on schedule-controlled behavior. *J Pharmacol Exp Ther*, *251*, 142–149.

13. Ginsburg, B. C., Kimmel, H. L., Carroll, F. I., Goodman, M. M., and Howell, L. L. (2005). Interaction of cocaine and dopamine transporter inhibitors on behavior and neurochemistry in monkeys. *Pharmacol Biochem Behav*, *80*, 481–491.

14. Cline, E. J., Scheffel, U., Boja, J. W., Carroll, F. I., Katz, J. L., and Kuhar, M. J. (1992). Behavioral effects of novel cocaine analogs: a comparison with in vivo receptor binding potency. *J Pharmacol Exp Ther*, *260*, 1174–1179.

15. Vaugeois, J. M., Bonnet, J. J., Duterte-Boucher, D., and Costentin, J. (1993). In vivo occupancy of the striatal dopamine uptake complex by various inhibitors does not predict their effects on locomotion. *Eur J Pharmacol*, *230*, 195–201.

16. Izenwasser, S., Terry, P., Heller, B., Witkin, J. M., and Katz, J. L. (1994). Differential relationships among dopamine transporter affinities and stimulant potencies of various uptake inhibitors. *Eur J Pharmacol*, *263*, 277–283.

17. Desai, R. I., Kopajtic, T. A., French, D., Newman, A. H., and Katz, J. L. (2005). Relationship between in vivo occupancy at the dopamine transporter and behavioral effects of cocaine, GBR 12909 [1-{2-[bis-(4-fluorophenyl)methoxy]ethyl}-4-(3-phenyl-propyl)piperazine], and benztropine analogs. *J Pharmacol Exp Ther*, *315*, 397–404.

18. Wood, D. M., and Emmett-Oglesby, M. W. (1988). Substitution and cross-tolerance profiles of anorectic drugs in rats trained to detect the discriminative stimulus properties of cocaine. *Psychopharmacology (Berl)*, *95*, 364–368.

19. Kleven, M. S., Anthony, E. W., and Woolverton, W. L. (1990). Pharmacological characterization of the discriminative stimulus effects of cocaine in rhesus monkeys. *J Pharmacol Exp Ther*, *254*, 312–317.

20. Broadbent, J., Michael, E. K., Riddle, E. E., and Apple, J. B. (1991). Involvement of dopamine uptake in the discriminative stimulus effects of cocaine. *Behav Pharmacol*, *2*, 187–197.

21. Cunningham, K. A., and Callahan, P. M. (1991). Monoamine reuptake inhibitors enhance the discriminative state induced by cocaine in the rat. *Psychopharmacology (Berl)*, *104*, 177–180.

22. Spealman, R. D. (1993). Modification of behavioral effects of cocaine by selective serotonin and dopamine uptake inhibitors in squirrel monkeys. *Psychopharmacology (Berl)*, *112*, 93–99.

23. Kleven, M. S., and Koek, W. (1998). Discriminative stimulus properties of cocaine: enhancement by monoamine reuptake blockers. *J Pharmacol Exp Ther*, *284*, 1015–1025.

24. Walsh, S. L., and Cunningham, K. A. (1997). Serotonergic mechanisms involved in the discriminative stimulus, reinforcing and subjective effects of cocaine. *Psychopharmacology (Berl)*, *130*, 41–58.

25. Spealman, R. D. (1995). Noradrenergic involvement in the discriminative stimulus effects of cocaine in squirrel monkeys. *J Pharmacol Exp Ther*, *275*, 53–62.

26. Schama, K. F., Howell, L. L., and Byrd, L. D. (1997). Serotonergic modulation of the discriminative-stimulus effects of cocaine in squirrel monkeys. *Psychopharmacology (Berl)*, *132*, 27–34.

27. Katz, J. L., Izenwasser, S., and Terry, P. (2000). Relationships among dopamine transporter affinities and cocaine-like discriminative-stimulus effects. *Psychopharmacology (Berl)*, *148*, 90–98.

28. Bergman, J., Madras, B. K., Johnson, S. E., and Spealman, R. D. (1989). Effects of cocaine and related drugs in nonhuman primates. III: Self-administration by squirrel monkeys. *J Pharmacol Exp Ther*, *251*, 150–155.

29. Wilcox, K. M., Lindsey, K. P., Votaw, J. R., Goodman, M. M., Martarello, L., Carroll, F. I., and Howell, L. L. (2002). Self-administration of cocaine and the cocaine analog RTI-113: relationship to dopamine transporter occupancy determined by PET neuroimaging in rhesus monkeys. *Synapse*, *43*, 78–85.

30. Wee, S., Carroll, F. I., and Woolverton, W. L. (2006). A reduced rate of in vivo dopamine transporter binding is associated with lower relative reinforcing efficacy of stimulants. *Neuropsychopharmacology*, *31*, 351–362.

31. Roberts, D. C., Phelan, R., Hodges, L. M., Hodges, M. M., Bennett, B., Childers, S., and Davies, H. (1999). Self-administration of cocaine analogs by rats. *Psychopharmacology (Berl)*, *144*, 389–397.

32. Kelley, A. E., and Lang, C. G. (1989). Effects of GBR 12909, a selective dopamine uptake inhibitor, on motor activity and operant behavior in the rat. *Eur J Pharmacol*, *167*, 385–395.

33. Heikkila, R. E., and Manzino, L. (1984). Behavioral properties of GBR 12909, GBR 13069 and GBR 13098: specific inhibitors of dopamine uptake. *Eur J Pharmacol, 103*, 241–248.

34. Baumann, M. H., Char, G. U., de Costa, B. R., Rice, K. C., and Rothman, R. B. (1994). GBR12909 attenuates cocaine-induced activation of mesolimbic dopamine neurons in the rat. *J Pharmacol Exp Ther, 271*, 1216–1222.

35. Czoty, P. W., Justice, J. B., Jr., and Howell, L. L. (2000). Cocaine-induced changes in extracellular dopamine determined by microdialysis in awake squirrel monkeys. *Psychopharmacology (Berl), 148*, 299–306.

36. Izenwasser, S., French, D., Carroll, F. I., and Kunko, P. M. (1999). Continuous infusion of selective dopamine uptake inhibitors or cocaine produces time-dependent changes in rat locomotor activity. *Behav Brain Res, 99*, 201–208.

37. Kunko, P. M., Loeloff, R. J., and Izenwasser, S. (1997). Chronic administration of the selective dopamine uptake inhibitor GBR 12,909, but not cocaine, produces marked decreases in dopamine transporter density. *Naunyn Schmiedebergs Arch Pharmacol, 356*, 562–569.

38. Melia, K. F., and Spealman, R. D. (1991). Pharmacological characterization of the discriminative-stimulus effects of GBR 12909. *J Pharmacol Exp Ther, 258*, 626–632.

39. Terry, P., Witkin, J. M., and Katz, J. L. (1994). Pharmacological characterization of the novel discriminative stimulus effects of a low dose of cocaine. *J Pharmacol Exp Ther, 270*, 1041–1048.

40. Witkin, J. M., Nichols, D. E., Terry, P., and Katz, J. L. (1991). Behavioral effects of selective dopaminergic compounds in rats discriminating cocaine injections. *J Pharmacol Exp Ther, 257*, 706–713.

41. Koetzner, L., Riley, A. L., and Glowa, J. R. (1996). Discriminative stimulus effects of dopaminergic agents in rhesus monkeys. *Pharmacol Biochem Behav, 54*, 517–523.

42. Roberts, D. C. (1993). Self-administration of GBR 12909 on a fixed ratio and progressive ratio schedule in rats. *Psychopharmacology (Berl), 111*, 202–206.

43. Tella, S. R., Ladenheim, B., Andrews, A. M., Goldberg, S. R., and Cadet, J. L. (1996). Differential reinforcing effects of cocaine and GBR-12909: biochemical evidence for divergent neuroadaptive changes in the mesolimbic dopaminergic system. *J Neurosci, 16*, 7416–7427.

44. Howell, L. L., and Byrd, L. D. (1991). Characterization of the effects of cocaine and GBR 12909, a dopamine uptake inhibitor, on behavior in the squirrel monkey. *J Pharmacol Exp Ther, 258*, 178–185.

45. Bergman, J., Madras, B. K., Johnson, S. E., and Spealman, R. D. (1989). Effects of cocaine and related drugs in nonhuman primates. III: Self-administration by squirrel monkeys. *J Pharmacol Exp Ther, 251*, 150–155.

46. Skjoldager, P., Winger, G., and Woods, J. H. (1993). Effects of GBR 12909 and cocaine on cocaine-maintained behavior in rhesus monkeys. *Drug Alcohol Depend, 33*, 31–39.

47. Stafford, D., LeSage, M. G., Rice, K. C., and Glowa, J. R. (2001). A comparison of cocaine, GBR 12909, and phentermine self-administration by rhesus monkeys on a progressive-ratio schedule. *Drug Alcohol Depend, 62*, 41–47.

48. Tella, S. R. (1995). Effects of monoamine reuptake inhibitors on cocaine self-administration in rats. *Pharmacol Biochem Behav, 51*, 687–692.

49. Glowa, J. R., Wojnicki, F. H., Matecka, D., Bacher, J. D., Mansbach, R. S., Balster, R. L., and Rice, K. C. (1995). Effects of dopamine reuptake inhibitors on food- and cocaine-maintained responding. I: Dependence on unit dose of cocaine. *Exp Clin Psychopharmacol*, *3*, 219–231.

50. Porrino, L. J., Migliarese, K., Davies, H. M., Saikali, E., and Childers, S. R. (1994). Behavioral effects of the novel tropane analog, 2β-propanoyl-3β-(4-toluyl)-tropane (PTT). *Life Sci*, *54*, L511–L517.

51. Villemagne, V. L., Rothman, R. B., Yokoi, F., Rice, K. C., Matecka, D., Dannals, R. F., and Wong, D. F. (1999). Doses of GBR12909 that suppress cocaine self-administration in non-human primates substantially occupy dopamine transporters as measured by [^{11}C] WIN35,428 PET scans. *Synapse*, *32*, 44–50.

52. Porrino, L. J., Davies, H. M., and Childers, S. R. (1995). Behavioral and local cerebral metabolic effects of the novel tropane analog, 2β-propanoyl-3β-(4-tolyl)-tropane. *J Pharmacol Exp Ther*, *272*, 901–910.

53. Hemby, S. E., Co, C., Reboussin, D., Davies, H. M., Dworkin, S. I., and Smith, J. E. (1995). Comparison of a novel tropane analog of cocaine, 2β-propanoyl-3β-(4-tolyl)tro-pane with cocaine HCl in rats: nucleus accumbens extracellular dopamine concentration and motor activity. *J Pharmacol Exp Ther*, *273*, 656–666.

54. Nader, M. A., Grant, K. A., Davies, H. M., Mach, R. H., and Childers, S. R. (1997). The reinforcing and discriminative stimulus effects of the novel cocaine analog 2β-propanoyl-3β-(4-tolyl)-tropane in rhesus monkeys. *J Pharmacol Exp Ther*, *280*, 541–550.

55. Dworkin, S. I., and Pitts, R. C. (1994). Use of rodent self-administration models to develop pharmacotherapies for cocaine abuse. *NIDA Res Monogr*, *145*, 88–112.

56. Birmingham, A. M., Nader, S. H., Grant, K. A., Davies, H. M., and Nader, M. A. (1998). Further evaluation of the reinforcing effects of the novel cocaine analog 2β-propanoyl-3β-(4-tolyl)-tropane (PTT) in rhesus monkeys. *Psychopharmacology (Berl)*, *136*, 139–147.

57. Lile, J. A., Morgan, D., Freedland, C. S., Sinnott, R. S., Davies, H. M., and Nader, M. A. (2000). Self-administration of two long-acting monoamine transport blockers in rhesus monkeys. *Psychopharmacology (Berl)*, *152*, 414–421.

58. Lile, J. A., Morgan, D., Birmingham, A. M., Wang, Z., Woolverton, W. L., Davies, H. M., and Nader, M. A. (2002). The reinforcing efficacy of the dopamine reuptake inhibitor 2β-propanoyl-3β-(4-tolyl)-tropane (PTT) as measured by a progressive-ratio schedule and a choice procedure in rhesus monkeys. *J Pharmacol Exp Ther*, *303*, 640–648.

59. Lile, J. A., Morgan, D., Birmingham, A. M., Davies, H. M., and Nader, M. A. (2004). Effects of the dopamine reuptake inhibitor PTT on reinstatement and on food- and cocaine-maintained responding in rhesus monkeys. *Psychopharmacology (Berl)*, *174*, 246–253.

60. Clarke, R. L., Daum, S. J., Gambino, A. J., Aceto, M. D., Pearl, J., Levitt, M., Cumiskey, W. R., and Bogado, E. F. (1973). Compounds affecting the central nervous system. 4,3β-phenyltropane-2-carboxylic esters and analogs. *J Med Chem*, *16*, 1260–1267.

61. Spealman, R. D., Goldberg, S. R., Kelleher, R. T., Goldberg, D. M., and Charlton, J. P. (1977). Some effects of cocaine and two cocaine analogs on schedule-controlled behavior of squirrel monkeys. *J Pharmacol Exp Ther*, *202*, 500–509.

62. Heikkila, R. E., Cabbat, F. S., Manzino, L., and Duvoisin, R. C. (1979). Rotational behavior induced by cocaine analogs in rats with unilateral 6-hydroxydopamine lesions of the substantia nigra: dependence upon dopamine uptake inhibition. *J Pharmacol Exp Ther*, *211*, 189–194.

63. Reith, M. E., Meisler, B. E., and Lajtha, A. (1985). Locomotor effects of cocaine, cocaine congeners, and local anesthetics in mice. *Pharmacol Biochem Behav*, *23*, 831–836.

64. Reith, M. E. (1986). Effect of repeated administration of various doses of cocaine and WIN 35,065-2 on locomotor behavior of mice. *Eur J Pharmacol*, *130*, 65–72.

65. Steketee, J. D. (1998). Repeated injection of GBR 12909, but not cocaine or WIN 35,065-2, into the ventral tegmental area induces behavioral sensitization. *Behav Brain Res*, *97*, 39–48.

66. D'Mello, G. D., Goldberg, D. M., Goldberg, S. R., and Stolerman, I. P. (1981). Conditioned taste aversion and operant behavior in rats: effects of cocaine, apomorphine and some long-acting derivatives. *J Pharmacol Exp Ther*, *219*, 60–68.

67. Boja, J. W., Cline, E. J., Carroll, F. I., Lewin, A. H., Philip, A., Dannals, R., Wong, D., Scheffel, U., and Kuhar, M. J. (1992). High potency cocaine analogs: neurochemical, imaging, and behavioral studies. *Ann N Y Acad Sci*, *654*, 282–291.

68. Cline, E. J., Terry, P., Carroll, F. I., Kuhar, M. J., and Katz, J. L. (1992). Stimulus generalization from cocaine to analogs with high in vitro affinity for dopamine uptake sites. *Behav Pharmacol*, *3*, 113–116.

69. Spealman, R. D., Bergman, J., Madras, B. K., and Melia, K. F. (1991). Discriminative stimulus effects of cocaine in squirrel monkeys: involvement of dopamine receptor subtypes. *J Pharmacol Exp Ther*, *258*, 945–953.

70. Jarbe, T. U. (1981). Cocaine cue in pigeons: time course studies and generalization to structurally related compounds (norcocaine, WIN 35,428 and 35,065-2) and (+)-amphetamine. *Br J Pharmacol*, *73*, 843–852.

71. Spealman, R. D., and Kelleher, R. T. (1981). Self-administration of cocaine derivatives by squirrel monkeys. *J Pharmacol Exp Ther*, *216*, 532–536.

72. Spealman, R. D., Bergman, J., and Madras, B. K. (1991). Self-administration of the high-affinity cocaine analog 2β-carbomethoxy-3β-(4-fluorophenyl)tropane. *Pharmacol Biochem Behav*, *39*, 1011–1013.

73. Schenk, S. (2002). Effects of GBR 12909, WIN 35,428 and indatraline on cocaine self-administration and cocaine seeking in rats. *Psychopharmacology (Berl)*, *160*, 263–270.

74. Reith, M. E., Meisler, B. E., Sershen, H., and Lajtha, A. (1986). Structural requirements for cocaine congeners to interact with dopamine and serotonin uptake sites in mouse brain and to induce stereotyped behavior. *Biochem Pharmacol*, *35*, 1123–1129.

75. Boja, J. W., Carroll, F. I., Rahman, M. A., Philip, A., Lewin, A. H., and Kuhar, M. J. (1990). New, potent cocaine analogs: ligand binding and transport studies in rat striatum. *Eur J Pharmacol*, *184*, 329–332.

76. Boja, J. W., Patel, A., Carroll, F. I., Rahman, M. A., Philip, A., Lewin, A. H., Kopajtic, T. A., and Kuhar, M. J. (1991). [^{125}I]RTI-55: a potent ligand for dopamine transporters. *Eur J Pharmacol*, *194*, 133–134.

77. Fleckenstein, A. E., Kopajtic, T. A., Boja, J. W., Carroll, F. I., and Kuhar, M. J. (1996). Highly potent cocaine analogs cause long-lasting increases in locomotor activity. *Eur J Pharmacol*, *311*, 109–114.

78. Cook, C. D., Carroll, F. I., and Beardsley, P. M. (1998). Separation of the locomotor stimulant and discriminative stimulus effects of cocaine by its C-2 phenyl ester analog, RTI-15. *Drug Alcohol Depend*, *50*, 123–128.

79. Weed, M. R., Mackevicius, A. S., Kebabian, J., and Woolverton, W. L. (1995). Reinforcing and discriminative stimulus effects of β-CIT in rhesus monkeys. *Pharmacol Biochem Behav*, *51*, 953–956.

80. Cline, E. J., Scheffel, U., Boja, J. W., Mitchell, W. M., Carroll, F. I., Abraham, P., Lewin, A. H., and Kuhar, M. J. (1992). In vivo binding of [^{125}I]RTI-55 to dopamine transporters: pharmacology and regional distribution with autoradiography. *Synapse*, *12*, 37–46.

81. Innis, R., Baldwin, R., Sybirska, E., Zea, Y., Laruelle, M., al-Tikriti, M., Charney, D., Zoghbi, S., Smith, E., and Wisniewski, G. (1991). Single photon emission computed tomography imaging of monoamine reuptake sites in primate brain with [^{123}I]CIT. *Eur J Pharmacol*, *200*, 369–370.

82. Neumeyer, J. L., Wang, S. Y., Milius, R. A., Baldwin, R. M., Zea-Ponce, Y., Hoffer, P. B., Sybirska, E., al-Tikriti, M., Charney, D. S., and Malison, R. T. (1991). [^{123}I]-2β-carbomethoxy-3β-(4-iodophenyl)tropane: high-affinity SPECT radiotracer of monoamine reuptake sites in brain. *J Med Chem*, *34*, 3144–3146.

83. Baldwin, R. M., Zea-Ponce, Y., Zoghbi, S. S., Laurelle, M., al-Tikriti, M. S., Sybirska, E. H., Malison, R. T., Neumeyer, J. L., Milius, R. A., and Wang, S. (1993). Evaluation of the monoamine uptake site ligand [^{123}I]methyl 3β-(4-iodophenyl)-tropane-2β-carboxylate ([^{123}I]β-CIT) in non-human primates: pharmacokinetics, biodistribution and SPECT brain imaging coregistered with MRI. *Nucl Med Biol*, *20*, 597–606.

84. Tolliver, B. K., and Carney, J. M. (1995). Locomotor stimulant effects of cocaine and novel cocaine analogs in DBA/2J and C57BL/6J inbred mice. *Pharmacol Biochem Behav*, *50*, 163–169.

85. Kimmel, H. L., Carroll, F. I., and Kuhar, M. J. (2001). Locomotor stimulant effects of novel phenyltropanes in the mouse. *Drug Alcohol Depend*, *65*, 25–36.

86. Carroll, F. I., Pawlush, N., Kuhar, M. J., Pollard, G. T., and Howard, J. L. (2004). Synthesis, monoamine transporter binding properties, and behavioral pharmacology of a series of 3β-(substituted phenyl)-2β-(3′-substituted isoxazol-5-yl)tropanes. *J Med Chem*, *47*, 296–302.

87. Carroll, F. I., Runyon, S. P., Abraham, P., Navarro, H., Kuhar, M. J., Pollard, G. T., and Howard, J. L. (2004). Monoamine transporter binding, locomotor activity, and drug discrimination properties of 3-(4-substituted-phenyl)tropane-2-carboxylic acid methyl ester isomers. *J Med Chem*, *47*, 6401–6409.

88. Kotian, P., Abraham, P., Lewin, A. H., Mascarella, S. W., Boja, J. W., Kuhar, M. J., and Carroll, F. I. (1995). Synthesis and ligand binding study of 3β-(4′-substituted phenyl)-2β-(heterocyclic)tropanes. *J Med Chem*, *38*, 3451–3453.

89. Kotian, P., Mascarella, S. W., Abraham, P., Lewin, A. H., Boja, J. W., Kuhar, M. J., and Carroll, F. I. (1996). Synthesis, ligand binding, and quantitative structure-activity relationship study of 3β-(4′-substituted phenyl)-2β-heterocyclic tropanes: evidence for an electrostatic interaction at the 2β-position. *J Med Chem*, *39*, 2753–2763.

90. Carroll, F. I., Howard, J. L., Howell, L. L., Fox, B. S., and Kuhar, M. J. (2006). Development of the dopamine transporter selective RTI-336 as a pharmacotherapy for cocaine abuse. *AAPS J*, *8*, E196–E203.

91. Kimmel, H. L., O'Connor, J. A., Carroll, F. I., and Howell, L. L. (2007). Faster onset and dopamine transporter selectivity predict stimulant and reinforcing effects of cocaine analogs in squirrel monkeys. *Pharmacol Biochem Behav*, *86*, 45–54.

92. Carroll, F. I., Fox, B. S., Kuhar, M. J., Howard, J. L., Pollard, G. T., and Schenk, S. (2006). Effects of dopamine transporter selective 3-phenyltropane analogs on locomotor activity, drug discrimination, and cocaine self-administration after oral administration. *Eur J Pharmacol*, *553*, 149–156.

93. Howell, L. L., Carroll, F. I., Votaw, J. R., Goodman, M. M., and Kimmel, H. L. (2006). Effects of combined dopamine and serotonin transporter inhibitors on cocaine self-administration in rhesus monkeys. *J Pharmacol Exp Ther*, *320*, 757–765.

94. Acri, J. B., Siedleck, B. K., and Witkin, J. M. (1996). Effects of benztropine on behavioral and toxic effects of cocaine: comparison with atropine and the selective dopamine uptake inhibitor 1-[2-(diphenylmethoxy)ethyl]-4-(3-phenyl-propyl)-piperazine. *J Pharmacol Exp Ther*, *277*, 198–206.

95. Newman, A. H., Allen, A. C., Izenwasser, S., and Katz, J. L. (1994). Novel 3α-(diphenylmethoxy)tropane analogs: potent dopamine uptake inhibitors without cocaine-like behavioral profiles. *J Med Chem*, *37*, 2258–2261.

96. Kline, R. H., Izenwasser, S., Katz, J. L., Joseph, D. B., Bowen, W. D., and Newman, A. H. (1997). 3′-Chloro-3α-(diphenylmethoxy)tropane but not 4′-chloro-3α-(diphenylmethoxy)-tropane produces a cocaine-like behavioral profile. *J Med Chem*, *40*, 851–857.

97. Katz, J. L., Izenwasser, S., Kline, R. H., Allen, A. C., and Newman, A. H. (1999). Novel 3α-diphenylmethoxytropane analogs: selective dopamine uptake inhibitors with behavioral effects distinct from those of cocaine. *J Pharmacol Exp Ther*, *288*, 302–315.

98. Tolliver, B. K., Newman, A. H., Katz, J. L., Ho, L. B., Fox, L. M., Hsu, K., Jr., and Berger, S. P. (1999). Behavioral and neurochemical effects of the dopamine transporter ligand 4-chlorobenztropine alone and in combination with cocaine in vivo. *J Pharmacol Exp Ther*, *289*, 110–122.

99. Tanda, G., Ebbs, A., Newman, A. H., and Katz, J. L. (2005). Effects of 4′-chloro-3α-(diphenylmethoxy)-tropane on mesostriatal, mesocortical, and mesolimbic dopamine transmission: comparison with effects of cocaine. *J Pharmacol Exp Ther*, *313*, 613–620.

100. Woolverton, W. L., Rowlett, J. K., Wilcox, K. M., Paul, I. A., Kline, R. H., Newman, A. H., and Katz, J. L. (2000). 3′- and 4′-Chloro-substituted analogs of benztropine: intravenous self-administration and in vitro radioligand binding studies in rhesus monkeys. *Psychopharmacology (Berl)*, *147*, 426–435.

101. Woolverton, W. L., Hecht, G. S., Agoston, G. E., Katz, J. L., and Newman, A. H. (2001). Further studies of the reinforcing effects of benztropine analogs in rhesus monkeys. *Psychopharmacology (Berl)*, *154*, 375–382.

102. Katz, J. L., Agoston, G. E., Alling, K. L., Kline, R. H., Forster, M. J., Woolverton, W. L., Kopajtic, T. A., and Newman, A. H. (2001). Dopamine transporter binding without cocaine-like behavioral effects: synthesis and evaluation of benztropine analogs alone and in combination with cocaine in rodents. *Psychopharmacology (Berl)*, *154*, 362–374.

103. Katz, J. L., Kopajtic, T. A., Agoston, G. E., and Newman, A. H. (2004). Effects of N-substituted analogs of benztropine: diminished cocaine-like effects in dopamine transporter ligands. *J Pharmacol Exp Ther*, *309*, 650–660.

104. Katz, J. L., Izenwasser, S., Kline, R. H., Allen, A. C., and Newman, A. H. (1999). Novel 3α-diphenylmethoxytropane analogs: selective dopamine uptake inhibitors with behavioral effects distinct from those of cocaine. *J Pharmacol Exp Ther*, *288*, 302–315.

105. Campbell, V. C., Kopajtic, T. A., Newman, A. H., and Katz, J. L. (2005). Assessment of the influence of histaminergic actions on cocaine-like effects of 3α-diphenylmethoxytropane analogs. *J Pharmacol Exp Ther*, *315*, 631–640.

106. Desai, R. I., Kopajtic, T. A., Koffarnus, M., Newman, A. H., and Katz, J. L. (2005). Identification of a dopamine transporter ligand that blocks the stimulant effects of cocaine. *J Neurosci*, *25*, 1889–1893.

107. Raje, S., Cornish, J., Newman, A. H., Cao, J., Katz, J. L., and Eddington, N. D. (2005). Pharmacodynamic assessment of the benztropine analogues AHN-1055 and AHN-2005 using intracerebral microdialysis to evaluate brain dopamine levels and pharmacokinetic/pharmacodynamic modeling. *Pharm Res*, *22*, 603–612.

108. Li, S. M., Newman, A. H., and Katz, J. L. (2005). Place conditioning and locomotor effects of N-substituted, 4',4''-difluorobenztropine analogs in rats. *J Pharmacol Exp Ther*, *313*, 1223–1230.

109. Wilson, M. C., Hitomi, M., and Schuster, C. R. (1971). Psychomotor stimulant self administration as a function of dosage per injection in the rhesus monkey. *Psychopharmacologia*, *22*, 271–281.

110. Bergman, J., Madras, B. K., Johnson, S. E., and Spealman, R. D. (1989). Effects of cocaine and related drugs in nonhuman primates. III: Self-administration by squirrel monkeys. *J Pharmacol Exp Ther*, *251*, 150–155.

111. Kleven, M. S., Kamenka, J. M., Vignon, J., and Koek, W. (1999). Pharmacological characterization of the discriminative stimulus properties of the phencyclidine analog, *N*-[1-(2-benzo[b]thiophenyl)-cyclohexyl]piperidine. *Psychopharmacology* (*Berl*), *145*, 370–377.

112. Mansbach, R. S., and Balster, R. L. (1993). Effects of mazindol on behavior maintained or occasioned by cocaine. *Drug Alcohol Depend*, *31*, 183–191.

113. Kleven, M. S., and Woolverton, W. L. (1993). Effects of three monoamine uptake inhibitors on behavior maintained by cocaine or food presentation in rhesus monkeys. *Drug Alcohol Depend*, *31*, 149–158.

114. Elmer, G. I., Brockington, A., Gorelick, D. A., Carrol, F. I., Rice, K. C., Matecka, D., Goldberg, S. R., and Rothman, R. B. (1996). Cocaine cross-sensitization to dopamine uptake inhibitors: unique effects of GBR 12909. *Pharmacol Biochem Behav*, *53*, 911–918.

115. Heikkila, R. E., Babington, R. G., and Houlihan, W. J. (1981). Pharmacological studies with several analogs of mazindol: correlation between effects on dopamine uptake and various in vivo responses. *Eur J Pharmacol*, *71*, 277–286.

116. Houlihan, W. J., Boja, J. W., Parrino, V. A., Kopajtic, T. A., and Kuhar, M. J. (1996). Halogenated mazindol analogs as potential inhibitors of the cocaine binding site at the dopamine transporter. *J Med Chem*, *39*, 4935–4941.

117. Kulkarni, S. S., Newman, A. H., and Houlihan, W. J. (2002). Three-dimensional quantitative structure-activity relationships of mazindol analogues at the dopamine transporter. *J Med Chem*, *45*, 4119–4127.

ERRATA

Dopamine Transporters: Chemistry, Biology, and Pharmacology
Edited by Mark L. Trudell and Sari Izenwasser
© 2008 John Wiley & Sons, Inc.   ISBN 978-0-470-11790-3

In Chapter 12:

On page 323, the full affiliation for **Dean F. Wong** is missing one department. This should properly read:

The Russell H. Morgan Department of Radiology and Radiological Science, and Department of Psychiatry, Department of Neuroscience, and Department of Environmental Health Sciences-Bloomberg School of Public Health, Johns Hopkins University, Baltimore, Maryland

On page 328, the first sentence in the second full paragraph should be replaced by the following:

The fluorophenyl tropane [^{11}C]WIN 35,428 (also known as CFT) was used for the first PET studies in normal human subjects [3] and then subsequently in patients with Parkinson's disease (PD) [4]. These probably represent the first human imaging of DAT with a fairly specific and higher affinity PET radioligand compared to radiolabeled cocaine itself, which preceded these radiolabeled tropanes (see below). Derivatives of these tropanes are now routinely used in human imaging studies.

This was not correct in the first printing of this book. We apologize for this error.

118. Houlihan, W. J., Kelly, L., Pankuch, J., Koletar, J., Brand, L., Janowsky, A., and Kopajtic, T. A. (2002). Mazindol analogues as potential inhibitors of the cocaine binding site at the dopamine transporter. *J Med Chem*, *45*, 4097–4109.

119. Houlihan, W. J., and Kelly, L. (2003). Assessment of mazindane, a pro-drug form of mazindol, in assays used to define cocaine treatment agents. *Eur J Pharmacol*, *458*, 263–273.

120. Baker, L. E., Riddle, E. E., Saunders, R. B., and Appel, J. B. (1993). The role of monoamine uptake in the discriminative stimulus effects of cocaine and related compounds. *Behav Pharmacol*, *4*, 69–79.

121. Lamb, R. J., and Griffiths, R. R. (1990). Self-administration in baboons and the discriminative stimulus effects in rats of bupropion, nomifensine, diclofensine and imipramine. *Psychopharmacology (Berl)*, *102*, 183–190.

122. Broadbent, J., Gaspard, T. M., and Dworkin, S. I. (1995). Assessment of the discriminative stimulus effects of cocaine in the rat: lack of interaction with opioids. *Pharmacol Biochem Behav*, *51*, 379–385.

123. Johanson, C. E., and Barrett, J. E. (1993). The discriminative stimulus effects of cocaine in pigeons. *J Pharmacol Exp Ther*, *267*, 1–8.

124. Kamien, J. B., and Woolverton, W. L. (1989). A pharmacological analysis of the discriminative stimulus properties of *d*-amphetamine in rhesus monkeys. *J Pharmacol Exp Ther*, *248*, 938–946.

125. Blitzer, R. D., and Becker, R. E. (1985). Characterization of the bupropion cue in the rat: lack of evidence for a dopaminergic mechanism. *Psychopharmacology (Berl)*, *85*, 173–177.

126. Jones, C. N., Howard, J. L., and McBennett, S. T. (1980). Stimulus properties of antidepressants in the rat. *Psychopharmacology (Berl)*, *67*, 111–118.

127. Bergman, J., Madras, B. K., Johnson, S. E., and Spealman, R. D. (1989). Effects of cocaine and related drugs in nonhuman primates. III: Self-administration by squirrel monkeys. *J Pharmacol Exp Ther*, *251*, 150–155.

128. Rush, C. R., Kollins, S. H., and Pazzaglia, P. J. (1998). Discriminative-stimulus and participant-rated effects of methylphenidate, bupropion, and triazolam in *d*-amphetamine-trained humans. *Exp Clin Psychopharmacol*, *6*, 32–44.

129. Kollins, S. H., MacDonald, E. K., and Rush, C. R. (2001). Assessing the abuse potential of methylphenidate in nonhuman and human subjects: a review. *Pharmacol Biochem Behav*, *68*, 611–627.

130. Shuster, L., Hudson, J., Anton, M., and Righi, D. (1982). Sensitization of mice to methylphenidate. *Psychopharmacology (Berl)*, *77*, 31–36.

131. Tyler, T. D., and Tessel, R. E. (1980). Norepinephrine uptake inhibitors as biochemically and behaviorally selective antagonists of the locomotor stimulation induced by indirectly acting sympathomimetic aminetic amines in mice. *Psychopharmacology (Berl)*, *69*, 27–34.

132. Yang, P. B., Swann, A. C., and Dafny, N. (2006). Chronic methylphenidate modulates locomotor activity and sensory evoked responses in the VTA and NAc of freely behaving rats. *Neuropharmacology*, *51*, 546–556.

133. Niculescu, M., Ehrlich, M. E., and Unterwald, E. M. (2005). Age-specific behavioral responses to psychostimulants in mice. *Pharmacol Biochem Behav*, *82*, 280–288.

134. Torres-Reverón, A., and Dow-Edwards, D. L. (2005). Repeated administration of methyl-phenidate in young, adolescent, and mature rats affects the response to cocaine later in adulthood. *Psychopharmacology (Berl)*, *181*, 38–47.

135. Brandon, C. L., Marinelli, M., Baker, L. K., and White, F. J. (2001). Enhanced reactivity and vulnerability to cocaine following methylphenidate treatment in adolescent rats. *Neuropsychopharmacology*, *25*, 651–661.

136. Brandon, C. L., and Steiner, H. (2003). Repeated methylphenidate treatment in adolescent rats alters gene regulation in the striatum. *Eur J Neurosci*, *18*, 1584–1592.

137. Izenwasser, S., Coy, A. E., Ladenheim, B., Loeloff, R. J., Cadet, J. L., and French, D. (1999). Chronic methylphenidate alters locomotor activity and dopamine transporters differently from cocaine. *Eur J Pharmacol*, *373*, 187–193.

138. Poncelet, M., Chermat, R., Soubrie, P., and Simon, P. (1983). The progressive ratio schedule as a model for studying the psychomotor stimulant activity of drugs in the rat. *Psychopharmacology (Berl)*, *80*, 184–189.

139. Emmett-Oglesby, M. W., Wurst, M., and Lal, H. (1983). Discriminative stimulus properties of a small dose of cocaine. *Neuropharmacology*, *22*, 97–101.

140. Li, S. M., Campbell, B. L., and Katz, J. L. (2006). Interactions of cocaine with dopamine uptake inhibitors or dopamine releasers in rats discriminating cocaine. *J Pharmacol Exp Ther*, *317*, 1088–1096.

141. Czoty, P. W., Makriyannis, A., and Bergman, J. (2004). Methamphetamine discrimination and in vivo microdialysis in squirrel monkeys. *Psychopharmacology (Berl)*, *175*, 170–178.

142. Griffiths, R. R., Findley, J. D., Brady, J. V., Dolan-Gutcher, K., and Robinson, W. W. (1975). Comparison of progressive-ratio performance maintained by cocaine, methylphenidate and secobarbital. *Psychopharmacologia*, *43*, 81–83.

143. Brady, J. V., and Griffiths, R. R. (1976). Behavioral procedures for evaluating the relative abuse potential of CNS drugs in primates. *Fed Proc*, *35*, 2245–2253.

144. Aigner, T. G., and Balster, R. L. (1979). Rapid substitution procedure for intravenous drug self-administration studies in rhesus monkeys. *Pharmacol Biochem Behav*, *10*, 105–112.

145. Johanson, C. E., and Schuster, C. R. (1975). A choice procedure for drug reinforcers: cocaine and methylphenidate in the rhesus monkey. *J Pharmacol Exp Ther*, *193*, 676–688.

146. Lile, J. A., Wang, Z., Woolverton, W. L., France, J. E., Gregg, T. C., Davies, H. M., and Nader, M. A. (2003). The reinforcing efficacy of psychostimulants in rhesus monkeys: the role of pharmacokinetics and pharmacodynamics. *J Pharmacol Exp Ther*, *307*, 356–366.

147. Schenk, S., and Izenwasser, S. (2002). Pretreatment with methylphenidate sensitizes rats to the reinforcing effects of cocaine. *Pharmacol Biochem Behav*, *72*, 651–657.

148. Schenk, S., and Partridge, B. (1999). Cocaine-seeking produced by experimenter-administered drug injections: dose–effect relationships in rats. *Psychopharmacology (Berl)*, *147*, 285–290.

149. Itzhak, Y., and Martin, J. L. (2002). Cocaine-induced conditioned place preference in mice: induction, extinction and reinstatement by related psychostimulants. *Neuropsychopharmacology*, *26*, 130–134.

150. Nomikos, G. G., Damsma, G., Wenkstern, D., and Fibiger, H. C. (1990). In vivo characterization of locally applied dopamine uptake inhibitors by striatal microdialysis. *Synapse*, *6*, 106–112.

151. Deutsch, H. M., Shi, Q., Gruszecka-Kowalik, E., and Schweri, M. M. (1996). Synthesis and pharmacology of potential cocaine antagonists. 2: Structure–activity relationship studies of aromatic ring–substituted methylphenidate analogs. *J Med Chem*, *39*, 1201–1209.

152. Wayment, H. K., Deutsch, H., Schweri, M. M., and Schenk, J. O. (1999). Effects of methylphenidate analogues on phenethylamine substrates for the striatal dopamine transporter: potential as amphetamine antagonists? *J Neurochem*, *72*, 1266–1274.

153. Deutsch, H. M., Ye, X., Shi, Q., Liu, Z., and Schweri, M. M. (2001). Synthesis and pharmacology of site specific cocaine abuse treatment agents: a new synthetic methodology for methylphenidate analogs based on the Blaise reaction. *Eur J Med Chem*, *36*, 303–311.

154. Meltzer, P. C., Wang, P., Blundell, P., and Madras, B. K. (2003). Synthesis and evaluation of dopamine and serotonin transporter inhibition by oxacyclic and carbacyclic analogues of methylphenidate. *J Med Chem*, *46*, 1538–1545.

155. Davids, E., Zhang, K., Tarazi, F. I., and Baldessarini, R. J. (2002). Stereoselective effects of methylphenidate on motor hyperactivity in juvenile rats induced by neonatal 6-hydroxydopamine lesioning. *Psychopharmacology (Berl)*, *160*, 92–98.

156. Gatley, S. J., Meehan, S. M., Chen, R., Pan, D. F., Schechter, M. D., and Dewey, S. L. (1996). Place preference and microdialysis studies with two derivatives of methylphenidate. *Life Sci*, *58*, L345–L352.

157. Schweri, M. M., Deutsch, H. M., Massey, A. T., and Holtzman, S. G. (2002). Biochemical and behavioral characterization of novel methylphenidate analogs. *J Pharmacol Exp Ther*, *301*, 527–535.

INDEX

Affinity labeling, 84–85. *See also* Irreversible
    binding
Aging, 14–15, 34, 334–335
Allele, 55–62
Aminorex, 313
Amphetamines, 74, 82, 103, 104, 309–311
Amino acid sequence, 5
Anatomy, 6–15
Attention-deficit/hyperactivity disorder (ADHD),
    39, 59–60
Autoradiography, 6, 8, 10–11, 32–34, 296
    DAT mapping, 32–34
    Parkinson's disease, 35

Bacterial leucine transporter, s*ee* LeuT
Benztropine, 172–173. *See also* Structure–activity
    relationships; Synthesis
    SAR 2-substituted benztropines, 189–191
    SAR 3-substituted benztropines, 175–186
    SAR 6/7-substituted benztropines,
        184, 186
Binding sites
    assay, 350–351
    benztropines, 203–204
    cocaine, 49–50, 83–8
    ligands, 83–87
    multiple sites, 31, 49, 251
    postmortem human brain, 30–34
    substrates, 83–87
    zinc, 81–82
Biotinylation, 368–372

Brain regions. *See also* Tissue preparation
    human, 15–17
    primate, 11–14
    rat, 10–11
Brain slices, 348–349, 357, 361, 363
Brasofensine, 127, 151. *See also* Parkinson's
    disease
Bupropion, 420–421

Calbindin-negative cell, 6
Caudate-putamen, 8. *See also* Binding sites
cDNA, *see* Cloning
Cell surface expression, 367–372
    subcellular fractionation, 367–368
    surface biotinylation, 368–372
β-CFT, *see* Autoradiography; PET; SPECT;
    Radioligands; WIN, 35, 428
Chimeras
    DAT/NET chimeras, 83
    human/bovine chimeras, 83
Chromosomes, 51–52
β-CIT, *see* Autoradiography; PET; SPECT;
    Radioligands; RTI-55
Clatherin, 368–369, 372
Cloning
    bovine, 5
    Caenorhabditis elegans, 5
    cDNA, 48–51
    genomic DNA, 51–55
    human, 5, 47–51
    monkey, 5

Dopamine Transporters: Chemistry, Biology, and Pharmacology. Edited by Mark L. Trudell
and Sari Izenwasser
Copyright © 2008 John Wiley & Sons, Inc.

Cloning (*Continued*)
  mouse, 5. *See also* Knockout mice
  rat, 5, 48
Cocaine, 126
  binding, 84–86, 126,
  comparison with benztropines, 196–203
  self-administration, 104–106
  structure–activity, 127
Cocaine abuse, 36–39
Comparative molecular field analysis (CoMFA),
    223, 255–256

DAT gene (*DAT1* or *DAT*), 47–48
  chromosomal position, 52
  structure, 52
DAT protein, 10–14
  amino acid sequence, 5, 49
  caudate-putamen, 8
  changes with age, 15, 34
  conformational changes, 86–87
  ultrastructural localization, 8–10
DAT structure. *See also* Posttranslational
    modifications
  glycosylation sites, 5, 49
  oligomerization, 81
  phosphorylation, 49
  primary sequence, 5, 74–75
  quaternary, 82–83
  tertiary, 81–82
  topology, 74–75
  transmembrane domains, 5
Disease states, 30, 35–40, 234. *See also*
    ADHD; Parkinson's disease
  Alzheimer's disease, 337
  Huntington's disease, 337
  Lesch–Nyhan syndrome, 337
  Tourette's syndrome, 337
Distribution
  cocaine abusers, 36–39
  human, 12, 15–17, 32, 36–39
  mRNA, 6–8, 34
  primate, 11–12
  rodent, 10–11
L-DOPA, 3
Dopamine release, 75, 305–308
  electrochemistry, 100–104, 363–364
  measurement, 358–360
Dopamine releasers
  amphetamines, 103–104,
      309–311
  MDMA analogs, 311–313
  others, 313
Dopamine synaptic markers, 34
Dopamine transport
  kinetics, 75

  mechanism, 75
  stoichiometry, 75
Dopamine uptake, measurement, 348, 357–358
Dopamine uptake inhibition
  behavioral activation, 113–115
  competitive, 30, 235–36
  fast onset, 110–113
  measurement, 357–358
  methylphenidate derivatives, 254
  noncompetitive, 235
  uncompetitive, 235
Dopamine uptake inhibitors, 235–237, 267–268
Drug discrimination
  3-aryl-heterotropanes, 296
  benztropines, 164
  3-phenyltropanes, 150

Electrochemistry
  DAT-associated currents, 364–367
  dopamine clearance, 360–363
  dopamine efflux, 363–364
Electrophoresis, 355–356
Endogenous monoamine transporter ligands,
      268–269, 308–309
Ephedrine, 306, 313
Exons, 52–56

Fast-scan cyclic voltammetry (FSCV), 98–10
  behavioral activation, 113–115
  cocaine self-administration, 104–106
  DAT-KO mice, 101–103
  effects of amphetamines, 103–104
  effects of cocaine, 100–103, 110–113
  electrode, 99
  fast onset, 110–113
  psycho stimulant reinforcement, 106–109
  regional dynamics, 109–110
Fen-phen, 315
Fenfluramine, 315
Fluorescent antibodies, 372
Fluorescence microscopy, 372–374
Fluorescent ligands, 148–149
FRET (fluorescence resonance energy transfer)
      microscopy, 81, 374–375
Functional assays, 357–367

GABA transporter
  gene, 51–53
  structural similarities, 4–5
GBR 12909, 212. *See also* Structure–activity
      relationships; In vivo studies
  GBR12909 derivatives, 212–219, 224, 237–238
GBR12935, 212
Gene expression 6–8, 53–55. *See also* Midbrain
      regions; Substantia nigra;
      Ventral tegmental area

Genes. *See also* DAT gene
  transporters, 51–52
Genetic analysis
  humans, 61–62
  nonhuman primates, 60–61
Genetics, 55–62
  ADHD, 59–60
  alcoholism, 57
  manic depression/bipolar disorder, 58
  neuropsychiatric disorders, 56–57
  PET, 335
  schizophrenia, 58
  smoking, 57
Genotype, 55–62. *See also* Genetics
Glycosylation
  functional role, 77–78
  patterns, 78
  sites, 78
Growth factors, 349, 352

Heterotropanes. *See also* Structure–activity
    relationships; Synthesis
  carbatropane synthesis, 276–277
  oxatropane synthesis, 278–279
  SAR 3-aryl-heterotropanes, 139–140,
    283–291
  thiatropane synthesis, 275–276
Histamine receptors, 175–182, 197
Histochemistry, 4, 6
Human Genome Organization (HUGO), 47
Hypothalamus, 7

Immunohistochemistry, 9–12, 356–357
Immunogold labeling, 9–10
Imaging
  in vitro, 32–33
  in vivo, 33
International Committee on Standardized
    Genetic Nomenclature for Mice, 47
Introns, 54–57
In vivo studies
  benztropines, 414–418
  bupropion, 420–421
  GBR 12909, 402–404
  mazindol, 418–420
  methylphenidate, 421–425
  3-phenyltopanes, 404–414
Irreversible binging, 147–149, 219

Kinase enzymes, 79–80, 82, 369–370
Knockout (KO) mice, 55, 235
  in vivo studies, 391–395

LeuT (bacterial leucine transporter)
  binding site, 76, 83–84

  crystal structure, 77
  relationship to DAT, 76–77
  transport mechanism, 77
Localization, 8–10
Locomotor activity. *See also* In vivo studies
  3-aryl-heterotropanes, 293–296
  benztropines, 192–194
  cocaine, 201–203
  3-phenyltropines, 150

Mazindol, 236–337
MDMA, 306, 311
Meperidine, 251
  derivatives, 251–252
Methamphetamine, 37, 74, 306
Methylphenidate. *See also* ADHD;
    Autoradiography; PET, Radioligands
  structure–activity studies, 252–255, 306
MDMA analogs, 311–313
Midbrain regions, 6–8. *See also* Substantia
    nigra; Ventral tegmental area
Midbrain structures, 6–8. *See also* Substantia
    nigra; Ventral tegmental area
MPP^+ (1-methyl-4-phenylpyridinium), 50, 356,
    357–359
MPTP (1-methyl-4-phenyl-1,2,3,6-
    tetrahydropyridine), 34, 50, 356
mRNA
  mRNA distribution, 6–8
  expression measurement, 351–354
Muscarinic receptors, 197–199
  benztropines, 175–191

Neuronal cultures, 350, 357
Neurotoxins, 34–35
Neurotransmission, 3–5, 29–30, 74,
    115–117, 348
Neurotransmitter sodium symporters (NSSs),
    74–77
Nicotine, 353
Nigral cell dendrites, 9
Norepinephrine release, 313–315
Norepinephrine transporter
  gene, 51–53
  structural similarities, 4, 74–75
Norepinephrine transporter selectivity
  amphetamines, 309–315
  benztropine derivatives, 175–191
  GBR12909 derivatives, 212–219, 224,
    237–238
  MDMA analogs, 311–313
  methylphenidate analogs, 252–255
  phenylpiperidines, 243–250
  3-phenyltropines, 127–142
  piperidines, 220–224, 238–243

Northern blotting, 352
Nucleus accumbens, 8
Nuclease protection assay, 354

Obesity, 181
Oligomerization
    FRET, 81
    homodimers, 81
    motif, 81
    tetramers, 81

Parkinson's disease, 17, 35–36, 50, 127,
        335–336
PET (positron emission tomography), 15–17,
        142–146, 296–297, 324–328
    ADHD, 337
    aging, 334
    genetic analysis, 61
    human postmortem brains, 34–39
    in vivo, 332
    neuropsychiatric diseases, 17, 337–338
    Parkinson's disease, 17, 335–336
    psychostimulant action, 338
    saturation binding parameters for DAT
        ligands, 332–334
    substance abuse, 338
Pharmacokinetics
    benztropines, 199–203
    radioligands, 332–334
Phasic neurotransmission, 115–117
Phenmetrazine, 313
Phenotypes, see Genetics
Phentermine, 315
Phenylpiperidines, 243–250
3-Phenyltropanes, see Radioligands; In vivo
        studies; PET; SPECT
    structure–activity relationships, 127–142
Phosphorylation
    CaMK, 79–80
    DAT trafficking, 79
    MAPK, 79–80
    phosphorylation sites, 79–80
    PKC activation, 79–80
Photoaffinity ligands, see Irreversible Binding
Piperidines, see Structure–activity relationships
    GBR12909 analogs, 237–238
Piperazines, see GBR12909; Structure–activity
    relationships
Place conditioning, benztropines, 196
Polymerase chain reaction, 354
Polymorphic sites, 55–60
Postmortem human brain, 30–39
Posttranslational modifications
    disulfide bonds, 78–79

glycosylation, 77–78
phosphorylation, 79–80
ubiquitylation, 80
PPT (2β-Propanoyl-3β-(4-tolyl)tropane), 7,
        102, 151
Protein, see DAT protein
Protein–protein interactions, 82–83
    α-synuclein, 82–83
    endocytosis motif, 83
    Hic-5, 82–83
    PICK1, 82–83
    protein kinase C (PKC), 82
    RACK1, 82–83
    syntaxin 1A, 82–83

QSAR (quantitative structure–activity
    relationships)
    GBR12909 analogs, 223
    piperidine analogs, 255–256

Radioligand binding assay, 350–351
Radioligands, 30–34, 142–146,
        328–332. See also Autoradiography; PET
        (positron emission tomography); Single
        photon emission computerized tomography
        (SPECT)
Ritalin, see Methylphenidate
RNAi, 375–377
RTI-55. See also 3-Phenyltropanes; Radioligands;
        Structure–activity relationships
    autoradiography, 32
    PET, 33, 35–38
RTI-76, 147
RTI-121, 10–12, 30–35. See also
    Autoradiography
RTI-336, 126, 150–151

SCAM (substituted cysteine accessibility method)
    TM2 analysis, 86
    TM7/TM8 region, 86–87
Self-administration studies
    benztropines, 195–196
    3-phenyltropanes, 150
Serotonin release, 315–316
Serotonin transporter
    gene, 51–53
    structural similarities, 6, 74–75
Serotonin transporter selectivity
    amphetamines, 309–311
    3-aryl-heterotropanes, 139–140, 283–291
    benztropine derivatives, 175–191
    GBR12909 derivatives, 212–219, 224,
        237–238
    MDMA analogs, 311–313

methylphenidate analogs, 252–255
phenylpiperidines, 243–250
3-phenyltropanes, 127–142
piperidines, 220–224, 238–243
Site-directed mutagenesis, 85–86, 203–204
SLC6 family, 47, 74, 348
SPECT (single-photon emission computed
    tomography) 15–17, 146, 324–328
in vivo, 332
neuropyschiatric diseases, 337–338
Parkinson's disease, 335–336
saturation binding parameters for DAT ligands,
    332–334
Structure–activity relationships (SAR)
amphetamines, 309–311
3-aryl-heterotropanes, 139–140, 283–291
benztropinamines, 184–188
cocaine, 12, 127
GBR12909 derivatives, 212–219, 224,
    237–238
MDMA analogs, 311–313
meperidine derivatives, 251–252
methylphenidate analogs, 252–255
phenylpiperidines, 243–250
3-phenyltropanes, 127–142
piperidines, 220–224, 238–243
ring-modified tropane derivatives, 140–141
2-substituted benztropines, 189–191
3-substituted benztropines, 175–186
6/7-substituted benztropines, 184, 186
Substantia nigra, 6–8, 10, 53
Synaptosomes, 349, 357–360, 368–369, 371
Synthesis
3-arylheterotropanes, 272–278
benztropinamines, 184–185

2-substituted benztropines, 189–192
3-substituted benztropines, 173–174
6/7-substituted benztropines, 184–185

Tesofensine, 127, 151
Tissue preparation, 349–350
Trafficking, 372–377
Transcription, 53–55, 61
Transmembrane domains, 84–86
Tropanes, *see* Benztropines; 3-Phenyltropanes
Tyrosine hydroxylase, 6, 34

Ubiquination, 369–370
Ubiquitylation
monoubiquitylation, 80
polyubiquitylation, 80
ubiquitylation sites, 80

Ventral tegmental area, 6–9, 10, 53
Voltammetry, *see* Electrochemistry;
    Fast-scan cyclic voltammetry (FSCV)

Western blotting, 355–356
WIN 35,065-2, 126–127. *See also*
    3-Phenyltropanes
[^{3}H]WIN 35,428, 10, 16, 130. *See also*
    3-Phenyltropanes; Radioligands
autoradiography, 10, 30–34
PET, 16

Zinc binding
anion conductance, 82, 365–366
coordination site, 81–82
conformational properties, 82